Light Agricultural
and Industrial
Structures

LIGHT AGRICULTURAL AND INDUSTRIAL STRUCTURES
Analysis and Design

G. L. Nelson
Agricultural Engineering Department
Ohio State University
Columbus, Ohio

H. B. Manbeck
Agricultural Engineering Department
Pennsylvania State University
University Park, Pennsylvania

N. F. Meador
Agricultural Engineering Department
University of Missouri
Columbia, Missouri

An **avi** Book
Published by Van Nostrand Reinhold Company
New York

AN AVI BOOK
(AVI is an imprint of Van Nostrand Reinhold Company Inc.)
Copyright © 1988 by Van Nostrand Reinhold Company Inc.

Library of Congress Catalog Card Number 87-37263

ISBN 0-442-26777-0

All rights reserved. No part of this work covered by the copyright hereon may be reproduced or used in any form or by any means—graphic, electronic, or mechanical, including photocopying, recording, taping, or information storage and retrieval systems—without written permission of the publisher.

Printed in the United States of America

Van Nostrand Reinhold Company Inc.
115 Fifth Avenue
New York, New York 10003

Van Nostrand Reinhold Company Limited
Molly Millars Lane
Wokingham, Berkshire RG11 2PY, England

Van Nostrand Reinhold
480 La Trobe Street
Melbourne, Victoria 3000, Australia

Macmillan of Canada
Division of Canada Publishing Corporation
164 Commander Boulevard
Agincourt, Ontario M1S 3C7, Canada

16 15 14 13 12 11 10 9 8 7 6 5 4 3 2 1

Library of Congress Cataloging-in-Publication Data

Nelson, G. L. (Gordon Leon), 1919–
 Light agricultural and industrial structures.

 "An AVI book."
 Bibliography: p.
 Includes index.
 1. Farm buildings—Design and construction.
 2. Structures, Theory of.   I. Manbeck, H. B.
 II. Meador, N. F.   III. Title.
 TH4911.N45 1988   690'.892   87-37263
 ISBN 0-442-26777-0

Contents

Preface	vii
1. Analysis and Design Concepts	1
2. Fundamental Concepts of Stress Analysis	7
3. Stress Analysis of Coplanar Statically Determinate Trusses	33
4. Bending Deformation	63
5. Analysis of Statically Indeterminate Coplanar Frames	91
6. Load Analysis	127
7. Fundamentals of Structural Connections	185
8. Structural Steel Design	209
9. Cold-Formed Steel Design	283
10. Light Timber Design	359
11. Reinforced Concrete Design	455
Index	545

Preface

This book is an outgrowth of a much earlier book, *Farm Structures*, by H. J. Barre and L. L. Sammet, published by John Wiley & Sons in 1950 as one of a series of textbooks in agricultural engineering sponsored by the Ferguson Foundation, Detroit, Michigan. *Light Agricultural and Industrial Structures: Analysis and Design* will be useful as an undergraduate student textbook for junior- or senior-level comprehensive courses on structural analysis and design in steel, wood, and concrete, and as a reference work for practicing engineers. Emphasis is on basic analysis and design procedures. The book should be useful in any country where there is a need to design structures for agricultural production and processing.

It is assumed that readers have had prerequisite course work in engineering mechanics and strength of materials as typically taught to undergraduate engineering students. The scope of this book is wide; it might be difficult for instructors and students to cover all of the chapters in a typical three credit-hour course. The instructor will need to assess his own situation and scheduling constraints. More or less time could be spent on chapters one through five, depending on the capability the students already have in analysis of statically determinate and indeterminate structures. Two to three weeks might then be allocated for study of each of the last six chapters dealing with design in steel, reinforced concrete, and wood.

We suggest that instructors help their students who may not have a major interest in structural analysis and design to understand that the content of this book is not limited in usefulness to light building structures. For example, students interested in farm machinery design must be able to analyze statically indeterminate welded steel frames used to a wide extent in farm machinery. Students whose primary interest is in soil and water engineering need capability in analysis and design of reinforced concrete culverts and related structures. Students interested in processing and food engineering will encounter structural analysis and design problems in relation to crop and food storage, handling, and transport facilities.

The timber, steel, and reinforced concrete chapters are based upon recent structural design specifications. However, it should be noted that design specifications are not static. Undoubtedly, one or more structural specifications will change with time. The authors suggest that the instructor may use the text to present basic concepts and methods of structural component design, but supplement the text by reference to the most up-to-date specification for each' structural material.

English and SI units have been used interchangeably throughout the book to reflect the present level of adoption of SI units. The timber and steel construction industries are moving slowly toward adoption of SI units. Reinforced concrete construction seems to be leading in adoption of SI units.

It is almost inevitable that errors in the text will come to light. We hope they are few in number. The authors will be indebted to anyone who identifies errors or the need for clearer meaning that could be remedied in future printings or editions.

1

Analysis and Design Concepts

Light structures are used for many agricultural and industrial needs. They are characterized by one- or two-story configuration, moderate spans, and light to moderate superimposed loads. Construction costs must be carefully controlled to maintain a profitable relationship between capital cost and income from use of the structures. Often, they are built according to standard plans and utilize mass-produced frames and other structural components.

A high degree of precision in structural analysis and design for light structures is often necessary. For example, a single standardized design may be used for several hundred or thousand buildings. This multiple use of one basic design can justify considerable analysis and design effort. A high degree of refinement is desirable for multiple-use plans for low-cost structures.

1.1 ANALYSIS AND DESIGN OBJECTIVES

One important and basic objective in structural analysis and design is to produce a structure capable of resisting all applied loads without undue deformation during its intended life. For some structures, preservation of the original shape may be a critical requirement. Even a minor amount of deflection or deformation may be objectionable. An example is a structural frame supporting a long run in a screw conveyor system. Here misalignment would cause excessive wear and operating difficulties. In other instances relatively large deformations may be permissible without impairing the usefulness of the structure. An example is a structural frame for a greenhouse enclosed with plastic sheeting. In all instances, the risk of collapse or destruction must be controlled by adequate analysis and design.

Production of an adequate design is the engineer's basic responsibility. The ubiquitous demand for economy, often severe in light structures, should not take precedence over structural adequacy. An owner seldom gives an engineer much credit for a structure that collapses or a machine frame that comes apart in the field, even though the structure or machine was inexpensive to build.

1.2 DESIGN PROCEDURE

Structural analysis and design are based on preliminary drawings previously prepared to meet functional requirements of the building. These drawings reveal gross dimensions and configuration of the building, and how space within the building will be used. Information on the preliminary drawings helps to establish spacing of frames, columns, and other primary components. Building width, height, and cross-section shape are also shown. The type of frame may be specified.

Structural design starts with *load analyses*. These are computations to estimate loads which the structure must carry. Information from the preliminary drawings plus knowledge of weights of building materials, wind forces, occupancy loads, pressures from stored materials, and other loads are used in preparing the load analyses. The objective is accurate estimates of the loads which may reasonably be expected to occur on the structure during its lifetime. These are the design loads.

Next, *stress analyses* are completed. These are computations to predict total stresses that will occur in the structural members when the design loads are applied. Total stress is the force resisted by a member subjected to axial or shear load; or bending moment resisted by a member subjected to bending. Unit stresses, such as pounds per square inch, either direct or bending stress, cannot be calculated until the cross-section properties of the members are known.

Usually a separate stress analysis is made for each kind of load. For example, in a simple shelter frame, separate analyses may be made for wind loads, dead loads, and snow loads. Then, the stresses produced by likely combinations of loads are investigated to determine which produce the critical stresses that control the design.

Structural design is done after the critical total stresses have been estimated. It includes selection of materials, member sizes, configurations, and fastenings for the load carrying parts of the building. For example, a beam is designed for bending by dividing the total bending moment obtained from the stress analysis by the allowable unit bending stress for the beam material to be used. The result is the section modulus, an index of the beam size and cross-section configuration required.

Structural design is often a "cut-and-try" process of successive selection, calculation, and refinement or modification of the original selection until satisfactory agreement is obtained between the selected size and the required size as revealed by design calculations. For example, in the general case of a statically indeterminate frame, a stress analysis cannot be made until a trial design is available to use for estimating stiffnesses of the frame members.

Final design drawings are prepared after completion of structural design. These, in combination with written specifications, should give all the information needed by the builder to erect an actual structure, which in all important respects is the same as the one created on paper by the designer. A novice in structural design is often at a loss to decide what is required in design drawings. The general requirement is to describe by drawings and specifications everything the builder needs to know beyond the dictates of standard practice or knowledge common to the trade, to erect the structure. Much helpful information can be gained by careful study of standard or typical drawings for the kind of structure to be designed.

An activity closely related to structural analysis and design is *checking* or *investigation*. This consists of verifying the adequacy of an existing design to carry specified loads; or in determining what stresses will result when specified loads are applied to a structure shown in an existing design. For example, various plan services offer complete construction plans for many kinds of light buildings. A structural designer or analyst often finds it necessary to check the structural design, because it may have been prepared for loading conditions that are not the same as for the locality or use being contemplated for the building.

1.3 DESIGN PHILOSOPHY

An engineer must make many decisions based partly on judgment and partly on calculations to complete a design for a building. Exercise of judgment requires assessment of the consequences and penalties of failure of the structure. If failure such as complete collapse would result in loss of human life, the designer makes conservative selections of allowable stresses. He selects live loads based on extreme conditions. He exercises every precaution to make the likelihood of failure exceedingly remote, even though these precautions may markedly increase the cost of the structure.

In contrast, the designer or owner of a simple shelter for cattle may tolerate a greater risk of failure to obtain worthwhile savings. Structural collapse might not be disastrous. If the designer is employed by a commercial organization or public service agency that publishes and distributes standard designs in large quantity, he must balance the extra cost imposed on all who use an overly conservative design against the potential penalty of failure in a few instances with a less conservative design.

The probability of failure always exists in any structure. The engineer's task is to produce a design whose probability of failure is compatible with consequences of failure and the added cost of making it less probable.

Engineering design is essentially an organized process of rendering judgment and making decisions. A skillful engineer utilizes all available means to support his judgment and decisions. In formalized courses, the emphasis is on analysis and design calculations, since these are among the most useful and powerful aids to judgment.

1.4 DESIGN GUIDES, AIDS, AND SHORTCUTS

The design of any structure requires many detailed computations. Some of these are of a routine nature. An example is computation of allowable bending moment for standard sizes, species, and grades of dimension lumber. Numerous tables and graphs are available to minimize or eliminate such routine and repeated computations.

Standard construction and assembly methods have evolved through experience and need for uniformity in the construction industry. These have resulted in standard details and standard components for building construction published in handbooks or guides.

Many designs are for structures that must meet local or area building codes. These often specify design loads, quality of materials, standard construction details, and other design and construction requirements. These must be met, and therefore serve as design guides.

Since structural design is often a "cut-and-try" process, the number of "cut-and-try" sequences often can be reduced by using information from previously executed designs for structures with comparable configurations and loadings. Therefore, designers should be familiar with the sources for standard plans and designs. One source with which designers of farm structures should be familiar is the Midwest Plan

1.4 Design Guides, Aids, and Shortcuts

Service (MWPS), Iowa State University, Ames, Iowa 50011. It is an official activity of 12 midwestern universities and the U.S. Department of Agriculture. The MWPS publishes design guides and building plans. This is done under the direction of agricultural engineers and consulting specialists. Particularly useful to farm structure designers are these MWPS publications:

i. *Handbook of Building Plans.*
   MWPS 20.
   1978 or later edition.
ii. *MWPS Structures and Environment Handbook.*
   MWPS 1.
   10th ed. 1980.
iii. *Professional Design Supplement*
   to the MWPS Structures and Environment Handbook.
   MWPS 17.
   5th ed. 1978.

2

Fundamental Concepts of Stress Analysis

In this chapter we will show how to apply concepts of stress, static equilibrium, and free body diagrams for stress analysis of coplanar statically determinate structures. A coplanar structure is one in which all the members, loads, and reactions are in the same plane. A statically determinate structure is one that can be analyzed by applying only equations of static equilibrium, without resorting to analysis of bending or axial deformation. The concepts in this chapter are fundamental to all stress analysis work. They should be understood clearly by engineers who design load-carrying frames or members. The nomenclature used in this chapter is listed at the end of the chapter.

2.1 PURPOSE OF STRESS ANALYSIS

A stress analysis is an organized set of computations used to calculate reactions and total stresses (bending moment, shear, direct stress) in a frame or member when specified loads are applied. For example, the stress analyst may be given the overall configuration of a frame for a machine or building and the loads to carry. His task is to compute the total stresses (bending, shear, direct) at critical parts of the frame, and the reactions, say, at the wheels for a machine frame or the foundation of a building frame. The results of the stress analysis are used to design the load-carrying members and joints in frames and other members.

The reactions need to be evaluated for designing the bearings, supports, or foundations. Total stresses are needed to compute member cross-section shape and size so unit stresses will not exceed some allowable or limiting value.

A complete stress analysis of a structural frame or member will give (1) type of stress, (2) magnitude, (3) direction, and (4) point of application, for each total stress which needs to be known for sizing the member, designing fastenings, or designing foundations. Results of the stress analysis can be displayed in tables, shear and bending moment diagrams, or free body diagrams.

It is important to distinguish between total stress and unit stress. A *total stress* is an internal force (shear, bending, or axial force) acting at a designated section in a structural element. For example, an external axial load applied to the ends of a member used as a column produces a total axial force or total stress within the column. For brevity, *stress* usually means *total stress*. The meaning is usually clear from the context.

Reactions are total forces transmitted at a point of attachment of one part of a frame to another, especially at a fastening or support such as a foundation. Examples are the reaction force transmitted from the heel of a truss onto a girder or other support element, or where a foundation supports the end of a column.

Total stresses and reactions may be expressed in lb force (lbf), kilo pounds (kip), or Newtons (N), when the stress or reaction is a shear or axial force. If the total stress is a bending force it may be expressed as foot pounds force (ft · lbf), inch pounds force (in. · lbf), or meter Newtons (m · N).

Unit stress is the stress per unit area of the section at which the unit stress is to be calculated. For example, the axial unit stress in a column is calculated by dividing the axial load applied to the column by the column cross-section area. The maximum unit bending stress at a section is calculated by dividing the total bending moment by the section modulus, an index of the cross-section geometry. Unit stress may be expressed as $lbf/in.^2$ (psi), $kip/in.^2$ (ksi), or pascals (Pa).

An *external load* is a force applied to a structural element from without. An example is a weight (the external load) applied at midspan of a girder.

The terms allowable stress and applied stress are used in stress analysis and design. *Allowable stress* is the maximum stress per unit area that can be resisted by a structural element without exceeding a value permitted by standards. *Applied stress* is the actual unit stress that occurs at a section due to applied loads. Generally, the applied stress at a section should not exceed the allowable stress.

2.2 KINDS OF STRESSES

The three kinds of stresses for which analyses are made in coplanar structures are (1) axial stress, (2) shear stress, and (3) bending stress.

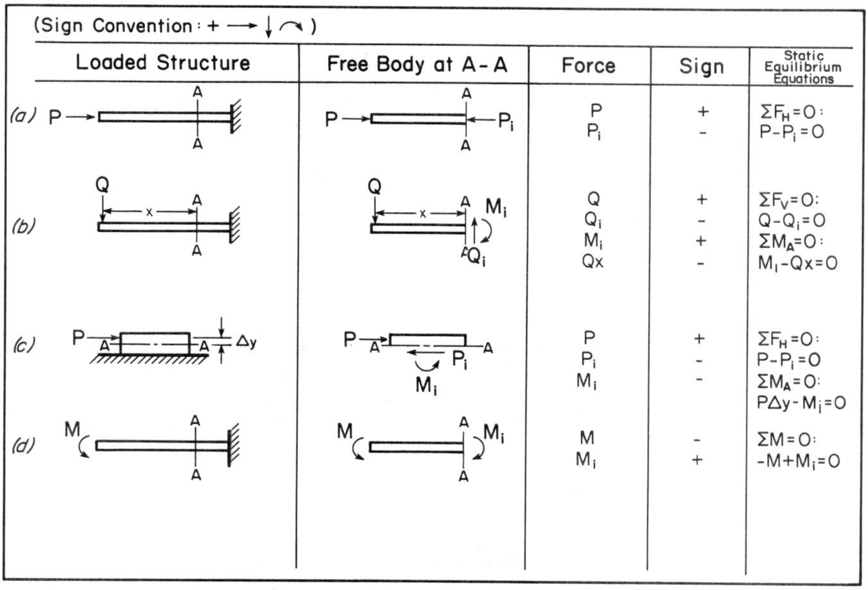

2.2 Kinds of Stresses

	Loaded Structure	Free Body at A-A	Force	Sign	Static Equilibrium Equations
(a)	(Sign Convention: $+ \rightarrow \downarrow \curvearrowright$)		P P_i	+ −	$\Sigma F_H = 0:$ $P - P_i = 0$
(b)			Q Q_i M_i Qx	+ − + −	$\Sigma F_V = 0:$ $Q - Q_i = 0$ $\Sigma M_A = 0:$ $M_i - Qx = 0$
(c)			P P_i M_i	+ − −	$\Sigma F_H = 0:$ $P - P_i = 0$ $\Sigma M_A = 0:$ $P\Delta y - M_i = 0$
(d)			M M_i	− +	$\Sigma M = 0:$ $-M + M_i = 0$

Fig. 2.1. Kinds of free body stresses in structural members.

Axial stress is the internal force at a transverse section of a member tending to either shorten (compressive axial stress) or elongate (tensile axial stress) the member parallel to its longitudinal axis. An example is given in Fig. 2.1a. Here, external axial force P applied to the end of the member produces internal axial stress (compression) P_i at Section A-A.

Shear stress is an internal force tending to displace the member on one side of a transverse section with respect to the member on the other side in a direction parallel to the plane of the section. Transverse shear is illustrated by Fig. 2.1b. Shear stress parallel to the longitudinal axis of a member is illustrated in Fig. 2.1c.

Bending stress is internal stress developed at a section to resist external forces tending to change the curvature of the longitudinal axis of a member loaded by external bending. Illustrations are given in Fig. 2.1b and d. In Fig. 2.1b, external bending moment due to external force produces bending moment M_i at section A-A. Internal bending moment M_i resists and is numerically equal but opposite in sense (clockwise vs. counterclockwise) to bending moment Qx in Fig. 2.1b or M in Fig. 2.1d. An applied bending moment at a section not accompanied by shear may be called a *couple*, because it can be produced by two equal and opposite coplanar forces equidistant from the

10 Fundamental Concepts of Stress Analysis

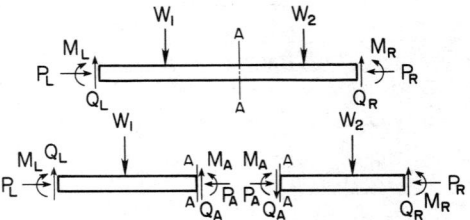

Fig. 2.2. Forces at a cut section.

member neutral axis. Bending moment M, Fig. 2.1d is an example of a couple. Bending moments at a section are in opposite directions on the cut faces to the right and left, respectively, at the section.

2.3 SIGNS AND SYMBOLS

To describe completely a load, stress, or reaction, we must give magnitude of the force, its direction, and point of application. For example, the required cross section size for a long, slender, axially loaded member depends on whether it is loaded in compression and must be designed as a column; or, on the other hand, loaded in tension, and designed as a tensile member. In stress analysis, the directions in which forces are transmitted affect other forces. An unambiguous method for specifying both magnitude and direction is a basic requirement for stress analysis.

    The magnitude of a force is given without ambiguity by a number followed by the kind of units of measure, for example, 1500 inch pounds of bending moment.

    The direction of a force is specified by pictorial symbols and algebraic signs.

    Pictorial symbols for direction are always used on free body diagrams of frames, members, or parts. They are unambiguous and almost self-explanatory.

    For example,

→   is a rightward force

↓   is a downward force

↻   is clockwise moment

Illustrations of their use are given in Figs. 2.1 and 2.2. Each case in

2.3 Signs and Symbols

Figs. 2.1 and 2.2 illustrates a loaded member and the internal stresses (forces) exposed at the cut section A-A. We note that the internal stresses in Fig. 2.2 are in opposite directions on the two faces at the cut section. This is always true.

Algebraic signs, plus (+) and minus (−), are used to denote directions in writing equations of static equilibrium. Mistakes with algebraic signs and algebraic manipulation seem to be the greatest single cause of wrong results in stress analysis.

By direction of a force, we mean rightward versus leftward, upward versus downward, or clockwise versus counterclockwise in the case of moments. An algebraic sign alone is ambiguous for specifying direction unless accompanied by a statement telling which direction is plus (+). This statement is called the sign convention. For a coplanar system of forces, we need three sign conventions; one each for two orthogonal lines of action (usually horizontal and vertical), and one for moments and couples.

The sign convention is an essential part of every stress analysis. It should be written down whenever an algebraic equation of static equilibrium of forces or moments is written. Otherwise, mistakes and confusion are apt to occur. The sign convention is usually written as a part of the statement that identifies the equilibrium equation.

For example,

ΣF_H +→: An equation summing forces in the horizontal direction (ΣF_H) follows. In that equation, the sign convention (+→) is plus (+) for rightward (→) forces. The minus sign is for leftward forces.

ΣF_V + ↓ : An equation summing forces in the vertical direction follows. In that equation, the plus sign is used for downward forces (↓).

ΣM_P ↻: An equation summing moments about point P follows. The plus sign is for moments in the clockwise direction ↻ about P.

In every equation or static equilibrium, there may be one or more direct forces of unknown magnitude and direction, and one or more couples or moments of unknown magnitude and direction. For example, we may have an equation with an unknown applied force Q_1, and an unknown internal bending moment M_1. The force Q_1 has a known line of action. The moment M_1 has a known point of application. The question is: Which algebraic sign should accompany unknown Q_1 and M_1, respectively, in equations of static equilibrium summing direct forces and moments?

The question is only partly resolved by writing down the sign convention for each equation. This only tells us what sign to use if we know the true direction of the force. We note that in algebraic equations

12 Fundamental Concepts of Stress Analysis

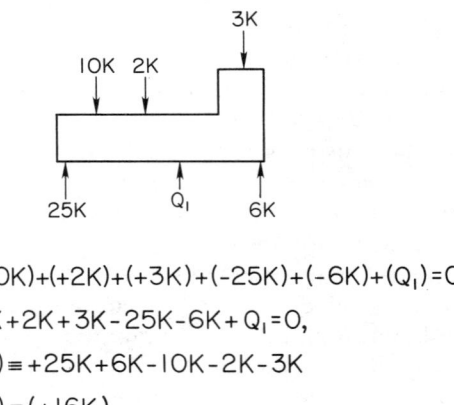

$\Sigma F_v + \downarrow$:

$(+10K)+(+2K)+(+3K)+(-25K)+(-6K)+(Q_1)=0,$
or
$10K+2K+3K-25K-6K+Q_1=0,$

$(Q_1) \equiv +25K+6K-10K-2K-3K$

$(Q_1) \equiv (+16K)$

Fig. 2.3. Application of sign convention.

for force equilibrium, we use the same algebraic signs for two completely different meanings, summation and direction. The use of the plus sign for summation is valid regardless of whether we are summing forces, apples, or the number of gold bars at Fort Knox.

The direction meaning of plus and minus is unique to the particular system with which we are dealing, in the present case, a force system. Therefore, the force Q_1 of unknown magnitude and direction is simply summed into the equation with a plus (+) sign without attempting to include a direction sign, since this is unknown. When the algebra is performed correctly to solve for Q_1 both the correct sign and numerical magnitude will emerge. This is illustrated in the example of Fig. 2.3. In the diagram, we show the symbol for the unknown force Q_1 directed upward, but this is only a guess, which may prove to be wrong. In this case, we can tell by inspection that it is wrong, but in more complex systems, we have to resort to solving algebraic equations with paper and pencil. The sign convention ($\Sigma F_V + \downarrow$) is established below the diagram of Fig. 2.3, and an equation of static equilibrium is written to sum vertical forces. We use two sets of signs in the equation. The sign inside each bracketed term has the direction meaning for that particular force. The plus (+) sign between bracketed terms has the summation meaning. When the brackets are eliminated, it is apparent that the summation sign, because it is always plus, has no effect on the direction sign.

This discussion may be summarized by the following rule: "Sum unknown forces into an equation of static equilibrium with a plus sign.

When the equation is solved, the correct direction sign according to the sign convention will emerge." When the equation in Fig. 2.3 is solved, $Q_1 = +16K$. The sketch should be corrected immediately to show Q_1 downward so that it is consistent with the results.

The same reasoning is used for unknown couples or moments whose directions, clockwise or counterclockwise, are unknown. They are written into the equation preceded by a plus sign for summation. When the equation is solved, the proper direction sign will emerge. Note that a positive force can produce either a positive or negative moment, depending on the location of the force with respect to the point about which moments are summed.

Whoever is writing an equation is free to choose the sign convention for that equation. Normally, we become so accustomed to using a particular convention that we may neglect to write it down each time, but this is not recommended.

2.4 EQUILIBRIUM OF FORCES

Building structures and parts must be designed to maintain integrity of shape and configuration after forces are applied. Otherwise the structure will collapse or develop other undesirable displacement. A frame, member, or other part of a structure that remains essentially at rest when forces are applied is said to be in static equilibrium. In general, forces acting on parts of a structure include (1) reactions, for example, the reaction force developed on a frame at its foundation; (2) external loads, for example, a pressure exerted by grain on the walls and floor of a bin; and (3) internal stresses, for example, bending and shear stresses in a beam. All of these appear as forces on a part of a structure or member when it is conceptually detached by cutting it loose from other parts as explained in section 2.5, Free Body Diagrams.

An additional requirement is dynamic equilibrium whenever significant inertial forces are generated due to acceleration of the structure. Examples are building frames subjected to blast or earthquakes; or machine frames subjected to vibration or shaking. In this book, we deal only with structures in which such inertial forces are insignificant.

Static equilibrium must exist for every part. For a coplanar system each part will be in static equilibrium if it satisfies the equations of static equilibrium. For a static coplanar system, there are three such equations.

14 Fundamental Concepts of Stress Analysis

$$\sum F_x = 0$$
$$\sum F_y = 0$$
$$\sum M_p = 0$$

where ΣF_x is the summation, taking direction into account, of all forces on the body in the x direction, ΣF_y is the summation of all forces in the y direction; and ΣM_p is summation of all moments and couples in the x–y plane tending to rotate the body about some coplanar point p. The x and y directions may be any two orthogonal directions. Some other convenient notation such as H or V for horizontal and vertical may be used instead of x and y.

These three equations of static equilibrium of forces and stresses may be written for any free body, frame, member, or part of a structure. These equations are independent, and give direct solutions for the directions and magnitudes of as many as three unknown coplanar independent forces.

If more than three independent, unknown forces exist, the others must be solved by some other method (such as an elastic analysis) or evaluated from other information. For example, a three-hinged frame has four independent reaction components, or two each at the two supports. Four independent equations are needed. Three equations of static equilibrium for the entire frame can always be written. A fourth equation can be written for the circumstance that moment is zero at the internal hinge. An equation summing moments, $\Sigma M = 0$, about the internal hinge can be written. The portion of the frame on either side of the hinge can be used.

Writing equations of static equilibrium for parts of structures must become second nature for a stress analyst, who should use a clear, logical system and notation when writing each equation. This will help to eliminate confusion and mistakes. First, a statement should be made to show which of the three equations is being written. Next, a symbol should be shown to denote the sign convention used. Then the complete equation is written using numerical values for known forces and letter symbols for unknown forces. Signs must agree with the sign convention adopted. All numerical values and letter symbols should be written on the left-hand side of the equation and equated to zero. For example, the three equations of static equilibrium for the part shown in Fig. 2.4 should be written:

$$\sum F_H = 0, \overset{+}{\rightarrow}: (H_L) - 30K - 10K = 0$$
$$\sum F_V = 0, +\uparrow: (V_L) + (V_R) - 5K = 0$$
$$\sum M_R = 0, \overset{+}{\curvearrowright}: (5V_L) - 3 \times 5K + (M_R) = 0$$

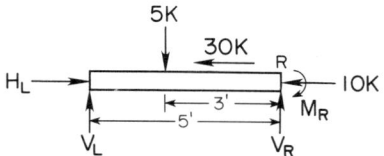

Fig. 2.4. Static equilibrium of forces.

which can be solved to get

$$H_L = + 40K$$
$$V_L = + 5K - V_R$$
$$V_R = + 2K + M_R/5$$

where H_L is the horizontal force at point L; V_L, the vertical force applied at point L; M_R, the moment applied at R; V_R, the vertical force at R.

An additional equation needed to solve for all four of the unknown forces must come from another source of information since, in general, only three independent equations of static equilibrium can be written for a coplanar system of forces on a free body.

The consistent use of an orderly notation as illustrated in each of the three foregoing equations will help to eliminate mistakes and blunders.

2.5 FREE BODY DIAGRAMS

Free body concepts and preparation of valid free body diagrams are basic to all structural analyses and design. To write equations of static equilibrium, the first step is to prepare a valid free body diagram. You can analyze stresses in any statically determinate structure if you can draw the valid free bodies, write the valid basic equations of static equilibrium, and then correctly perform the arithmetic and algebra. Free body diagrams are used extensively in analysis of statically indeterminate structures.

A free body is a schematic diagram of either all or a part of a structure. It shows the point of application, direction, and magnitude of all known forces; and the point of application and line of action for unknown forces. Diagrammatically and conceptually, a free body must be completely detached or cut loose from any physical connection or

16   Fundamental Concepts of Stress Analysis

support from any other part of the structure. Forces shown on the free body must include all stresses which were acting internally at sections where the part is cut loose, and all external loads or reactions on the part. Symbols show direction, point of application, and line of action. Supplementary data with each symbol includes an identifying letter for unknown forces, or, when known, the numerical value of the force. For stresses or reactions whose directions are unknown and must be determined by analysis, the positive direction is in effect assumed as previously explained. When the equations of static equilibrium are solved, the correct sign will be revealed. Then, the free body diagram should be corrected immediately.

A free body diagram is either valid or invalid. A valid free body diagram is one which includes all the forces correctly displayed on the free body. An invalid free body has one or more forces which are omitted, incorrectly displayed, or inconsistent in direction with adjoining free bodies. Of course, invalid free bodies are worse than useless because they lead to incorrect equations of static equilibrium and incorrect results.

A free body diagram may include an entire frame, member, part of a member, or joint. The portion to include in a free body is determined by making sections to create free bodies wherever reactions or internal stresses are to be computed.

To prepare free body diagrams, you must follow the key requirements and procedures:

1. The free body must be completely cut loose at sections where you desire to analyze stresses or reactions.
2. The free body must be *completely cut loose* from the rest of the structure or part. This typically requires two or more cuts for a member to obtain one free body.
3. All internal stresses of known or unknown magnitude at cut sections must be shown as forces acting at the cut on the free body.
4. All external forces (loads, reactions) on the free body must be shown. A common mistake in free body analysis is failure to include all external forces on the part and all internal stresses at cut sections.
5. The free body should include a finite, even if small, part of the structure. For a free body of a joint, a portion of each member meeting at the joint should be shown. Then the stresses acting in each member can be shown without unduly crowding the sketch. This is illustrated in Fig. 2.5. We understand that conceptually,

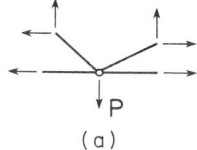

(a)

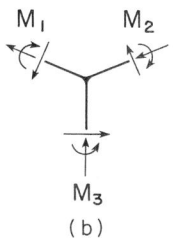

(b)

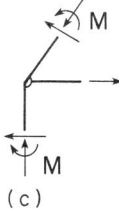

(c)

Fig. 2.5. Joint free bodies.

the lengths of the members included in the joint free body approach zero so that shear stresses, if any, in the members develop zero moment about the joint. Therefore, in Fig. 2.5b it is known immediately that

$$M_1 + M_2 - M_3 = 0$$

since the parts shown meeting at the joint approach zero length. In Fig. 2.5c two members are joined by a rigid connection. The intermediate member is joined by a hinge to the other two members. We note that any member terminating in a hinge at a joint can have only axial forces unless the member also has both bending and shear forces acting on it. Otherwise, the member would rotate about the pinned joint. Thus, in Fig. 2.5a the resultant of the horizontal and vertical stress components in each member must be collinear with the member.

18 Fundamental Concepts of Stress Analysis

After the free body diagram has been drawn clearly and correctly, the sign convention should be written. Then the corresponding equations of static equilibrium are written with care to ensure that they are consistent with the free body and the sign convention.

2.6 SUPPORT CONDITIONS

Every structure resting on the ground must have suitable supports to transmit loads on the structure into the ground. The kinds of forces the supports can resist depend on the support configuration. Supports can be arranged to resist only one or any combination of the three possible kinds of forces, viz., horizontal, vertical, and moment, in a coplanar system. This is done by appropriate combinations of three basic, idealized support elements. These, and their pictorial symbols are:

Frictionless hinge

Frictionless roller

Immovable base

The frictionless hinge can transmit forces in any coplanar direction but not moment. The frictionless roller can transmit forces normal to but not parallel to the direction of rolling. Also, it can transmit a couple applied to a plate resting on the rollers. Conceptually, such a roller can consist of a frictionless four-wheeled dolly, with the wheels confined to roll within a channel-shaped pair of tracks. The dolly resists a couple because the wheels cannot be disengaged from the track. The immovable base transmits all forces into the ground without displacement or rotation. In reality, the immovable base can only be approximated by a well-designed foundation.

The seven possible combinations of these support elements are illustrated in Fig. 2.6. In real structures, any of these combinations may occur, although (a), (d), and (g) are encountered more frequently than the others in typical foundations and supports. Transverse Q forces applied to a member, supported as in (b) and (d) of Fig. 2.6, will cause rotation of the member if the distance Δy is finite. We prevent this by conceptually shrinking Δy to approach zero. Then the supports can only transmit the forces shown because they are applied to the hinge pin.

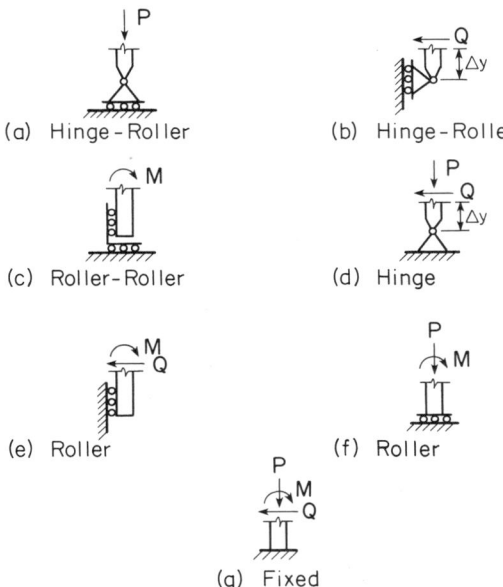

Fig. 2.6. Support conditions.

In actual construction, it is impossible to build completely frictionless hinges or rollers, or completely immovable bases. By careful design and construction they can be approached as closely as necessary. In some cases, this may require an actual roller, for example. In other cases, a hinge-roller may be approximated by a connection which yields fairly readily, up to a point. For example, wooden trussed rafters are normally held down by metal clips at the trussed rafter supports. In some instances, only toenailing may be used. Such a connection can yield rather freely by a small amount without destroying the joint when either horizontal or vertical forces or moments are applied. After a small amount of yielding, the connection develops significant resistance to horizontal movement.

Support conditions assumed or specified in analysis and design can be altered in the real building due to carelessness in construction or unexpected natural effects. For example, the designer may wish to utilize resistance to uplift that can be developed by the end of a pole or rigid frame member extending into the ground; or by a rigid attachment between a reinforced concrete column and a concrete pier extending into the ground. If the ground shrinks away from the pole due to natural drying, uplift resistance may be almost completely lost

20 Fundamental Concepts of Stress Analysis

and condition (c) or (e) may be approximated; see Fig. 2.6. If a careless builder neglects to provide reinforcement continuity from a concrete pier into the column, resistance to moment and uplift may be lost. Such changes may transform the stable structure conceived and analyzed by the designer into an unstable structure in actual use, with disastrous consequences.

The stress analyst must understand clearly both the idealizations and practical realities of the support conditions for the structure he is analyzing. He should consider the possibility that construction errors or natural causes may transform the designed supports to those which allow the structure to become unstable and in danger of collapse; or statically indeterminate and stressed differently, compared to an original statically determinate support system.

2.7 APPLICATION OF FREE BODY CONCEPTS

Figures 2.7 and 2.8 illustrate how to prepare free body diagrams and write accompanying equations of static equilibrium for stress analysis of coplanar, statically determinate frames. The frame, Fig. 2.7a, is loaded by one horizontal and one vertical load. Its statically determinate supports include a hinge on an immovable base for the left column and a hinge-roller on an immovable base at the right column. All joints, other than at the supports, are rigid. Sections A–L, Fig. 2.7a, are made at locations where the analyst wants to compute internal stresses. Additional or alternative sections might be chosen, if more detailed knowledge of internal stresses is desired.

A free body of the *entire frame* is prepared first, as illustrated in Fig. 2.7b, to compute reactions. Notice that the three free body equations in Fig. 2.7b assume that the unknowns, V_A, Q_A, and V_L, are in the positive directions. Subsequent calculations will reveal the correct directions. Then, free body diagrams for parts of the frame of interest to the analyst are prepared. Equations of static equilibrium are written for each free body and solved to determine internal stresses at sections as illustrated in Fig. 2.8. This procedure, called *free body analysis,* starts with a part that includes, at one end, known reaction forces, and proceeds to the next adjoining part, until all free bodies have been analyzed. For each free body, the directions of the internal stresses exposed at the cut sections should be compatible with the same cut sections shown on an adjoining free body. Adjoining ends of two free bodies at a common section always have corresponding internal stresses in opposite directions. For example, the left column is known to be in

2.7 Application of Free Body Concepts   21

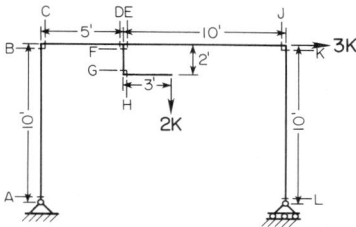

(a) Frame Configuration and Loads

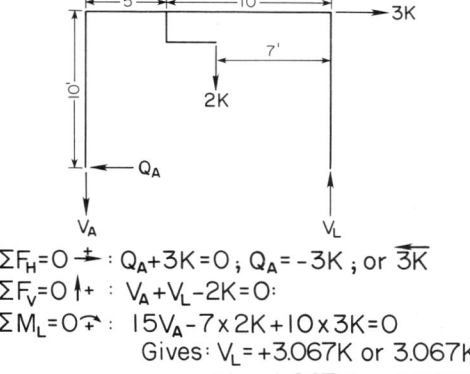

$\Sigma F_H = 0 \xrightarrow{+}: Q_A + 3K = 0; Q_A = -3K;$ or $\overleftarrow{3K}$
$\Sigma F_V = 0 \uparrow +: V_A + V_L - 2K = 0:$
$\Sigma M_L = 0 \curvearrowright: 15 V_A - 7 \times 2K + 10 \times 3K = 0$
   Gives: $V_L = +3.067K$ or $3.067K \uparrow$
   $V_A = -1.067K$ or $1.067K \downarrow$

(b) Frame Free Body Diagram and Static Equilibrium Equations

Fig. 2.7. Frame and its free body analysis.

tension because the free body analysis of the entire frame, Fig. 2.7b, reveals that the left vertical reaction component pulls down on the left column. Therefore, in Fig. 2.8a the tension force V_B must pull up to produce tension in the left member. In Fig. 2.8b, the same force must pull down on the corner joint, because the column is in tension.

One end or cut section of each free body will have stresses that are unknown in direction and magnitude until the static equilibrium equations for that free body have been solved. The positive direction is assumed and the stress simply summed into the equation. When the true direction is revealed, the free body diagram should be corrected immediately.

Note again that, regardless of what direction is shown for an unknown stress on a free body diagram, the symbol for that stress or load is preceded by a plus (+) for summation of forces. For example, in the equation for ΣF_H in Fig. 2.8a, the sign preceding the unknown Q_B is

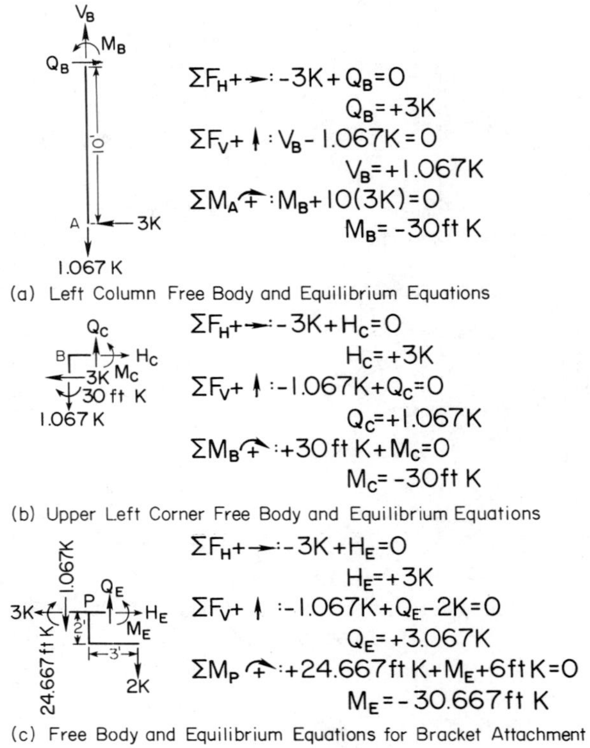

22 Fundamental Concepts of Stress Analysis

(a) Left Column Free Body and Equilibrium Equations

$\Sigma F_H + \rightarrow : -3K + Q_B = 0$
$Q_B = +3K$
$\Sigma F_V + \uparrow : V_B - 1.067K = 0$
$V_B = +1.067K$
$\Sigma M_A \curvearrowleft : M_B + 10(3K) = 0$
$M_B = -30 \text{ ft K}$

(b) Upper Left Corner Free Body and Equilibrium Equations

$\Sigma F_H + \rightarrow : -3K + H_C = 0$
$H_C = +3K$
$\Sigma F_V + \uparrow : -1.067K + Q_C = 0$
$Q_C = +1.067K$
$\Sigma M_B \curvearrowleft : +30 \text{ ft K} + M_C = 0$
$M_C = -30 \text{ ft K}$

(c) Free Body and Equilibrium Equations for Bracket Attachment

$\Sigma F_H + \rightarrow : -3K + H_E = 0$
$H_E = +3K$
$\Sigma F_V + \uparrow : -1.067K + Q_E - 2K = 0$
$Q_E = +3.067K$
$\Sigma M_P \curvearrowleft : +24.667 \text{ ft K} + M_E + 6 \text{ ft K} = 0$
$M_E = -30.667 \text{ ft K}$

Fig. 2.8. Free body diagrams of frame parts.

plus (+) regardless of the sign convention for F_H. This plus sign stands for summation in the equation, rather than the unknown direction of Q_B.

The sign preceding unknown moments on a free body diagram corresponding to internal bending stresses is always plus (+) in an equation summing moments. For example, Fig. 2.8c shows the free body for the bracket attachment. The unknown moment M_E is assumed to be in the counterclockwise direction on the free body. This is the negative direction according to the sign convention adopted for that equation. Nevertheless, M_E is preceded by a plus (+) for summation sign in the equation. Solving the equation indicates that M_E is negative according to the sign convention adopted.

When equations have been solved for the magnitude and direction of unknown forces, couples, or moments, the correct sign should always be shown, plus (+) or minus (−). Otherwise, uncertainty will arise

2.8 Shear and Moment Diagrams

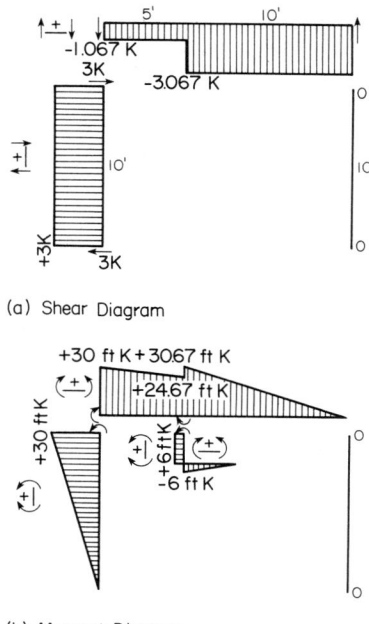

(a) Shear Diagram

(b) Moment Diagram

Fig. 2.9. Shear and moment diagrams for frame members.

later whether the sign is plus, or whether the analyst simply forgot to write down the correct sign.

2.8 SHEAR AND MOMENT DIAGRAMS

Shear and moment diagrams for structural components with bending forces are useful and valuable aids to the designer. They tell at a glance where maximum and minimum shear and bending occur, what their magnitudes are, and how they vary.

A shear diagram is a graph of internal transverse shear at all points along a member such as a beam. A moment diagram is a graph of internal bending moment at all points along a member. Figure 2.9 illustrates shear and moment diagrams for the members of the frame analyzed in Fig 2.7, except for shear in the small bracket.

24 Fundamental Concepts of Stress Analysis

A nonambiguous sign convention must be adopted for shear and moment diagrams. A nonambiguous sign convention is a small, free body symbol, shown by the shear or moment diagram as illustrated in Fig. 2.9. The symbol is a small free body element with internal shear, moment, or direct stress at the ends of the element and a plus sign to indicate the sign convention, as follows:

Shear: ↑ —+— ↓

Moment: ⤴ —+— ⤴

Direct stress: ← —+— →

The opposite sign conventions would be just as valid.

2.9 PREPARATION OF SHEAR AND MOMENT DIAGRAMS

Shear and moment diagrams for a member are prepared from a free body diagram of the member after all unknown forces (stresses, reactions, and loads) have been evaluated. It is usually convenient to draw the moment diagram below or adjoining the shear diagram. For horizontal members, construction of the shear diagram customarily begins at the left end of a member. The known shear there is plotted to a suitable scale above or below the baseline according to the sign convention adopted. Next, shear is plotted at other significant points along the member, such as at point loads, or where the character of distributed loads changes. Free body diagrams should be used to evaluate shear at these points.

Then, connecting lines are drawn to show how shear varies between these points. For example, shear V_X due to uniformly distributed gravity load on a horizontal member is

$$V_X = V_L - wX$$

where V_L is shear at the left end, w is the load intensity, and X is distance measured to the right from V_L. This is the equation for a straight line with slope $-w$. For a linearly varying load intensity, shear varies according to the expression for a second-degree curve:

$$V_X = V_L - wX - aX^2/2$$

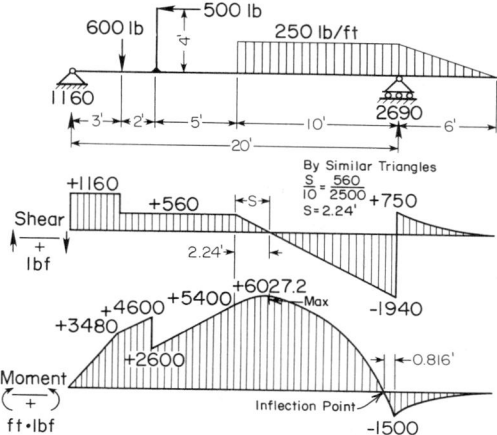

Fig. 2.10. Shear and moment diagrams for a horizontal member.

where w is load intensity at the left end, and a is increase in load intensity per unit length of member. At each point load normal to the axis of the member, shear changes abruptly by the amount of the load.

The moment diagram is constructed by free body analysis from information in the shear diagram. Additional information needed includes the bending moment at one end, and any bending moment or couples introduced from other members attached at intermediate points. An example is the bending moment introduced into horizontal member C–J from the small bracket attached at F in Fig. 2.7a. Such a moment causes a step change in the moment diagram as shown in Fig. 2.9b. It is seldom necessary to evaluate the moment at all points. A connecting line is drawn according to the way that the moment is known to vary between critical points where moment has been computed. For example, the moment in a beam carrying only point loads varies linearly between the loads. If the load is uniformly distributed, the moment varies as a second degree curve, or parabola. Fig. 2.10 illustrates these concepts.

When preparing bending moment diagrams, a basic relationship between the shear in the member and the change in bending moment M, due to shear, can expedite preparation of moment diagrams. This relationship is derived in Fig. 2.11. The relationship is valid for members carrying both distributed and point loads, even though the expres-

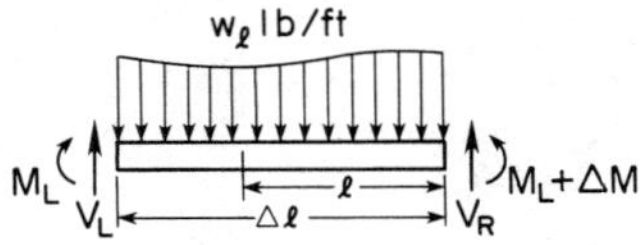

26 Fundamental Concepts of Stress Analysis

$$\Sigma M_R = 0, \curvearrowright:$$
$$M_L - (M_L + \Delta M) + V_L(\Delta \ell) - \int_0^{\Delta \ell} w_\ell \, \ell d\ell = 0$$

Divide by $\Delta \ell$:
$$-\frac{\Delta M}{\Delta \ell} + V_L - \frac{\ell}{\Delta \ell}\int_0^{\Delta \ell} w_\ell \, d\ell = 0$$

Let $\Delta \ell \to 0$:

Then: $\int_0^{\Delta \ell} w_\ell \, d\ell \to 0$

so: $\frac{\Delta M}{\Delta \ell}\Big|_{\Delta \ell \to 0} = \frac{dM_L}{d\ell} = V_L$

or $V_L = \frac{dM_L}{d\ell}$

in General:
$$dM = V \, d\ell$$

and $\Delta M \Big|_{\ell_1}^{\ell_2} = \int_{\ell_1}^{\ell_2} V \, d\ell$

or $M_{\ell_2} = M_{\ell_1} + \int_{\ell_1}^{\ell_2} V \, d\ell$

Fig. 2.11. Moment–shear relationships.

sion relating shear to distance along the member is not differentiable at a point load. Integration, or the corresponding arithmetical computation of area enclosed by the shear diagram, must be carried out within each length interval between point loads. This discussion may be paraphrased: The change in bending moment between two points along the beam is equivalent to the area of the shear diagram between two points, provided that no external moments are introduced between the points.

When a couple or bending moment is introduced from another member at some intermediate point between l_1 and l_2, an abrupt or step change occurs in moment at the point of attachment of the other member, without any corresponding change in the area enclosed by the shear diagram. The basic relationship between shear and bending moment in the member is valid, however, if applied along the length interval between introduction of couples or moments from connecting members.

An example of preparation of shear and moment diagrams for a member carrying several kinds of loads is shown in Fig. 2.10. The reader should verify the analysis of internal shear and moment, and the shear and moment diagrams.

2.10 INFLECTION POINTS

In some instances, a designer may need to know where the internal moments in a member become zero. Locations of zero internal moments are *inflection points*. An example is in the design of reinforced concrete beams. At or near inflection points, the tensile reinforcement location must be changed so that the beam will be adequately reinforced against tensile stresses.

Inflection points are identified by inspection of the moment diagram. For example, Fig. 2.10 reveals that the beam has an inflection point a short distance to the left of the right support. If the moment diagram were drawn to scale, the distance to the left of the support could be measured. Also, free body analysis that takes into account that the internal bending moment is zero at the inflection point can be used. Free body analysis to locate the inflection point is illustrated in Fig. 2.12. The free body includes the portion of the beam between the right end and the inflection point at unknown distance X from the right-hand support.

Note: The internal bending moment is zero at the inflection point. Write free body equation:

$\Sigma M_L = 0, \curvearrowleft+$

$$-2690X + 250(X)(X/2) + 250 \times 3 \times (X + 2) = 0$$

reduces to:

$$X^2 - 15.52X + 12.00 = 0$$

which gives $X = 0.816$ ft. Therefore, the inflection point occurs at 0.816 ft left of support.

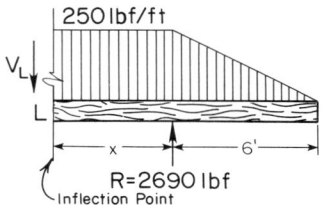

Fig. 2.12. Free body analysis for inflection point, beam illustrated in Fig. 2.10.

PROBLEMS

For each of the frames or beams in Figs. 2.13–2.21, compute the reactions and draw the shear and moment diagrams. Use free body analysis. Show your sign conventions. Double check the reactions before proceeding to analysis of internal stresses. Specify the position of each inflection point.

For the problems of Figs. 2.15, 2.16, and 2.17, convert to SI units and specify the shear and moment diagrams in SI units.

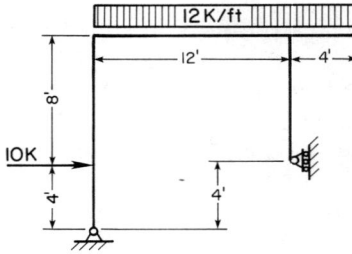

Fig. 2.13. Rigid frame with overhang and an unusual support.

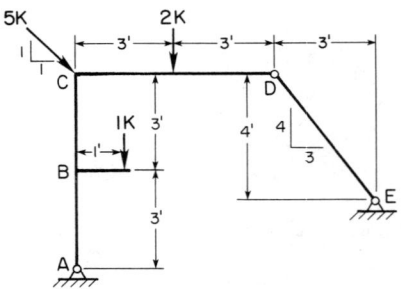

Fig. 2.14. Three-hinged rigid frame.

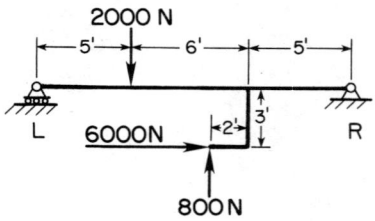

Fig. 2.15. Beam with a bracket.

Problems 29

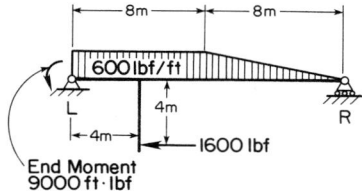

Fig. 2.16. Beam with distributed loads and a bracket.

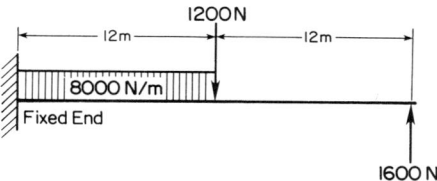

Fig. 2.17. Cantilevered beam with distributed and concentrated loads.

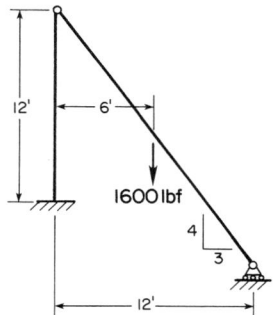

Fig. 2.18. Two-hinged triangular frame.

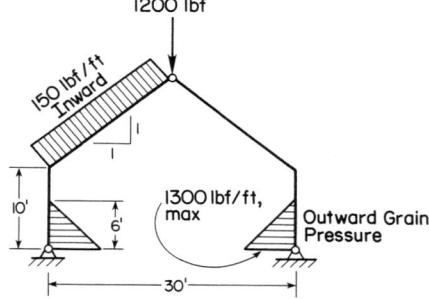

Fig. 2.19. Three-hinged gable frame.

30 Fundamental Concepts of Stress Analysis

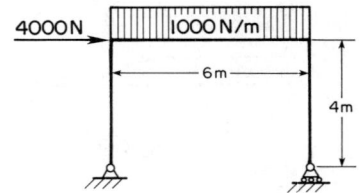

Fig. 2.20. Rectangular frame with uniformly distributed load.

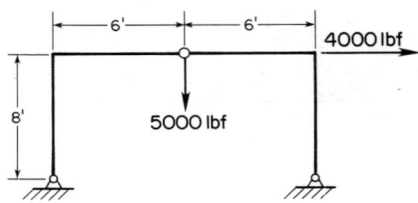

Fig. 2.21. Three-hinged rectangular frame.

NOMENCLATURE FOR CHAPTER 2*

a	Linear rate of increase in distributed load intensity on a member lbf/ft^2 (N/m^2)
F_h	Horizontal force, lbf (N)
F_v	Vertical force, lbf (N)
F_x	Force in x direction, lbf (N)
F_y	Force in y direction, lbf (N)
H	Horizontal load, lbf (N)
k, or kip	Applied load, 1000 lbf (N)
ksi	Unit stress, kilo pounds force per sq in. (Pa)
l	Distance measured along a member, in. or ft (cm or m)
lbf	Pounds force (N)
m	Meters (ft)
M	Applied bending moment, lbf · ft (N · m)
N	Force, Newtons
psi	Unit stress or pressure, lbf/in.^{2} (Pa)
P	Applied force, lbf (N)

*Quantities such as F, H, M, P, Q, V, W, Δ may be subscripted to show location, direction, or point of application.

Q	Applied force, usually transverse, lbf (N)
Pa	Pressure or unit stress, pascals
V	Vertical force or load lbf (N)
w	Uniformly distributed load, lbf/ft (N/m)
W	Concentrated load, lbf (N)
Δ	Small increment of length or distance, in. (cm)

3

Stress Analysis of Coplanar Statically Determinate Trusses

In this chapter we will explain basic concepts of truss configuration, how to compute stresses in the members of idealized statically determinate trusses, and how to compute truss deflections.

Trusses are versatile and have wide application in buildings and machines. Examples are hitch and frame assemblies for field machines, frames for buildings, and supports for long conveying systems. Every engineer who designs machines or building structures should have clear concepts of how trusses are arranged, and how to compute stresses in truss members. The nomenclature used in this chapter is listed at the end of the chapter.

3.1 CONCEPTS AND DEFINITIONS

Truss. We define a two-dimensional truss as a coplanar arrangement of bars (members) linked together by pinned joints, and intended to form a stable configuration under any general system of loads. Member center lines are concurrent at the joints. Loads are applied only at the joints. Some truss configurations are unstable, as will be explained later. Ideally, the pinned joints are frictionless and offer no resistance to rotation of the members about the pin. The longitudinal axes of all members and the lines of action of the loads are all in the same plane.

A truss ideally has a support system of three independent reaction forces which are neither all parallel nor all coincident. This makes the supports stable and statically determinate.

3.1.1. Stresses in Truss Members

Only axial forces exist in members of an idealized truss. A stress analysis of a truss consists in computation of the reaction forces at the

34 Stress Analysis of Coplanar Statically Determinate Trusses

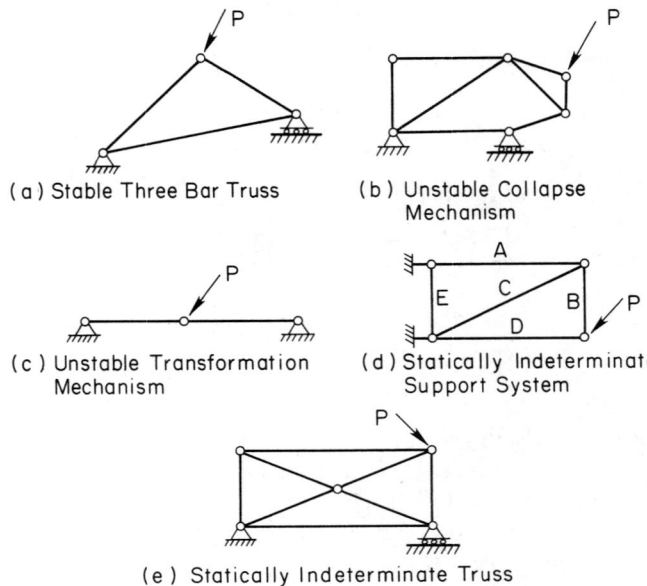

Fig. 3.1. Basic truss concepts.

supports and magnitude and character (tension or compression) of the axial force in each member. The computed values are used by designers to determine the size of each member, to design the fastenings at the joints, and to design the supports.

Shear and bending can exist in truss members if loads with transverse components are applied to members between joints, or if restraint against free rotation occurs at joints. These possibilities are ruled out in an ideal truss. Therefore, the basic problem dealt with here is how to compute axial stresses in idealized truss members.

3.1.2 Stable Configuration

A stable truss is one which has a stable configuration. A stable configuration is a system of bars linked together in such a way that it cannot behave as a mechanism under a general system of loads. A basic, three-bar, stable truss is illustrated in Fig. 3.1a. An unstable truss is a mechanism, i.e., a coplanar system of bars linked together in a way that permits unrestrained rotation, at least initially, of the bars about their joints when load is applied. The two kinds of mechanisms that can occur are (1) a collapse mechanism and (2) a transformation mechanism.

A collapse mechanism is illustrated in Fig. 3.1b. When load is applied, collapse into a stable mode occurs accompanied by severe structural damage.

The other, a transformation mechanism, is illustrated in Fig. 3.1c. Note that the supports are pinned and statically indeterminate. They develop no reaction to load until a slight rotation of the two members has occurred. This transforms the mechanism to a statically indeterminate system. The horizontal reaction components and axial stress in the bars depend on the elastic properties of the bars.

In contrast, the basic three-bar truss of Fig. 3.1a is stable, and is not a mechanism. Unstable configurations should be shunned by engineers because they are always in danger of sudden collapse.

Unstable collapse mechanisms are sometimes used inadvertently in buildings and other engineering structures. Collapse may not occur immediately when the structure is loaded because of fortuitous constraints such as continuity of a member through a joint where ideally a hinge occurs; or fortuitous constraint at the supports. Later, slightly higher loads may destroy the constraint, followed by sudden collapse of the structure. For example, this author has seen a collapsed roof which had been supported by a prototype of the truss configuration shown in Fig. 3.5d. The structure had carried the roof loads for a number of years because the upper chord members were continuous through the joints at 1, instead of pinned as idealized in Fig. 3.5d. At the time of collapse, heavy snowfall had occurred. The high bending stresses produced by the snowload caused bending failure at the joints 1–1 in the top chord members. The truss collapsed, dropping large quantities of snow and the wreckage of the roof into the building interior. Not only was the building virtually destroyed, but the contents were subsequently damaged by water from the snow melt. Designing unstable structures is one of the cardinal sins of engineering practice.

Unstable transformation mechanisms in trusses should also be avoided. Transformation to a statically indeterminate mode can be accompanied by very high member stresses. Members or joints may rupture, accompanied by collapse of the structure.

3.1.3 Static Determinacy

A statically determinate configuration is one for which all member stresses can be computed by applying equations for static equilibrium of forces. A statically indeterminate configuration is one for which all member stresses cannot be computed by statics alone, but also depend on the elastic characteristics of the members. Many desirable truss configurations are statically indeterminate.

A truss may be carried on statically indeterminate supports. An example is shown in Fig. 3.1d. Each support can develop two reaction components for a total of four, or one more than can be computed by the three general equations of static equilibrium. Therefore, the magnitude of the vertical reactions depends on the axial stiffness of truss member E, as well as the amount the reactions move when load is applied. Stresses in members A, B, C, and D are statically determinate.

Correctly designed statically indeterminate trusses need not be avoided. Engineers should learn to recognize them, and when the need arises, learn how to make stress analyses for them. In any statically indeterminate structure, whether the indeterminacy arises from the arrangement of the members, or supports, or both, stresses can be changed significantly by movement of the supports, or by temperature strain in the members. Analysis of statically indeterminate trusses is not covered in this book.

A fundamental characteristic of a statically indeterminate truss is that *no* member stresses can occur without superimposed loads and reactions. Thus, shortening or elongation of some member in a statically determinate truss due to temperature strain, for example, will not produce stresses in any of the members. A statically indeterminate truss, in contrast, can be stressed in the absence of superimposed loads and reactions. For example, temperature strain in one member produces stresses in all members of a statically indeterminate truss. This is an important characteristic which we will utilize later in the analysis of certain complex truss configurations.

3.2 STATICALLY DETERMINATE TRUSS TYPES

Statically determinate trusses can be divided into types on the basis of bar arrangement, as follows:

1. Simple
2. Compound
    a. Compounded by three-bar linkages
    b. Compounded by pin and bar linkages
3. Complex
    a. Statically determinate form
    b. Critical form

The method of stress analysis depends partly on the type of truss. Concepts of bar arrangement for the three types are based on the stable three-bar truss in Fig. 3.1a on stable, statically determinate supports.

3.2 Statically Determinate Truss Types

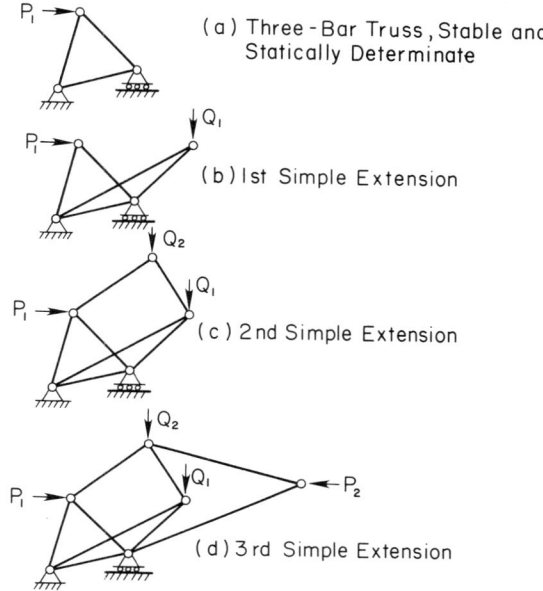

Fig. 3.2. Development of a simple truss.

Every simple truss has at least one, and every compound truss at least two stable three-bar truss configurations. A complex truss may or may not possess one or more three-bar trusses.

3.2.1 Simple Truss

A simple truss consists of a stable three-bar truss augmented by a succession of one or more simple extensions. A simple extension consists of two noncollinear bars linked together at one point, and with the other two ends linked to any two points on an existing simple truss. The development of a simple truss is illustrated in Fig. 3.2, starting with a basic three-bar truss. An indefinite number of simple extensions can be added. The simple truss of Fig. 3.2d is not particularly simple in outward appearance, but is simple by definition because it can be developed starting with a basic three-bar truss augmented with simple extensions. Most practical simple trusses will have a more orderly appearance.

We note in Fig. 3.2 that simple trusses may have one or more members that cross others. No connections (neither pinned nor rigid) between members occur at these crossings. The only connections are the pinned joints denoted symbolically by small circles. Quadrilateral enclosures, for example, as produced by the second simple extension in Fig. 3.2c, may occur in simple trusses. After the first simple extension, the original three-bar truss may lose its identity. Two stable, statically determinate support elements are assumed for the initial three-bar simple truss. After the test, the reactions are restored to their original locations.

It is easy to demonstrate that all simple trusses are stable and statically determinate. The original three-bar truss is stable and determinate. The first simple extension is statically determinate because only two members with a total of only two unknown axial forces are added. These can be computed by a free body analysis of the added joint. The extension is stable because the added joint is linked with two stable points on the original three-bar truss, forming a stable, three-point arrangement. Analogous reasoning can be applied to each simple extension used to produce the final simple truss configuration.

A simple truss always has bars equal in number to $2J - 3$, where J denotes the number of pinned joints in the truss. This is true because, starting with a three-bar truss and adding simple extensions, the progression of joint count J will be 3, 4, 5, 6, 7, ..., J; and the corresponding progression of bar count will be 3, 5, 7, 9, 11, ..., $2J - 3$. Each bar has an unknown axial stress. Therefore, $2J - 3$ unknown bar stresses exist in the members of a simple truss. Also, there are three unknown reaction forces in a statically determinate support system. The total number of unknown forces is therefore $(2J - 3) + 3$, or $2J$, i.e., twice the number of joints. Each of the J joints can be isolated as a free body, and two independent equations of static equilibrium, $\Sigma F_x = 0$ and $\Sigma F_y = 0$ written for each joint for a total of $2J$ independent equations. This is equal to the number of unknown stresses plus reaction forces. Therefore, we can always obtain a set of independent, simultaneous linear equations equal in number to the unknown bar forces and reactions. These will always have unique, nontrivial solutions because, as was demonstrated earlier, every simple truss is stable and statically determinate.

3.2.2. Compound Truss

A compound truss is a nonsimple truss obtained by properly interconnecting two simple trusses. Compound trusses may have unstable configurations. Interconnections can be either three nonparallel and noncoincident bars, each connecting a different pair of joints on the two simple trusses, as in Fig. 3.3a, or a joint shared by two simple trusses

3.2 Statically Determinate Truss Types

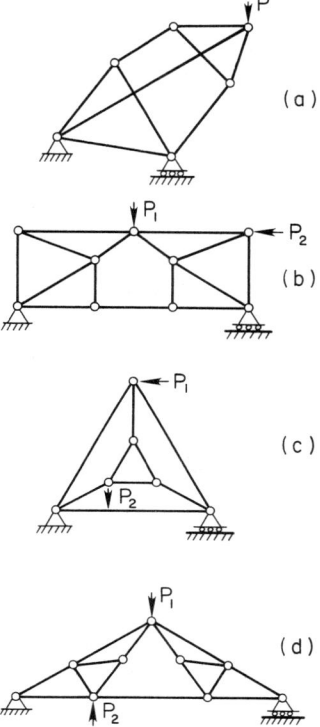

Fig. 3.3. Compound trusses.

plus a bar connecting two other points, one on each of the two simple trusses, as in Fig. 3.3b or d. These three links, or the common joint plus one link, may be regarded as reaction elements that furnish supports for one simple truss from the other. The three-bar interconnection is stable if, and only if, the three bars are neither all parallel nor their lines of action all coincident, i.e., with their lines of action, extended, meeting at a point. The interconnection shown in Fig. 3.3c is unstable because the three links joining the outer and inner simple trusses are coincident at the center of the inner configuration, and can develop initially no restraint against rotation of the inner simple truss with respect to the outer one. The interconnection shown in Fig. 3.5f is unstable because the three interconnecting bars are all parallel, and can develop no initial restraint against motion normal to the parallel bars. The compound configuration with common joint plus bar connection is always stable, provided only that the bar connection does not cross the common joint between the two simple trusses. Thus, the interconnections in Figs. 3.3b and d are stable. A compound truss is stable if and only if the interconnections between the simple trusses are stable.

The number of bars in a compound truss is always $2J - 3$. This is demonstrable as follows: Let J_1 and J_2 denote the number of joints in the two simple trusses. Each simple truss will have $(2J_1 - 3)$ and $(2J_2 - 3)$ bars, respectively. When the two are joined together by three bars, the total bars will be $(2J_1 - 3) + (2J_2 - 3) + 3$, or $2J - 3$, where J is the total joint count in the compound truss. When interconnected by a common joint plus a bar, the total bars will be $(2J_1 - 3) + (2J_2 - 3) + 1$, or $2(J_1 + J_2) - 5$. Since one joint is shared, the total number of joints J in the compound truss will be $J = J_1 + J_2 - 1$, or $J_1 + J_2 = J + 1$. Substituting $J + 1$ for $J_1 + J_2$ gives, again, $2J - 3$ for the total number of bars.

A compound truss can be transformed to a simple truss by relocating one member. The reader should inspect the compound trusses in Fig. 3.3 to verify this. The basic compound truss consisting of a pair of simple trusses with compound interconnections can be augmented by compound interconnections to another simple or compound truss. To transform such a multiply compounded truss to a simple truss will require $n - 1$ bar relocations, where n is the number of simple trusses in the multiply compounded truss.

3.2.3 Complex Trusses

A complex truss is a nonsimple, noncompound truss that can be transformed to either a simple or compound truss by relocating one member. For example, the truss of Fig. 3.4a is neither simple nor compound. Relocation of one of the diagonal bars as in Fig. 3.4b transforms the complex truss to a simple truss. The complex truss of Fig. 3.4e can be transformed to a compound truss by relocating member m, then to a simple truss by relocating member n as shown.

Since a complex truss has the same number of bars as a simple truss connecting the same number of joints, it is statically determinate unless it possesses a critical form. Then, it is an unstable mechanism that becomes statically indeterminate when load is applied. The critical form concept will be explained in a later section.

3.3 TESTS FOR STABILITY AND DETERMINACY

Before starting a stress analysis of a truss, the configuration should be analyzed for stability and static determinacy. If unstable, the truss

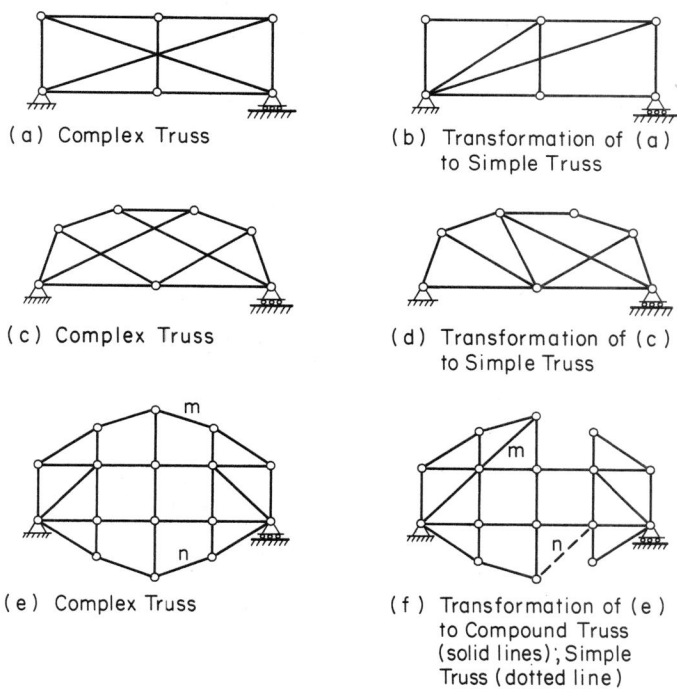

Fig. 3.4. Complex trusses.

is dangerous, and the configuration should be modified to produce stability. We have seen that a simple truss is always stable. A compound truss is stable if the interconnections between the simple trusses are stable. A complex truss is stable if it does not possess a critical form, explained later. Therefore, the best test for stability and determinacy is development of a paper-and-pencil sketch to reproduce the truss configuration, starting with a three-bar truss contained somewhere in the configuration. Then add simple extensions until the arrangement of the truss in question is matched. If it cannot be matched, the truss is compound, unstable, or complex. If the truss being analyzed has a missing bar needed to complete the configuration, it is unstable. If the truss is compound, the two simple trusses will have to be interconnected by appropriate stable connections. If the truss is complex, one bar relocation will transform it either to a simple or compound truss.

Another test for stability sometimes proposed is to count the number J of the joints, and compute $2J - 3$. If the number of bars is less than $2J - 3$, the truss is certainly unstable. If the number of bars is equal to $2J - 3$, the test is ambiguous, and no valid conclusions can be drawn

42 Stress Analysis of Coplanar Statically Determinate Trusses

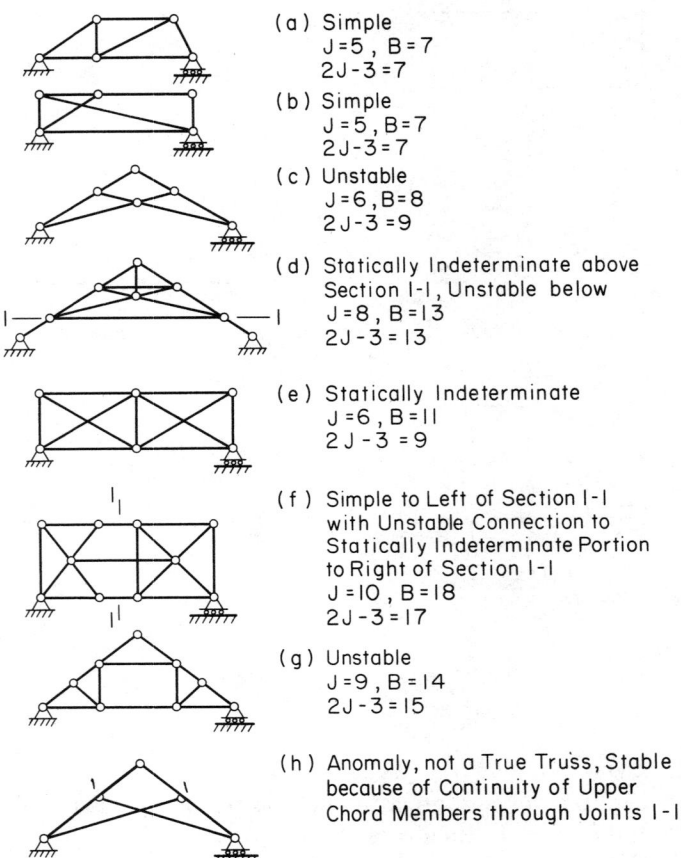

Fig. 3.5. Variations in truss configurations.

without examining the configuration in more detail. For example, the truss of Fig. 3.5d has $2J - 3$ bars, but it is a combination of unstable and statically indeterminate bar arrangements. For a simple truss, the test is always valid, but a simple truss is always stable and statically determinate, so the test is superfluous for a simple truss. The fact that a truss has $2J - 3$ bars does not prove it is a simple truss.

The analyst must know whether a truss is statically determinate or indeterminate. If the number of bars is greater than $2J - 3$, the truss has at least one bar that is statically indeterminate. All simple and compound trusses are statically determinate. Complex trusses are statically determinate unless they include a critical form. This produces indeterminacy when the truss is loaded.

If the truss in question matches a simple, compound, or complex truss, except that it has one or more added bars, it is statically indeterminate. The extra bars are called redundants.

A number of stable, unstable, and statically indeterminate trusses are illustrated in Fig. 3.5. The reader should analyze their configurations in the light of the concepts of truss configuration. For each truss, the joint count J and bar count B are given, and $2J - 3$ computed. The reader should note that the comparison of $2J - 3$ with B is sometimes ambiguous.

The foregoing discussion of truss configuration and its relationship to stability and determinacy is based on the premise that the supports are stable and determinate. This is true if the supports develop three independent reaction elements that are not all parallel. For example, a simple truss on a hinged support at one joint and a roller support at another joint has stable and statically determinate supports.

A truss with a statically determinate bar arrangement becomes a statically indeterminate system if supported on statically indeterminate reactions; for example, the truss of Fig. 3.1d. Stresses in bars A, B, C, and D are statically determinate, but not E. Bar E can be discarded. Then the truss and the reactions are stable and statically determinate.

3.4 STRESS ANALYSIS

The first step in truss stress analysis is to determine whether the truss configuration and supports are stable and statically determinate. If so, the next step is to compute the reactions by a free body analysis of the loaded truss. Then, the stresses in the truss members are computed by free body analyses of each truss joint. This sometimes is called the method of joints. Free body analysis of portions that include more than one joint can be used as an alternative or supplemental method. This is called the method of sections. The analyst should use the method that is more convenient for him. Both may be used in combination, if desired.

3.4.1 Simple Truss

A simple truss can be analyzed by starting at a joint carrying an external load or reaction and where only two members meet. A free body analysis of each such joint is made, based on $\Sigma F_x = 0$ and $\Sigma F_y = 0$, to solve for the stresses in the two bars. Then, proceed to an adjoining

44 Stress Analysis of Coplanar Statically Determinate Trusses

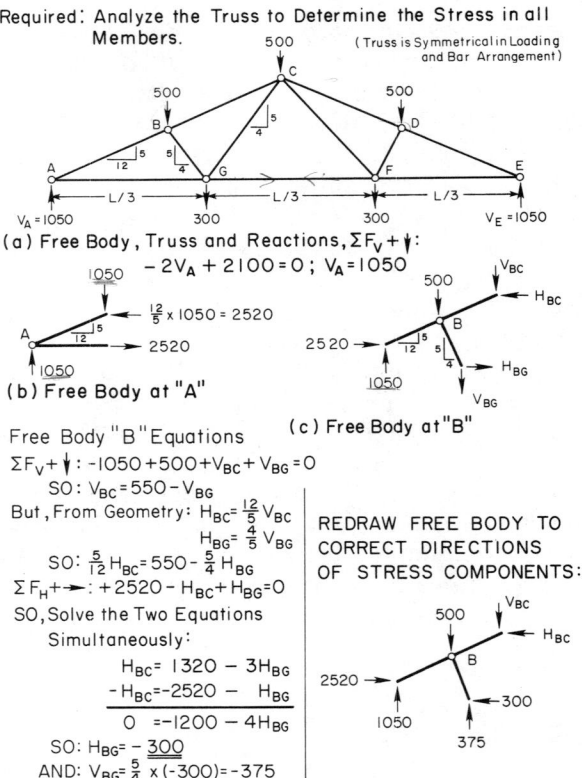

Fig. 3.6. Stress analysis of a simple truss.

joint. If more than three members meet at a joint, all but two of the bar stresses need to have been determined by analyses of other joints. An example of free body analysis of joints for a simple truss is given in Fig. 3.6.

After the horizontal and vertical components of the stress in each bar have been determined, the total resultant axial stress is computed by

$$S = \sqrt{H^2 + V^2}$$

where S is the total axial stress, H is the horizontal component, and V is the vertical component. Finally, the stresses should be tabulated for use in design computations as in Table 3.1.

$\Sigma F_V +\!\downarrow: -1050 - 375 + 500 + V_{BC} = 0$
SO : $V_{BC} = +\underline{925}$
AND : $H_{BC} = \frac{12}{5} V_{BC} = +\underline{2220}$
CHECK : $\Sigma F_H +\!\rightarrow: +2520 - 300 - 2220 \equiv 0$; Checks.
$\Sigma F_V +\!\downarrow: +500 + 925 - 1050 - 375 \equiv 0$; Checks.

(d) Free Body at "C"

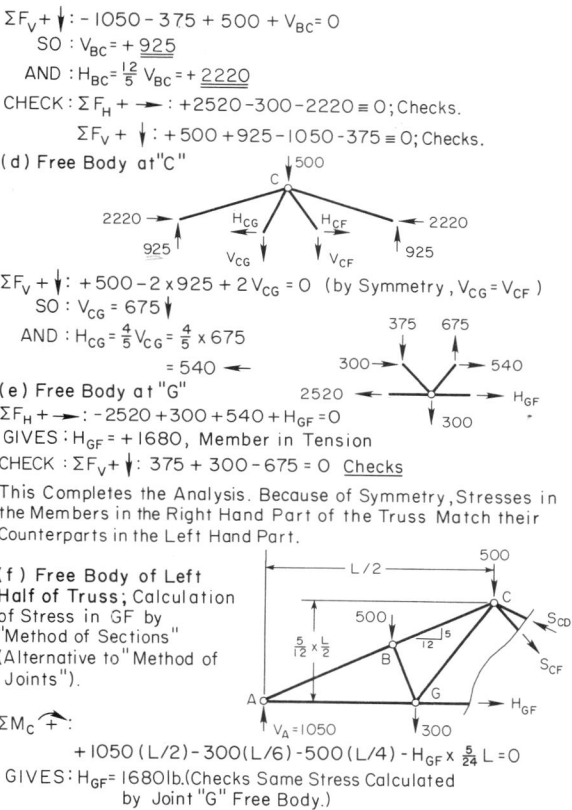

$\Sigma F_V +\!\downarrow: +500 - 2 \times 925 + 2 V_{CG} = 0$ (by Symmetry, $V_{CG} = V_{CF}$)
SO : $V_{CG} = 675 \downarrow$
AND : $H_{CG} = \frac{4}{5} V_{CG} = \frac{4}{5} \times 675$
$= 540 \leftarrow$

(e) Free Body at "G"
$\Sigma F_H +\!\rightarrow: -2520 + 300 + 540 + H_{GF} = 0$
GIVES : $H_{GF} = +1680$, Member in Tension
CHECK : $\Sigma F_V +\!\downarrow: 375 + 300 - 675 = 0$ Checks

This Completes the Analysis. Because of Symmetry, Stresses in the Members in the Right Hand Part of the Truss Match their Counterparts in the Left Hand Part.

(f) Free Body of Left Half of Truss; Calculation of Stress in GF by "Method of Sections" (Alternative to "Method of Joints").

$\Sigma M_C \curvearrowright:$
$+1050 (L/2) - 300(L/6) - 500(L/4) - H_{GF} \times \frac{5}{24} L = 0$
GIVES : $H_{GF} = 1680$ lb.(Checks Same Stress Calculated by Joint "G" Free Body.)

Fig. 3.6. Continued.

In stress analysis of a truss, it is important to identify correctly whether a member is in tension (T) or compression (C), because the design of the member depends on the character of the stress. For example, if a member is in compression, it is designed as a column. The character of the stress can be determined by inspection of a joint-free body that contains the ends of the members in question. If the resultant of the member stress components pulls away from the joint, the member is in tension. If the resultant pushes toward the joint, the member is in compression.

The most difficult part of a free body analysis of simple trusses is the arithmetic. Because of the considerable amount of arithmetic cal-

Table 3.1. Member Stresses in Truss of Fig. 3.6.

Member	H	V	Axial Stress	Compression, C or Tension, T
AB	2520	1050	2730	C
BC	2220	925	2405	C
CD	2220	925	2405	C
DE	2520	1050	2730	C
EF	2520	0	2520	T
FG	1680	0	1680	T
GA	2520	0	2520	T
GB	300	375	480	C
GC	540	675	864	T
CF	540	675	864	T
FD	300	375	480	C

culation that must be performed, many opportunities for mistakes occur. After each free body analysis has been completed, it should be double-checked by resumming the computed forces and the applied joint loads in vertical and horizontal directions. Otherwise, undetected mistakes will be carried from one free body to the next, and any semblance of a valid analysis is lost.

In free body analysis of trusses, it is usually best to write the horizontal and vertical components of the member stresses in terms of the sides of the member slope triangle, as shown in Fig. 3.6. This avoids computing trigonometric functions of angles.

Great care should be exercised in drawing free bodies to ensure that member stresses are consistent from one free body to the next. If stresses in a certain member on one free body correspond to, say, tension, the stresses in the same member on an adjacent free body must also correspond with tension. If calculations reveal that the actual stress in a member is opposite to that assumed in the free body, the free body should be corrected immediately.

When symmetry occurs, it can be utilized to simplify the calculations as in Fig. 3.6, free body at C.

3.4.2 Compound Truss

If only joint-free bodies are used in analysis of a compound truss, a joint will always be encountered where there are more than two unknown bar stresses, regardless of the sequence of joints selected. To eliminate this impasse use a free body section of the truss that cuts the connections between the two simple trusses. Analysis of the portion of the truss as a free body will reveal the forces in the connections. Then the analyses through each of the remaining simple trusses can be completed by the method of joints or sections, as desired.

3.4.3 Complex Truss

Except for peripheral simple extensions, all joints in a complex truss have three or more members, and all section free bodies cut four or more members. Therefore, neither the method of joints nor the method of sections, alone or in combination, can be used to progressively evaluate the stresses. Instead, one of the following alternative methods can be used:

Method A *(Simultaneous Equations).* Write $2J$ independent equations of static equilibrium by applying $\Sigma X = 0$, $\Sigma Y = 0$ to free bodies for the J joints. Solve these simultaneously for the unknown bar stresses and external reactions. This is laborious to do by hand calculations.

Method B *(Undetermined Member Stress).* Select one member at a joint where only three members meet. Designate the stress in that member by some symbol, say, S. Compute the stresses in the other members at that joint in terms of S. Then proceed to the next joint. Compute member stresses at that joint in terms of S. Continue until a joint is reached where two independent expressions, each in terms of S, are available for the stress in one member. Solve for S; then compute numerical values for member stresses already evaluated in terms of S; then evaluate the other member stresses.

This method is illustrated in Fig. 3.7. The complex truss is symmetrical and symmetrically loaded. The unknown components of the stress in member AB are designated by S, since member AB has a one-to-one slope. Then, free body analyses are made successively of joints A, B, and C. By symmetry, it is known that the stress in CF equals the stress in AD. The stress in AD was obtained in terms of S by analysis of joint A. The stress in CF was obtained in terms of S by analysis of joint C. The horizontal components of the stress in these two symmetrical members, AD and CF, are equated to evaluate S. Then, all stresses can be computed. Had the truss been nonsymmetrical, either in configuration or loading, the free body analysis of joints to get member stresses in terms of S would have to continue through joint D. This would yield an expression in terms of S for the stress in AD, which could be equated to the expression for stress in AD obtained at joint A. This equation would be solved for S.

48 Stress Analysis of Coplanar Statically Determinate Trusses

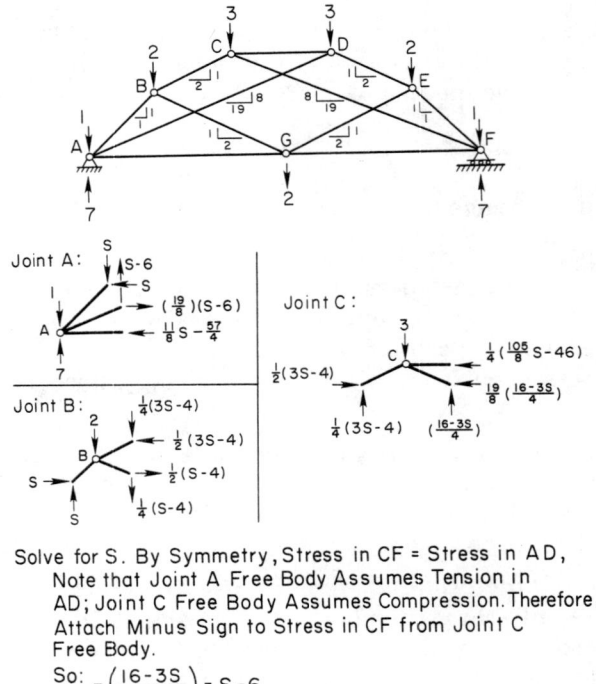

Solve for S. By Symmetry, Stress in CF = Stress in AD, Note that Joint A Free Body Assumes Tension in AD; Joint C Free Body Assumes Compression. Therefore, Attach Minus Sign to Stress in CF from Joint C Free Body.

So: $-\left(\dfrac{16-3S}{4}\right) = S-6$

Gives: $S = \underline{8}$

Fig. 3.7. Complex truss analysis.

Method C (Collinear Members). If a complex truss has two collinear members at a three-member joint, the stress in the noncollinear member can be determined directly by free body analysis of that joint. Then, the known stress in that member can be used in a free body analysis of the joint at the other end of the member, and all member stresses can be evaluated immediately. This is illustrated in Fig. 3.8, where members A and B are collinear.

3.4.3.1 Critical Form. By definition, complex trusses have the proper number of bars for a stable truss configuration. This is not a sufficient test for stability, because critical orientation of some bars may create instability. Complex trusses that include such orientation are said to have critical form. We define critical form as an orientation of bars in a complex truss which produces instability due to a transformation mechanism.

3.4 Stress Analysis 49

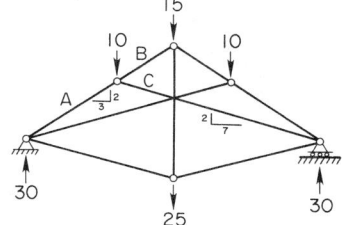

Free Body, Joint AB:

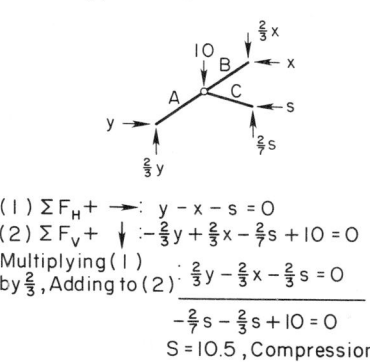

(1) $\Sigma F_H +$ →: $y - x - s = 0$
(2) $\Sigma F_V +$ ↓: $-\frac{2}{3}y + \frac{2}{3}x - \frac{2}{7}s + 10 = 0$

Multiplying (1)
by $\frac{2}{3}$, Adding to (2): $\frac{2}{3}y - \frac{2}{3}x - \frac{2}{3}s = 0$

$$-\frac{2}{7}s - \frac{2}{3}s + 10 = 0$$

$S = 10.5$, Compression

Fig. 3.8. Complex truss with collinear members at a joint.

The instability disappears when the truss is loaded and the truss become statically indeterminate. Some members may be seriously overstressed. In some instances, critical form can be identified by inspection. In others, apparently stable complex trusses possess critical form.

Figure 3.9 illustrates a complex truss in which the critical form can be identified by inspection. It is a modification of the truss analyzed in Fig. 3.7 by merely realigning members *AB* and *BC* to make them collinear, doing the same for *DE* and *EF*, then adding a simple exten-

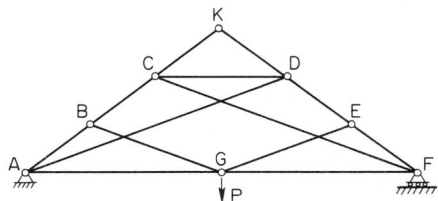

Fig. 3.9. Unstable complex truss.

50 Stress Analysis of Coplanar Statically Determinate Trusses

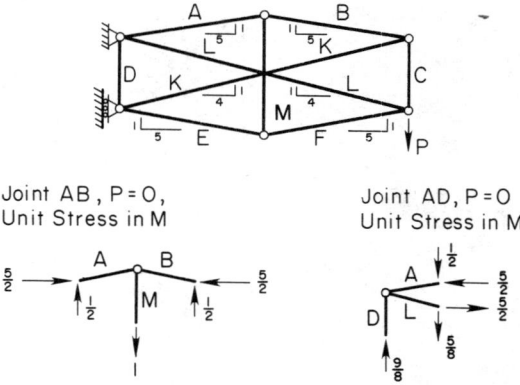

Fig. 3.10. Complex truss-positive test for critical form.

sion to K. When load P is applied, no restraint can be developed in member BG or EG until some rotation of triangle ABG about A and FEG about F has occurred.

Figure 3.10 illustrates a complex truss in which the critical form may not be readily apparent. When loaded by P, the stresses in members A, B, E, and F will be equal in magnitude and have the same character. Stresses in members K and L can develop no torque about their intersection. Therefore, rotation occurs about the point of intersection of the three internal members when P is applied. After deformation has occurred so that the bars no longer have symmetrical orientations, a resisting torque about P is developed, but some of the member stresses are apt to be high and may rupture the members.

3.4.3.2 Statical Indeterminacy. Because critical form is not always apparent by inspection, engineers need a "foolproof" test to identify it. We use the fact that a complex truss with critical form is transformed to a statically indeterminate truss when load is applied. Therefore, if we can show that a complex truss is statically indeterminate when loaded, we have sufficient proof that it has critical form. If it is not statically indeterminate when loaded, it does not have critical form.

A convenient test for statical indeterminacy is to apply an internal stress to one of the members. Conceptually, this amounts to replacing one of the members by a tie rod equipped with a turnbuckle. We tighten the turnbuckle to apply, say, a unit tensile stress in the tie rod. Then we attempt to analyze the stresses in the other members due to this unit tensile stress only, without any external load. If a consistent stress

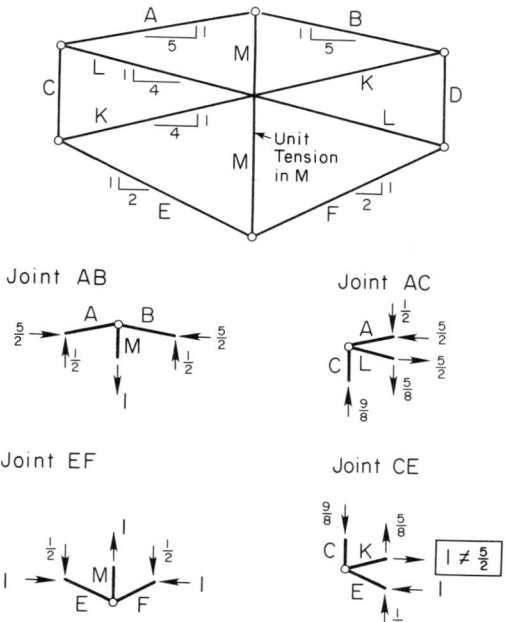

Fig. 3.11. Complex truss-negative test for critical form.

analysis is possible, the truss is statically indeterminate and possesses critical form. If a consistent stress analysis is impossible, the truss is statically determinate when loaded, and does not possess critical form.

This test is illustrated in Fig. 3.10. To expedite the test, apply a unit tensile stress to the internal vertical member M by tightening the turnbuckle. All external reactions and loads such as P are removed. Free body analyses of joints AB and AD, together with symmetry, show that a consistent stress analysis is possible. Therefore, the truss does have critical form, and is a potentially dangerous configuration.

In contrast, the complex truss of Fig. 3.11 does not have critical form. We apply the unit tensile load to member M. Utilizing free body analyses, we arrive at the free body for joint CE. By symmetry, we conclude that stresses in members K and L must be identical. However, the free body of joint AC shows that the horizontal component of the stress in L is $\frac{5}{2}$; but the free body at joint CE shows it to be unity. This inconsistency proves that the truss is statically determinate, and does not possess critical form.

3.5 REAL VERSUS IDEAL TRUSSES

An ideal truss has frictionless hinges at the joints, the external loads and reactions are applied only at the joints, none of the members are continuous through a joint, and the system is coplanar. Real trusses used in buildings or machines always depart from these idealizations to some degree. For example, the joints may be made with gusset plates to which the members are welded, riveted, or glued. Because of practical assembly requirements, the members may not all be exactly coplanar. Not all the axes of members meeting at a joint may be coincident at that joint. Roof trusses may have distributed loads along the upper and lower chords. The reactions may be statically indeterminate. Some of these differences can be ignored, and the real truss analyzed as an equivalent ideal truss. Idealization of rigid or semirigid joints by frictionless hinged joints is usually tolerable in a truss. The bending and shear stresses that occur in the members because the joints are actually rigid may be of secondary importance compared to the axial stresses computed on the assumption that the joints are pinned. These stresses due to joint rigidity are called secondary stresses. The stresses occurring with ideal joints are called primary stresses.

3.6 GRAPHICAL STRESS ANALYSIS

Direct stresses in trusses can be computed by graphically constructing a force polygon for the applied loads and member stresses. This force diagram is called a Maxwell diagram. It is essentially a graphical analysis of free bodies. No new concepts are involved. It has advantages of speed and self-checking characteristics, particularly if stresses due to several different loadings are to be investigated for one-truss configuration. Drafting equipment is required but the tedious arithmetic required in free body analyses is eliminated. For an explanation of the mechanical details of graphical analysis, see Michalos and Wilson (1965, pp. 56–59); or Parker and Ambrose (1982, chapter 7).

3.7 TRUSS SHAPES

An almost endless variety exists of truss configurations, both in profile and arrangement of internal members. Some have the name of the originator, for example, Pratt, Howe, and Fink. Some typical shapes

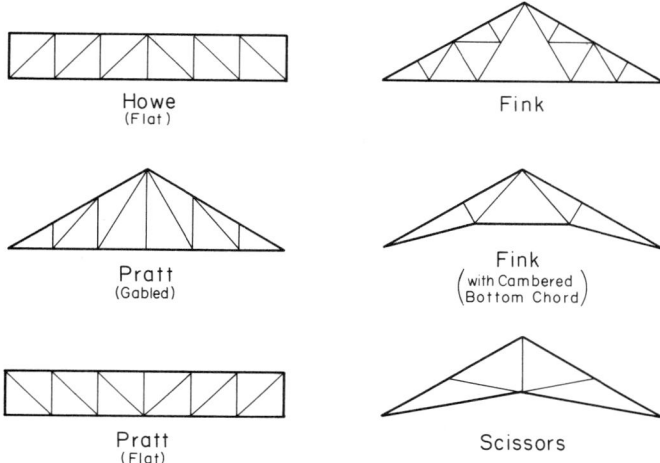

Fig. 3.12. Typical truss configurations.

are shown in Fig. 3.12. Sometimes unstable arrangements are proposed or created, e.g., for the scissors truss by omitting the vertical member. Another example is omitting one of the diagonals in a flat Howe or flat Pratt configuration to provide space for duct work or other mechanical and electrical requirements. The engineer should be able to detect potentially dangerous, proposed truss configurations.

3.8 COMMERCIAL SYSTEMS

Computer-based commercial design systems are available for analysis, design, cost estimates, and fabrication procedures for a wide variety of truss configurations and loading patterns. A customer, e.g., an architect or builder, can call the service to input span, loading, material specifications, and other pertinent data. The customer quickly receives an optimized design and associated cost information. Availability of computer-based analysis and design systems does not obviate the need for engineers to know how to analyze stresses in trusses. Engineers need to be able to investigate proposed trusses for stability and adequacy. If the engineer is not capable of doing this, he has no basis for creative designs or detecting faulty designs. Engineers, of course, should utilize but not be at the mercy of computer-based analysis and design systems.

3.9 TRUSS DEFLECTION

In many instances, truss deflection needs to be controlled to prevent unsightly sagging, misalignment of equipment supported by the truss, faulty operation of doors and windows, water ponding on roofs supported by flat profile trusses, and other manifestations of unsatisfactory performance. Truss deflection can occur from axial deformation of truss members due to member axial stress, slippage at joints due to inadequate joint design, and uneven temperature change.

Trusses of steep to moderate gable profile ordinarily are quite stiff because of a relatively large depth-to-span ratio. Flat profile trusses may have larger deflections.

3.9.1 Real and Ideal Trusses

Real trusses may have greater stiffness compared to their ideal counterparts. Real truss joints (gusset plates, toothed plate connectors, welded connections) typically resist relative rotation of members, whereas the ideal truss has pinned connections. Truss members may be continuous through a joint, instead of terminating with pinned ends.

Because of these and other differences between the real truss and its ideal counterpart, calculated deflections due to superimposed loads on the idealized truss are approximations to deflections of the actual truss. The most accurate information on truss deflection probably is obtained by load-testing of prototype trusses. Nevertheless, designers should be able to calculate deflections of ideal trusses to ascertain that truss performance criteria, such as limitations on maximum deflection, are met. Graphical and analytical methods are available for estimating truss deflection. We will present only an analytical method based on principles of virtual work.

3.9.2 Virtual Work

The principle of virtual work, first presented by the Italian engineer Alberto Castigliano in 1873, is based on the concept of virtual displacements developed by John Bernoulli in 1717.

The objective is to obtain a method for calculating the deflection at a specified joint or joints in specified directions (usually horizontal and vertical) for a truss under a specified pattern of loads. The member sizes and moduli of elasticity must be specified.

The basic approach can be described with reference to a typical statically determinate truss in Fig. 3.13. The loads $P_1, P_2, \ldots, P_5$ are the

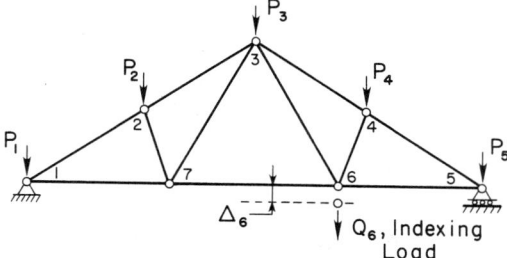

Fig. 3.13. Indexing load applied to truss.

actual loads for which deflection is to be determined. Suppose we wish to calculate the vertical deflection at joint 6. (The same basic procedure would be used for any other joint.) Then we apply a vertical indexing load Q_6 at joint 6. For convenience, we assign a value of unity to the indexing load. The load is applied in the direction of the desired deflection, in this illustration, downward at joint 6. (This indexing load is sometimes referred to as an imaginary or virtual load of unity.) With the indexing load in place, we now apply the regular loads, $P_1, \ldots, P_5$. After the P_i loads have been applied, each member has experienced axial deformation due to those loads which can be calculated based on the axial stress on the member, its length, cross-section area, and modulus of elasticity. In general,

$$\delta_{ij} = S_{ij} l / AE$$

where δ_{ij} is axial deformation, in. (cm), in member ij when subjected to axial stress, S_{ij}, lbf (N), due to P_i loads; S_{ij}, axial stress in member ij, lbf (N), due to P_i loads; l, member length, in. (cm); A, member cross-section area, in.^{2} (cm^2); E, member modulus of elasticity, lbf/in.^{2} (N/cm^2).

At the same time, each member will be subjected to a stress s_{ij}, due to the indexing load Q_i. This stress can be calculated by the same procedure as for the stresses due to the P_i loads. Conceptually, this indexing stress "rides along" with the axial deformation produced by the P_i loads. The work done, or energy accumulated in a specified member by the s_{ij} stress moving through deformation δ_{ij}, is W_m, expressed by

$$W_m = (\delta_{ij})(s_{ij}) \quad \text{in.} \times \text{lbf}$$

The total amount of work done on all the members is the sum of the

56 Stress Analysis of Coplanar Statically Determinate Trusses

work done on each member by the s_{ij} stresses moving through δ_{ij}. This work generated by the s_{ij} stresses in the members must be equal to the work done by the indexing load moving through deflection Δ_i at joint i, in this case, joint 6. This work W_i is expressed by

$$W_i = (\Delta_i)(Q_i) \quad \text{in.} \times \text{lbf}$$

where Δ_i is deflection at joint i due to unit indexing load in direction of that load applied at joint i, in. (cm); and Q_i, unit indexing load, lbf (N). By conservation of energy in an ideal system,

$$W_i = \Sigma W_m$$

where W_m is the work done by each of the member stresses s_{ij} moving through δ_{ij} or

$$(\Delta_i)(Q_i) = \Sigma[(\delta_{ij})(s_{ij})]$$

rearranging,

$$\Delta_i = \Sigma \left[(\delta_{ij}) \left(\frac{s_{ij}}{Q_i} \right) \right] \tag{3.1}$$

The term s_{ij}/Q_i is the dimensionless ratio of the stress generated in member ij to the load Q_i applied at a specified joint and generating stress s_{ij}. As long as the truss deforms in a linearly elastic manner, stress s_{ij} is directly proportional to load Q_i. For convenience, a unit load is used for Q_i in calculations.

By substituting into Eq. (3.1) for δ_{ij}, we have

$$\Delta_i = \Sigma \left[\frac{(S_{ij}l)}{AE} \frac{(s_{ij})}{Q_i} \right] \tag{3.2}$$

Application of Eq. (3.2) for calculating the deflection in a specified direction and at a specified joint is done as follows:

1. Calculate stresses S_{ij} and the corresponding axial deformations in the members of the truss due to the applied loads P_i.
2. Calculate the S_i stresses in the members of the truss due to an indexing load Q_i of unity applied at the joint where deflection is to be calculated. The indexing load must be applied in the direction for which deflection is to be calculated, usually vertically,

horizontally, or both. (A separate set of calculations must be made for each direction.)
3. Calculate, for each member, the product of the axial deformation due to the applied P_i loads times the axial stress in that member due to an indexing load Q_i of unity.
4. The summation for all the members of the products obtained in (3) will be the deflection Δ_i of joint i in the specified direction.

A sign convention must be established for the axial deformations due to the applied loads and the indexing load, based on whether axial shortening or elongation occurs. It is convenient to use plus ($+$) for axial elongation (tension) and minus ($-$) for axial shortening (compression).

An example of calculation of deflection in a truss of just a few members follows.

Example. Calculate the vertical and horizontal deflection of the peak of the truss of Fig. 3.14.

Solution. The calculations are arranged in Table 3.2. The stresses S_{ij}, s_{ijv}, and s_{ijh} entered in Table 3.2 are obtained by free body analysis for member stresses of the truss. One analysis is made for S, the member stresses due to the applied loads, P_1, P_{2H}, and P_{2V}. The other two analyses are made for member stresses s_v and s_h due to (a) a vertical indexing load of unit value applied to the peak, and (b) a horizontal indexing load of unity applied to the peak, where deflection is to be calculated.

The final values for peak deflection are

$$\Delta_h = 0.232 \text{ in. } (0.589 \text{ cm}), \rightarrow \text{horizontal displacement of peak}$$

$$\Delta_v = 0.310 \text{ in. } (0.787 \text{ cm}), \downarrow \text{vertical displacement of peak}$$

The same procedure would be followed for any other joint. Care must be exercised to use consistent units throughout the analysis.

Note that in this example, the modulus of elasticity was introduced in the final calculation for deflection. This is valid only if all members have the same E value. Otherwise, the calculation for each member would include that member's E value.

Table 3.2. Calculations of Deflections in Truss of Fig. 3.14.[a]

Member ij	Length l, ft	A, in.²	S_{ij}, lbf	s_{ijv}/Q_{iv}	s_{ijh}/Q_{ih}	$\dfrac{S_{ij}l\, s_{ijv}}{A\, Q_{iv}}$	$\dfrac{S_{ij}l\, s_{ijh}}{A\, Q_{ih}}$
AB	25	2	−1250	−0.833	−0.625	+13,015	+ 9765
BC	25	3	−2750	−0.833	−0.625	+19,089	+14,323
CD	20	3	+2200	+0.667	+0.500	+ 9782	+ 7333
DA	20	3	+2200	+0.667	+0.500	+ 9782	+ 7333
DB	15	2	+ 900	0.000	0.000	0	0
						+51,668	+38,754

Vertical deflection of peak = $\dfrac{51{,}668}{2 \times 10^6}$ = 0.02583 ft or 0.310 in. (0.787 cm) ↓

Horizontal deflection of peak = $\dfrac{38{,}754}{2 \times 10^6}$ = 0.01938 ft or 0.232 in. (0.589 cm) →

[a] Plus (+) denotes tension in member; minus (−) denotes compression.

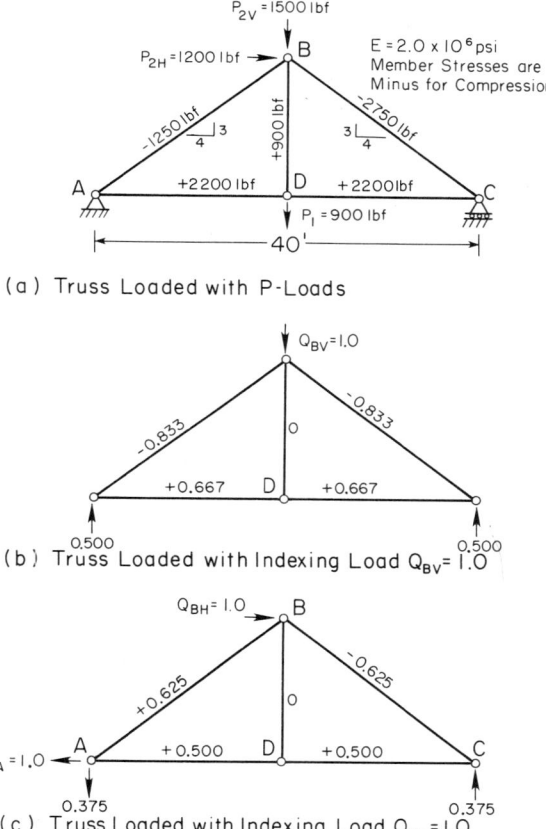

Fig. 3.14. Truss deflection calculations.

PROBLEMS

3.1. Compute the axial stresses in the trusses shown in Fig. 3.15a, b, c, d, e, and f. Tabulate the stress in each member. Indicate the character of each member stress: T, tensile or C, compression.

3.2. Compute the deflection in millimeters (horizontal and vertical components) of the peak joint and the two lower chord joints for the truss in Fig. 3.15a. Indicate direction of the deflections. $E = 12{,}400 \times 10^6$ pascals. Members are 2 in. × 8 in. for all members.

60 Stress Analysis of Coplanar Statically Determinate Trusses

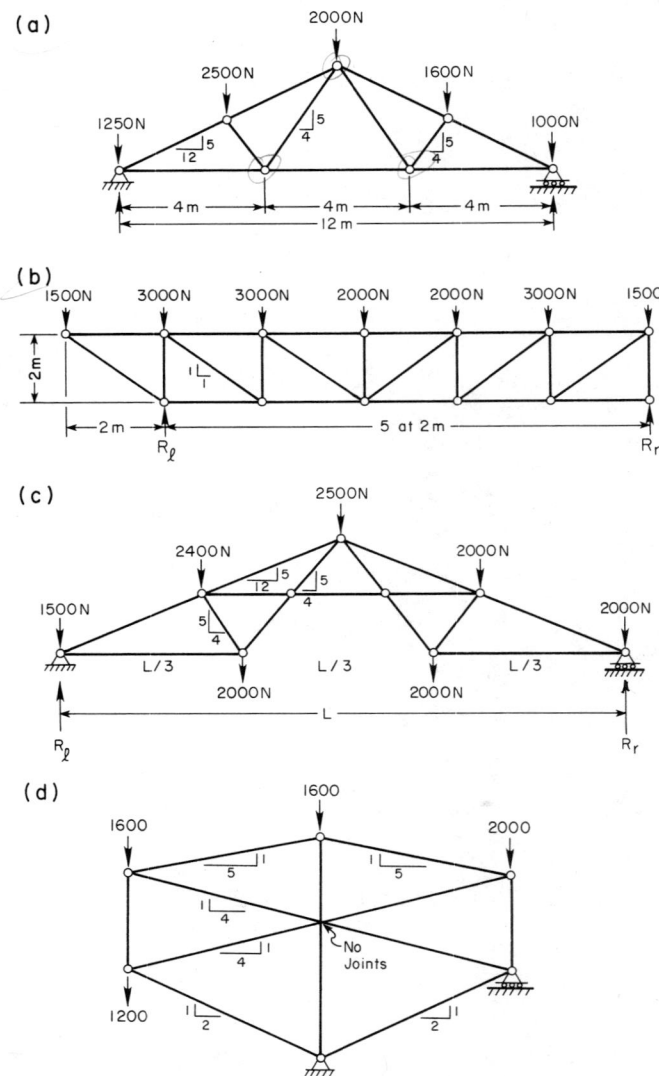

Fig. 3.15. Problems.

3.3. Compute the vertical deflection at the lower joint at the outboard end of the truss shown in Fig. 3.15e. $E = 2.4 \times 10^6$ psi. Assume the lengths are in feet and the loads are in pounds force. The member cross sections are 20 in.^{2} for the diagonals, 8 in.^{2} for the vertical members, and 15 in.^{2} for the horizontal members.

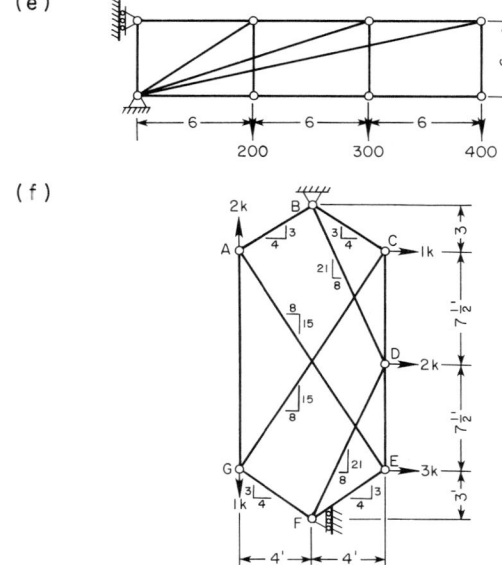

Fig. 3.15. Continued.

NOMENCLATURE FOR CHAPTER 3*

A	Truss member cross-section area, in.^{2} (cm^2)
B	Number of bars (members) in a truss
E	Modulus of elasticity, lbf/in.^{2} (pascals)
F_x, F_y	Force in x direction, y direction, lbf (N)
H, V	Horizontal (vertical) component of truss member stress, lbf (N)
J	Number of pinned joints in a truss
l	Truss member length, ft or in. (m, cm)
P_i	External load applied at a truss joint i, lbf (N)
Q_i	Unit indexing load applied at a truss joint i, lbf (N)
Q_{iv}, Q_{ih}	Vertical or horizontal indexing load, lbf (N) at joint i
S	Axial stress in a truss member lbf (N)
S_{ij}	Axial stress in truss member ij due to applied P loads, lbf (N)
s_{ij}	Truss member ij axial stress generated by unit indexing load, lbf (N)

Note: Quantities such as F, H, V, P, Q, and S may be subscripted to indicate point of application or direction.

s_{ijh}, s_{ijv} Truss member ij axial stess generated by horizontal indexing load, h, or vertical load, v, lbf (N)
W_i Work equivalent of unit indexing load Q_i acting through deflection Δ_i, in. · lbf (J)
W_m Work equivalent of s_{ij} member stress acting through distance δ_{ij}, in. · lbf (J)
δ_{ij} Axial deformation of truss member ij due to applied loads P_i, in. (cm)
Δ_{ih} Deflection of truss joint i in a horizontal direction, in. (cm)
Δ_{iv} Deflection of truss joint i in vertical direction, in. (cm)
s_h Horizontal component of member stress generated by a horizontal indexing load of unity, lbf (N)
s_v Vertical component of member stress generated by a vertical indexing load of unity, lbf (N)

REFERENCES

Michalos, J., and Wilson, E. N. 1965. *Structural mechanics and analysis*. Macmillan, New York.

Parker, H., and Ambrose, J. 1982, *Simplified design of building trusses for architects and builders*. Wiley, New York.

4

Bending Deformation

The purposes of this chapter are to develop the concepts of bending and of bending deformation and their computation. This is of practical usefulness. Also, bending deformation is the basis for analysis of statically indeterminate frames and members.

The amount of bending is often critical in deciding whether or not members or frames are adequate to carry superimposed loads. Excessive deformation of machine frames, for example, may cause misalignment and wear in moving parts. In building frames, it may cause unsightly sagging of the building, or faulty operation of doors and windows. The nomenclature used in this chapter is listed at the end of the chapter.

4.1 DEFINITIONS

For later reference, some important definitions related to bending deformation are given here. These will be illustrated in subsequent pages.

Bending. Change in curvature of the elastic line at a point due to applied bending effect.

Elastic line. Longitudinal neutral axis of an elastic member, as seen in the plane of bending.

Curvature. Rate of change of direction of the elastic line with respect to distance measured along the elastic line.

Direction of the elastic line. Direction of the tangent to the elastic line at a specified point, measured with respect to a set of reference axes.

Bending effect, β. An effect such as bending moment or nonuniform temperature change causing bending. In linear bending, the amount of bending is linearly and directly proportional to the bending effect. Bending effect may vary according to some function of distance along the elastic line; i.e., $\beta = f(l)$ or $\beta(l)$.

Bending deformation. Change in configuration (shape, position) of the elastic line due to bending. Deformation includes rotation and displacement of the elastic line.

Rotation. Change in direction of the elastic line due to bending. Also, rotation at a hinged joint or reaction.
  (i) *Relative rotation:* Rotation at a point measured relative to the direction of an elastic line at some specified point.
  (ii) *Absolute rotation:* Rotation measured with respect to direction of a fixed reference line, i.e., one which does not rotate when bending occurs.

4.2 CURVATURE AND BENDING

The basic concept of bending is change in curvature of the elastic line of a member. The configuration of the elastic line is described by position and orientation of a tangent to the line at some specified point, and its curvature at all other points. The concept of curvature is illustrated in Fig. 4.1a. Tangents defining the direction of the elastic line at P and P' form angle θ. The average change in direction of the tangent per unit distance along the line is $\theta/\Delta l$. Now we conceptually shrink Δl, and cause P' to approach P. As Δl approaches zero, the average rate of change of curvature becomes the derivative at P of θ with respect to l. We define the curvature at point P as $d\theta/dl$. For example, a straight line has zero curvature; and a circular arc has constant curvature. For other curves, the curvature varies from point to point.

Based on this concept of curvature, we define bending as change in curvature of the elastic line at a point on the elastic line. Some bending effect must be produced by the environment, or "outside world," such as temperature change or applied forces, for example. The concept of bending is illustrated in Fig. 4.1. For generality, we show an elastic line that is slightly curved before bending. The concepts are applicable to initially straight or initially curved members.

A tangent to the elastic line at P' (Fig. 4.1a) before bending makes angle θ with the tangent at P. Bending effect β applied along segment Δl of the elastic line causes θ to change to θ' (Fig. 4.1b). As Δl approaches

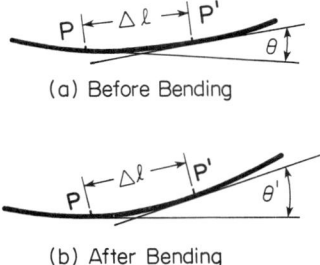

Fig. 4.1. Bending concept.

zero, the curvature before bending becomes $d\theta/dl$ at P; and $d\theta'/dl$ after bending. We define bending effect $\beta(l)$ at point P as

$$\beta(l) = \frac{d\theta'}{dl} - \frac{d\theta}{dl}$$

or change in curvature of the elastic line at a point. Note that depending on loading and configuration, $\beta(l)$ may be constant or may vary as some function of distance along the elastic line. $\beta(l)$ has the dimension of change in curvature per unit length, or L^{-1}. For bending due to bending moment M applied to a member with modulus of elasticity E and cross-section moment of inertia I,

$$\beta(l) = M/EI$$

The product EI is sometimes called the stiffness modulus. Therefore, bending is simply the change in curvature at a point due to bending effects. The units of bending are radians per unit length.

4.3 BENDING ROTATION

In general, bending effect β varies along the elastic line, and can be related to distance l measured along the elastic line by some function, $\beta(l)$, i.e., $d\theta/dl = \beta(l)$. Total change in direction, or angle change, $\Delta\theta$, of the elastic line due to bending applied along Δl is the integral of $\beta(l)dl$, or

$$\Delta\theta = \int_{\Delta l} \beta(l)\, dl$$

66 Bending Deformation

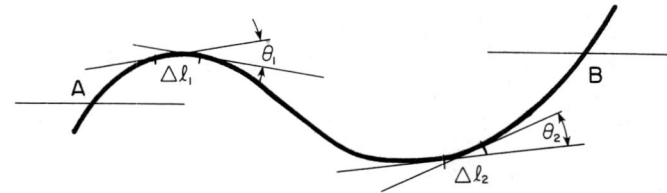

(a) Elastic Line Before Bending

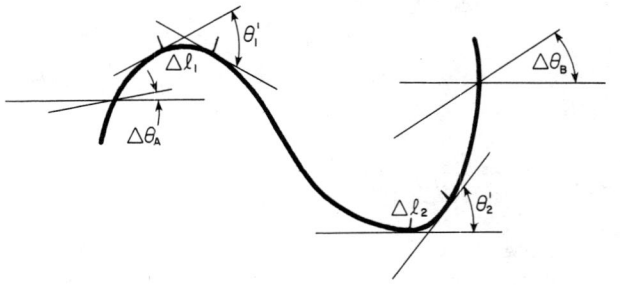

(b) Elastic Line After Bending

Fig. 4.2. Effects of bending on elastic line.

Bending applied to the elastic line changes its configuration by causing rotations and displacements. The member or frame represented by the elastic line rotates and is displaced by the same amounts as the elastic line.

Elastic line rotation is illustrated in Fig. 4.2, greatly exaggerated. An elastic line is shown before bending, Fig. 4.2a, and after bending, Fig. 4.2b. Usually we are interested in rotation of a frame or member at specified points such as A and B in Fig. 4.2. To indicate rotation at A and B relative to a fixed reference system, we conceptually solidly attach reference lines to the elastic line at A and B. Since the reference lines are attached to the elastic line, they rotate with it. One could initially orient the reference lines to coincide with horizontal or vertical directions before bending. Next, bending effect, $\beta_1(l)$, is introduced along segment Δl_1, and $\beta_2(l)$ along segment Δl_2. This produces the changed configuration shown in Fig. 4.2b. θ_1 has increased to θ_1', and θ_2 to θ_2'. We use the notation $\Delta\theta$ to denote these angle changes and write

$$\Delta\theta_1 = \theta_1' - \theta_1 = \int_{\Delta l_1} \beta_1(l)\, dl$$

and

$$\Delta\theta_2 = \theta_2' - \theta_2 = \int_{\Delta l_2} \beta_2(l)\, dl$$

Angle changes $\Delta\theta_1$ and $\Delta\theta_2$ produce absolute rotations $\Delta\theta_A$ and $\Delta\theta_B$. Since there are no hinges between A and B, the rotation at B relative to A is $\Delta\theta_B - \Delta\theta_A$, and

$$\Delta\theta_B - \Delta\theta_A = \Delta\theta_1 + \Delta\theta_2$$

In the general case,

$$\Delta\theta_B - \Delta\theta_A = \int_A^B \beta(l)\, dl \tag{4.1}$$

We emphasize the idea that the difference, $\Delta\theta_B - \Delta\theta_A$, is relative rotation. The absolute rotation at A or B depends on the location of the point where the elastic line is fixed against absolute rotation. For example, if the elastic line is fixed against rotation at A, Fig. 4.2, $\Delta\theta_A$ is zero and $\Delta\theta_B$ is the sum of all rotations between A and B. If the elastic line is fixed at some other point between Δl_1 and Δl_2, $\Delta\theta_A$ and $\Delta\theta_B$ have some other values. To compute absolute rotations at designated points, we must know the absolute rotation of some other specified point, say, at an end of the elastic line, or of some intermediate point.

Because integration or summation of bending effects along an elastic line in general gives only relative rotation, we now introduce a double subscript notation for relative rotation to distinguish it from absolute rotation. Thus, $\Delta\theta_{BA}$ means "relative rotation of the elastic line at B with respect to the elastic line at A, or $\Delta\theta_B - \Delta\theta_A$." The magnitudes of $\Delta\theta_{BA}$ and $\Delta\theta_{AB}$ are equal, but have opposite senses. We use a single subscript to denote absolute rotation. For example, $\Delta\theta_A$ means "absolute rotation at A."

A tacit assumption throughout the foregoing explanation of bending rotation is that no hinges exist between A and B. A hinge is a discontinuity at which $d\theta/dl$ becomes undefined when bending effect is ap-

68 Bending Deformation

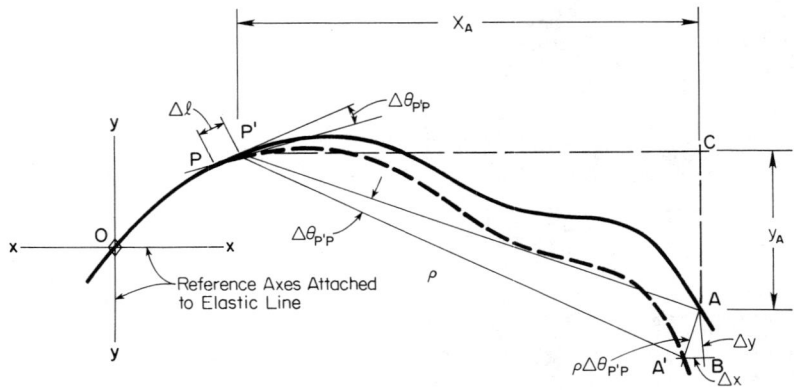

Fig. 4.3. Geometry of small displacements of the elastic line.

plied because a hinge has no resistance to bending. Therefore, bending cannot be transmitted through a hinge. However, hinges may coincide with A and B, Fig. 4.2, and integration or summation can be carried out between hinges.

4.5 BENDING DISPLACEMENTS

The preceding section dealt with relative changes in direction of the elastic line due to bending. Next we develop an analysis for relative displacements due to bending. This is based on the geometry of small displacements, and used later to compute deformations of frames and members subjected to bending.

Displacement effects are illustrated in Fig. 4.3. For generality, a nondescript shape is shown, with bending deformation greatly exaggerated for clarity. Our objective is to understand and analyze how some point A on the elastic line moves when bending is applied along segment PP' of the elastic line. First, we establish reference directions by conceptually fixing an orthogonal pair of reference axes to the elastic line, say, at O. We observe that these axes are only fixed to the elastic line at O, not necessarily fixed in absolute terms. If the elastic line rotates at O, *the reference axes rotate the same amount as the elastic line.*

Next, we construct chord ρ from P' to A. This chord is conceptually fixed to the elastic line at P', and rotates with the elastic line at P'. Next, bending is applied along Δl. This produces rotation $\Delta\theta_{P'P}$ of the elastic line and chord ρ at P'. Point A is displaced to A' due to $\Delta\theta_{P'P}$.

4.5 Bending Displacements

Neglecting as a second order effect the small displacement of P' relative to P over Δl, the displacement AA' is $\rho(\Delta\theta_{P'P})$. Its components in the reference directions are Δy and Δx. The position of A with respect to P' is described by distances x_A and y_A. Because corresponding sides are orthogonal, small triangle $AA'B$ is similar to large triangle $P'AC$. Based on the geometry of similar triangles

$$\frac{\Delta y}{x_A} = \frac{(\rho)(\Delta\theta_{P'P})}{\rho}$$

or

$$\Delta y = (x_A)(\Delta\theta_{P'P})$$

Similarly,

$$\Delta_x = (y_A)(\Delta\theta_{P'P})$$

Now we conceptually shrink Δl causing it to approach zero length and P' to approach P. Then $\Delta\theta_{P'P}$ becomes $d\theta$, Δy becomes dy, and Δx becomes dx. We write

$$dy = x_A \, d\theta$$

$$dx = y_A \, d\theta$$

Since $d\theta = \beta(l) \, dl$, we can rewrite the foregoing equations:

$$dy = x_A \beta(l) \, dl$$

$$dx = y_A \beta(l) \, dl$$

We integrate these to compute the x- and y-displacement components due to bending effect applied throughout a specified portion of an elastic line, say, between points A and O, Fig. 4.3, as follows:

$$\Delta y_{AO} = \int_O^A x_A \, d\theta = \int_O^A x_A \beta(l) \, dl \qquad (4.2)$$

and

$$\Delta x_{AO} = \int_O^A y_A \, d\theta = \int_O^A y_A \beta(l) \, dl \qquad (4.3)$$

It is important to see that Δy_{AO} and Δx_{AO} are displacements of A measured with reference to the direction of axes fastened to and rotating with the elastic line at point O. Therefore, we must distinguish between relative and absolute displacements. Relative and absolute displacement of a point, say, A, Fig. 4.3, are the same *only when* the elastic line and the member it represents are fixed against rotation and displacement where the reference axes are attached. In general, relative and absolute displacements are not the same.

The subscript notation we use for displacement is analogous to the one previously specified for rotations. A single subscript denotes absolute displacement. For example, Δy_A means "absolute displacement of point A in the y direction." A double subscript denotes relative displacement. The first subscript designates the displaced point. The second denotes the point at which the reference axes are attached to the elastic line. For example, Δy_{AO} means "bending displacement in the y direction of point A relative to point O measured with respect to x–y axes attached to the elastic line at O." As previously explained, we measure displacements either in the original or rotated reference direction, and ignore second order difference because the angle is so small.

The limits of integration in Eqs. (4.1), (4.2), and (4.3) must be consistent with the subscripts, otherwise ambiguity in sign conventions will occur. The first subscript should designate the upper limit of integration. Integration proceeds along the elastic line from the point where the reference axes are attached to the point for which we compute rotation or displacement. The subscripts on distances x and y must correspond with the displaced point. It is important to note that distances x_A and y_A are measured from where bending occurs to the point whose displacement due to that bending is being calculated. The distances x_A and y_A are regarded as scalar quantities, so no sign convention is needed for them.

4.6 SIGN CONVENTIONS FOR BENDING DEFORMATION

The analyst must establish a sign convention to use in analysis of bending deformation. For rotation, signs denote sense, i.e., clockwise versus counterclockwise. For Δx, signs denote rightward versus leftward displacement; and for Δy, upward versus downward displacement. The choice of which sense or direction to call positive is optional. One should establish and denote the sign convention at the beginning of each analysis.

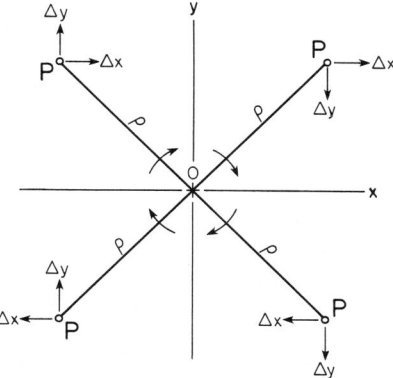

Fig. 4.4. Effect of location of a point on direction of displacements.

Figure 4.4 illustrates how directions and consequently signs of relative displacements Δx and Δy depend on the direction of relative rotation and the position of the displaced point P relative to the reference point O. For example, clockwise relative rotation of ρ causes rightward relative displacement of a point P above the reference point O and downward displacement of P to the right of the reference point O.

4.7 DEFORMATION DUE TO BENDING MOMENT

Equations (4.1), (4.2), and (4.3) express relative rotation and displacement of the elastic line as a function of bending effect $\beta(l)$, without defining the nature of the bending effect. Usually the bending effect is bending moment produced by forces on the frame or member. Now we relate bending moment to bending deformation. This will enable us to compute deformations due to specified bending moments.

Conceptually, amount of bending $d\theta/dl$, caused by bending moment M, varies directly with the magnitude of M and inversely with the stiffness or resistance to bending of the member. Bending stiffness is defined as the product of the elastic modulus of the member material and the moment of inertia of the member cross section. The equation relating these quantities, developed in all elementary books on strength of materials, is

$$\frac{d\theta}{dl} = \frac{M}{EI}$$

where $d\theta/dl$ is bending, radians per inch (cm), produced by bending

72 Bending Deformation

moment at a specified section; M, bending moment, lbf · in. (N · m), acting at the section; E, modulus of elasticity of the member material, lbf per sq. in. (Pa); I, moment of inertia of member cross section, in.^{4} (cm^4).

One limitation introduced by the foregoing equation is that the curvature of the member before bending must be small enough that the effects of curvature on internal stress distribution when bending is applied are negligible compared to a straight member. If the ratio of initial radius of curvature (reciprocal of curvature) to member cross-section depth is greater than approximately 20, the effects of initial curvature in stress distribution may be neglected. Another limitation arises if the modulus of elasticity is regarded as constant when bending is applied. This implies that unit stress varies linearly with strain throughout the range of applied bending. Some materials used for members to resist bending only approximately conform to this assumption.

With these limitations noted, we write for bending effect $\beta(l)$ its equivalent M/EI. Then we rewrite Eqs. (4.1), (4.2), and (4.3) as follows:

$$\Delta\theta_{AB} = \int_B^A \frac{M}{EI}\, dl \tag{4.4}$$

$$\Delta y_{AB} = \int_B^A x_A \frac{M}{EI}\, dl \tag{4.5}$$

$$\Delta x_{AB} = \int_B^A y_A \frac{M}{EI}\, dl \tag{4.6}$$

Note that bending effect $B = \dfrac{M}{EI}$ can vary due to variations along the member in M, E, or I.

Equations (4.4), (4.5), and (4.6) are the basic relationships for computing bending rotation and displacements at A relative to reference axes fixed to a specified point B on the elastic line. We note that distance l is measured *along* the elastic line, *not* in the x or y directions, unless these happen to coincide with the elastic line.

4.8 APPLICATION OF DEFORMATION EQUATIONS

Two applications of the deformation equations are (1) computation of deformations when the bending effect is known; and (2) stress analysis

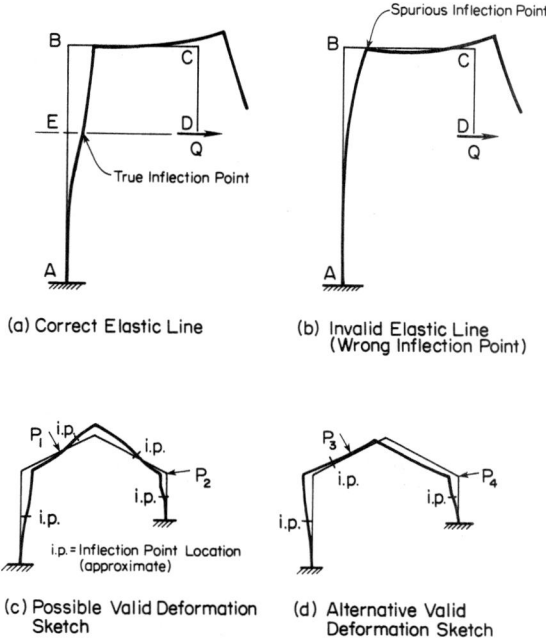

Fig. 4.5. Valid and invalid elastic line sketches.

of statically indeterminate members and frames subjected to specified loads and reactions. We will deal only with the first application in Chapter 4.

A first and essential step in computing deformations is to sketch the elastic line as deformed by applied forces. The sketch is the basis for writing the deformation equations. It should be compatible with the geometry of the undeformed member or frame, and applied bending effects. A frequent mistake is to inadvertently introduce an inflection point at a rigid corner in a frame. Shown in Fig. 4.5 is a correctly drawn elastic line sketch, Fig. 4.5a, and one incorrectly drawn with a spurious inflection point at a rigid joint, Fig. 4.5b. The shape of the deformed elastic line in the incorrect sketch is incompatible with the applied force which produces zero bending moment (inflection point) at E. Figures 4.5c and d illustrate two different but valid deformation sketches of a hingeless frame.

The second step is to choose the location of axes for referencing relative rotations and displacements. If the member has at least one fixed support (no rotation or translation), the reference axes should be

74 Bending Deformation

located there. (Recall that these relative reference axes are conceptually fixed to the elastic line and rotate with it.) Then relative and absolute rotations and displacements will be the same. If the member or frame has only hinged supports, the reference axes can be attached to the elastic line at one of the hinges.

The third step is to establish the sign convention for rotations and displacements. This can be done by showing symbols, for example:

$$+: \rightarrow \quad \downarrow \quad \curvearrowright$$

The choice of positive sense and directions is optional with the analyst.

The fourth and final step is to write the deformation equations, based on the sketch of the deformed elastic line; then compute the desired deformations.

These procedures are illustrated in the following examples. At present, we will not be concerned with the forces that produce bending moment and bending effect. For simplicity, we use the symbol β for bending effect, $\beta(l) = M/EI$.

Example 4.1. Cantilevered Member. *Required:* Compute the absolute end rotation and displacement of the cantilever member shown in Fig. 4.6 when subjected to a constant bending effect β due to an end moment.

Solution. Figure 4.6 shows the deformed elastic line, sign convention, deformation equations, and the computed deformations. The reference axes are attached to the elastic line at A where the end of the cantilever is fixed. Then, relative deformations computed with respect to these axes are also absolute deformations.

We note that Δx_{BA} is zero, since for the undeformed member y_B is everywhere zero. This illustrates the important point that distances x_B and y_B are measured with the elastic line in its undeformed position, regardless of whether the reference axes are fixed, as in the present example, or rotate, as do axes at a hinged support. This latter case will be illustrated in another example.

Example 4.2. Cantilevered Member. *Required:* Compute the end rotation and displacement at C of the cantilevered member in Fig. 4.7. The member is subjected to constant bending β throughout AB and zero bending throughout BC.

Solution. The deformed elastic line, sign convention, deformation

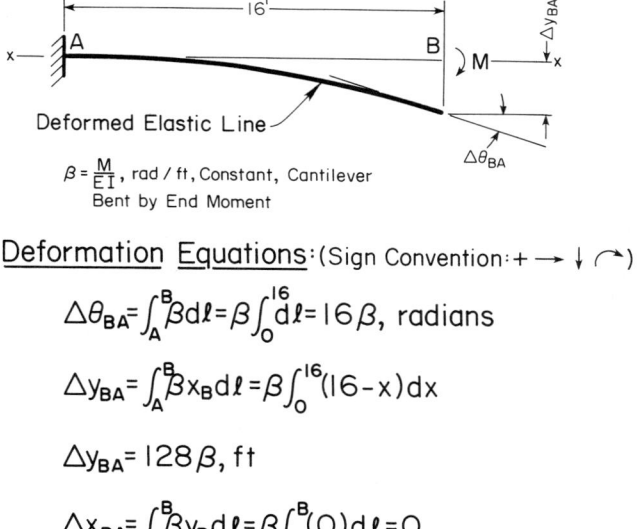

Fig. 4.6. Deformation of cantilever bent by end moment.

equations, and the computed values are given in Fig. 4.7. We note that integration is performed along portion AB, from A to B since bending effect is zero along BC. However, distance x_c is measured to point C, whose displacement Δy_{CA} we wish to compute.

Example 4.3. Cantilevered Frame. *Required:* Compute the absolute rotation and displacements at end D for the frame of Fig. 4.8. The frame is subjected to a constant bending effect β applied in the direction shown.

Solution. The assumed valid deformed elastic line, sign convention, deformation equations, and calculations are shown in Fig. 4.8. It is convenient to integrate separately for each segment of a frame. The limits for each integration correspond with the ends of the segment. Distances x and y are always measured from the point on the frame where bending occurs to the point P whose displacement we compute. For example, the y distance for bending effect in segment BC is constant at 4 in. The x distance for bending in segment CD is 3 in. minus $3/5 l$ in., with integration from $l = 0$ in. at point C to $l = 5$ in. at D.

This example illustrates how the position of the displaced point D

76 Bending Deformation

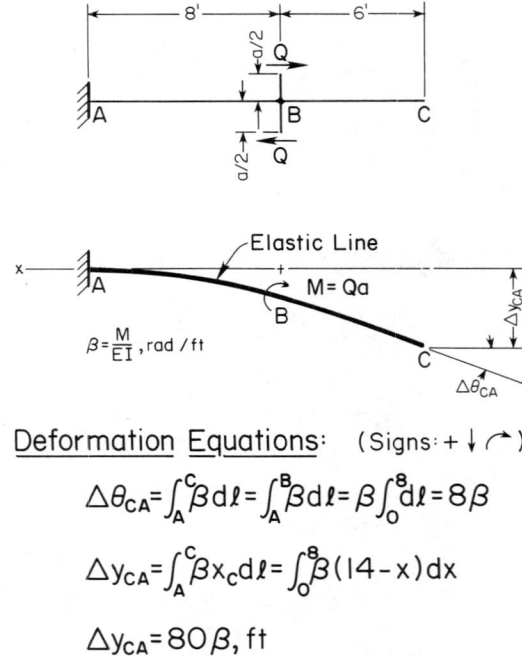

Deformation Equations: (Signs: $+\downarrow \curvearrowright$)

$$\Delta\theta_{CA} = \int_A^C \beta \, d\ell = \int_A^B \beta \, d\ell = \beta \int_0^8 d\ell = 8\beta$$

$$\Delta y_{CA} = \int_A^C \beta x_c \, d\ell = \int_0^8 \beta(14-x) \, dx$$

$$\Delta y_{CA} = 80\beta, \text{ ft}$$

Fig. 4.7. Deformation of cantilever bent by intermediate couple.

relative to the reference axes influences its direction of displacement. The applied counterclockwise bending effect in column segment AE, below point E, displaces D to the left. The same bending effect along EB, above point E, displaces D to the right. This can be verified conceptually by noting that the tip of a chord, FD, for example, fixed to segment AE at point F, moves leftward and upward when counterclockwise bending and rotation of AB occur. The tip of chord GD, for example, moves rightward and upward when counterclockwise bending and rotation of AB occur. This is the effect previously illustrated in Fig. 4.4. To take this effect into account, it is convenient to write separate integrals for Δx_{DA} for the two portions AE and EB.

Example 4.4. Two-Hinged Frame. *Required:* Write the deformation and other equations needed to compute absolute rotations at A and D for the two-hinged frame of Fig. 4.9. Bending effect $\beta(l)$ varies as a function of position along the frame.

Solution. Figure 4.9 shows the deformed elastic line in two different

4.8 Application of Deformation Equations 77

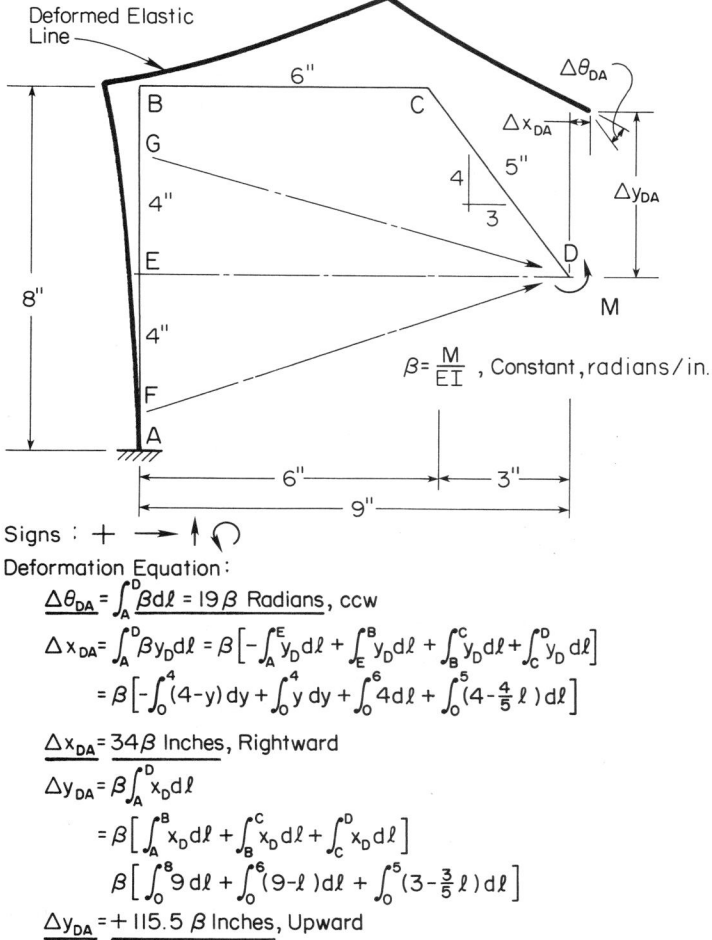

Signs : $+ \longrightarrow \uparrow \circlearrowleft$

Deformation Equation:

$\Delta\theta_{DA} = \int_A^D \beta \, d\ell = 19\beta$ Radians, ccw

$\Delta x_{DA} = \int_A^D \beta y_D \, d\ell = \beta \left[-\int_A^E y_D \, d\ell + \int_E^B y_D \, d\ell + \int_B^C y_D \, d\ell + \int_C^D y_D \, d\ell \right]$

$= \beta \left[-\int_0^4 (4-y) \, dy + \int_0^4 y \, dy + \int_0^6 4 \, d\ell + \int_0^5 (4 - \tfrac{4}{5}\ell) \, d\ell \right]$

$\underline{\Delta x_{DA} = 34\beta}$ Inches, Rightward

$\Delta y_{DA} = \beta \int_A^D x_D \, d\ell$

$= \beta \left[\int_A^B x_D \, d\ell + \int_B^C x_D \, d\ell + \int_C^D x_D \, d\ell \right]$

$\beta \left[\int_0^8 9 \, d\ell + \int_0^6 (9-\ell) \, d\ell + \int_0^5 (3 - \tfrac{3}{5}\ell) \, d\ell \right]$

$\underline{\Delta y_{DA} = +115.5\beta}$ Inches, Upward

Fig. 4.8. Cantilevered bracket bent by end moment.

positions, I and II. In position I, the deformed elastic line has been detached at D and swung counterclockwise to make AB vertical at A (as in the unloaded configuration), while preserving the bent shape of the elastic line. For this position, $\Delta\theta_{DA}$ and ΔX_{DA} are computed by integration.

The second, final position of the elastic line is produced by swinging the entire frame clockwise so that it can be reattached at D. Absolute rotation $\Delta\theta_A$ is calculated by $\Delta X_{DA}/h$ or $\Delta Y_{DA}/s$.

78   Bending Deformation

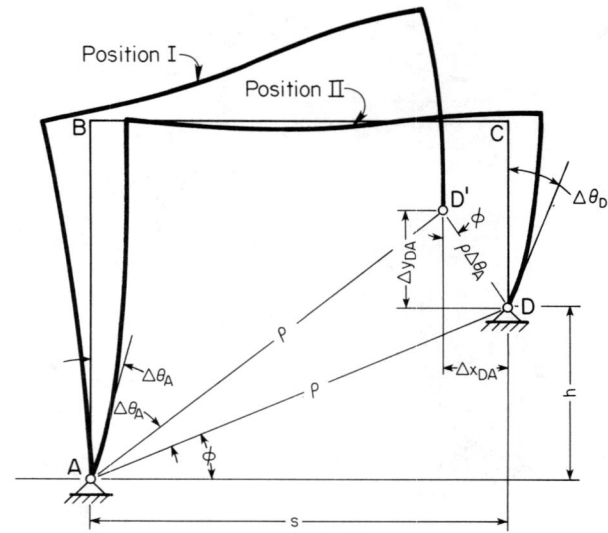

Deformation Equations: (Sign Convention: $+ \rightarrow \downarrow \curvearrowright$)

(1) $\Delta\theta_{DA} = \int_A^D \beta_\ell d\ell$

(2) $\Delta x_{DA} = \int_A^D \beta_\ell y_D d\ell = \rho(\Delta\theta_A)\sin\phi = h \times \Delta\theta_A$

(3) $\Delta y_{DA} = \int_A^D \beta_\ell x_D d\ell = \rho(\Delta\theta_A)\cos\phi = s \times \Delta\theta_A$

Geometric Equations:

(4) From (2), $\Delta\theta_A = \Delta x_{DA}/\rho \sin\phi = \Delta X_{DA}/h$

(5) $\Delta\theta_{DA} = \Delta\theta_D - \Delta\theta_A$; $\Delta\theta_D = \Delta\theta_{DA} + \Delta\theta_A$

(6) From (2) & (3), $\Delta x_{DA} = \Delta y_{DA} \tan\phi$

Fig. 4.9. Deformation of a two-hinged frame, supports at unequal heights.

Finally, the absolute rotation at D is calculated from the expression

$$\Delta\theta_{DA} = \Delta\theta_D - \Delta\theta_A$$

We note that in two-hinged frames with the supports at different elevations,

$$\Delta X_{DA} = \Delta Y_{DA} \tan\phi$$

Therefore, if the supports are at the same elevation, even though ab-

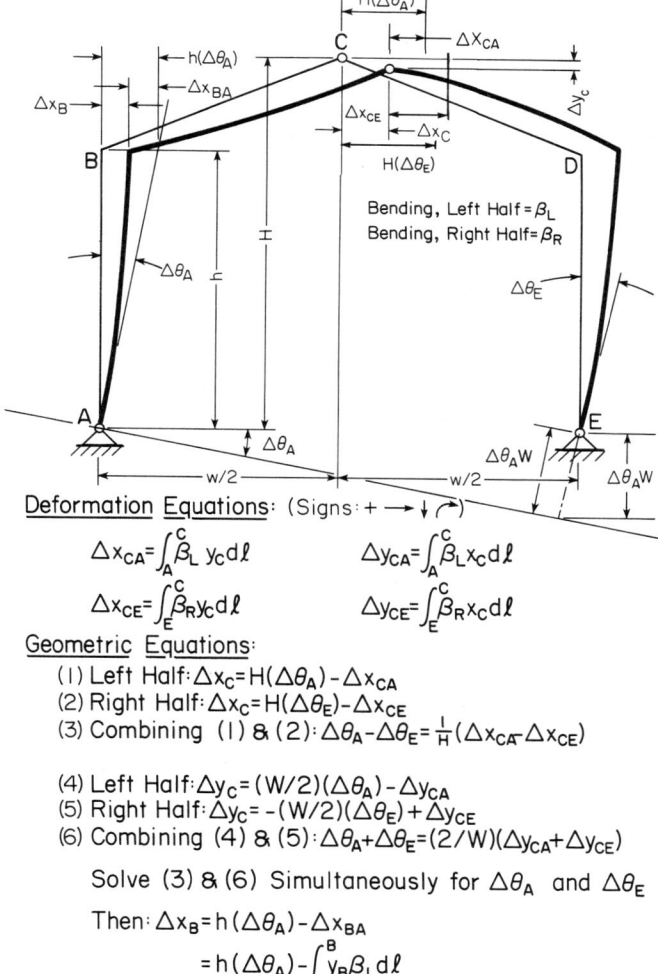

Fig. 4.10. Deformation of a three-hinged frame.

solute rotation occurs at both supports, $\phi = 0$ and $\Delta X_{DA} = 0$, as in Fig. 4.10.

Example 4.5. Three-Hinged Frame. *Required:* Write equations for the absolute rotations, $\Delta\theta_A$ and $\Delta\theta_E$ at the supports and the absolute displacement of the left haunch, ΔX_B, for the three-hinged frame of

80 Bending Deformation

Fig. 4.10. The right and left halves of the frame are subjected to bending moments β_R and β_L, respectively. These vary in a definable manner along the members.

Solution. The deformed elastic line is shown in Fig. 4.10. We note that because of the internal hinge at C,

$$\Delta\theta_{EA} \neq \Delta\theta_E - \Delta\theta_A$$

Instead, we use the condition that absolute displacements at point C due to deformation of the right and left halves of the frame, respectively, must be equal. This produces two equations, one for ΔX_c and one for Δy_c, which can be solved for $\Delta\theta_A$ and $\Delta\theta_B$.

Finally, the absolute displacement of the left haunch is computed as the sum of the rotation and bending displacements in the left column.

The equations are written out in detail in Fig. 4.10.

4.9 AREA–MOMENT ANALOGIES

Integration of the deformation equations can lead to cumbersome calculations, especially when a member is subjected to bending effect β_l that varies along the member. For example, the bending effect in a cantilever beam with an end load varies linearly with distance from the unsupported end,

$$\beta_l = \frac{M_l}{EI} = \frac{Pl}{EI}$$

For either constant or varying bending effects, calculations are simplified by comparing the integrals in the deformation equations to areas and moments of areas, as illustrated in Fig. 4.11. Here, the elastic line is shown for a member that has some general curvilinear configuration in the xy plane. We wish to compute deformation of end B with respect to the elastic line at A. The bending effect diagram is superimposed on the elastic line, with ordinates drawn in the z direction. The centroid of the bending effect diagram occurs at $\overline{X}_B$ and $\overline{Y}_B$ located with respect to end B, where we wish to compute deformation with respect to A. As illustrated by Eq. (1) in Fig. 4.11, relative rotation of end B is numerically equal to the area enclosed between the bending effect diagram and the elastic line. The displacements of end B in the

4.9 Area–Moment Analogies

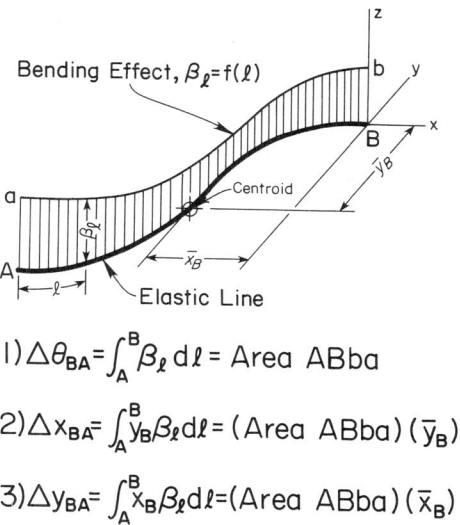

1) $\Delta\theta_{BA} = \int_A^B \beta_\ell \, d\ell = \text{Area ABba}$

2) $\Delta x_{BA} = \int_A^B \bar{y}_B \beta_\ell \, d\ell = (\text{Area ABba})(\bar{y}_B)$

3) $\Delta y_{BA} = \int_A^B \bar{x}_B \beta_\ell \, d\ell = (\text{Area ABba})(\bar{x}_B)$

Fig. 4.11. Generalized area–moment analogy.

x and y directions are equal to the products of the area and the appropriate centroid distances. Therefore, the end displacements are analogous to static moments of the area taken with respect to the end whose displacement we wish to compute. These analogies have led to the name "area–moment" equations for Eqs. (4.4), (4.5), and (4.6), and their counterparts in Fig. 4.11, where $\beta_l = M_l/EI$. Equation (4.4) is sometimes called the first area–moment equation, and Eqs. (4.5) and (4.6) the second area–moment equations.

The generalized example of the area–moment analogy just given had the elastic line lying in the xy plane. The ordinates of the bending effect diagram are plotted in the z plane, and can be regarded as vectors whose directions define the plane of bending, i.e., the vectors are normal to the xy plane. A conventional approach has the bending effect diagram plotted in the xy plane. This latter approach may in some cases be easier to conceptualize, depending on individual perceptions. The meaning of the direction of the bending effect diagram ordinates becomes somewhat ambiguous in the conventional approach, particularly when the member is curvilinear. Murphy (1950, p. 253) developed a "Conjugate Frame Analogy" in which the bending effect diagram ordinates are normal to the plane of the frame. The analyst should use the analogy that seems to him to be easier to apply (see Murphy, Glenn. 1950. *Similitude in Engineering*. Ronald Press, New York).

82 Bending Deformation

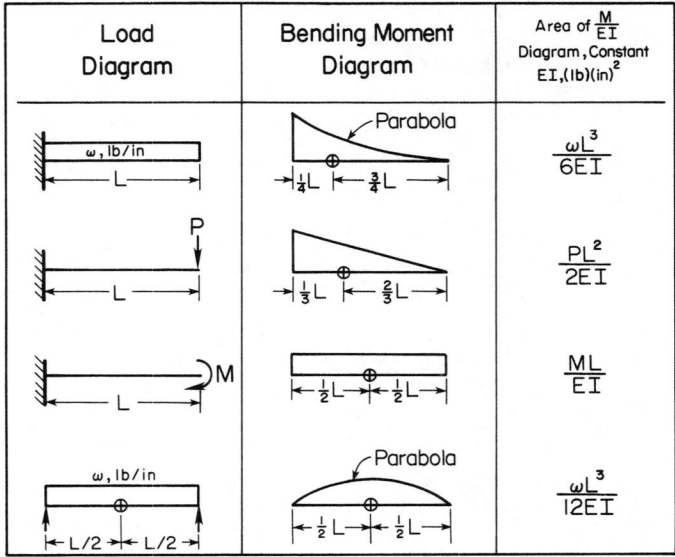

Fig. 4.12. Basic bending moment diagrams.

When the area–moment concept is used in analysis, mathematical integration can usually be replaced by arithmetic computations of geometrically commonplace areas such as rectangles, triangles, and parabolas. For example, the bending effect diagram for a cantilever beam with an end load is a right triangle whose area is $Pl^2/2EI$, as shown in Fig. 4.12. The centroid distance is $2/3\ L$. For members bent by uniform distributed loads, the bending effect diagram is a parabola. For members bent by a constant bending effect, such as due to an end moment, the bending effect diagram is a rectangle, Fig. 4.12.

The area–moment analogies are helpful in recalling how to write the deformation equations, and how to calculate deformation. However, one should avoid blindly substituting the analogy for a basic understanding of the deformation equations.

The examples that follow illustrate how the area–moment analogies can be used in computing deformation.

Example 4.6. Cantilevered Bracket. *Required:* Compute the absolute rotation and displacement of end C of the bracket shown in Fig. 4.13 due to an end vertical load of 500 lbf.

Solution. The deformed elastic line, sign convention, moment dia-

4.9 Area–Moment Analogies 83

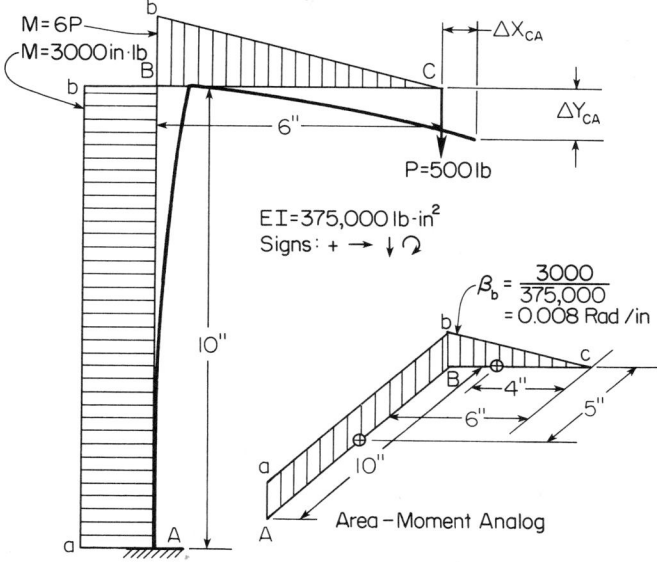

$$\Delta\theta_C = \Delta\theta_{CA} = +(10)(0.008) + (\tfrac{1}{2})(0.008)(6) = +0.104 \text{ radians}$$

$$\Delta X_C = \Delta X_{CA} = +(0.08)(5) + (0.024)(0) = +0.40 \text{ in.}$$

$$\Delta Y_C = \Delta Y_{CA} = +(0.08)(6) + (0.024)(4) = +0.576 \text{ in.}$$

Fig. 4.13. Area–moment analogy, cantilevered bracket.

gram, and area–moment analog are shown in Fig. 4.13. We select A as the reference point, since no rotation or displacement occurs there. We note that end C is displaced both vertically and horizontally. Bending in AB produces both vertical and horizontal displacement at C, but bending in BC produces only vertical displacement, since y_c is everywhere zero for BC.

The area–moment equations and the numerical calculations are given in Fig. 4.13. Signs are assigned to the rotation and bending effects based on the sketch of the deformed elastic line and the chosen sign convention. Rotation $\Delta\theta_{CA}$ is the sum of the areas BbC and $AabB$, divided by EI. The bending effects in AB and BC both cause clockwise rotation at end C with respect to fixed end A, and therefore are given positive signs, consistent with the chosen sign convention.

The displacements are calculated as the sum of the moments of the

84 Bending Deformation

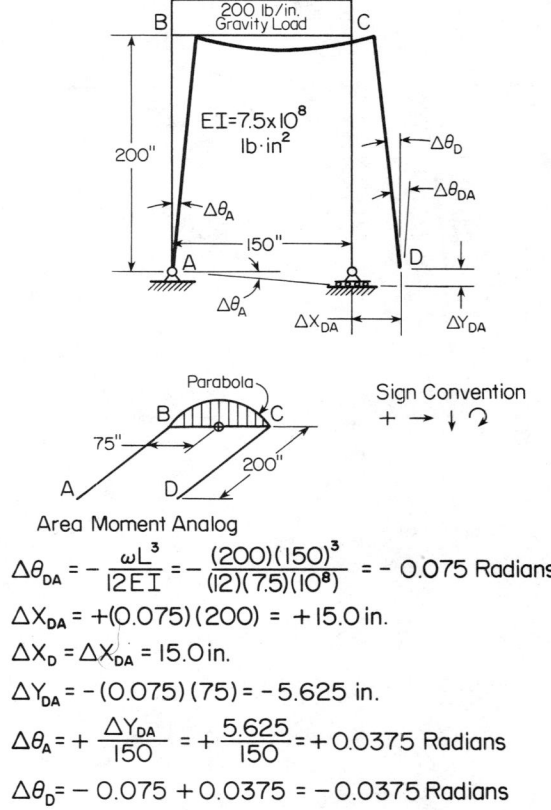

Fig. 4.14. Area–moment analogy, simply supported frame.

areas about C. For ΔY_{CA} the moment of BbC is area Bbc times its centroid distance from C, 4 in.; and the moment of $AabB$ is area $AabB$ times its centroid distance, 6 in. For ΔX_{CA}, similar calculations are made, using $\overline{Y}_C$, which for BbC is 0 in. and for $AabB$ is 5 in.

Example 4.7. Simply Supported Frame. *Required:* Find the absolute rotations and displacements at end D of the simply supported rigid frame shown in Fig. 4.14.

Solution. The deformed elastic line, sign convention, and area–moment analog sketch are shown in Fig. 4.14. We note no bending can occur in the columns. Bending occurs only in the girder. The bending

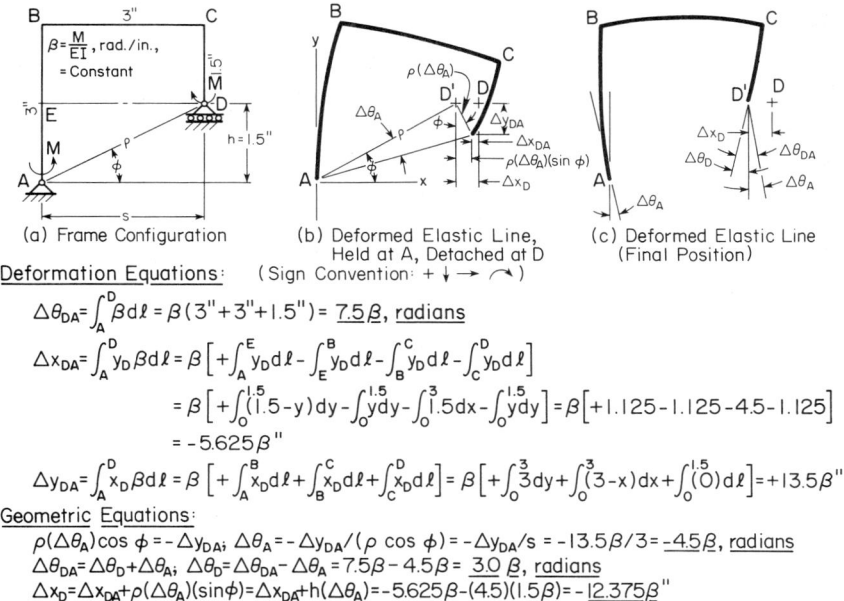

Fig. 4.15. Frame subjected to constant bending effect.

moment is the same as for a simply supported beam with a uniformly distributed load, shown in Fig. 4.12. The area of the M/EI diagram is $wL^3/12EI$.

The deformation calculations are given in Fig. 4.14. We note that, because the supports have the same elevation, ΔX_{DA} and ΔX_D are the same, even though rotation of the reference axes occurs at A. We neglect the difference between ΔX_{DA} and $\Delta X_{DA} \cos \theta$, because θ is small and $\cos \theta$ is virtually unity. Note that $\Delta Y_D = 0$.

Example 4.8. Simply Supported Frame, Supports at Unequal Heights. The frame shown in Fig. 4.15a is subjected to a constant bending moment M throughout the frame. Joint A is pinned. Joint D is pinned to a roller which allows displacement only in a horizontal direction. Equal and opposite end moments M are "cranked" into the frame just above the pin-connected supports so that no reactions are transmitted into the supports. These moments applied to the frame produce a constant bending effect β throughout the frame.

Required. Calculate in terms of β: $\Delta\theta_{DA}$, Δy_{DA}, Δx_{DA}, $\Delta\theta_A$, Δx_D.

86 Bending Deformation

The final deformed elastic line configuration can be achieved by a two-step process of analyzing the elastic line. First clamp the frame at A so that no rotation can occur there. Next release end D and introduce end moment M at D. This will produce end moment M also, at clamped end A and the configuration of Fig. 4.15b. With the elastic line in this position, calculations for the displacements of Fig. 4.15b can be made. Finally, unclamp the frame at A while maintaining moment M at A and D. Swing the frame CCW through angle $\Delta\theta_A$ about end A to place end D at its final position reattached to the roller at D. In this position, calculations can be made for the rotations shown in Fig. 4.15c.

The deformations are:

$\Delta\theta_{DA} = +7.5\beta$ radians
$\Delta y_{DA} = +13.5\beta$ radians
$\Delta x_{DA} = -5.63\beta$ in.
$\Delta\theta_A = -4.5\beta$ radians
$\Delta x_d = -12.38\beta$ in.

4.10 ELASTIC LINE DEFORMATION SKETCHES

A valid sketch of the deformed elastic line for a frame is a basic part of every deformation analysis. The main purpose of the sketch is to show how joints move and member curvatures change when the frame is loaded. It guides the analyst in writing valid and consistent deformation equations.

The sketch cannot be drawn to any particular scale of displacements because they are not known quantitatively when the sketch is made. Furthermore, displacements and changes in curvature are exaggerated for clarity. Otherwise, they would be practically invisible on a sketch of ordinary size.

Because the sketch is made before any deformations are calculated, assumptions have to be made sometimes for the directions of certain displacements and curvature changes. The sketch is then developed consistent with these assumptions. Subsequent calculations will reveal the correct directions.

The following suggestions will help in making deformation sketches for rigid frames consisting of straight members.

Start with a scale drawing of the undeformed elastic line. It must correspond with the actual frame support conditions, location of hinged joints, and position and direction of applied loads. First, apply the loads and translate the joints into new positions without bending the members. Second, resketch the members to show bending curvature between the joints.

To first translate the joints, conceptually replace rigid joints and supports by hinges that are spring loaded to prevent completely free rotation and collapse of the frame. Next apply the loads of the joints. If some or all of the original loads are intermediate between joints, apply equivalent loads at the joints. Next allow the loads to force the frame into a new position achieved by rotation at the joints. At this stage, the frame behaves like a mechanism of bars linked together by pinned, spring-loaded joints. Angles between members meeting at a joint will change slightly. It is important to preserve the original lengths of all members as we shift the joints into new positions.

To complete the sketch, redraw all the original rigid joints at their new locations. *This requires that the original angles between members be restored at each rigid joint but not at originally pinned joints or supports.* The original directions of members with fixed supports must be restored at the fixed supports.

Finally, resketch the straight elastic lines to transform them to smoothly curved lines that enter each joint tangent to the joint angle. At each fixed support, the elastic line must enter tangent to its original direction. This may require an inflection point in the elastic line between some joints. A common mistake is to inadvertently introduce inflection points to coincide with rigid joints. Theoretically, such coincidence can occur, but is unlikely in a real frame. Also, restore to their original intermediate positions any loads previously relocated as equivalent joint loads. This restoration may change the number of inflection points in the member carrying the intermediate loads. Corresponding corrections are made in the sketch.

After some practice based on valid concepts, one can visualize and sketch the deformed elastic line for a rigid frame directly without resorting to all of the intermediate steps that have been described.

PROBLEMS

4.1 Refer to Fig. 4.15. Write the equations for ΔX_{BA}, ΔX_{CD}, Δy_{CD}, and θ_D. (Express bending effect as β, assumed constant throughout the frame.)

4.2 Refer to Fig. 4.16. Write the deformation equations for calculating ΔX_{BA}, ΔX_B, and Δy_{DA}. Assume member stiffness is constant throughout the frame.

88 Bending Deformation

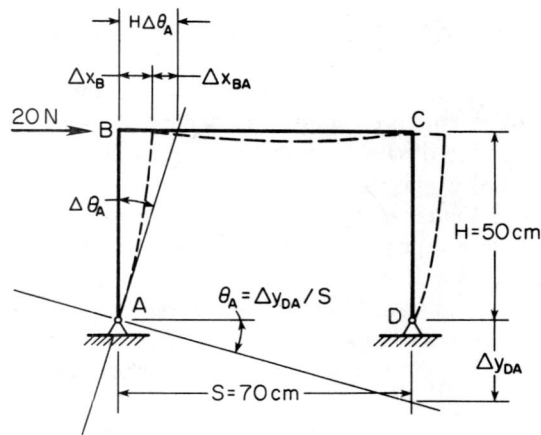

Fig. 4.16. Two-hinged drawbar.

4.3 Refer to Fig. 4.10. The frame is loaded by a concentrated downward load at point C instead of the generalized bending effects β_L and β_R. Draw the deformed elastic line. Write the deformation equations for solving for ΔX_B, ΔX_D, and Δy_C.

4.4 Refer to Fig. 4.17. Draw the deformed elastic line for the one-hinged frame. Write the deformation equations for calculating ΔX_B. Joints A, C, and D are rigid. Joint B is pinned.

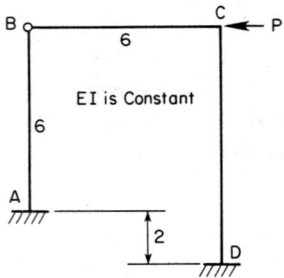

Fig. 4.17. One-hinged frame, Problem 4.4.

4.5 Refer to Fig. 4.18. Sketch the deformed elastic line. Write equations for computing Δy_E.

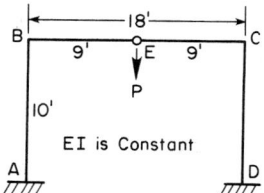

Fig. 4.18. One-hinged frame, Problem 4.5.

4.6 Draw a valid sketch of the deformed elastic line for each of the frames of Fig. 4.19.

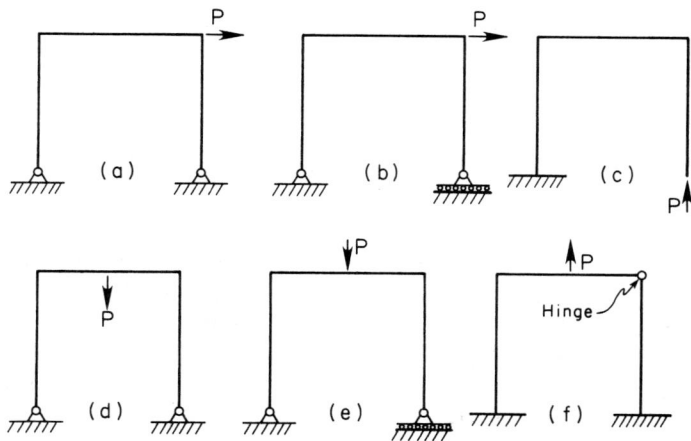

Fig. 4.19. Frames for sketching deformed elastic curve, Problem 4.6.

4.7 The frame of Fig. 4.20 has a constant EI value of 2×10^6 lbf·ft². Compute the horizontal and vertical displacement of points 1 and 2.

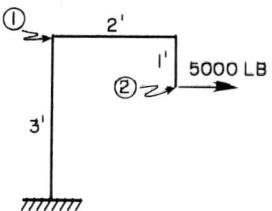

Fig. 4.20. Frame for Problem 4.7.

4.8 The frame of Fig. 4.21 has an EI value of 5×10^5 lbf · ft². Compute the internal pinned joint displacement when a downward load of 2500 lbf is applied at the pin.

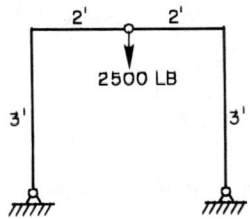

Fig. 4.21. Frame for Problem 4.8.

NOMENCLATURE FOR CHAPTER 4
E Modulus of elasticity of a member, lbf/in.² (pascals)
I Cross-section moment of inertia of a member, in.⁴ (cm⁴)
M Bending moment applied to a member, in. · lbf (cm · N)
P Concentrated load applied to a member, lbf (N)
l Distance along the elastic line, in. (cm)
β Bending effect, radians per unit length along elastic line
θ Direction of a tangent to the elastic line, radians, with respect to some reference direction
Δ increment, e.g., $\Delta\theta$ is an incremental change in direction of the elastic line
w uniformly distributed load, lbf/in. (N/cm)

5

Analysis of Statically Indeterminate Coplanar Frames

5.1 CHARACTERISTICS

Statically indeterminate coplanar frames have reactions and internal stresses that cannot be determined only by applying the three basic equations of static equilibrium. In general, indeterminacy exists when there are more unknown reaction or stress components applied to a frame or member than the number of independent equations of static equilibrium that can be written. For example, the reactions from the supports of a two-hinged arch are statically indeterminate because the supports can develop a total of four independent reaction components (a horizontal and a vertical force at each hinge), but only three independent equations of static equilibrium can be written. In this case there is one *redundant* reaction component, or one *degree of indeterminacy*.

The nomenclature used in this chapter is listed at the end of the chapter. A frame can have only statically indeterminate reactions, or only statically indeterminate member stresses, or both. Examples are shown in Fig. 5.1.

Many engineering structures are statically indeterminate. Some examples are (1) frames on at least two supports but with fewer than three hinges; (2) welded joint machine frames; (3) concrete culverts and drop inlet structures used in soil and water conservation; and (4) slabs or panels continuous over more than two supports. Because coplanar statically indeterminate frames and members are encountered so frequently in structures and machines, an engineer should be able to analyze them to determine reactions, member stresses, and displacements.

92   Analysis of Statically Indeterminate Coplanar Frames

Analysis of statically indeterminate frames is accomplished by investigating how the frame deforms when loads are applied. The deformation depends on the elastic properties of the frame. Deformation includes bending of members, rotation of joints, and translation of joints.

The two methods of statically indeterminate analysis presented here include (1) the area–moment method; and (2) the moment distribution method.

5.2 AREA–MOMENT METHOD

5.2.1 Area–Moment Concepts

The area–moment method is useful for frames that have only two supports, such as the frames in Fig. 5.1. The area–moment method can become cumbersome to use for more complex frames with many supports.

The area–moment method is based on the area–moment theorems. These should be thoroughly understood for three reasons. (1) They are directly applicable in the analysis of many indeterminate frames. (2) They are the basis for relationships used in other analytical methods. (3) They develop an understanding of how loaded frames deform.

A basic problem in analysis of a statically indeterminate frame is usually to compute the unknown reaction components. When these have been determined, moments and shears in other parts of the frame can be computed. Because the frame is statically indeterminate, equations in addition to those for static equilibrium must be written so that the total number of independent equations is equal to the total number of reaction components. In general, the additional equations are written based on the geometry of the frame and how it deforms at specific points when loads are applied. These equations will include terms containing unknown forces at the reactions. These equations (static equilibrium equations plus displacement geometry equations) are solved simultaneously to yield values of the reactions.

In Chapter 4, the three area–moment equations were developed to describe the geometry of deformation of coplanar frames. These equations are the basis for analysis of statically indeterminate coplanar frames. The three equations are

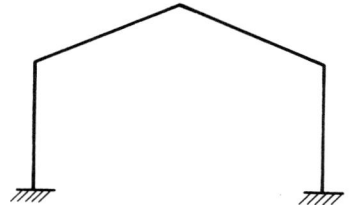

(a) Reactions Statically Indeterminate, Member Stresses Determinate

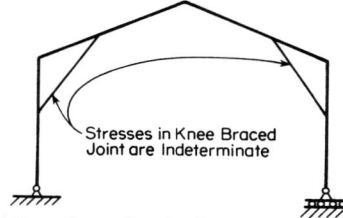

(b) Reactions Statically Determinate, Member Stresses Indeterminate

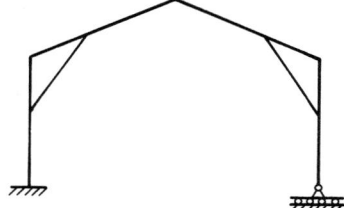

(c) Reaction and Member Stresses Statically Indeterminate

Fig. 5.1. Statically indeterminate rigid frame.

$$\Delta \theta_{AB} = \int_B^A \frac{m\,ds}{EI} \tag{5.1}$$

$$\Delta x_{AB} = \int_B^A y_A \frac{m\,ds}{EI} \tag{5.2}$$

$$\Delta y_{AB} = \int_B^A x_A \frac{m\,ds}{EI} \tag{5.3}$$

where $\Delta\theta_{AB}$ is the rotation in radians of the elastic line at A relative to the elastic line at B; $\Delta\theta_{BA}$ is the rotation of the elastic line at B relative to the elastic line at A; m is the internal bending moment in in. · lbf at s; s is distance along the elastic line; ds is differential increment distance;

94 Analysis of Statically Indeterminate Coplanar Frames

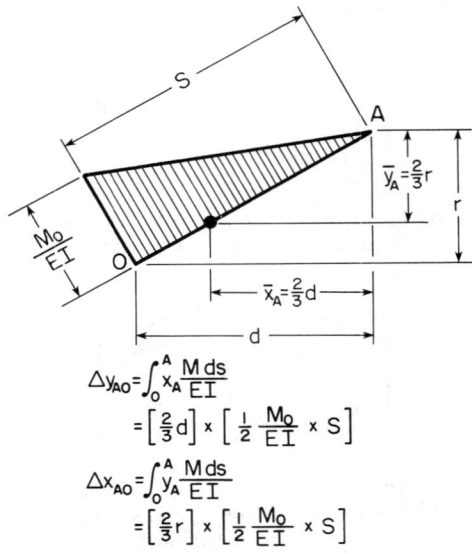

Fig. 5.2. Computation of integrals for area–moment analysis.

E is the modulus of elasticity of the member in lbf per square inch; Δx_{AB} is the bending displacement in the X direction of point A on the elastic line relative to a set of x–y reference axes fixed to the elastic line at B; Δy_{AB} is the bending displacement in the y direction of point A with respect to the reference axes fixed to the elastic line at point B; y_A is distance in the y direction between the point where bending moment m is applied to point A on the elastic line; x_A is distance in the x direction from point A on the elastic line to the point where bending moment is applied; and I is the cross-section moment of inertia of the member in in.[4].

Equation (5.1) is sometimes referred to as the first area–moment equation. Equations (5.2) and (5.3) may be referred to as the second area–moment equations. The integral in the first [Eq. (5.1)] area–moment equation represents the area divided by EI under the moment diagram between the points A and B. The integrals in Eqs. (5.2) and (5.3) represent the moment about point A of the area divided by EI. An illustration of the computation scheme is given in Fig. 5.2. The cross-hatched area is the area under the bending moment diagram, O to A.

5.2.2 Area–Moment Applications

The application of area–moment equations to analysis of statically indeterminate frames will be illustrated by writing them in generalized

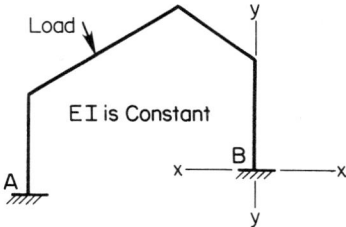

Fig. 5.3. Hingeless rigid frame.

form for frames with two supports and hingeless, one-hinged, and two-hinged configurations. Then numerical examples will be worked out.

5.2.2.1 Hingeless Frame. Figure 5.3 illustrates a hingeless gable frame with fixed end supports at different elevations. Since EI is constant, for all parts of the frame, EI need not be included in the computations. Reference axes can be attached to the frame either at A or B, since no rotation of the frame occurs at the supports. The three area–moment equations written for this frame are

$$\Delta\theta_{AB} = \int_B^A m \, ds = 0$$

$$\Delta x_{AB} = \int_B^A y_A m \, ds = 0$$

$$\Delta y_{AB} = \int_B^A x_A m \, ds = 0$$

These three equations plus the three static equilibrium equations give a total of six equations that can be solved for the six reaction components, three at each support.

5.2.2.2 One-Hinged Frame. A frame with only one hinge at 0 or B as in Fig. 5.4 has only two redundants because the condition that bending moment M is zero at the hinge can be used to write

$$\Sigma M_0 = 0$$

If the hinge is between the supports as in Fig. 5.4a, two more equations can be written based on the condition that

$$\Delta x_{0A} = \Delta x_{0B}$$

and

$$\Delta y_{0A} = \Delta y_{0B}$$

96 Analysis of Statically Indeterminate Coplanar Frames

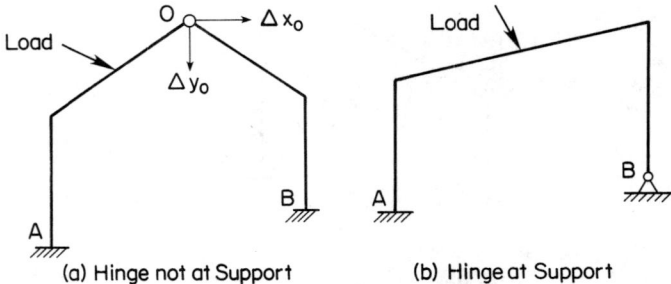

Fig. 5.4. One-hinged rigid frame: (a) hinge not at support and (b) hinge at support.

The reference axes are placed at A and B, and do not rotate because the ends of the frame are fixed.

From the area–moment equations for Δx and Δy, the two foregoing equations give

$$\int_A^0 y_0 \frac{m\,ds}{EI} = \int_B^0 y_0 \frac{m\,ds}{EI}$$

and

$$\int_A^0 x_0 \frac{m\,ds}{EI} = \int_B^0 x_0 \frac{m\,ds}{EI}$$

If the hinge occurs at a reaction as at B in Fig. 5.4b, B does not move with respect to axes at A, so the equations of zero displacement become

$$\Delta x_{BA} = 0 = \int_A^B y_B \frac{m\,ds}{EI}$$

and

$$\Delta y_{BA} = 0 = \int_A^B x_B \frac{m\,ds}{EI}$$

It is important to carefully choose the reference point to which distances x and y are measured if there are hinges as in Fig. 5.4a and b. For example, in Fig. 5.4b frame rotation occurs at B, so

$$\Delta x_{AB} \neq 0$$
$$\Delta y_{AB} \neq 0$$

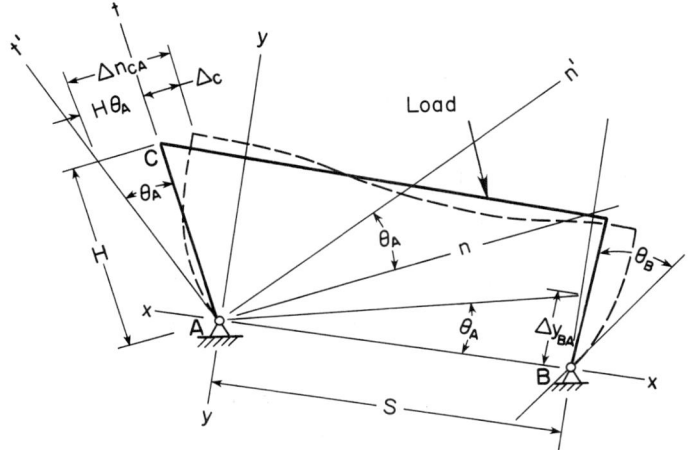

Fig. 5.5. Two-hinged frame, hinges at supports.

because the reference axes at B rotate, and A and B are at different elevations.

5.2.2.3 Two-Hinged Frame. A two-hinged frame has only one redundant reaction component. In the general case, shown in Fig. 5.5, the supports are not at the same elevation. One of the reference axes should be chosen to pass through both reactions as in Fig. 5.5. Then the gross configuration of the frame can be established with respect to these axes. The area–moment equation that can be written with respect to these axes is either

$$\Delta x_{BA} = 0 = \int_A^B y_B \frac{m\ ds}{EI}$$

or

$$\Delta x_{AB} = 0 = \int_B^A y_A \frac{m\ ds}{EI}$$

Deflections of the frame can be computed by first evaluating the absolute rotations, θ_A and θ_B, of the frame at the reactions, by the following expressions:

98 Analysis of Statically Indeterminate Coplanar Frames

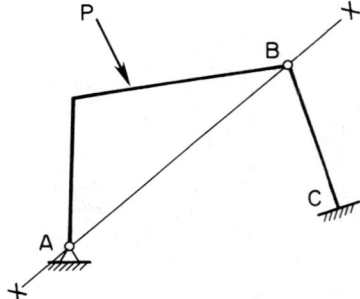

Fig. 5.6. Two-hinged frame, with internal hinge.

$$\theta_A = \frac{\Delta y_{BA}}{S} = \frac{1}{S}\int_A^B x_B \frac{m\,ds}{EI}$$

$$\theta_B = \frac{\Delta y_{AB}}{S} = \frac{1}{S}\int_B^A x_A \frac{m\,ds}{EI}$$

Then, deflections can be computed with respect to nt axes. For example, in Fig. 5.5,

$$\Delta n_{CA} = \int_A^C t_C \frac{m\,ds}{EI}$$

written with respect to an axis t that includes member AC and a perpendicular axis n. Then, the absolute deflection Δ_C of point C is

$$\Delta_C = \Delta n_{CA} - H\theta_A$$

This deflection is normal to member AC.

If one of the hinges is an internal hinge (not at a support) as in Fig. 5.6, the equation that can be written is

$$\Delta x_{BA} = \Delta x_{BC}$$

because the frame does not rotate at C. The x axis must be established through the two hinges.

5.2.3 Examples of Analysis by Area–Moment Theorems

The general procedure for applying the area–moment equations in the analysis of reactions of coplanar statically indeterminate frames is as follows:

1. Determine the number of redundant reaction components. Draw a valid sketch of how the frame deforms when loaded.
2. Write in generalized form the appropriate area–moment equations. As many equations must be written as redundant reaction components. Careful consideration must be given to specifying the location of the reference axes and the effects of rotation of these axes.
3. Draw free-body diagrams for the parts of the frame used in writing the area–moment equations.
4. Draw moment diagrams for each part. For clarity, draw an individual diagram for each load or reaction component. Compute the values of the integrals in the area–moment equations by a suitable tabulation routine. Notice that each part of the frame will have unknown reaction components (shear, direct stress, end moment) applied to each end of the part. A valid moment diagram can be drawn for moments about either end. Choose the end which will include the fewer number of unknown forces.
5. Substitute numerical values from the tabulation into the generalized area–moment equations.
6. Use the area–moment equations and equations of static equilibrium to solve for the redundant reaction components.

Deflections can be investigated in either statically determinate or indeterminate frames by direct application of the area–moment theorems after the reactions have been determined.

The following numerical examples illustrate how to apply the area–moment equations to determine reactions and deflections in statically indeterminate coplanar frames with two supports.

5.2.3.1 One-Hinged Frame (Numerical Example). Figure 5.7 shows

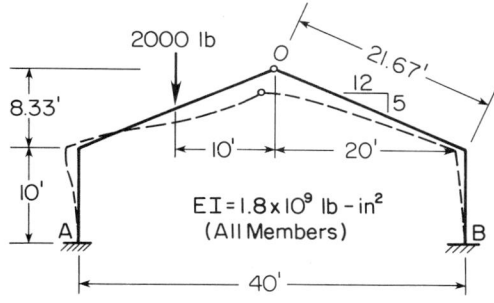

Fig. 5.7. One-hinged rigid frame.

100 Analysis of Statically Indeterminate Coplanar Frames

a symmetrical, one-hinged frame carrying a single concentrated load. The stiffness, EI, is the same for all parts of the frame. The problem is to determine (a) reactions at A and B; and (b) deflection of 0.

Solution. There are a total of six reaction components to be determined. Four equations from statics can be written (three for static equilibrium at the reactions, and one for the condition of zero bending moment at the hinge). Two additional equations must be written by applying the area–moment equations. These two equations are

$$\Delta x_{0A} = \Delta x_{0B}$$

$$\Delta y_{0A} = \Delta y_{0B}$$

where

$$\Delta x_{0A} = \int_A^0 y_0 \frac{m\,ds}{EI}$$

$$\Delta x_{0B} = \int_B^0 y_0 \frac{m\,ds}{EI}$$

$$\Delta y_{0A} = \int_A^0 x_0 \frac{m\,ds}{EI}$$

$$\Delta y_{0B} = \int_B^0 x_0 \frac{m\,ds}{EI}$$

It should be noted that, in the present example,

$$\Delta \theta_{AB} \neq 0$$

and

$$\int_B^A x_A \frac{m\,ds}{EI}$$

$$\int_B^A y_A \frac{m\,ds}{EI}$$

are not defined because of the hinge at 0. EI is zero at the hinge, and

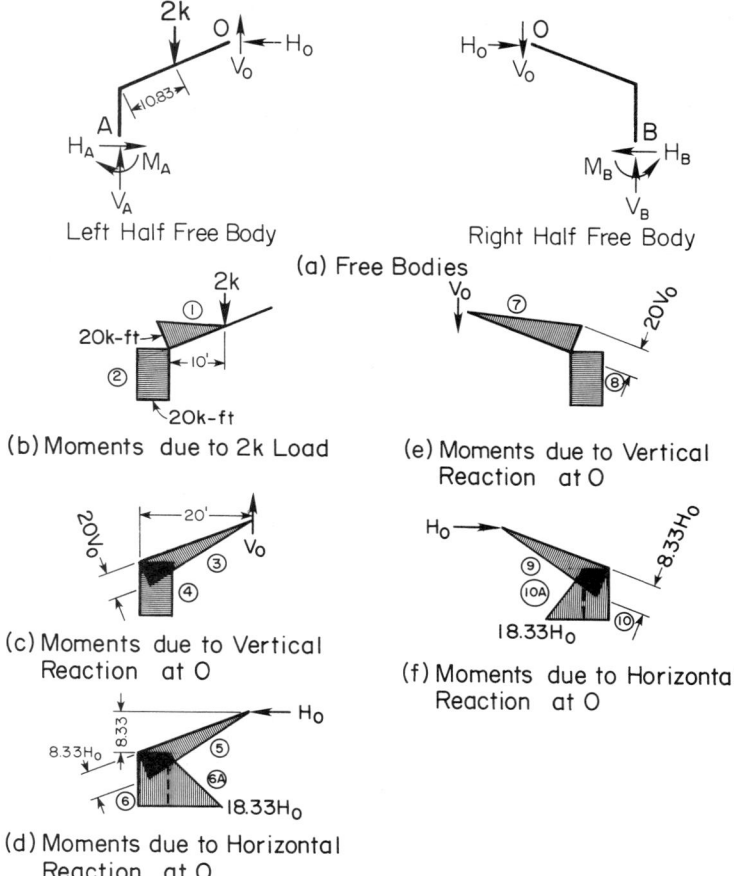

Fig. 5.8. (a) Free body and (b)–(f) moment diagrams for the frame in Fig. 5.7.

a discontinuity occurs in $d(\Delta\theta)/ds$ there. Integration through the hinge is not valid, and m/EI is not defined at the hinge.

The deformed frame is shown by the dashed line configuration in Fig. 5.7. Free body diagrams of the right and left parts of the frame are shown in Fig. 5.8a. Moment diagrams for the $2K$ load and each reaction component at 0 are shown in Figs. 5.8b–f. Note that some simplification in computation is obtained by drawing the moment diagrams for the effects of reactions at the hinge instead of at the supports, although the latter procedure would be valid. The hinge has only two reactions components, but there are three at each support.

Table 5.1. Analysis for One-Hinged Building Frame of Figs. 5.7 and 5.8.[a]

Part (Fig. 5.8)		$\int m\, ds$ (k−ft²)	$\overline{X}_0$ (ft)	$\overline{Y}_0$ (ft)	$\overline{X}_0 \int m\, ds$ (k−ft³)	$\overline{Y}_0 \int m\, ds$ (k−ft³)
Left half	1	108.33	16.67	6.94	−1805.86	+751.81
	2	200.00	20.00	13.33	−4000.00	+2666.00
	3	216.67V_0	13.33	5.56	+2888.20V_0	−1204.69V_0
	4	200.00V_0	20.00	13.33	+4000.00V_0	−2666.00V_0
	5	90.28H_0	13.33	5.56	+1203.43H_0	−501.96H_0
	6	83.33H_0	20.00	13.33	+1666.60H_0	−1110.79H_0
	6A	50.00H_0	20.00	15.00	+1000.00H_0	−750.00H_0
Right half	7	216.67V_0	13.33	5.56	−2888.20V_0	−1204.69V_0
	8	200.00V_0	20.00	13.33	−4000.00V_0	−2666.00V_0
	9	90.28H_0	13.33	5.56	+1203.43H_0	+501.96H
	10	83.33H_0	20.00	13.33	+1666.60H_0	+1110.70H_0
	10A	50.00H_0	20.00	15.00	+1000.00H_0	+750.00H_0
				Summation, left half	−5805.86	+3417.81
					+6888.20V_0	−3870.69V_0
					+3870.03H_0	−2362.75H_0
				Summation, right half	−6888.20V_0	−3870.69V_0
					+3870.03H_0	+2362.75H_0

[a] Sign convention for displacements: $+\rightarrow$, $+\uparrow$.

The values of the integrals in the area–moment theorems are computed by the tabulation scheme shown in Table 5.1. The values tabulated under column $\int m\, ds$ are the areas of the moment diagrams shown in Fig. 5.8. In this example the stiffness, EI, of the frame is the same for all members. Therefore the numerical stiffness value need not be taken into account when solving for reactions. In general, stiffnesses of different members would not all be the same. The $\int m\, ds$ in Table 5.1 would then be replaced by $\int m\, ds/EI$ for each member. The values of $\overline{X}_0$ and $\overline{Y}_0$ are taken from the moment diagrams as previously explained.

An appropriate sign convention must be established for the values in Table 5.1 under $\overline{X}_0 \int m\, ds$ and $\overline{Y}_0 \int m\, ds$. The sign convention used here is (+) for effects which tend to move point 0 to the right or upward. For example, it is apparent that the $2k$ load whose moment diagram is shown in Fig. 5.8b, tends to bend member A0 so point 0 moves to the right. Therefore, the values of $\overline{Y}_0 \int m\, ds$ corresponding to the moment diagram parts (1) and (2) in Fig. 5.8b have plus (+) signs as in the last column of Table 5.1. The downward force, V_0, on the right half of the frame, Fig. 5.8e, tends to bend the right half of the frame so point 0 moves to the left and downward. Accordingly, minus (−) signs are given to the values of $\overline{X}_0 \int m\, ds$ and $\overline{Y}_0 \int m\, ds$ corresponding to parts (7) and (8) in Table 5.1.

The analysis for the reactions at A and B is completed by writing two equations that involve the unknown reaction components H_0 and V_0. H_0 and V_0 are evaluated. The reaction components at the supports are computed by free body equations.

Compute reactions:

Write

$$\Delta X_{0A} = \Delta X_{0B}$$

and

$$\Delta Y_{0A} = \Delta Y_{0B}$$

But

$$\Delta X_{0A} EI = \int_A^0 y_0\, m\, ds = \Sigma \left(\overline{Y}_0 \int m\, ds, \text{left half} \right)$$

$$\Delta X_{0B} EI = \int_B^0 y_0\, m\, ds = \Sigma \left(\overline{Y}_0 \int m\, ds, \text{right half} \right)$$

$$\Delta Y_{0A} EI = \int_A^0 x_0\, m\, ds = \Sigma \left(\overline{X}_0 \int m\, ds, \text{left half} \right)$$

$$\Delta Y_{0B} EI = \int_B^0 x_0\, m\, ds = \Sigma \left(\overline{X}_0 \int m\, ds, \text{right half} \right)$$

So, inserting numerical values we have

$$-5805.86 + 6888.20 V_0 + 3870.03 H_0 = -6888.20 V_0 + 3870.03 H_0$$

and

$$+3417.81 - 3870.69 V_0 - 2362.75 H_0 = -3870.69 V_0 + 2362.75 H_0$$

Solving these gives

$$V_0 = +0.421 \text{ kip, upward, left half, as assumed}$$

$$H_0 = +0.723 \text{ kip, leftward, left half, as assumed}$$

104 Analysis of Statically Indeterminate Coplanar Frames

Next, by free body analysis, left half of frame is

$$V_A = 1.579 \text{ kip (7023 N), upward}$$
$$H_A = 0.723 \text{ kip (3216 N), rightward}$$
$$M_A = 1.673 \text{ kip-ft (2268 N} \cdot \text{m), cw}$$

By free body analysis, right half of frame is

$$V_B = 0.421 \text{ kip (1873 N), upward}$$
$$H_B = 0.723 \text{ kip (3216 N), leftward}$$
$$M_B = 4.833 \text{ kip-ft (6553 N} \cdot \text{m), ccw}$$

Next, compute ΔX_0 as follows:

$$\Delta X_0 = \Delta X_{OB} = \frac{\int_B^0 Y_0 m \, ds}{EI} = \frac{\sum_B^0 \left[\overline{Y}_0 \int_B^0 m \, ds \right]}{EI}$$

$$= \frac{(2362.75 H_0 - 3870.69 V_0) \times 1000 \times 1728}{1.8 \times 10^9}$$

$$\Delta X_0 = 0.075 \text{ in. (0.1905 cm), rightward}$$

Next, compute ΔY_0:

$$\Delta Y_0 = \Delta Y_{OB} = \frac{\sum_B^0 \left[\overline{X}_0 \int m \, ds \right]}{EI}$$

$$\Delta Y_0 = \frac{(-6888.20 V_0 + 3870.03 H_0) \times 1000 \times 1728}{1.8 \times 10^9}$$

$$\Delta Y_0 = -0.098 \text{ in. or } 0.098 \text{ in. (0.249 cm), downward}$$

The free body analysis results compared with the assumed elastic line deformation, Fig. 5.7, reveal: a) the fixed end moment at B is actually counterclockwise instead of clockwise as first assumed; b) pt. O actually moves rightward instead of leftward as assumed in Fig. 5.7; c) no inflection point occurs in column AA. The student should verify this and correct the elastic line to conform with the results of the analysis.

This is given to the reader as an exercise on deformation of the elastic line.

5.2.3.2 Two-Hinged Drawbar Frame (Numerical Example). Figure 5.9 is a two-hinged drawbar frame. The problem is to compute (a) the reactions; and (b) the movement of points B and C when the 5 kip horizontal load is applied at B.

Solution. The free body diagram for the frame is shown in Fig. 5.9b. A sketch of the deformed elastic line is shown in Fig. 5.10. The

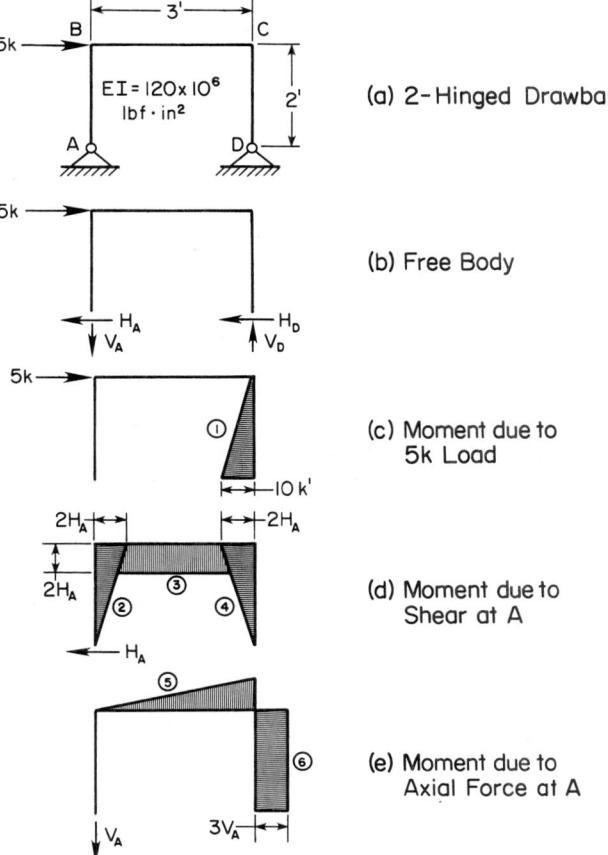

Fig. 5.9. Analysis diagram for two-hinged drawbar.

106 Analysis of Statically Indeterminate Coplanar Frames

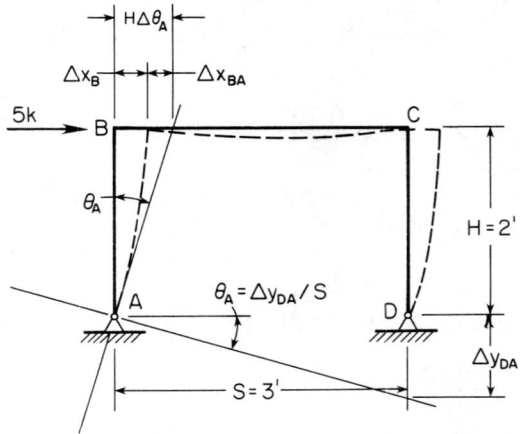

Fig. 5.10. Geometry of deformed two-hinged drawbar.

reactions have only one redundant. Only one area–moment equation is needed. The supports cannot move laterally and are at the same elevation. Therefore,

$$\Delta x_{DA} = \int_A^D y_D \frac{m\,ds}{EI} = 0$$

Note that, although the frame rotates at A and D, the foregoing equation is valid because the difference between unity and $\cos \theta_A$ is negligible, where θ_A is the rotation at A. In other words, the x distance from D to the y axis on the frame at A is the same before and after loading, neglecting second order effects. The numerical computations for the reactions are given in Table 5.2.

The geometry for the computation of horizontal deflection of B and C is illustrated in Fig. 5.10. This deflection is due to two effects: frame rotation at A; and bending of member AB. Frame rotation θ_A is obtained by first evaluating Δy_{DA} by the second area–moment equation. Then θ_A can be computed by the relationship

$$\theta_A = \Delta y_{DA}/S$$

as illustrated in Fig. 5.10. Movement of B due to bending relative to the frame axis at A is

5.2 Area–Moment Method

$$\Delta x_{BA} = \int_A^B y_B \frac{m \, ds}{EI}$$

Finally, the absolute movement of B due to rotation and bending is

$$\Delta x_B = H\theta_A - \Delta x_{BA}$$

The computations for this analysis are worked out in Table 5.2. The movements of B and C are equal.

Table 5.2. Analysis for Two-Hinged Drawbar Frame of Figs. 5.9 and 5.10.

Part (Fig. 5.9)	$\int m \, ds$ (k – ft^2)	$\overline{X}_D$ (ft)	$\overline{Y}_D$ (ft)	$\overline{X}_D \int m \, ds$ k – ft^3	$\overline{Y}_D \int m \, ds$ k – ft^3
1	10.0	0.000	0.667	0.000	+ 6.667
2	$2.0H_A$	3.000	1.333	+ $6.000H_A$	+ $2.667H_A$
3	$6.0H_A$	1.500	2.000	+ $9.000H_A$	+ $12.000H_A$
4	$2.0H_A$	0.000	1.333	0.000	+ $2.667H_A$
5	$4.5H_A$	1.000	2.000	– $4.500V_A$	– $9.000V_A$
6	$6.0V_A$	0.000	1.000	0.000	– $6.000V_A$
				+ $15.000H_A$	+ 6.667
				– $4.500V_A$	+ $17.333H_A$
					– $15.000V_A$

Compute reactions:

Write $\Sigma M_D = 0$, ↻: $-3V_A + 5 \times 2 = 0$; $V_A = \underline{\dfrac{10}{3}}$ kip ↓

Also $\Delta x_{DA} \times EI \int_A^D y_D m \, ds = 0$

or $+ 6.667 + 17.333 H_A - 15.000 V_A = 0$
gives $H_A = \underline{2.500}$ kip leftward (11,120 N)

Compute θ_A: ↻
Write $\theta_A = \dfrac{\Delta y_{DA}}{S}$

but $\Delta y_{DA} \times EI = 15.000 H_A - 4.500 V_A = 22.500$ kip-ft^3

gives $\Delta y_{DA} = \dfrac{22.500 \times 1728 \times 1000}{120 \times 10^6} = \underline{0.324}$ in. (0.823 cm)

so $\theta_A = \dfrac{0.324 \text{ in.}}{36 \text{ in.}} = 0.009$ rad

Compute ΔX_B (cf. Fig. 5.10):
Write $\Delta x_B = H\theta_A - \Delta x_{BA}$

But $\Delta x_{BA} = \dfrac{\int_A^B y_B \, m \, ds}{EI} = \dfrac{2H_A \times 1.333}{EI} = \dfrac{2 \times 2.5 \times 10^3 \times 1.333 \times 1728}{120 \times 10^6}$

$\Delta x_{BA} = 0.096$ in. (0.244 cm)
So $\Delta x_B = 24 \times 0.009 - 0.096 = \underline{0.120 \text{ in., to right}}$ (0.305 cm).

108 Analysis of Statically Indeterminate Coplanar Frames

The foregoing problem illustrates the difference between absolute and relative rotations of a frame. The first area–moment equation integrated from A to B would give the relative rotation of the frame at A with respect to the frame at B. This is not the same as the absolute rotation at A because the frame also rotates at B. Absolute rotation at A must be computed by application of the second area–moment equations.

An effect ignored in the foregoing examples is length change due to elastic axial strain in the members. It is assumed that the members are the same length before and after loading. The length change in members produced by stresses can usually be ignored.

5.2.3.3 Application to Arches. The area–moment equations can be used to analyze arches of curvilinear configuration. The general approach is the same as for frames with rectilinear members. The area–moment integrals can be evaluated graphically by approximating the actual curved configuration by short straight line segments. Then the integrals can be evaluated for each segment. A scale drawing of the arch should be used to measure distances instead of calculating these analytically. Too few segments give a poor approximation to the actual arch configuration. Too many segments require too many computations. The selection of an adequate segment length is a matter of judgment.

5.2.3.4 Application to Fixed End Members. The area–moment equations can be applied to calculate bending moments at fixed ends of members due to loads on the member, or displacements and rotations at the ends. For example, consider case G in Table 5.3. The problem is to obtain the fixed end moments that occur when the left end is rotated through angle θ_A. The first area–moment equation is applicable,

$$\theta_A = \Delta\theta_{AB} = \int_B^A \frac{m\,ds}{EI}$$

The analysis for this case is worked out in Fig. 5.11. The student should verify the other cases in Table 5.3 by application of the area–moment equations.

5.3 MOMENT DISTRIBUTION

Analysis of statically indeterminate structures by moment distribution is rapid, self-checking, and can be carried to any desired degree of

Table 5.3. Fixed End Members.

Case	Load and Support Configuration	End Moment (Absolute Values) Left M_L	Right M_R
A	Uniform load w, lb/ft over span ℓ, both ends fixed	$\dfrac{w\ell^2}{12}$	$\dfrac{w\ell^2}{12}$
B	Concentrated load P at distance a from left, b from right, span ℓ	$\dfrac{Pab^2}{\ell^2}$	$\dfrac{Pa^2 b}{\ell^2}$
C	Partial uniform load w over length a from left end, span ℓ	$M_L: \dfrac{wa}{12\ell}(6\ell^2 - 8a\ell + 3a^2)$ $M_R: \dfrac{wa^2}{12\ell}(4\ell - 3a)$	
D	Triangular load, zero at left, w at right, span ℓ	$\dfrac{w\ell^2}{30}$	$\dfrac{w\ell^2}{20}$
E	Support settlement δ, both ends fixed, span ℓ	$\dfrac{6EI\delta}{\ell^2}$	$\dfrac{6EI\delta}{\ell^2}$
F	Support settlement δ, left end pinned, right end fixed, span ℓ	0	$\dfrac{3EI\delta}{\ell^2}$
G	Rotation θ at left, both ends fixed, span ℓ	$\dfrac{4EI\theta}{\ell}$	$\dfrac{2EI\theta}{\ell}$
H	Rotation θ at left, left fixed, right pinned, span ℓ	$\dfrac{3EI\theta}{\ell}$	0

110 Analysis of Statically Indeterminate Coplanar Frames

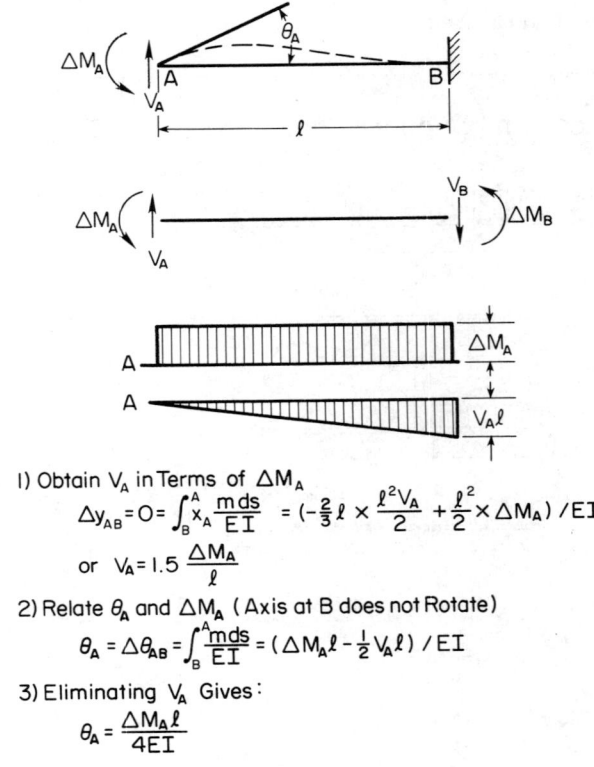

Fig. 5.11. Application of area–moment theorems to beam analysis.

precision. If no joint translations occur, moment distribution does not require solution of simultaneous equations. Therefore it is often a more convenient method to use for frames with many supports. In contrast, the area–moment method can become cumbersome to use when many simultaneous equations must be solved for several supports. The moment distribution computation routine is patterned after the mechanical behavior of the loaded frame. Therefore, it should appeal to engineers who prefer an analytical method which readily portrays physical behavior.

5.3.1 Moment Distribution Concepts

The basic concepts of moment distribution are illustrated by analysis of the four-member frame in Fig. 5.12. The outer ends of the members

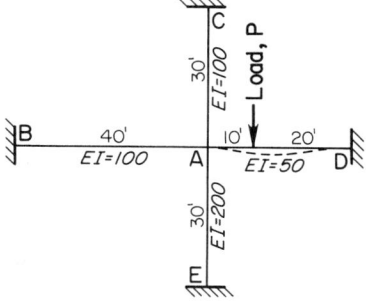

(a) Frame Characteristics

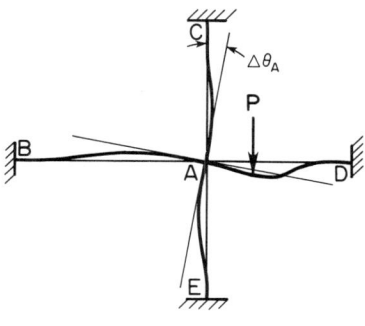

(b) Deformation Due to Load   (continued)

Fig. 5.12. Four-member frame for analysis by moment distribution.

are fixed. The members are connected at A by a rigid joint. Joint A can only rotate, not translate, because the members are mutually perpendicular and held at the outer ends. The problem is to analyze the frame by moment distribution to obtain the bending moments that occur at the ends of all four members when load P is applied.

First, imagine that a large, immovable clamp is temporarily applied at joint A. This completely prevents rotation of A when the members are loaded. A is said to be "locked" by the clamp.

Next, load P is applied. It produces deformation of member AD as shown in Fig. 5.12a. With A locked, AD is identical to an ordinary fixed end member, and fixed end bending moments, or FEMs, can be computed by a standard formula from Table 5.3. Moments at joint A in the locked position are shown in Fig. 5.12c.

Next, imagine that the clamp is removed to unlock A. Joint A can now rotate through some angle $\Delta\theta_A$ as shown in Figs. 5.12b and d. The A ends of all members rotate through the same angle, $\Delta\theta_A$, because of

112 Analysis of Statically Indeterminate Coplanar Frames

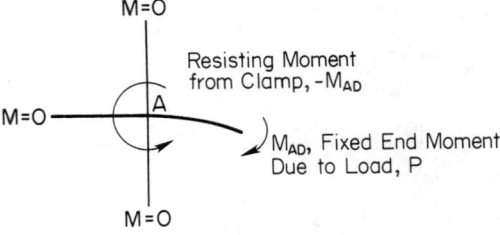

(c) Moments at Joint A in Locked Condition

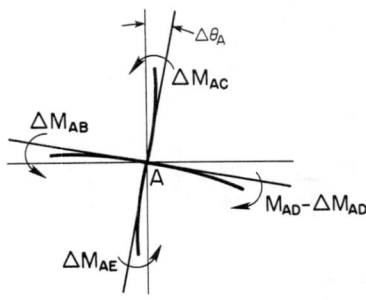

(d) Moments at Joint A in Unlocked Condition

Fig. 5.12. Continued.

the rigid joint at A. The rotation changes the bending moments in the ends of the members at A. The change in end moment in each member depends on the EI value of the member, its length, and whether the opposite or far end is fixed or hinged.

5.3.1.1 Stiffness Factor. A stiffness factor k can be defined for each member that takes into account these factors:

$$k = \Delta M/\Delta \theta \text{ ft} \cdot \text{lbf/radian}$$

where k is stiffness factor, units of bending moment at end, per unit end rotation, $\Delta M/\Delta \theta$, ft · lbf/radian; ΔM is bending moment change at the end of the member, ft · lbf; and $\Delta \theta$ is rotation at the end, radians.

In general, the stiffness factor will be different for each end of a member. ΔM and $\Delta \theta$ are measured at the end for which k is defined. Stiffness factors can be evaluated from the relationships between bending moment and angular rotation for cases G and H in Table 5.3. When the opposite end of the member is fixed (case G)

$$k = \frac{\Delta M}{\Delta \theta} = \frac{4EI}{l}$$

When the opposite end is pinned (case H)

$$k = \frac{\Delta M}{\Delta \theta} = \frac{3EI}{l}$$

The change, ΔM in bending moment produced in the A end of each member when joint A is unlocked can be calculated by analysis of the free body, Fig. 5.12d. By $\Sigma M = 0$ applied to joint A,

$$\Delta M_{AB} + \Delta M_{AC} + \Delta M_{AD} + \Delta M_{AE} = M_{AD}$$

Notice that the shear at the A end of each member, although present, is not shown because, conceptually, the ΔMs are applied at joint A, and the end shears are zero distance from joint A. Thus, end shear at end A produces no moment at end A.

The FEM, M_{AD}, is unbalanced when the clamp is removed until the joint rotates enough that opposite moments totaling M_{AD} are developed in the ends of the members. Each ΔM is called a *distributed moment* because it is a share of the unbalanced moment distributed to each member at the joint. Substituting $\Delta M = k\, \Delta \theta$ for each of the terms on the left-hand side of the preceding equations:

$$k_{AB}\, \Delta \theta_A + k_{AC}\, \Delta \theta_A + k_{AD}\, \Delta \theta_A + k_{AE}\, \Delta \theta_A = M_{AD}$$

The proportion $\Delta M/M_{AD}$ of the unbalanced moment resisted at each member end, using end AB as an example, is

$$\frac{\Delta M_{AB}}{M_{AD}} = \frac{k_{AB}\, \Delta \theta_A}{k_{AB}\, \Delta \theta_A + k_{AC}\, \Delta \theta_A + k_{AD}\, \Delta \theta_A + k_{AE}\, \Delta \theta_A}$$

and

$$\frac{\Delta M_{AB}}{M_{AD}} = \frac{k_{AB}}{k_{AB} + k_{AC} + k_{AD} + k_{AE}} = \frac{k_{AB}}{\Sigma k}$$

5.3.1.2 Distribution Factor.
Next, define $r_{AB} = k_{AB}/\Sigma k$. The term r_{AB} is called the *distribution factor* for end A of member AB. The subscript denotes the member and end. For example, AB denotes end A of member AB. As the words *distribution factor* imply, r_{AB} when multiplied by the unbalanced moment at a joint, in this case M_{AD}, gives the share of the unbalanced bending moment distributed to the end AB. By similar analysis for ends AC, AD, and AE,

$$r_{AC} = \frac{k_{AC}}{\Sigma k}$$

$$r_{AD} = \frac{k_{AD}}{\Sigma k}$$

$$r_{AE} = \frac{k_{AE}}{\Sigma k}$$

where $\Sigma k = k_{AB} + k_{AC} + k_{AD} + k_{AE}$.

5.3.1.3 Carry-over Moment.
The last step is to compute the bending moment at the outer or opposite end of each member when joint A is unlocked. The moment at the other end is called the *carry-over moment*, abbreviated COM, because, loosely speaking, it is an effect transmitted or carried from end A to the opposite end.

The COM for a member with the opposite end fixed can be evaluated from the relationship for case G in Table 5.3. By comparing bending moments at the right and left ends, it is seen that

$$\frac{M_R}{M_L} = -\frac{1}{2}$$

Therefore, bending moment applied to one end of a member produces one half that moment with opposite sign at the other end, if this other end is fixed.

The sign convention used here is based on interpretation of the direction (clockwise, cw, or counterclockwise, ccw) the moment in question tends to rotate the joint. Here, a positive sign is used if the moment tends to turn any joint, either internal or at a fixed support in a clockwise direction. Therefore, the original FEMs of member AD (Fig. 5.12) are positive at end A and negative at end D. The carry-over moment will always have the opposite sign of the distributed moment, as inspection of Table 5.3, case G, will reveal. A counterclockwise rotation

5.3 Moment Distribution

(1) Computation of Distribution Factors:

Member	EI	Length	$k=\frac{4EI}{\ell}$	$r=\frac{k}{\Sigma k}$
AB	100	40	10.00	0.176
AC	100	30	13.33	0.235
AD	50	30	6.67	0.118
AE	200	30	26.67	0.471
		Σk	56.67	1.000

(2) Computation of Fixed End Moments from Table 5.3:

$$FEM_{AD} = \frac{Pab^2}{\ell^2} = \frac{2 \times 10 \times 400}{900} = 8.88 \text{ k-ft}$$

$$FEM_{DA} = \frac{Pa^2b}{\ell^2} = \frac{2 \times 100 \times 20}{900} = 4.44 \text{ k-ft}$$

(3) Moment Distribution:

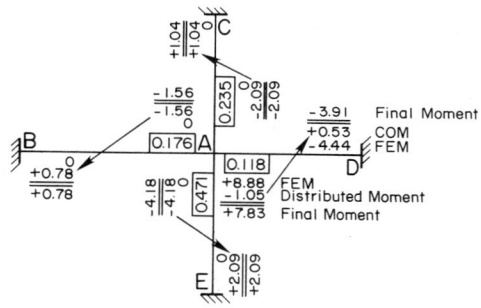

Fig. 5.13. Moment distribution computations.

of the left-end joint tends to rotate the right-end joint in a clockwise direction.

5.3.2 Example, Four-Member Rigid Frame

The foregoing principles are illustrated by a numerical example using the frame of Fig. 5.12a. The problem is to calculate the end moments developed in both ends of each member when a load P of 2 kip is applied to member AD.

1. Compute stiffness factors k and distribution factors r as shown in Fig. 5.13. The r values are entered in the diagram of the frame in Fig. 5.13. For example, the distribution factor for end A of member AC is 0.235.
2. Compute fixed end moments for member AD due to the 2 kip load.

116 Analysis of Statically Indeterminate Coplanar Frames

These are evaluated by case B, Table 5.3, and entered as shown in Fig. 5.13. For example, the FEM for end A of member AD is +8.88 ft kip. The FEMs for the other members are, of course, zero, since they are not loaded while joint A is clamped or locked. A sign convention must be adopted and used in moment distribution. The sign convention used here is *plus* (+) for moments which tend to turn any joint *clockwise*. This sign convention is used for FEMs, distributed moments, and COMs.

3. Unlock joint A. Distribute the unbalanced moment of +8.88 ft kip. The share of the unbalanced moment distributed to each member at the joint is r for that end multiplied by the total unbalanced moment. Note that the sign for the distributed moments is always the opposite of the sign for the total unbalanced moment which, in this case, is the FEM at end AD. The distributed moments are entered in Fig. 5.13 following the unbalanced FEM.
4. Carry over each distributed moment from the A end of each member to the opposite end. The sign of the COM is always the opposite of the distributed moment.
5. Add the moments at each end to obtain the actual end moments in the loaded structure.
6. By free body analysis of each member, compute end shears and draw shear and moment diagrams.

5.3.3 Multijoint Structures

The foregoing analysis dealt with a frame with only one joint which could rotate but not translate. In the general case, a frame has several joints, some of which can translate as well as rotate. The following example illustrates moment distribution for a frame with more than one joint. The joints can only rotate, not translate.

5.3.3.1 Example, Multijoint Structures. Figure 5.14 depicts a frame with two internal joints, B and C, which can rotate. Joint translation is prevented because each joint is held by two members in mutually perpendicular directions and with the outer ends restrained against translation. As a result, the only translation that can occur is due to axial strain. This is neglected as a second order effect. The problem is to compute the end moments for all members. The EI values for the members are given in Fig. 5.14. The analysis starts with calculation of stiffness factors, distribution factors, and FEMs as shown in Fig.

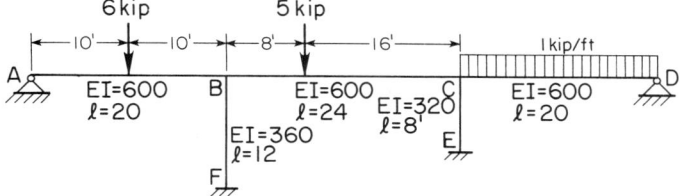

(1) **Computation of Distribution Factors:**

	End	EI/ℓ	k	r=k/Σk
Joint B	BA	30	×3=90	0.290
	BC	25	×4=100	0.323
	BF	30	×4=120	0.387
			Σk=310	1.000
Joint C	CB	25	×4=100	0.286
	CD	30	×3=90	0.257
	CE	40	×4=160	0.457
			Σk=350	1.000

(2) **Computation of Fixed End Moments** (cf. Table 5.3, Case A and Case B):

$$M_{BA}: \frac{6 \times 20}{8} = 15.00 \text{ ft-k}$$

$$M_{BC}: \frac{5 \times 8 \times 256}{576} = 17.78 \text{ ft-k}$$

$$M_{CB}: \frac{5 \times 64 \times 16}{576} = 8.89 \text{ ft-k}$$

$$M_{CD}: \frac{1 \times 400}{12} = 33.33 \text{ ft-k}$$

Fig. 5.14. Continuous frame for analysis by moment distribution.

5.14. Note that the stiffness factors for members *BA* and *CD* are $3EI/L$, since the opposite ends are hinged (Table 5.3, case H).

The distribution factors and FEMs are entered in the frame diagram; see Fig. 5.15. The joints are then unlocked and the unbalanced moment at each joint is distributed according to the distribution factors. This completes the first cycle. Note that the initial unbalanced moment at each joint is the sum taking sign into account of the FEMs for the ends of all members meeting at that joint. One half of each distributed moment is carried over to the opposite end. Joints *A* and *D* do not receive COMs, since hinged ends cannot resist bending. The COMs to joints *B* and *C* again unbalance them. The unbalanced moment at each joint is again distributed. This completes the second cycle. Each moment distribution cycle after the first includes (a) carrying over; (b)

118 Analysis of Statically Indeterminate Coplanar Frames

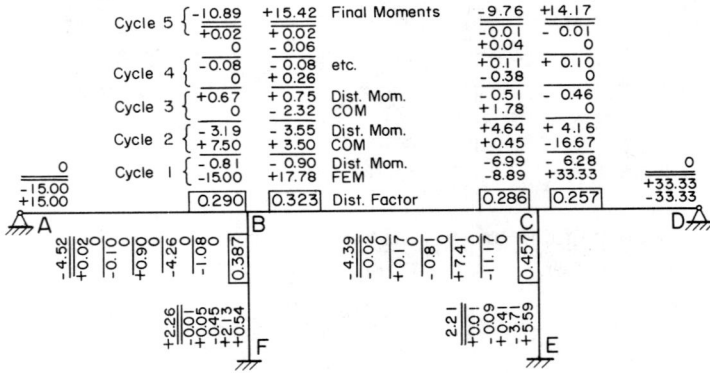

Fig. 5.15. Moment distribution analysis for continuous frame.

computing the unbalanced moment at each joint; and (c) distributing the unbalanced moment to the ends of the members meeting at each joint. In each cycle, the sum of the distributed moments at each joint must be equal but of opposite sign to the unbalanced moment.

The COM normally becomes smaller with each new cycle. When the distributed moments become negligibly small, no more cycles are needed, and the computations stop. In the example of Fig. 5.15, five cycles have been illustrated. The calculations could be continued through additional cycles if desired to obtain greater precision. The final end moment is the algebraic sum of the original FEM, the distributed moments, and the COMs at each member end; or, simply, the algebraic sum of each column of figures. The algebraic sum of all the final end moments at each internal joint should, of course, be zero. If not, a computation error has occurred.

Many statically indeterminate frames have pinned connections at the supports or at internal joints. A pinned connection cannot resist any moment. The moment distribution must take this into account by calculations for k, the stiffness factor for each member that has a pinned end connection. The k value for a member with a pinned end is $3\ EI/l$, compared to a member with the opposite end fixed, for which the stiffness is $4\ EI/l$. For example, Fig. 5.14, joint B, member end BA has a stiffness factor of $3\ EI/l$ because the opposite end, A, is pinned instead of fixed. This is reflected in the distribution factor, r. The initial fixed end moments due to the applied loads are calculated assuming that all pinned as well as the fixed joints are temporarily fixed against rotation.

The moment distribution process begins with unlocking the joints, then calculating the unbalanced moment and the distributed moment at each joint. The pinned joints are also unlocked. The original FEMs at the pinned joints are balanced with an end moment opposite in sense but equal in magnitude to the original FEM with the pinned joint locked. For example, pinned joint A, Fig. 5.15, has an original FEM of +15.00 ft.k. When unlocked, the balancing moment is −15.00 ft.k. One half of the balancing moment, +7.50 ft.k is carried over to end B, and becomes part of the unbalanced moment at joint B, cycle 2.

After this initial balancing of pinned end joints, joint A remains pinned and no further effect of pinned end A occurs. The pinned end can resist no moment in its locked condition. No further COM is generated at A. The same is true for joint B.

Notice again that the stiffness factor for a member with one end fixed and the other pinned is 3 EI/l instead of 4 EI/l for a member with both ends fixed.

5.3.4 Joint Translation

The foregoing moment distribution methods assume that only joint rotations, not joint translations (sometimes called "sidesway") occur when the frame is loaded. The configuration and loading arrangements of many frames are such that some joints translate as well as rotate. For example, the knee joints of a symmetrical gable-shaped frame spread apart under gravity loads. As a result, moment distribution analysis assuming no joint translation is modified to produce a spurious load for static equilibrium when reactions and internal stresses are calculated. To correct for this spurious load a second moment distribution is completed for the frame deformed by only an assumed load of arbitrary value and opposite in direction with respect to the spurious load. Fixed end moments and stiffness factors corresponding to joint translation are computed by case E or F, Table 5.3, depending on whether the member is fixed at both ends or only one end. Finally, the moments of the first distribution are corrected by the ratio of the spurious load to the opposition load from the second distribution. The corrections for joint translation are usually quite small. (It should be noted that the area–moment method inherently takes into account joint translation.) Further exposition of sidesway corrections can be found in most textbooks on statically indeterminate structures.

5.4 OTHER METHODS FOR ANALYSIS OF STATICALLY INDETERMINATE FRAMES

Other methods for analysis of statically indeterminate structures are available. They all require solution of a set of simultaneous algebraic equations. The number in the set is equal to the degree of static indeterminacy. For example, a coplanar frame supported on four fixed-end reactions, with no internal hinges for which $\Sigma_m = 0$ could be written, is indeterminate to degree 9 and would require some analytical method to produce a set of nine simultaneous linear equations. The coefficients in these equations are used to set up a matrix which is then solved on a computer.

One might ask why it is thought necessary to understand the fundamentals of analysis of statically indeterminate structures because computer programs are available to do the otherwise tedious work. Engineers need to have a basic understanding of the response of a load-bearing system to forces, displacements, member properties, and changes in geometry, even though hand calculations can be eliminated for solution of the system. Otherwise the analyst will be at the mercy of the computer and unable to use it creatively.

Some additional methods for analysis of statically indeterminate frames and members include the following:

1. *Slope–deflection method.* Simultaneous equations are written in terms of the slopes and deflections at the ends of the members. The number of independent equations is equal to the degree of statical indeterminacy.

2. *Force (equilibrium) method.* A statically determinate solution is assumed. Then the discrepancies in geometry resulting from the assumed solutions are computed and corrected. This produces a set of simultaneous equations equal in number to the degree of indeterminacy. A coefficient matrix can be written and solved, typically by a high-speed computer.

3. *Displacement (geometrical) method.* Instead of assuming a statically determinate solution, a solution that satisfies the geometry of the deformed frame is assumed and corrected for the discrepancies in static equilibrium. A set of linear equations is obtained and solved by matrix algebra on a high-speed computer.

4. *Column analogy.* The equations for the axial stress in a short column loaded by an eccentric force are analogous to the equations for stresses in a statically indeterminate arch or other ring-shaped structure. A detailed development of the column analogy can be found in Michalos, James, and Wilson, Edward N., *Structural Mechanics and Analysis.* Macmillan, New York. 1965.

PROBLEMS

5.1 Compute the reactions at the supports for the frame of Fig. 5.16. Draw the shear and moment diagram for each member. Joints A, B, and D are rigid.

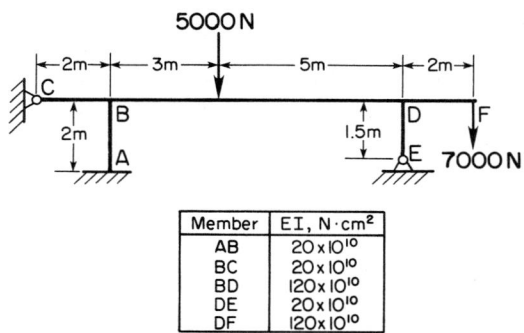

Member	EI, N·cm²
AB	20×10^{10}
BC	20×10^{10}
BD	120×10^{10}
DE	20×10^{10}
DF	120×10^{10}

Fig. 5.16. Statically indeterminate frame, Problem 5.1.

5.2 Compute the reactions at the supports for the frame of Fig. 5.17. Draw the shear and moment diagram for each member. Joints B, C, and D are rigid.

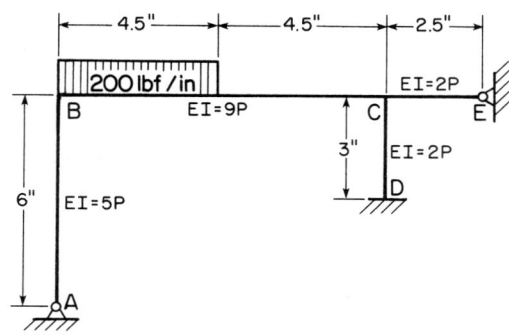

Fig. 5.17. Statically indeterminate frame, Problem 5.2.

5.3 Compute the reactions at the supports for the frame of Fig. 5.18. Draw the shear and moment diagrams for each member. Calculate the maximum deflection of member BC.

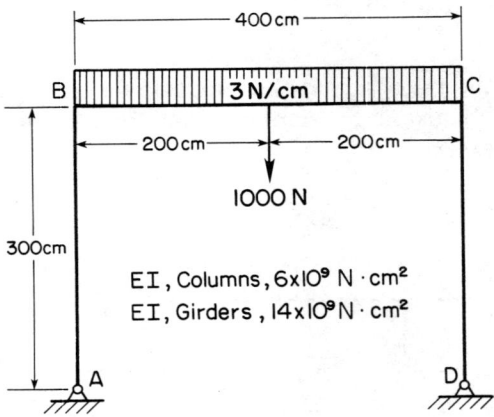

Fig. 5.18. Two-hinged frame, Problem 5.3.

5.4 Calculate the reactions at the supports for the drawbar frame of Fig. 5.19. Draw the shear and moment diagrams for each member. Calculate the vertical and horizontal displacements at point Q.

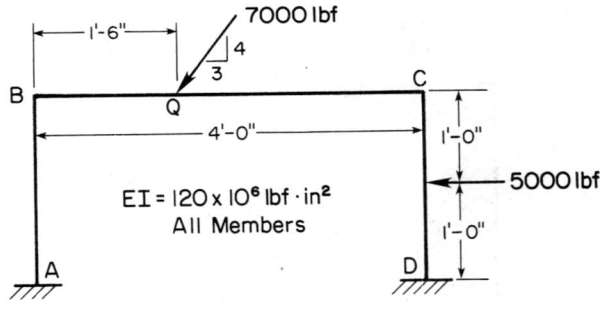

Fig. 5.19. Hingeless frame, Problem 5.4.

5.5 Calculate the shear and moment at the ends of each member for the buried road culvert of Fig. 5.20. Draw the shear and moment diagrams for each member.

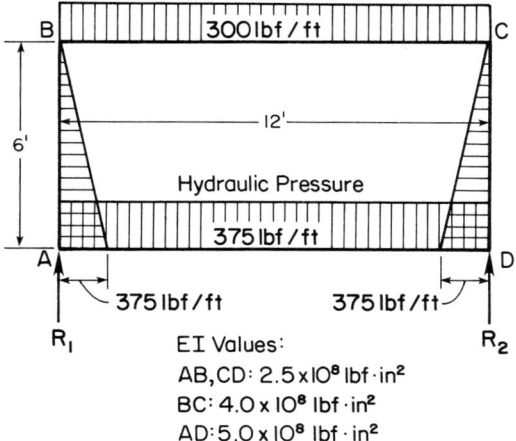

Fig. 5.20. Box culvert, Problem 5.5.

5.6 Compute the reaction at the supports for the frame of Fig. 5.21. Compute the deformation (absolute displacement, absolute rotation) at joints B, C, and D.

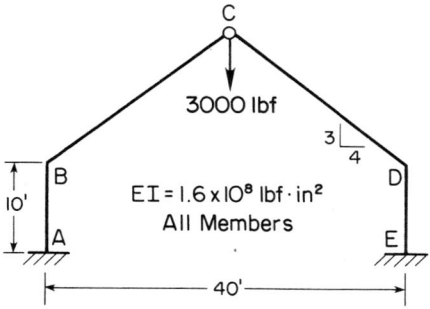

Fig. 5.21. One-hinged gable frame, Problem 5.6.

5.7 The R/C frame for an agricultural products warehouse is configured and loaded as shown in Fig. 5.22. Compute the reactions. Draw the shear and moment diagram for each member.

124 Analysis of Statically Indeterminate Coplanar Frames

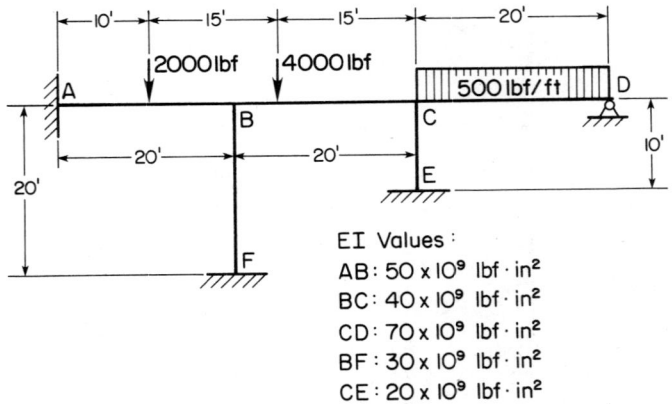

Fig. 5.22. Warehouse, Problem 5.7.

5.8 An all welded tool bar frame is configured and loaded as shown in Fig. 5.23. Calculate the shear and moment at the end of each member. (Neglect the axial strains induced by direct stresses in the members.)

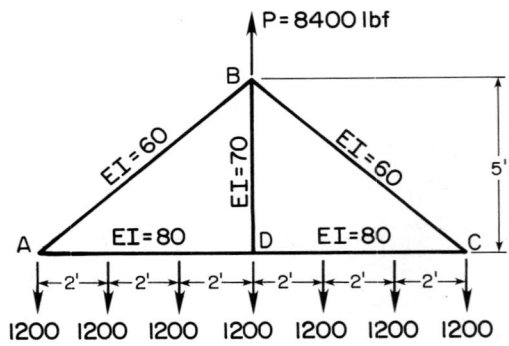

Fig. 5.23. Welded tool bar frame, Problem 5.8.

5.9 Verify the relationships shown for each of the cases of Table 5.3.
5.10 Calculate the reactions for the frame of Fig. 5.21. The frame is loaded by a wind load of 150 lbf/lineal ft of the left half of the frame. The concentrated load at the ridge is removed.

NOMENCLATURE FOR CHAPTER 5

E	Modulus of elasticity, lbf/sq in. (pascal)
I	Flexural member cross-section moment of inertia, in.^{4} (cm^4)
kip	Force or reaction, thousands of pounds force
M	Meeting moment
m	Bending moment at designated position on elastic line, lbf · ft (N · m)
s, l	Length increment measured along the elastic line, in. (cm)
$\overline{X}_A$	Distance in x direction from centroid of bending moment diagram to point A (cf. Fig. 5.2), in. (cm)
x_A	Distance in x direction between point A on the elastic line to point where bending effect is applied, causing displacement at A (cf. Fig. 4.3)
Δx_{AB}	Bending displacement in x direction of point A on the elastic line relative to point B where a set of x–y reference axes are attached to the elastic line in. (cm)
$\overline{Y}_A$	Distance in y direction from centroid of bending moment diagram to point A in. (cm) (cf. Fig. 5.2)
y_A	Distance in y direction between point A on the elastic line and the point of application of the bending effect, in. (cm) (cf. Fig. 4.3)
Δy_{AB}	Displacement in y direction of point A on the elastic line relative to point B where a set of reference axes are attached to the elastic line, in. (cm)
$\Delta\theta_{AB}$	Rotation of elastic line at A relative to elastic line at B

Note: Consistent units must be used in calculations of forces and deformations. The following equivalents will be useful in converting from English units to SI units, and vice versa.

1 ft = 0.3048 m
1 ft · lbf = 1.3558 Newton meter (N · m)
1 ft · kip = 1355.8 Newton meter (N · m)
1 in. = 2.540 cm
1 in. · lbf = 0.1130 Newton meter (N · m)
1 in.^{4} (cross-section moment of inertia) = 41.62 cm^4
1 kip = 4448 N
1 kip/sq in. = 6.895 × 10^6 Pa
1 lbf = 4.448 Newton (N)
1 psi = 6895 pascal (Pa)
1 sq ft = 0.0929 meter^2 (m^2)

6

Load Analysis

Often design loads are specified by a building code or possibly by a professional standard. The designer must check to see if the design is supposed to conform to one code or if it can be based on the best available information from a variety of sources. This chapter does not follow one code or standard. The American National Standards Institute (ANSI), Building Officials and Code Administrators (BOCA), and Uniform Building Code (UBC) publications have been used to illustrate the factors affecting loads.

Design load selection is important because loads are the foundation of design. The designer has an opportunity to assume conservative or unconservative loads according to his engineering judgment. Loads are frequently used as the point to interject the designer's judgment because often the designer has a better feel for the magnitude of the loads than for design stresses or procedures.

The load assumption may be modified in some manner to take into account the severity of failure. Severity of failure is usually measured in terms of possible loss of human life. Code writers recognize the probability of loss of human life during collapse by classifying the structures according to the use and therefore according to the risk involved. Classifications such as assembly areas, mercantile, business, storage, residence, and animal housing have associated probabilities of loss of life during collapse.

Structures may also be classified according to the permanence required and the financial losses that can occur upon collapse. Codes recommend loads for permanent and temporary structures and provide for design load selection to account for various magnitudes of financial loss upon failure. Technological advances in agriculture have caused rapid obsolescence of agricultural buildings in the past. This factor has caused designers to consider these buildings less permanent than houses, for example, and to choose design loads accordingly.

Many agricultural structures are designed for lighter loads than most other structures, because the comparative severity of failure, in terms of loss of human life, is considered small. This assumption should, however, be examined when designing each structure. Higher loads should be used if the structure is likely to be occupied by people a high percentage of the time, such as for produce processing buildings, or if the structure is likely to contain items for which a collapse would cause a heavy financial burden on the owner. As a measure of the relative magnitude of assumed loads, agricultural structures, where loss of life is not likely, have been designed on a 25-year recurrence interval for snow loads while residences have been designed on a 50-year recurrence interval. Assembly areas have been designed on a 100-year recurrence interval.

6.1 LOAD CLASSIFICATION

Loads are usually classified as dead loads, live loads, or natural loads. Dead loads are those loads that do not change in magnitude or location during the life of the structure. Dead load includes the weight of the structure plus permanently installed equipment.

Live loads are those loads that occur through use of the structure. People walking on the floor, weight of materials stored on a warehouse floor, pressure of grain against the sides of the bin, and loads imposed by vehicles crossing a bridge are all examples of live loads.

Natural loads are those loads that occur due to nature. Snow, wind, and earthquake loads are the usual natural loads considered.

6.2 COMBINATION OF LOADS

The design process must consider those combinations of loads that can be expected to act together. The most severe combination of loads will vary depending upon the member being considered. ANSI Standard A58.1-1982 suggests the following load combinations be investigated for allowable stress designs:

1. Dead
2. Dead + [Live + (Roof live or snow or rain)]   0.75
3. Dead + [Wind or Earthquake]   0.75

4. Dead + [Live + (Roof live or snow or rain) + (Wind or Earthquake)] 0.75

6.3 DEAD LOADS

Dead load estimation presents a dilemma in that the dead load is needed to design a structure but the dead load is not known until the structure is designed. Fortunately, the dead load of a structure is usually low when compared to the total load capacity. Often, a dead load assumption based on experience is acceptable. Dead loads are often assumed to be zero where experience is not helpful in estimating the dead load and then one or two iterations usually establishes the dead load and the total design load to an acceptable accuracy.

The designer is encouraged to be conservative in dead load estimates because dead loads can increase over the life of the structure. Wood and similar materials can absorb moisture; a second roof, floor, or wall covering can be installed; additional insulation added; or permanently installed equipment can increase the dead load. Dead load tables such as Table 6.1 are normally conservative.

The ratio of dead to total load for a structure varies with the construction material used, member type (tension, column, or beam), and the extent to which other structural dead loads are being supported by the member under study. To illustrate some of these variations and yet establish that the variation is remarkably small in well-designed structural members, consider the following examples:

Example 6.1. Consider a 2.44-m (8-ft) long concrete slat with a cross section as shown in Fig. 6.1. The 1983 edition of *Structures and Environmental Handbook* (Midwest Plan Service) gives the live load capacity of the slat as 189 lb/ft. The dead load per slat is calculated as follows:

Dead load

$= (178 \text{ mm})(127 \text{ mm} + 100 \text{ mm}) \left(\frac{1}{2}\right) (2440 \text{ mm})(2.40 \text{ mg/mm}^3)$

$= 118 \text{ kg}$

Live load capacity $= (189 \text{ lb/ft})\left(1.49 \dfrac{\text{kg/m}}{\text{lb/ft}}\right)(2.44 \text{ m}) = 687 \text{ kg}$

Total load capacity $= 118 + 687 \text{ kg} = 805 \text{ kg}$

Dead load/total load capacity $= (118 \text{ kg}/805 \text{ kg})100 = 14.7\%$

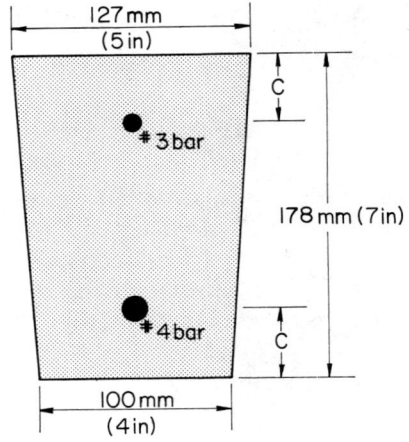

Fig. 6.1. Example 6.1.

Table 6.1. Dead Loads for Design Purposes.

Item	Specific Gravity[a]	kg/m²	lb/ft²
Basic Building Materials			
Clay, damp and plastic	1.76		
Concrete			
stone aggregate	2.31		
vermiculite and perlite aggregate, nonstructural	0.40–0.80		
light aggregate, load-bearing	1.12–1.68		
reinforced, stone aggregate	2.40		
Earth, dry and loose	1.22		
damp and packed	1.54		
wet and packed	1.92		
Gravel, dry, loose	1.65		
dry, packed	1.81		
wet, loose	1.92		
Limestone, crusher run	1.52		
38 to 50 mm (1.5 to 2 in.)	1.36		
above 50 mm (2 in.) grade	1.28		
Metals, rolled steel	7.85		
cast aluminum	2.64		
Plastics, medium density polyethylene	0.93		
polyvinylchloride	1.39		
fiberglass-reinforced polyester sheet	1.55		
Sand, dry, loose	1.52		
wet, packed	1.92		

Table 6.1. Continued.

Item	Specific Gravity	kg/m²	lb/ft²
Stone, granite	2.80		
limestone and marble	2.64		
slate	2.80		
sandstone	2.36		
Wood (12% moisture)			
Douglas fir	0.54		
fir, commercial, white	0.43		
hemlock	0.46		
oak, white	0.75		
pine, southern	0.61–0.67		
pine, white	0.43		
softwood lumber			
2 × 4—305 mm (1 ft) o.c.		7.3	1.5
2 × 4—406 mm (16 in) o.c.		5.4	1.1
2 × 4—610 mm (2 ft) o.c.		3.4	0.7
2 × 6—305 mm (1 ft) o.c.		11.0	2.3
2 × 6—610 mm (2 ft) o.c.		5.4	1.1
2 × 8—305 mm (1 ft) o.c.		15.0	3.0
2 × 10—305 mm (1 ft) o.c.		19.0	3.9
2 × 12—305 mm (1 ft) o.c.		23.0	4.7
Trusses, wood, span 6 to 18 m (20 to 60 ft), spacing 0.6 to 2.4 m (2 to 8 ft) psf = 1 + 0.1 (span, ft)/(spacing, ft) kg/m² = 4.9 + 0.5 (span, m)/(spacing, m)			
Roof, Wall and Floor Coverings			
Asbestos cement board, corrugated roofing or shingles		20	4
Asbestos cement board, flat, per 25 mm (1 in.) thickness	1.4	35	7.3
Asphalt shingles		12	2.5
Brick, 100 mm (4 in.)		171	35
Composition roofing, 5-ply felt and gravel		29	6
Concrete block, 100 mm (4 in.) hollow, stone aggregate		146	30
102-mm (4-in.) hollow, lightweight aggregate		98	20
203-mm (8-in.) hollow, stone aggregate		268	55
208-mm (8-in.) hollow, lightweight aggregate		186	38
203-mm (8-in.) solid, stone aggregate		327	67
Corrugated steel, 28 gauge		3.9	0.8
Corrugated aluminum, 0.53 mm (0.021 in.)		2.0	0.4
Glass, plate	2.58		
Gypsum board			
127 mm (1/2 in.)	0.8	10	2.1
159 mm (5/8 in.)		13	2.6
Particle boards, fiberboard			
sheathing, 127 mm (1/2 in.)	0.31	3.9	0.8
flake board, 127 mm (1/2 in.)	0.72	9.2	1.9
hardboard, per 25-mm (1-in.) thickness	1.16	29.0	6.0

(*continued*)

Table 6.1. Continued.

Item	Specific Gravity	kg/m²	lb/ft²
Plywood			
9.5 mm (3/8 in.)	0.57	5.4	1.1
127 mm (1/2 in.)	0.57	7.3	1.5
Insulation			
Blanket and batt			
fiberglass per 25-mm (1-in.) thickness		0.29	0.06
fiberglass, sheet or board per 25-mm (1-in.) thickness		0.39–0.88	0.08–0.18
Mineral wool, per 25-mm (1-in.) thickness		0.24	0.05
Foamed plastics			
polystyrene, extruded, per 25-mm (1-in.) thickness		0.93	0.19
polystyrene, molded, per 25-mm (1-in.) thickness		0.39	0.08
polyurethane, foil covered, per 25-mm (1-in.) thickness		0.88	0.18
urea formaldehyde, per 25-mm (1-in.) thickness		0.39	0.08
Loose fill (settled density)			
cellulose, per 25-mm (1-in.) depth	0.045	1.12	0.23
fiberglass, per 25-mm (1-in.) depth	0.035	0.88	0.18
mineral wool, per 25-mm (1-in.) depth	0.05	1.17	0.24
vermiculite, per 25-mm (1-in.) depth	0.13	3.32	0.68

[a]Specific gravity of $1.0 = 1.0$ Mg/m³ or 1.0 mg/mm³ (62.4 lb/ft³).

Example 6.2. The uniform load carrying capacity of a W6 × 20 A36 steel beam with a 16-ft span is 766 lb/ft according to the *Manual of Steel Construction* (American Institute of Steel Construction, 1980). The percent dead load is calculated as follows:

Dead load = (16 ft)(20 lb/ft) = 320 lb

Total load = (16 ft)(766 lb/ft) = 12,256 lb

Dead load/total load = (320 lb/12,256 lb)100 = 2.6%

If the load were a concentrated load at the center of the span, the minimum load would be 6,350 lb and the percent dead load would be:

Dead load/total load = (320 lb/6350 lb)100 = 5.0%

Example 6.3. Consider the truss shown in Fig. 6.2. The dead load is estimated as follows:

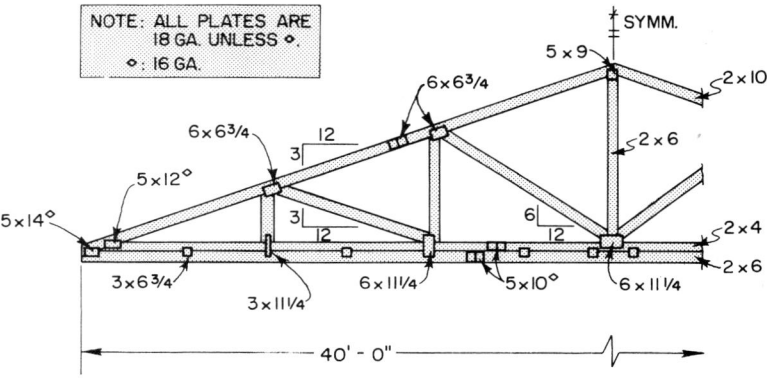

Fig. 6.2. Example 6.3.

Top chord = (1.03 × 40 ft)(3.9 lb/ft) = 161 lb

Lower chord = (40 ft)(1.5 + 2.3) lb/ft = 152 lb

Web members = (48 ft)(1.5 lb/ft) = 72 lb

18-gauge plates = (1,440 in.^{2}/144 in.^{2}/ft^2) × 2 lb/ft^2 = 20 lb

16-gauge plates = (920 in.^{2}/144 in.^{2}/ft^2) × 3.3 lb/ft^2 = 21 lb

Total dead load = 426 lb

Total design load (at normal load duration)

= (26 psf)(40 ft)(7.5 ft) = 7800 lb

Dead load/total design load = (426 lb/7800 lb) 100 = 5.5%

If the whole roof structure, including the truss, is considered then the percent dead load is changed considerably. Assume the roof consists of 28 gauge steel roofing, 2 × 4 purlins 2 ft o.c. (on center), the truss, 2 × 4 2-ft o.c. ceiling furring strips, $\frac{1}{2}$-in. plywood ceiling, and 6-in. fiberglass batt insulation. Then the dead load (D.L.) is calculated as follows:

Roofing	0.8 psf
Purlins	0.7 psf
Truss (426 lb/40 ft × 7.5 ft)	1.4 psf
Furring strips	0.7 psf
Insulation	0.4 psf
Total D.L.	4.0 psf

Total load (T.L.)   26.0 psf
D.L./T.L. = (4.0 psf/26.0 psf)   100 = 15.4%

6.4 LIVE LOADS

Live loads are those loads that occur through use of the structure. Sometimes these live loads, such as a bulk milk tank, can be accurately estimated in magnitude and location. In other instances, however, the magnitude and location of loads can vary so much that it seems impossible to predict them. In the latter cases, experience has shown that certain live loads, as shown in Tables 6.2, 6.3, and 6.4, do not necessarily predict the live loads that will occur but use of these loads has resulted in adequate designs unless unusual circumstances are present. For example, consider the floor load on the first floor of a dwelling. Table 6.2 gives the minimum design live load as 195 kg/m^2 (40 lb/ft^2). This design load has proven adequate for all but unusual circumstances even when the location and size of furniture is not known.

Table 6.2. Minimum Uniformly Distributed Live Floor Loads.[a]

Occupancy or Use	kg/m^2	lb/ft^2
Assembly halls and other places of assembly		
fixed seats	293	60
movable seats and platforms	488	100
Balcony (exterior)	488	100
one- and two-family dwellings only	293	60
Corridors		
first floor	488	100
other floors[b]		
Dining rooms and restaurants	488	100
Fire escapes (on multi- or single-family residential bldgs. only)	195	40
Garages (passenger cars only)	244	50
Manufacturing		
light	610	125
heavy	1220	250
Office buildings		
offices	244	50
lobbies	288	100
corridors, above first floor	390	80
file and computer rooms[c]		
Residential		
multifamily houses		
private apartments	195	40
public rooms	488	100
corridors	390	80
dwellings		
first floor	195	40

Table 6.2. Continued.

Item	kg/m³	lb/ft²
second floor and habitable attics	146	30
uninhabitable attics[d]	98	20
Reviewing stands and bleachers	488	100
Sidewalks, vehicular driveways, and yards subject to trucking	1220	250
Stairs and exitways	488	100
Storage warehouse		
light	610	125
heavy	1220	250
Stores		
retail		
first-floor rooms	448	100
upper floors	366	75
wholesale	610	125

[a] Abstracted from *The BOCA Basic General Building Code/1975* (Building Officials & Code Administrators International, Inc., 1975). For the design of members supporting more than 14 m² and the live load is less than 488 kg/m² the live load can be reduced a percentage R as given by $R = 0.86(A - 14m)$ where A is the area supported in m². The maximum reduction is 40% for members receiving load from one level only and 60% for others but not more than $R = 23.1(1 + D/L)$ where D is the dead load supported and L is the live load supported. Reduction shall not apply to places of public assembly or garage.

[b] Same as occupancy served, accept as indicated.

[c] Require heavier loads, based upon anticipated occupancy.

[d] Live load is applied to joists or to bottom chords of trusses or trussed rafters only in those portions of attic space having a clear height of 1.07 m (42 in.) or more between joist and rafter in conventional rafter construction; and between bottom chord and any other member in trusses or trussed rafter construction. However, joists or the bottom chords of trusses or trussed rafters shall be designed to sustain the imposed dead load or 49 kg/m² (10 lb/ft²) whichever is greater, uniformly distributed over the entire span.

A further ceiling dead load reduction to a minimum of 24 kg/m² (5 lb/ft²) or the actual dead load, whichever is greater, may be applied to joists in conventional rafter construction or to the bottom chords of trusses or trussed rafters under either or both of the following conditions:

If the clear height is not over 76 cm (30 in.) between joist and rafter in conventional construction and between the bottom chord and any other member for trusses or trussed rafter construction.

If a clear height of greater than 76 cm (30 in.) as defined in preceding item a does not exist for a horizontal distance of more than 30 cm (12 in.) along the member.

Table 6.2 gives the minimum uniformly distributed live load to be used in design of floors. Most of these buildings are nonagricultural but the table is included for occasional use and for comparison purposes.

Table 6.3 gives the recommended design floor live load for agricultural buildings. Table 6.4 gives concentrated loads to be considered if actual loads are not known. Unless otherwise specified these concentrated loads are to be applied over 0.23 m² (2.5 ft²). In cases where a uniform load and a concentrated load apply to the same design, consider each load separately and select the most critical load for each part of the design.

Table 6.3. Recommended Design Floor Live Loads for Agricultural Buildings.[a]

	Solid Floor and Floor Support		Slat per Unit Length	
	kg/m²	lb/ft²	kg/m	lb/ft
Beef cattle				
calves to 135 kg (300 lb)	244	50	223	150
feeders, breeders	488	100	372	250
Dairy cattle				
calves to 135 kg (300 lb)	244	50	223	150
mature	488	100	372	250
stall area	293	60	372	250
maternity or hospital pen	244	50	223	150
Swine[b]				
to 25 kg (50 lb)	171	35	74	50
90 kg (200 lb)	244	50	149	100
180 kg (400 lb)	317	65	223	150
225 kg (500 lb)	342	70	253	170
Sheep				
feeders	195	40	149	100
eves, rams	244	50	179	120
Horses	488	100	372	250
Turkeys	146	30	37	25
Chickens[c]				
floor houses	98	20	22	15
Greenhouses	244	50	—	—
Manure (per unit depth)	1040	65	—	—
Shops, storages, vehicles[d]				

	Suspended Load[f]	
	kg/m	lb/ft
Chickens, suspended cages[e] per length of cage row		
full stair step (double deck, no dropping boards)	112	75
modified stair step (double deck, with dropping boards)	164	110
modified stair step (triple deck, with dropping boards)	223	150

[a]Abstracted from ASAE Engineering Practice: EP378. 2, *Agricultural Engineers Yearbook of Standards* (American Society of Agricultural Engineers, 1983).
[b]For floors that are outdoors, add snow load on the ground. Increase solid floor live load 25% for floors supporting crowded animals (e.g., crowding pen, dairy holding pen with automatic gate, handling alleys near loading chute).
[c]Where slats are interconnected between supports so that three or more slats must deflect together, design each span of slat between support and interconnection, and between two interconnections, for the recommended load per unit length of slat. Design the full span of each slat for one-half the recommended load per unit length.
[d]*Cage loads.* Base design loads for floor-supported cages.

Table 6.3. Footnotes Continued.

^{e}*Storage loads:* Calculate the design load for product storage on the basis of individual mass but no less that 488 kg/m^2 (100 lb/ft^2). *Vehicle loads:* For vehicle traffic on a manure tank lid, use 4540 kg (10,000 lb) axle load, or a loading equivalent of two 2270 kg (5000 lb) concentrated loads 1.2m (4 ft) apart and oriented any direction on the tank cover. *Vehicle storage (uniformly distributed):* The minimum design load on a floor area used for farm machinery with traffic limited to access and egress should be 730 kg/m^2 (150 lb/ft^2), except where the area will be occupied by either loaded farm trucks or large farm tractors (those with a mass exceeding 5900 kg (13,000 lb) including mounted equipment). In such cases the design load should be 975 kg/m^2 (200 lb/ft^2).
^{f}*Suspended loads.* Loads for suspended poultry cages are based on 4-row (double deck) or 6-row (triple deck) cages with two birds per 0.2 m (8 in.) cage, or three birds per 0.3 m (12 in.) cage, and 0.05 m (2 in.) of manure accumulated on dropping boards under upper cages.

Table 6.4. Concentrated Live Loads.

Location	kg	lbs
Greenhouse roof bars, purlins, rafters[a]	45	100
Manufacturing		
light[b]	908	2000
heavy[b]	1362	3000
Office floors[a,b,c]	908	2000
accessible truss lower chord panel point or ceiling joist over manufacturing, processing, repair area[c]	908	2000
Slat, design for uniform load or concentrated load[d]	115	250
farrowing pen[d]	115	250
Stair tread on area of 2600 mm^2 (4 in.^{2}) of tread[b,c]	136	300
Stores		
retail[b]	908	2000
wholesale[b]	1362	3000
Vehicle		
passenger car on area of 1290 mm^2 (20 in.^{2})[b,c]	908	2000
tractor and implements on 200 × 300 mm (8 × 12 in.) per wheel[d]	2270	5000
loaded trucks *not* exceeding 9075 kg (20,000 lb) gross weight on 100 × 510 mm (4 × 20 in.) per wheel[d]	3632	8000
loaded trucks exceeding 9075 kg (20,000 lb) gross weight on 150 × 560 mm (6 × 22 in.) per wheel[d]	5448	12000

[a]Abstracted from *The BOCA Basic Building Code/1981*, (Building Officials & Code Administrators International, Inc., 1981).
[b]Abstracted from *Uniform Building Code* (International Conference of Building Officials, 1976).
[c]Abstracted from ANSI A58.1-1982 (American National Standards Institute, 1982).
[d]Abstracted from *ASAE Engineering Practice* EP378.2 (American Society of Agricultural Engineers, 1983).

Roof minimum live load recommendations are also applied to roofs as shown in Table 6.5. This live load specification is to allow for loads that occur during construction or during repair. Notice also that the load is less with increasing roof slope and area supported. This decrease in load intensity as the area increases can also be noted in Table 6.3 where slat loads are much higher per unit area than loads on floor support beams which support much greater floor areas. This is because a peak load can occur on a small area but it may be physically impossible or highly unlikely for the same peak load to occur over a much larger area.

6.5 SNOW LOADS

The snow load of interest to the building designer is the maximum snow load on the roof during the useful life of the structure. Available data, however, is usually snow depth or weight of snow on the ground at weather stations in the area. Difficulties arise in converting snow depth to snow load, translating weather station data to the building site, and then using the ground snow load to estimate the maximum load that will occur on the roof. The design roof snow load is predicted by the following equation:

$$\begin{aligned} S &= S_g I C_s C_t \\ &= S_g I C_r C_\alpha (\text{drift coefficient})\, C_t \end{aligned} \quad (6.1)$$

where S_g is ground snow load; I, importance factor; C_s, ground-to-roof load factor combining C_r, C_α, and the drift coefficient; C_t, heat loss factor; C_r, ground to roof load factor; C_α, roof slope factor.

6.5.1. Ground Snow Load (S_g)

Long-term ground snow records in the United States have been taken by the National Weather Service (NWS) and also by the Soil Conservation Service (SCS). Many NWS stations only record ground snow depth; however, since about 1952 most first-order weather stations have recorded ground snow load. The SCS records snow depth and weight to predict flooding of certain rivers or water availability. SCS data collection sites were selected for the preceding objectives and therefore do not give uniform nationwide coverage, but the data is useful as a supplement to other data. NWS ground snow load data is too sparse and the length of record is too short to make sufficiently accurate snow load maps in certain areas. For these regions, NWS

6.5 Snow Loads   139

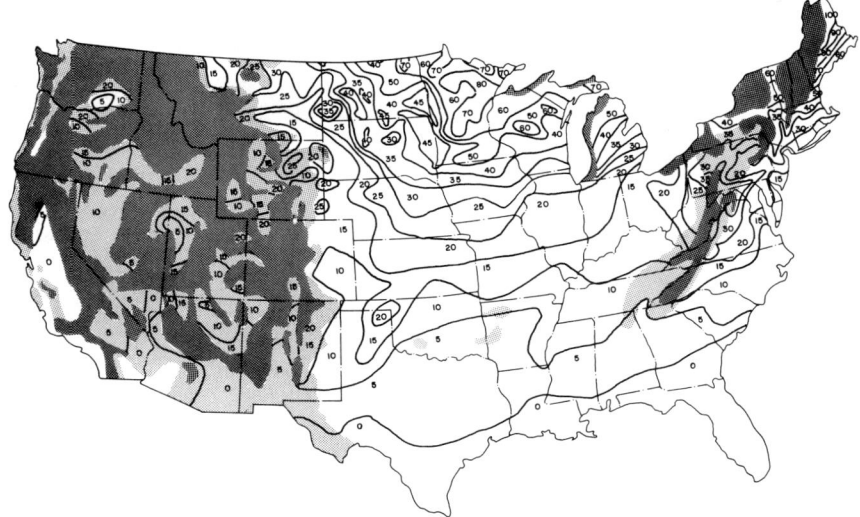

Fig. 6.3. Fifty-year recurrence interval ground snow map in pounds per square foot (S_g). Dark shaded areas are where the variation precludes mapping at this scale. Lighter shaded areas are where available data should be applied with caution. (Load in kg/m² = 4.88 lb$_m$/ft².) From ANSI Standard A58.1-1982.

ground snow load data, NWS snow depth data translated to ground snow load data by nearby NWS density data, and SCS load data are used to establish ground snow loads. The ANSI Standard A58.1-1982 ground snow load map, shown in Fig. 6.3, is based on both NWS and SCS data.

In instances where only snow depth is known, difficulties arise in predicting snow load because snow densities vary. As snow depth in-

Table 6.5. Roof Minimum Live Loads, kg/m² (psf).[a]

Roof Slope	Tributary Loaded Area for Any Structural Member m² (ft²)		
	0 to 18.6 (200)		>55.7 (600)
<4 rise in 12 run	98 (20)		59 (12)
>12 rise in 12 run	59 (12)		59 (12)

For loads between the above slope and area values:
  $98(1.20 - 0.0107A)(1.2 - 0.05F) > 59$ kg/m²
  $20(1.20 - 0.001A)(1.2 - 0.05F) > 12$ psf
where F is rise per 12 run; A is tributary loaded area.
[a]Abstracted from ANSI Standard A58.1-1982 (American National Standards Institute, 1982).

creases density increases due to snow compressibility. Melting and refreezing, rain on the snow, as well as variations in new fallen snow density also alter the snowpack average density. Snow specific densities of 0.05 to 0.2 for freshly fallen snow, 0.2 to 0.4 for snowpacks, and 0.7 to 0.8 for snowpacks in mountainous regions have been reported.

6.5.1.1 Importance Factor (*I*). As mentioned in a previous section, the designer may want to increase or decrease the design load because of the severity of failure or because of building permanence. Some codes use snow load maps with different probabilities of occurrence or different recurrence intervals (*RI*). The snow load map in Fig. 6.3 has an annual probability of occurrence of 0.02 or 50-year recurrence interval. ANSI Standard A58.1-1982 uses importance factors to convert the 50-year recurrence interval map to 25-year or 100-year values. The importance factors (*I*) used are as follows:

All buildings not in the categories listed below (50-year RI)	1.0
Assembly areas for 300 people or more	1.1
Essential facilities—hospitals, police stations, disaster center (100-year RI)	1.2
Low human life hazard facilities—agricultural buildings, temporary facilities (25-year RI)	0.8

6.5.2 Snow Loads in Mountainous Regions

In mountainous regions the usual snow load maps cannot show the extreme variation in snow load on the usual scale maps. It has been found more satisfactory to modify the map by relating the loads to a reference elevation. This greatly simplifies the map in most cases. To calculate the snow load at any location, the map snow load at the reference elevation is modified to the load at the site elevation by a functional relationship between snow load and elevation.

The functional relationship is usually of the form:

$$S_g = C_1 H^2 + C_2 H + S_0 \qquad (6.2)$$

where C_1 and C_2 are constants; H, elevation above sea level; S_0, snow load at a reference elevation.

The reference elevation sometimes assumed is sea level but more commonly it is 300 to 600 m (980 to 1970 ft).

Figure 6.4 shows 1969 snow load data presented by Schaerer (1970) from several locations near the coast of British Columbia, Canada. The

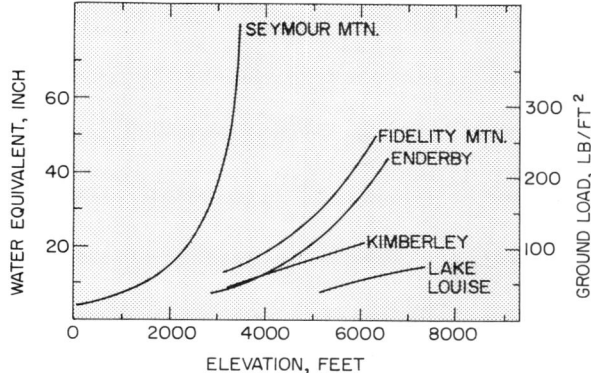

Fig. 6.4. Ground snow load versus elevation for several locations near the coast of British Columbia, Canada, in 1969.

Seymore Mountain location is near the coast and in a wet area. Fidelity mountain and Kimberly represent areas of less and less rainfall. Notice that the relationship is linear as rainfall decreases; i.e., C_1 in Eq. (6.2) is less significant.

Figure 6.5 shows the ground contour map and the normalized snow load map for Latah County, Idaho. As can be seen, the normalized snow

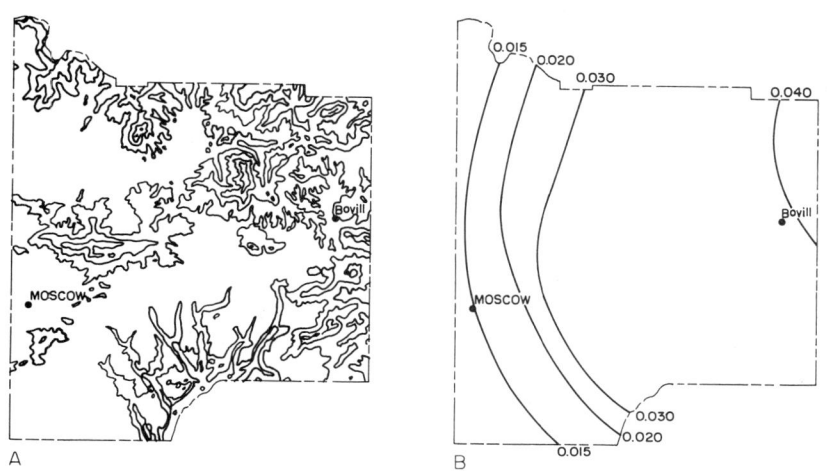

Fig. 6.5. (A) Ground contour map of elevation above sea level for Latah County, Idaho. (B) typical normalized snow load map for Latah County, Idaho. From Rusten et al. (1980).

loads do not follow the ground contours but a more general weather pattern. The relationship between ground snow load and elevation proposed by Rusten et al. (1980) for this county is

$$S_g = (\text{map normalized snow load, psf/ft})(\text{elevation, ft})$$

This equation assumes sea level as the reference elevation and a linear relationship ($C_1 = 0$).

To select design snow loads in mountainous regions with the present data available, the designer must consult local building code enforcement agencies and weather bureaus for sources of pertinent information. Future codes may use normalized snow maps and specify the load–elevation relationship in mountainous regions.

6.5.3. Ground-to-Roof-Snow-Load Relationship (C_r)

Wind during snow deposition, wind removing previously deposited snow, radiant energy, and snow melt due to radiant energy or heat loss through the roof all are factors which can reduce roof snow loads as compared to the ground snow load. The factor for converting the ground snow load (S_g) to the load on a flat roof is C_r. The value of C_r probably changes with climate. In cold, high snow regions, the maximum roof snow load is a function of the weather the entire winter. Snowfall after snowfall add to the roof snow load and increase the snowpack density. Wind and radiant energy reduce the snow load and C_r is reported to vary from 0.5 to 0.8. In more moderate climates where snow does not stay on the roof for very long, the maximum snow load often occurs as the result of a single snowfall, ice storm, or rain absorbed into snow and C_r could approach 1.0 depending on the associated wind. These two examples represent the variety of conditions that result in maximum snow loads on structures. Most codes assume one C_r for all regions, and the designer should know that this could be an unconservative assumption in low snow load regions. Minimum snow loads are sometimes written in the code to account for this factor.

Wind is probably the major factor in the determination of C_r. Ostavnov and Rosenberg (1966) stated that snow is removed from a snowpack at velocities over 6 ms^{-1} (13.5 mph) and during snowfalls at velocities over 3–4 ms^{-1} (6.5–9.0 mph). Ostavnov proposed that C_r is related to velocity by the following equation:

$$C_r = 1.24 - 0.13 \overline{V}_w \quad \text{when } \overline{V}_w \text{ is in ms}^{-1}$$

or

$$C_r = 1.24 - 0.29 \overline{V}_w \quad \text{when } \overline{V}_w \text{ is in mph}$$

where $\overline{V}_w$ is the average winter wind velocity.

On the basis of Canadian studies, both the 1970 National Building Code of Canada and ANSI Standard A58.1-1972 recommended a $C_r = 0.8$ for sheltered roofs. For roofs fully exposed to winds of sufficient velocity to remove snow and that have no projects such as parapet walls that stop snow from blowing off, a $C_r = 0.6$ is recommended. The ANSI Standard A58.1-1982 defines C_r as follows:

$$\begin{aligned}
C_r &= 0.7 C_e \\
&= (0.7)(0.8) = 0.56 \text{ for windy exposed areas} \\
&= (0.7)(0.9) = 0.63 \text{ for windy areas with little shelter} \\
&= (0.7)(1.0) = 0.70 \text{ for areas where wind cannot} \\
&\quad \text{be counted on to remove snowfall} \\
&= (0.7)(1.1) = 0.77 \text{ low wind areas where buildings or} \\
&\quad \text{trees shelter the roof*} \\
&= (0.7)(1.2) = 0.84 \text{ in dense conifer forest areas}
\end{aligned}$$

where C_e is an exposure factor.

6.5.4 Roof Slope Effect (C_α)

Roof slope also has an effect on the roof snow load. Most codes do not reduce the flat roof loading for roof slope angles less than 20 or 30≃. For slopes greater than 20 or 30≃, the flat roof snow load is linearly decreased with roof slope angle to 0 load at 60≃ or 70≃. ANSI Standard A581.1-1982 introduced the combined effect of less snow deposited on steep slopes and the potential for snow sliding off the roof as follows:

---

*Shelter is provided by obstructions within 10 h_o of the roof where h_o is the obstruction height above roof level.

$C_\alpha = (70 - \alpha)/55$ for $15° < \alpha < 70°$ and warm slippery roof which would allow snow to slide off the eaves

$C_\alpha = (70 - \alpha)/40$ for $30° < \alpha < 70°$ and warm not slippery roof or cold slippery roof which would allow snow to slide off the eaves

$C_\alpha = (70 - \alpha)/25$ for $45° < \alpha < 70°$ cold, not slippery roof

where α is the roof angle measured in degrees from horizontal. The slope reduction coefficient C_α is bounded by zero and one, therefore $C_\alpha = 1.0$ for any angle below the minimum and $C_\alpha = 0$ for any angle above $70°$.

6.5.5 Snow Drifting

Snow is deposited on roof areas where the velocity is below the deposition velocity for the snowflakes, or ice particles in that storm. In areas of high velocity and turbulence, snow is not deposited and previously deposited snow may be removed. It is important to consider drifting because of the high probability that drifting will occur and because the resulting snow distribution is often the critical loading.

Most codes are in general agreement on the snow distributions to be checked. Figure 6.6 is a summary of the distributions required by the ANSI Standard A58.1-1982. These distributions take into account the reduced snow on the roof compared to the ground, roof slope effects and drifting.

Snow may slide off a sloped roof onto a lower roof. Consider the lower roof to accumulate all the snow (balanced load condition) that can slide off the upper roof and remain on the lower roof.

6.5.6 Effects of Heat Loss Through the Roof (C_t)

With proper conditions, heat loss through the roof can reduce the roof snow load. The heat loss must be great enough to melt the snow and allow it to flow off the roof. Often the water from the melted snow will flow to a colder area of the roof, such as an overhang, and then freeze. Ice accumulations in this area can cause roof leaks and damage to gutters but usually are not critical structural loads. The amount of load reduction due to melting depends upon the inside temperature, ambient temperature, roof insulation value, and time. Restraint must

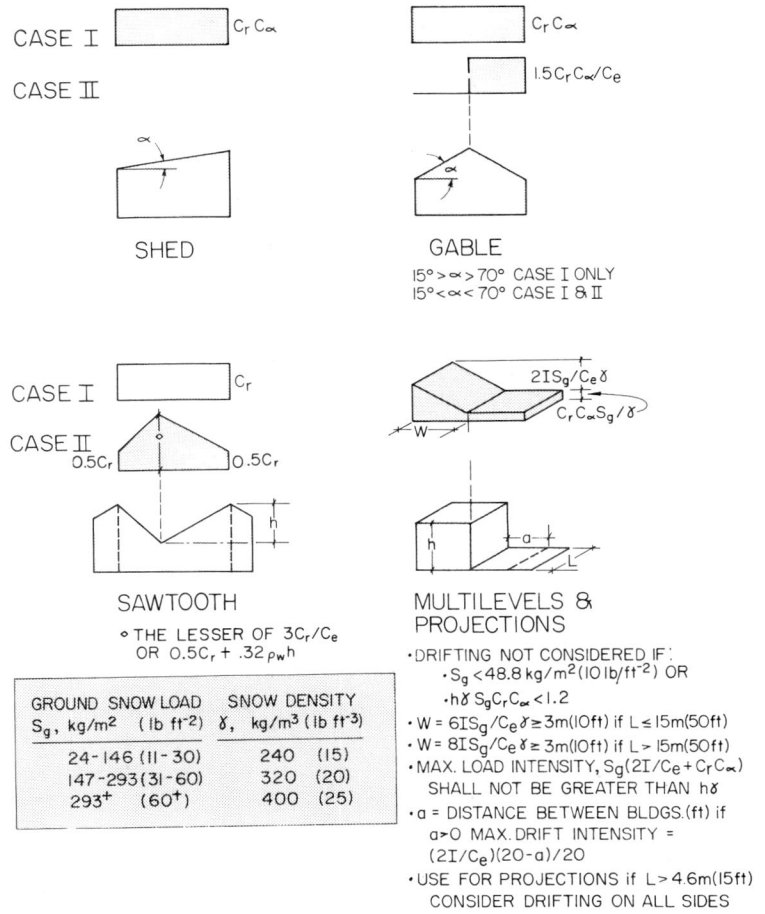

Fig. 6.6. Snow load distributions required by ANSI Standard A58.1-1982. Symbols a, n, and L defined in this figure apply only to this figure.

be exercised in reducing snow load due to heat loss because assumptions made about inside temperatures and insulation values may change during the life of the structure.

The ANSI Standard A58.1-1982 contains the following heat loss factors:

$$\text{Heated structure} \qquad C_t = 1.0$$

$$\text{Structure kept above freezing} \qquad C_t = 1.1$$

Unheated structure $C_t = 1.2$

6.5.7 Snow Load Summary

Ground snow loads (S_g) are obtained from snow load maps such as Fig. 6.3 or from local code enforcement agencies when snow load maps do not give sufficient information. This ground snow load is modified to obtain the roof snow load (S) as follows:

$$S = S_g I C_s C_t$$

where I is the importance factor; C_t, heat loss factor; $C_s = C_r C_\alpha$ (drift coefficient); C_r, ground to flat roof conversion factor; C_α, roof slope factor.

C_s, which includes drift coefficients as well as C_r and C_α, is given in Fig. 6.6. C_t is given in the previous section.

6.6 WIND

Wind forces on buildings result from the building in the path of the wind altering the wind velocity and direction. If the wind velocity is completely converted to pressure, then the pressure exerted by the wind is

$$\text{Pressure} = \gamma V^2/2g_c \quad (6.3)$$

where γ is air mass density, kg m^{-3} (lb ft^{-3}); V, air velocity, ms^{-1} (ft s^{-1}); g_c, gravitational constant, 1 ms^{-2} kg N^{-1} (32.17 ft s^{-2} lb lbf^{-1}).

Using standard values for the air density and the usual dimensions,

$$\text{Pressure} = 1.22 \text{ kg/m}^3 \times V^2/2g_c$$
$$= 0.61 V^2 \text{ N/m}^2 \text{ or Pa,}$$

or (6.4)

$$\text{Pressure} = \frac{(1/13.05)(5280 \text{ ft mi}^{-1}/3600 \text{ s hr}^{-1})^2 V^2}{(2)32.17 \text{ ft s}^{-2} \text{ lb lb}_f{}^{-1}}$$

$$= 0.00256 V^2 \text{ in } p_f\text{sf where } V \text{ is in mph}$$

As the wind changes direction and goes around the building it causes different pressures (positive—i.e., above atmospheric pressure; negative—i.e., below atmospheric) to act on the building surfaces. Usually these pressures are related to the velocity pressure by a wind surface coefficient. The pressure acting on a building surface is described by the design wind velocity and a pressure coefficient; i.e.,

$$p = f(V_z\, c)$$

where V_z is the design wind velocity at height z; and c is the pressure coefficient.

6.6.1 Wind Storms

The design wind velocity usually arises from a variety of wind storms depending on the area of the United States. Along the Atlantic and Gulf coasts of the United States, the design velocity winds result from hurricanes. They are storms that can last for several days with top winds approaching 44.7 ms^{-1} (100 mph). Hurricane winds are generally accompanied by rain that can increase the air mass density and therefore the pressure.

Tornadoes are cylindrically shaped rotating clouds extending downward from a thunderstorm cloud base to ground level. Winds in the intense whirl or vortex range from 45 to 100 ms^{-1} (100 to 250 mph) and the atmospheric pressure at the core is 10 to 25% lower than immediate surroundings. Analysis of the path of destruction from tornadoes indicates that most tornado paths are 45 to 130 m wide, 1.5 to 10 km long, with an average area of destruction of about 2.2 km^2 (0.85 mi^2).

Figure 6.7 gives the observed yearly tornado frequency per 25,900 km^2 (10,000 mi^2). This map shows where tornadoes are most or least likely to occur. That frequency along with the average area of destruction for a tornado can be used to estimate the recurrence interval. For example, consider the recurrence interval for an area that has 9 tornadoes per 25,900 km^2 (10,000 mi^2) per year when the damage per tornado covers 2.2 km^2 (0.85 mi^2) area. If the probability of occurrence is uniform over the area, then

$$\text{Probability of occurrence} = \frac{(9 \text{ tornadoes/yr})(2.2 \text{ km}^2/\text{tornado})}{25,900 \text{ km}^2}$$

$$= 0.00076$$

and

$$\text{Recurrence interval} = 1/0.00076$$

$$= 1308$$

148 Load Analysis

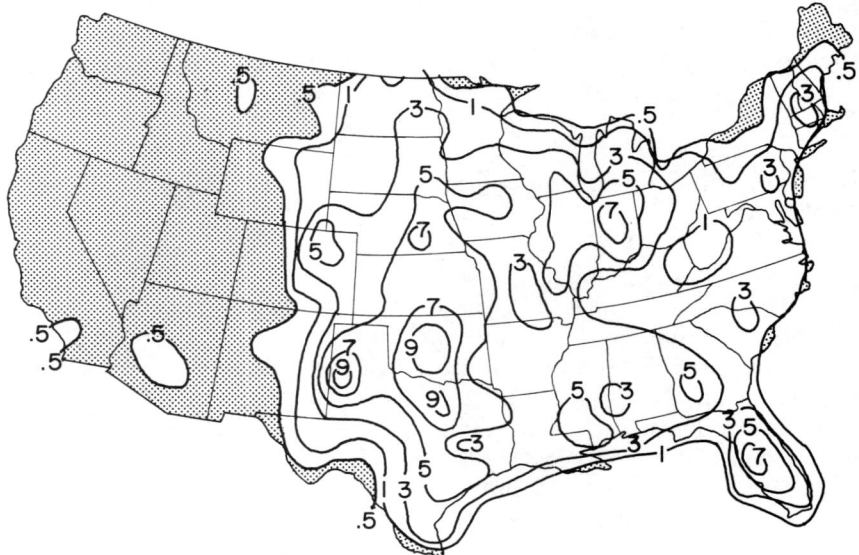

Fig. 6.7. Observed tornado frequency per 25,900 square kilometers (10,000 square miles) per year. From U.S. Department of Commerce, National Oceanic and Atmospheric Administration.

The recurrence interval for a specific site in the area where tornadoes are most likely to occur is over 1000 years. The usual conclusion is that it is uneconomical to design most farm buildings to resist tornadoes because the wind force is extreme and the probability of a tornado hitting a specific building is quite low.

At some locations there can be unusual winds that must be taken into account in design. Such areas are shown as special wind areas on basic wind velocity maps.

6.6.2 Design Wind Velocity (V_z)

Wind velocities are influenced by the atmospheric pressure gradient parallel to the earth surface (resulting in a wind at gradient wind velocity if no obstructions reduce the wind velocity) and the earth surface roughness. At the ground surface the velocity is considered to be zero and increases for several kilometers upward to the gradient wind velocity. The mathematical description of this variation in wind speed as originally suggested by Davenport is

$$V_z = V_G(z/H_G)^a \tag{6.5}$$

where V_z is wind velocity at height z; V_G, gradient wind velocity; H_G, height to the gradient wind velocity; a, exponent dependent on ground roughness; z, height above ground surface where V_z is calculated.

ANSI Standard A58.1 suggested the following values of H_G and a:

Terrain	H_G m (ft)	a	
Center of large cities and very rough terrain	455 (1500)	1/3	Exposure category A[a]
Suburban areas, wooded areas and rolling terrain	365 (1200)	1/4.5	Exposure category B[a]
Flat open country	275 (900)	1/7	Exposure category C[a]
Flat coastal areas <460 m (1500 ft) inland	213 (700)	1/10	Exposure category D[a]

[a]Exposure categories are defined later in this section.

Wind speed from a map such as in Fig. 6.8, however, is not the gradient wind speed but is the wind speed at 10 m (33 ft) above the ground in open terrain (exposure C). To calculate wind speed at another height (z) and in another type of terrain, first obtain the design wind velocity at the location in question V_{map}. Then

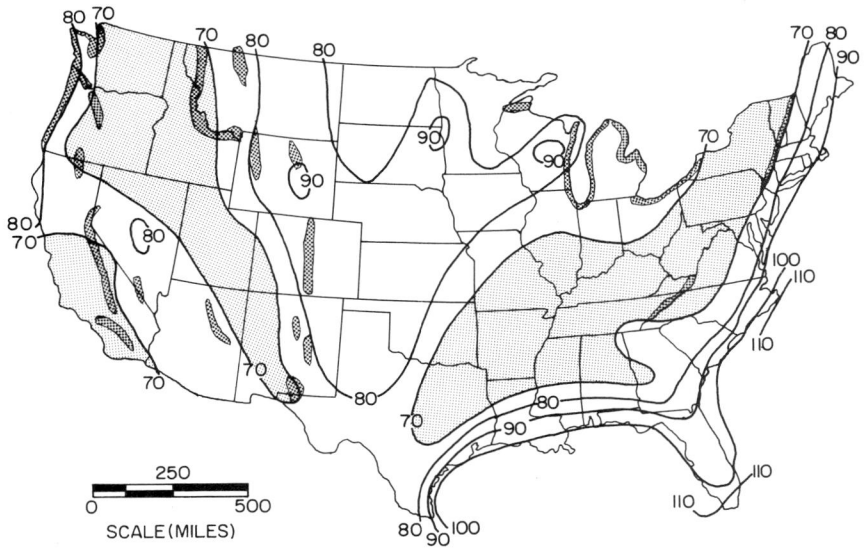

Fig. 6.8. Fifty-year recurrence interval basic wind speed map in miles per hour. Values are fastest-mile speeds at 33 ft (10 m) above ground for exposure category C. Linear interpolation between wind speed contours is acceptable. To convert to 25-year recurrence interval, multiply basic wind speed by 0.95. To convert to 100-year recurrence interval, multiply basic wind speed by 1.07.

150 Load Analysis

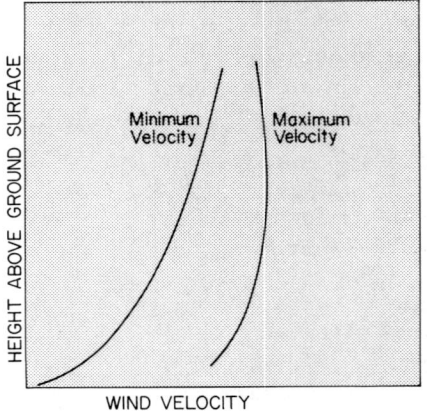

Fig. 6.9. Schematic showing increased wind stability with height above ground surface.

$$V_G = V_{map}/(z/H_G)^a = V_{map}/(10/275)^{1/7}$$

where $z = 10$ m for V_{map}; $H_G = 275$ m; $a = 1/7$ for exposure C.

Then for another exposure and height the design velocity (V_z) is

$$V_z = \frac{V_{map}(z/H_G)^a}{(10/275)^{1/7}}$$

or (6.6)

$$V_z = 1.61 V_{map}(z/H_G)^a$$

If wind velocity is measured at a point, it can be seen to vary significantly with a frequency of about 1 sec. One cause of velocity variation is turbulence caused by surface roughness. The smoother the surface, the less turbulence and more stable the wind velocity. This turbulence also diminishes with height above the ground surface. This decreased turbulence is shown schematically in Fig. 6.9.

To obtain design wind speed, velocity is averaged over a time period and the maximum average value used. The shorter the averaging period, the closer the maximum will be to the highest velocity wind gust. However, below some time period length, the design wind speed may occur for such a short duration that many structures would not be engulfed in that gust and would not react to that gust alone. If a long time period is selected, the design velocity determined could certainly be exceeded for a long enough time period for the building to react and therefore would underestimate the wind load. Cohen and Vellozzi (1968) gave the following factors to convert to a 1-hr time average velocity:

$$V_{1\text{ sec}} = 1.55 \times V_{1\text{ hr}}$$

$$V_{10\text{ sec}} = 1.43 \times V_{1\text{ hr}}$$

$$V_{1\text{ min}} = 1.25 \times V_{1\text{ hr}}$$

$$V_{10\text{ min}} = 1.06 \times V_{1\text{ hr}}$$

Code writers do not agree on the time period to use in determining the design wind speed. The time period has varied from 2–3 seconds in Australia to 1 hr in Canada. The ANSI Standard A58.1-1982 uses the "fastest mile" time period to develop the design or basic wind speed map in Fig. 6.8. The "fastest mile" time period is the shortest time necessary for a particle of air to travel 1 mi. ANSI Standard A58.1-1982 also recommends a gust factor be included to account for short-term high velocity winds.

Selection of gust factors should take into account the factors which affect wind gustiness and also building characteristics which make it susceptible or resistant to wind gusts. Most agree that buildings with high natural frequencies (≥ 10 cps) are not as susceptible to gusts and therefore can be designed with a gust factor approaching 1.0. Buildings with low natural frequencies (0.1 to 0.2 cps) are not as stable, therefore gusts are more critical and the gust factor can approach 3.6.

The natural frequency of a building is related to the dimensions, mass, stiffness, and damping characteristics. There are, however, several approximate formulas for natural frequency. One such equation is given in the Uniform Building Code (1976) as follows:

$$\text{Frequency} = 20\sqrt{d}/h_n \text{ cps} \tag{6.7}$$

where d is structure depth in the direction of the wind, ft; h_n is structure height, ft.

Gust factors used in the ANSI Standard A.58.1-1982 for the design of relatively stable buildings are shown in Table 6.6. If the building is tall and narrow, more detailed analysis is recommended.

6.6.3 Wind Pressures on Buildings

The basic wind speed, given in Fig. 6.8, must be related to the exposure and height of the building to be designed. Building height is usually

Table 6.6. Gust Response Factors G_H.[a]

Height, z, m (ft)	Exposure			
	A	B	C	D
0–4.5 (15)	2.36	1.65	1.32	1.15
6 (20)	2.20	1.59	1.29	1.14
7.5 (25)	2.09	1.54	1.27	1.13
9 (30)	2.01	1.51	1.26	1.12

[a] For the main structure, z is the mean roof height. For cladding, z is the cladding height.

taken as the mean roof height. Exposure categories are described in the ANSI Standard A58.1-1982 as

Exposure A: Large city centers with at least 50% of the buildings in excess of 21 m (70 ft) tall for the greater of 805 m (0.5 mi) or 10 times the building height (h) upwind.

Exposure B: Urban, suburban, wooded areas or other areas where numerous closely spaced obstructions the size of single family dwellings are upwind for a distance of 457 m (1500 ft) or 10 times the building height, whichever is greater.

Exposure C: Open terrain with scattered obstructions less than 9.1 m (30 ft) or flat open country.

Exposure D: Flat unobstructed coastal areas exposed to wind flowing over large bodies of water. This area shall extend inland 457 m (1500 ft) or 10 times the building height.

When the velocity is determined for the building exposure and height, the velocity is converted to a basic velocity pressure, q_z as follows:

$$q_z = 0.61[1.61V_{map}(z/H_G)^\alpha]^2 \quad \text{in Pa when } V_{map} \text{ in ms} \quad (6.8)$$
$$= 0.00256[1.61V_{map}(z/H_G)^\alpha]^2 \quad \text{in } p_f sf$$

or as tabulated in Table 6.7.

6.6.4 Pressure Coefficients (c)

The pressure on any part of the building is related to the basic wind pressure (q_z) by a pressure coefficient c. This coefficient is empirical and is usually determined by wind tunnel tests on models. The coefficient is negative if below atmospheric pressure and positive if above. In general, wind pressure on a gable-roofed building is as shown in Fig. 6.10 and general statements about wind pressures can be made as follows: (1) The more wind that is blocked by a building, the closer

Table 6.7. Basic Wind Pressures, q_z in Pa lb$_f$ ft^{-2}.[a]

V_{map}, m/s (mph)	Height z, m (ft)	Exposure			
		A	B	C	D
31.3 (70)	0–4.5 (15)	71 (1.5)	219 (4.6)	478 (10.0)	716 (15.0)
	6 (20)	86 (1.8)	250 (5.2)	519 (10.8)	759 (15.8)
	7.5 (25)	100 (2.1)	276 (5.8)	554 (11.6)	793 (16.6)
	9 (30)	113 (2.4)	299 (6.2)	583 (12.2)	823 (17.2)
35.8 (80)	0–4.5 (15)	94 (2.0)	287 (6.0)	626 (13.1)	937 (19.6)
	6 (20)	113 (2.4)	326 (6.8)	679 (14.2)	992 (20.7)
	7.5 (25)	131 (2.7)	360 (7.5)	724 (15.1)	1038 (21.7)
	9 (30)	148 (3.1)	391 (8.2)	763 (15.9)	1076 (22.5)
40.2 (90)	0–4.5 (15)	118 (2.5)	362 (7.6)	789 (16.5)	1181 (24.7)
	6 (20)	143 (3.0)	412 (8.6)	857 (17.9)	1251 (26.1)
	7.5 (25)	166 (3.5)	455 (9.5)	913 (19.1)	1308 (27.3)
	9 (30)	187 (3.9)	493 (10.3)	962 (20.1)	1357 (28.3)
44.7 (100)	0–4.5 (15)	146 (3.0)	448 (9.4)	976 (20.4)	1461 (30.5)
	6 (20)	176 (3.7)	509 (10.6)	1059 (22.1)	1547 (32.3)
	7.5 (25)	205 (4.3)	562 (11.7)	1129 (23.6)	1618 (33.8)
	9 (30)	231 (4.8)	609 (12.7)	1189 (24.8)	1678 (35.0)
49.2 (110)	0–4.5 (15)	176 (3.7)	543 (11.3)	1182 (24.7)	1770 (37.0)
	6 (20)	214 (4.5)	617 (12.9)	1283 (26.8)	1874 (39.2)
	7.5 (25)	248 (5.2)	681 (14.2)	1368 (28.6)	1960 (40.9)
	9 (30)	280 (5.8)	738 (15.4)	1441 (30.1)	2033 (42.5)

[a]Gust factor is not included.

the windward wall coefficient approaches 1.0 and the more negative are the coefficients on the roof and end walls; (2) the windward roof pressure coefficient is as low as -1.0 on flat roofs, increases to 0.0 at about 30° roof slope, and is positive above 30°; (3) winds parallel to the ridge cause negative pressures over the entire roof and often are the critical situation for roof uplift; (4) open sections of the building cause internal pressures of the approximate magnitude as would have occurred on the external surface of the removed section.

For a rectangular enclosed building with a gable roof, most codes are in agreement that the pressure coefficients should be approximately as follows:

Wind Perpendicular to the Ridge
Windward wall	$+0.8$	(0.5 to 0.8)
Leeward roof	-0.7	(-0.4 to -0.7)
Leeward wall	-0.5	(-0.4 to -0.6)
End or side walls	-0.7	

Wind Parallel to the Ridge
Windward wall	$+0.8$

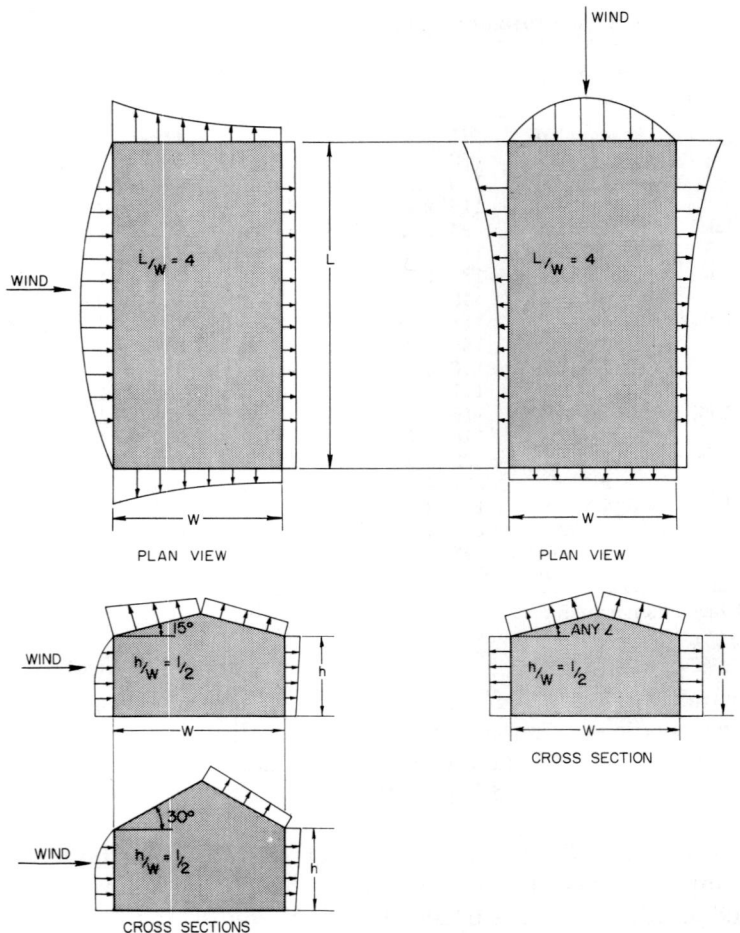

Fig. 6.10. General wind pressure patterns. Symbols h, L, and W defined in this figure apply only to this figure.

Roof	−0.7
End or side walls	−0.7

However, there seems to be much less agreement about the windward roof pressure coefficient when the wind is perpendicular to the ridge. Most agree that the coefficients are positive for roof slopes above 40°, zero somewhere between 20° and 40°, and −0.6 to −1.0 for flat roofs. Table 6.8 gives the 1978 ASAE Standard for windward roof pressure

Table 6.8. Shape Coefficients c for External Wind Loads on Single Span Gable-Type Building—Totally Enclosed.[a]

H/W	Windward Roof Coefficient Roof Slope						
	1:12	2:12	3:12	4:12	5:12	6:12	7:12
0.10	−0.34	−0.24	−0.13	−0.03	0.05	0.12	0.19
0.15	−0.51	−0.35	−0.20	−0.05	0.05	0.12	0.19
0.20	−0.60	−0.47	−0.27	−0.06	0.05	0.12	0.19
0.25	−0.60	−0.59	−0.34	−0.08	0.05	0.12	0.19
0.30	−0.60	−0.60	−0.41	−0.18	0.01	0.08	0.16
0.35	−0.60	−0.60	−0.47	−0.26	−0.07	0.05	0.12
0.40	−0.60	−0.60	−0.53	−0.33	−0.15	0.01	0.09
0.45	−0.60	−0.60	−0.57	−0.39	−0.22	−0.06	0.05
0.50	−0.60	−0.60	−0.60	−0.44	−0.29	−0.14	0.00
0.60	−0.60	−0.60	−0.60	−0.49	−0.34	−0.20	−0.06
0.70	−0.60	−0.60	−0.60	−0.53	−0.39	−0.25	−0.13
0.80	−0.60	−0.60	−0.60	−0.57	−0.43	−0.30	−0.18
0.90	−0.60	−0.60	−0.60	−0.60	−0.47	−0.35	−0.23
1.00+	−0.60	−0.60	−0.60	−0.60	−0.51	−0.39	−0.28

From ASAE Standard S288.3 *1978–79 Agricultural Engineers Yearbook*.
[a]For designing trusses, columns, rigid frames, and other main members.

coefficients. This standard represents a moderate estimate of the pressure coefficients, i.e., between the extreme values used by various codes.

The ANSI Standard A58.1-1982 recommends calculation of interior pressure as follows:

For main frames

$$p_i = q_h (GC_{pi}) \qquad (6.9)$$

where p_i is internal pressure; q_n is pressure calculated at *average roof height*; GC_{pi} is gust factor (G_H) multiplied by the building internal pressure coefficient $= 0.75$ *and* -0.25 if the openings in one wall exceed all other wall openings by 10% and the openings in all other walls do not exceed 20% of the total area of the one wall.

For components and cladding less than 18 m (16 ft) high:

$$p_i = q_h(GC_{pi}) \qquad (6.10)$$

where p_i and GC_{pi} are as defined for Eq. (6.9); q_h is pressure calculated at *average roof height* and *exposure C*.

Most codes agree on the coefficients to be used for arched roof buildings. The roof is usually divided into the windward quarter, center half, and leeward quarter. The coefficients shown in Table 6.9 are those accepted by the ANSI Standard A58.1-1982.

Load Analysis

Table 6.9. External Pressure Coefficients for Arched Roofs.

	Rise-to-Span Ratio, $f/w = r$	Windward Quarter	Center Half	Leeward Quarter
Roof on elevated structure	$0 < r < 0.2$	-0.9	$-0.7 - r$	-0.5
	$0.2 \leq r < 0.3$	$(1.5r - 0.3)^a$	$-0.7 - r$	-0.5
	$0.3 \leq r < 0.6$	$(2.75r - 0.7)$	$-0.7 - r$	-0.5
	$0.3 < r < 0.6$	$(2.75r - 0.68)$	$-0.7 - r$	-0.5
Arch springing from ground	$0.3 < r < 0.6$	$1.4r$	$-0.7 - r$	-0.5

^{a}Alternate coefficient $6r - 2.1$ shall also be used.

For towers the pressure coefficient depends upon the slenderness of the tower and its surface roughness. The net pressure coefficients that act on the vertical projected area, [height (h) × diameter or dimension perpendicular to the wind] of towers, chimneys, tanks, and similar structures, as given by ANSI A58.1 1982 are shown in Table 6.10.

6.6.5 Pressures on Components and Cladding

As the area supported by a member is decreased, the greater the probability is that the whole area will be subjected to a more extreme pressure. On large areas the structure is counted on to integrate the high and low pressures and therefore the design pressure is an average. For this reason, the absolute value of design pressure coefficients usually increase as the area supported decreases.

The ANSI Standard A58.1-1982 recommends that the pressure for components and cladding be calculated by the equation

$$p = q_h(G_H C_p) - p_i \qquad (6.11)$$

where p is wind pressure acting on components or cladding; q_h, wind

Table 6.10. Pressure Coefficients for Chimneys, Towers, Tanks, and Similar Structures.

Shape	Type Surface		H/D	
		1	7	25
Square	All	1.3	1.4	2.0
Round, $D\sqrt{q_z} < 2.5$	Moderately smooth	1.3	1.4	2.0
	Rough (D'/D^a 0.02)	0.5	0.6	0.7
	Very Rough (D'/D^a 0.08)	0.8	1.0	1.2

From ANSI Standard A58.1-1982.
$^aD'$ is the depth of protruding elements in ft; D is the diameter or least horizontal dimension in ft.

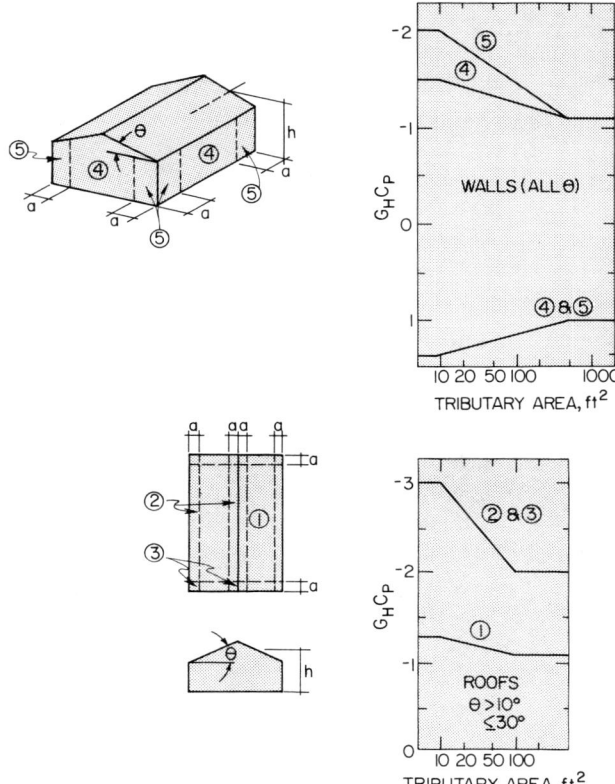

Fig. 6.11. Pressure coefficients for roofs and walls. From ANSI Standard A58.1-1982. $a = 10\%$ minimum width or $0.4h$, whichever is smaller, but not less than either 4% of minimum width or 3 ft. h = mean roof height, in feet, except that eave height may be used when $\theta \leq 10°$. θ = roof slope from horizontal, in degrees. Symbols a, h, and θ defined in this figure apply only to this figure.

pressure at mean roof height *using exposure C*; $G_H C_p$, given in Fig. 6.11; p_i, internal pressure as given in preceding section.

6.6.6 Wind Load Summary

The wind load for the main frame is given by

$$p = q_z G_H C \tag{6.12}$$

where z is mid-roof height for windward roof, leeward roof, and leeward wall, z is height to point where q_z is calculated for windward wall; $q_z = (0.61)[1.61\ V(z/H_G)^a]^2$ in Pa, $q_z = 0.00256\ [1.61\ V\ (z/H_G)^a]^2$ in psf or refer to Table 6.7; G_H is gust response factor from Table 6.6; c is pressure coefficient from pressure coefficient section and Tables 6.8, 6.9, and 6.10.

The wind load on components and cladding is given by

$$p = q_h\ (GC_p) - p_i \tag{6.13}$$

where q_h is wind pressure at mean roof height using exposure C; GC_p, gust factor and pressure coefficient given in Fig. 6.11; p_i, internal pressure [see Eq. (6.9)].

Example. Wind Load Problem. What is the design wind pressure acting on a 12-m wide, 30-m long and 4-m eave height machinery storage building located near Kansas City, MO? The roof slope is 3.5 in 12. One side of the building is 80% open. The site exposure category is C. See Fig. 6.12.

Solution. Design wind velocity

$$V_{map} = (76\ \text{mph})(0.447\ \text{ms}^{-1}/\text{mph})$$
$$= 34\ \text{ms}^{-1}\ \text{(interpolated value)}$$

Design height

$$h = 4\ \text{m} + (3\ \text{m})(3.5/12) = 4.875$$

Basic pressure at

$$z = h,\ q_z = (0.61)[1.61\ V_{map}(z/H_g)^a]^2$$
$$= (0.61)[(1.61)(34)(4.875/275)^{1/7}]^2$$
$$= 577\ \text{Pa}$$

Gust factor $G_H = 1.31$ (interpolated value).
External pressure

$$p = q_z G_H c$$

at 4.875 m,

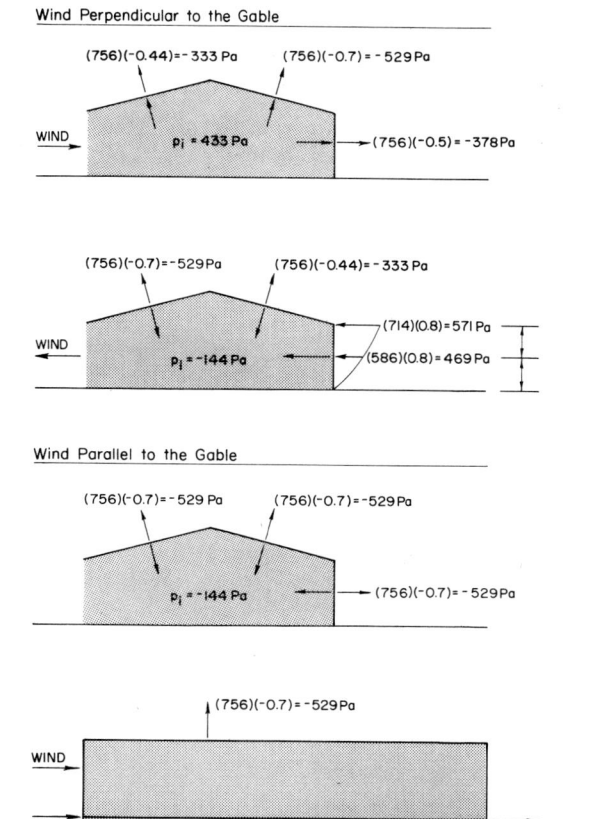

Fig. 6.12. Example—Wind Load Problem.

$$p = (577)(1.31)c = 756c \text{ Pa}$$

at 4m,

$$p = (0.61)[(1.61)(34)\,(4.0/275)^{1/7}]^2(1.31)c = 714c \text{ Pa}$$

at 2 m,

$$p = (0.61)[(1.61)(34)\,(2.0/275)^{1/7}]^2(1.31)c = 586c \text{ Pa}$$

Internal pressure

$$p_i = q_n\,(GC_{pi})$$
$$= (577)(0.75) = 433 \text{ Pa}$$

or

$$p_i = (577)(-0.25) = -144 \text{ Pa}$$

For components and cladding: Pressures for the five areas of the building shown in Fig. 6.11 are as follows:

$$p = q_h\,(GC_p) = (577 \text{ Pa})(GC_p)$$
$$p_i = q_h(GC_{pi}) = 433 \text{ or } -144 \text{ Pa}$$

Assume the tributary area for members being designed is 0.93 m² (10 ft²):

Area 1: $\quad p = (577 \text{ Pa})(-1.3) - 433 = -1183$ Pa (outward)

Area 2 & 3: $p = (577 \text{ Pa})(-3.0) - 433 = -2164$ Pa (outward)

Area 4: $\quad p = (577 \text{ Pa})(1.4) + 144 = 952$ Pa (inward)

and

$\quad\quad\quad\quad\quad p = (577 \text{ Pa})(-1.5) - 433 = -1298$ Pa (outward)

Area 5: $\quad p = (577 \text{ Pa})(1.4) + 144 = 952$ Pa (inward)

and

$\quad\quad\quad\quad\quad p = (577 \text{ Pa})(-2.0) - 433 = -1587$ Pa (outward)

6.7 LOADS EXERTED BY CONFINED LIQUIDS AND GRANULAR MATERIALS

During the lives of many agricultural structures, loads may be imposed on the structures by stored materials. The loads may be from a liquid such as water, fuel, or liquid fertilizer; from granular materials such as grains, fertilizers, or gravel; or from a mixture of granular and liquid materials such as silage or soil. The magnitude of the load can be large compared to the natural loads, and the nature of the granular and granular–liquid load is not well defined; therefore careful attention to these loads is required and conservative design is recommended.

6.8 LIQUIDS

Liquids at rest, in this discussion, cannot resist a shear force and therefore the forces on any particle of liquid must be the same in all directions. If the force were less in any direction, flow would occur. The vertical pressure exerted at any point in liquid is equal to the pressure exerted by the liquid mass above that point. Thus, the expression for the pressure exerted by a fluid in any direction at a vertical distance z below the surface is

$$p_n = \gamma z (g/g_c) = \frac{lb}{ft^3} \cdot ft \left(\frac{32.17 \, ft/s^2 \, lb_f}{321.7 \, ft/s^2 \, lb} \right) = \frac{lb_f}{ft^2} \quad (6.14)$$

where γ is liquid mass density, kg m^{-3} (lb ft^{-3}); z, vertical distance below the liquid surface, m (ft); g, the acceleration of gravity, $g = 9.8$ ms^{-2} (32.17 ft s^{-2}); $g_c = 1$ m s^{-2} kg N^{-1} (32.17 ft s^{-2} lb lb$_f^{-1}$).

When this liquid pressure is confined by a boundary, the pressure exerted is perpendicular to that boundary. Since the pressure is perpendicular to the boundary, no forces are exerted along the boundary. Tension or compression forces within the wall, however, can occur due to a curved boundary and weight of the boundary wall.

Example. Fluid Pressure Problem. What pressure does gasoline exert on the walls of a tank shown in Fig. 6.13?

Solution. Gasoline exerts pressure on the tank walls according to Eq. (6.14)

$$p_n = \gamma z (g/g_c)$$

where $\gamma = 673$ kg m^{-3}; $g = 9.8$ m s^{-2}; $g_c = $ m s^{-2} kg N^{-1}.
Therefore,

$$p_n = (673 \text{ kg m}^{-3})(9.8 \text{ m s}^{-2}/1\text{N}^{-1} \text{ kg m s}^{-2}) z$$

$$= 6595 \, z, \quad \text{N m}^{-2} \text{ or Pa}$$

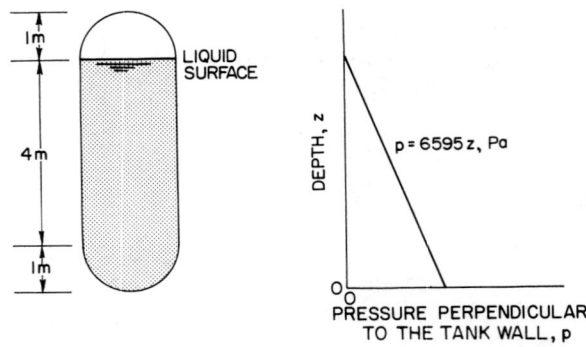

Fig. 6.13. Gasoline tank and the pressure exerted by the gasoline on the tank walls in Example—Fluid Pressure Problem.

The pressure perpendicular to the tank wall is a linear equation with depth as shown in Fig. 6.13.

6.9 GRANULAR MATERIALS WITHOUT BRIDGING

Granular materials in this discussion are materials, such as shelled corn, consisting of individual particles that are small with respect to the size of the storage; the particles exhibit insignificant cohesion or adhesion. This granular material, unlike liquids, does possess some resistance to shear forces and therefore the pressure at a point is not the same in all directions. A measure of the ability of the granular mass to resist shear forces can be demonstrated by the stable conical shape assumed by the mass when the material is discharged at one point and allowed to fall onto the floor. The steeper the cone angle, the greater the shear resistance, and therefore less pressure exerted by the material on any vertical boundary.

For a cohesionless material the shearing resistance at failure is given by Coulomb's equation

$$\tau/\sigma = \tan \phi \qquad (6.15)$$

where τ is shear stress on the failure plane; σ, compression stress perpendicular to the failure plane; ϕ, angle of internal friction. This angle ϕ can be determined by direct shear tests but is often approximated by the emptying angle of repose. The emptying angle of repose represents the minimum angle of internal friction.

6.9 Granular Materials Without Bridging

To investigate the conditions of failure in a cohesionless granular mass, consider the condition of equilibrium for an element within the granular mass (see Fig. 6.14). The stresses acting on the vertical and horizontal sides, σ_1 and σ_2, are principal stresses. The angle α is any arbitrary angle. Summing the horizontal and vertical forces for an element of unit thickness yields

$$\Sigma F_x = 0 = \sigma \, ds \sin \alpha - \tau \, ds \cos \alpha - \sigma_3 \, ds \sin \alpha$$
$$\Sigma F_x \, 0 = \sigma \sin \alpha - \tau \cos \alpha - \sigma_3 \sin \alpha \quad (6.16)$$
$$\Sigma F_z = 0 = -\sigma \, ds \cos \alpha - \tau \, ds \sin \alpha + \sigma_1 \, ds \cos \alpha$$
$$\Sigma F_z \, 0 = -\sigma \cos \alpha - \tau \sin \alpha + \sigma_1 \cos \alpha \quad (6.17)$$

Solving Eqs. (6.16) and (6.17) for τ and σ yields

$$\tau = (\sigma_1 - \sigma_3)\sin \alpha \cos \alpha = \frac{1}{2}(\sigma_1 - \sigma_3)\sin 2\alpha \quad (6.18)$$

$$\sigma = \sigma_1 \cos^2 \alpha + \sigma_3 \sin^2 \alpha = \frac{1}{2}(\sigma_1 + \sigma_3) + \frac{1}{2}(\sigma_1 - \sigma_3)\cos 2\alpha \quad (6.19)$$

These equations specify stresses τ and σ on a section at any angle α from the horizontal. These parametric equations plot as a circle on a τ–σ graph as shown in Fig. 6.15. For example, the stress on a plane

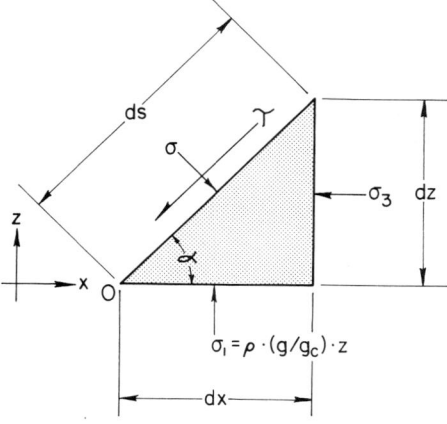

Fig. 6.14. Free body diagram of an element in a granular mass.

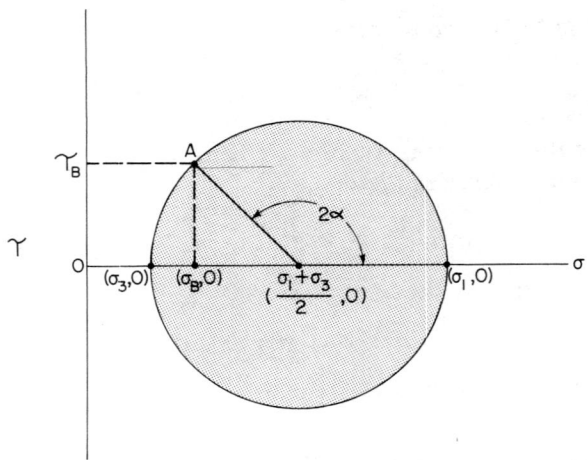

Fig. 6.15. The τ–σ diagram of the stresses on a plane in the granular mass.

at an angle α from the horizontal is represented by the point A at angle 2α from the σ axis. Stresses on the plane are σ_B and τ_B. Coulomb's equation is illustrated by straight lines at angles $\pm \phi$ from horizontal on the $\sigma - \tau$ graph as shown in Fig. 6.16. When the stress circle falls between the line $\pm \tau/\sigma = \tan \phi$, the material is stable. When the stress on any plane is such that $\tau/\sigma \geqslant \pm \tan \phi$, failure occurs along that plane.

Consider the stress circle A in Fig. 6.16. In this case the maximum and minimum stress are such that the material is stable. However,

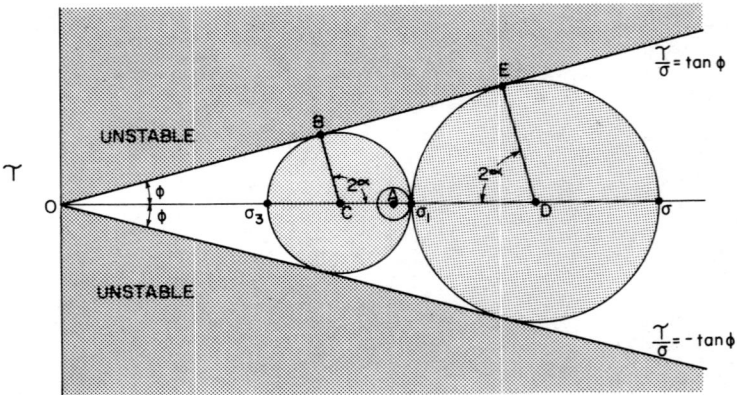

Fig. 6.16. The τ–σ diagram for the Rankine active and passive cases.

6.9 Granular Materials Without Bridging

consider the case where σ_1 remains constant as σ_3 is reduced. The stress circle enlarges until it touches the failure criteria line at B and failure occurs along a plane inclined at an angle α_f from the horizontal. Since the circle is tangent to the line $\tau/\sigma = \tan\phi$, the angle at B between the tangent and the radius is 90°. The angle OCB is $90° - \phi$ and the angle $2\alpha_f$ is

$$2\alpha_f = 180° - (90° - \phi) = 90° + \phi$$

$$\alpha_f = 45° + \phi/2$$

Therefore, the angle between the horizontal and the failure angle is $45° + \phi/2$.

The stress σ_3 is determined as follows:

$$\overline{BC}/\overline{OC} = \sin\phi$$

but

$$\overline{BC} = (\sigma_1 - \sigma_3)/2$$

and

$$\overline{OC} = (\sigma_1 + \sigma_3)/2$$

Therefore

$$\sin\phi = (\sigma_1 - \sigma_3)/(\sigma_1 + \sigma_3)$$

Rearranging the terms

$$\sigma_3/\sigma_1 = (1 - \sin\phi)/(1 + \sin\phi) \qquad (6.20)$$

By trigonometric identities it can be shown that

$$(1 - \sin\phi)/(1 + \sin\phi) = \tan^2(45° - \phi/2) \qquad (6.21)$$

Therefore

$$\sigma_3/\sigma_1 = \tan^2(45° - \phi/2)$$

$$\sigma_3 = \sigma_1 \tan^2(45° - \phi/2)$$

$$\sigma_1 = \gamma z \, (g/g_c)$$

$$\sigma_3 = \gamma z \, (g/g_c) \tan^2(45° - \phi/2) \qquad (6.22)$$

Referring again to Fig. 6.16, the material could also fail if σ_1 is kept constant while σ_3 is increased until the stress circle again touches the failure criteria line at E. This time

$$2\alpha_f = 90° - \phi$$
$$\alpha_f = 45° - \phi/2$$

The stress σ_3 is determined as follows:

$$\overline{DE}/\overline{OD} = \sin \phi$$
$$1/2\,(\sigma_3 - \sigma_1) / 1/2\,(\sigma_3 + \sigma_1) = \sin \phi$$

and

$$\sigma_3/\sigma_1 = (1 + \sin \phi)/(1 - \sin \phi) = \tan^2(45° + \phi/2)$$
$$\sigma_3 = \sigma_1 \tan^2(45° + \phi/2)$$
$$\sigma_3 = \gamma z\,(g/g_c) \tan^2(45° + \phi/2) \tag{6.23}$$

The first case, where σ_3 is reduced until failure occurs, is referred to as the Rankine active state. This problem was first solved by Rankine in 1857. It is termed the active state because as σ_3 is reduced, gravity forces are active in the failure of the mass to maintain σ_3 as a minimum pressure. If a vertical wall confining the material moves away from the material, gravity force on the material causes the mass to fail and move to the wall.

The second case where σ_3 is increased to failure is known as the Rankine passive case. Here the gravity force does not assist but resists the failure. The passive case represents the maximum pressure the granular mass can exert on the wall if the mass is free to fail in one direction.

This discussion has been concerned with the stresses in the granular mass but the real interest is the force exerted by the granular material on the confining structure. This force is considered to be the same as the material stress at the boundary, and frictional forces are considered to be related to the normal boundary forces by the friction coefficient. The forces on the vertical wall confining a granular mass are summarized as follows:

Active Rankine case:

$$p_h = \gamma z (g/g_o) \tan^2(45° - \phi/2) \quad (6.24)$$

$$p_v = p_h \tan \mu \quad (6.25)$$

Passive Rankine case:

$$p_h = \gamma z (g/g_c) \tan^2(45° + \phi/2) \quad (6.26)$$

$$p_v = p_h \tan \mu \quad (6.27)$$

where p_h is horizontal pressure on the wall, Pa; p_v, vertical pressure on the wall, Pa; μ, friction angle between the wall and the granular material.

The force on any horizontal surface is the weight of the granular material over the surface minus any frictional force. The frictional forces do not reduce the vertical loads very far from the wall, therefore the effect of these frictional forces reducing vertical forces on horizontal surfaces is usually ignored and the vertical pressure on the floor is given as follows:

$$\text{At the floor } p_{fv} = \gamma z (g/g_c) \quad (6.28)$$

Rankine active state is assumed to exist when the granular mass moves to the wall such as when a bin is filled. The active state exerts the minimum force on the walls. If there is any expansion of the grain, such as when the moisture increases or when material flowing in the bin forms a shear surface which causes the grain mass to expand, or if granular mass restrains the bin wall such as when the bin temperature drops rapidly, then the state of the grain mass is somewhere between the active and passive state.

Notice that the horizontal force on the wall is linearly related to z, the same as it is for a fluid. Sometimes the terms $\rho \tan^2(45° - \phi/2)$ are grouped in the active case and called the equivalent fluid density. The force on the wall is then given as

$$p_h = \gamma_e (g/g_c) z \quad (6.29)$$

where $\gamma_e = \gamma \tan^2(45° - \phi/2)$ is the equivalent fluid density, kg m^{-3}.

Properties of granular material necessary for the calculation of pressures exerted by these materials are given in Table 6.11.

Table 6.11. Properties of Granular Materials.

Product	Specific Gravity[a] (bulk), ρ/ρ_water	Angle of Repose, φ_min Empt.	Fill	Static Coefficient of Friction[b] Galv. Steel (tan μ)	Steel Troweled Concrete (tan μ)	Douglas Fir Lumber, Perp. to Grain (tan μ)	Ref.
Barley							
eastern	0.64	28	16	0.34	0.62	0.41	c
western	0.69	28	16	0.34	0.62	0.41	c
Coal							
anthracite	0.83	27		0.32	0.70	0.70	d
Corn							
ear	0.45					0.62	d
shelled	0.77	27	16	0.37	0.64	0.38	c
Flaxseed	0	25	14				c
Crushed stone	1.92	40					
Grain, sorghum	0.72	33	20				d
Hay, chopped	1.7 in. (10% n.c.)			0.4		0.45	e
Oats, central	0.56	32	18	0.59	0.44	0.30	c
Rice, rough	0.58	36	20				c
Phosph. rock	1.25	35					
Rye	0.74	26	17				c
Sand, dry	1.60	30		0.32	0.58	0.58	d
Soybeans	0.74	29	16	0.20	0.55	0.44	c
Wheat				0.33	0.68	0.50	c
hard red spring	0.83	28	17				
hard red winter	0.82	27	16				
soft red winter	0.78	27	16				
durum	0.83	26	17				

[a]At maximum bulk density expected.
[b]At high storage moisture contents for grains. From ASAE Paper 63-628 for grain.
[c]Midwest Plan Service 1983. *Structures and Environment Handbook*. 11th ed. Midwest Plan Services, Ames, IA.
[d]Ketchum, M. S. 1919. *The Design of Walls, Bins, and Grain Elevators*. McGraw-Hill Book Company, New York.
[e]Christensen, M. 1952. Investigation of Vertical Hay Feeder for Cattle and Sheep. Unpublished M.S. Thesis, Michigan State University, East Lansing, MI.

Example. Problem. What is the equivalent fluid density of grain sorghum?

Solution

$$\gamma_e = \gamma \tan^2(45° - \phi/2)$$
$$= 0.72 \text{ Mg m}^{-3} \tan^2(45° - 33/2)$$
$$= 212 \text{ kg m}^{-3}$$

6.9 Granular Materials Without Bridging

Example. Granular Pressure Problem. What is the active case total vertical force per linear foot at the bottom of the wall of a 6.4-m (21-ft) diameter bin if the grain depth is 4.9 m (16 ft)? The grain stored is shelled corn. The bin wall is corrugated galvanized steel.

Solution. The vertical pressure at any depth z is

$$p_v = p_h \tan \mu = \gamma z (g/g_c) \tan^2(45° - \phi/2) \tan \mu \qquad (6.30)$$

Assuming that no variables in Eq. (6.30) are functions of z, then integrating with respect to z, the total force V, is given by

$$V_t = \int p_v \, dz = \int \gamma z \, (g/g_c) \tan^2(45° - \phi/2) \tan \mu \, dz$$
$$= (1/2) \gamma z^2 \, (g/g_c) \tan^2 (45° - \phi/2) \tan \mu \qquad (6.31)$$

From Table 6.11 $\gamma = 770$ kg m^{-3} (48 lb ft^{-3}), $\phi = 27°$, and $\tan \mu = 0.37$, therefore

$$V_t = (1/2)(770 \text{ kg m}^{-3})(4.9 \text{ m})^2 (9.8 \text{ kg N}^{-1}) \tan^2[45° - (27°/2)](0.37)$$
$$= 12.59 \text{ kN m}^{-1}$$

What would the active horizontal pressure be at the bottom of the wall?

$$p_h = \gamma \, z (g/g_c) \tan^2(45° - \phi/2)$$
$$= (770 \text{ kg m}^{-3})(4.9 \text{ m})(9.8 \text{ kg N}^{-2}) \tan^2[45° - (27°/2)]$$
$$= 13.89 \text{ kN m}^{-2}$$

or

$$p_h = \frac{(48 \text{ pcf})(16 \text{ ft})(32.17 \text{ ft s}^{-2}) \tan^2[45° - (27/2)]^2}{(32.17 \text{ ft s}^{-2} \text{ lb lb}_f^{-1})} = 288 \; p_f sf$$

What is the passive pressure at the bottom of the wall?

$$p_h = \gamma z (g/g_c) \tan^2(45° + \phi/2)$$
$$= (770 \text{ kg m}^{-3})(4.9 \text{ m})(9.8 \text{ kg N}^{-1}) \tan^2[45° + (27°/2)]$$
$$= 98.46 \text{ kN m}^{-2}$$
$$= \frac{(48 \text{ pcf})(16 \text{ ft})(32.17 \text{ ft s}^{-2}) \tan^2[45° + (27/2)]^2}{(32.17 \text{ ft s}^{-2} \text{ lb lb}_f^{-1})} = 2045 \; p_f sf$$

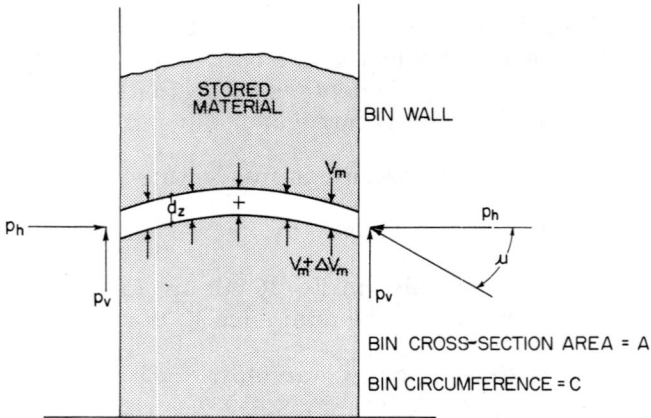

Fig. 6.17. Equilibrium of an infinitesimal arch in a granular storage. Used in the derivation of Janssen's equation.

If the surface of the granular mass is inclined at an angle α above the horizontal, then the horizontal pressure on a vertical wall is given by

$$p_h = \gamma z (g/g_c) \cos^2 \phi / [1 + (\sin \phi \sin(\phi - \alpha)/\cos \alpha)^{.5}]^2 \quad (6.32)$$

where α must be less than ϕ.

6.10 GRANULAR MATERIALS WITH BRIDGING

In 1895 Janssen conducted experiments with model bins in which he measured the force of the granular material on the wall and on the floor. He found that when the depth-to-diameter ratio reached two and more, a higher portion of the weight of additional granular material was supported by the wall as compared to the floor. He theorized that the only way that could happen was for arches to be formed in the granular mass.

To derive the equations that describe the horizontal and vertical pressure on the walls, Janssen isolated a thin layer within the mass (Fig. 6.17) and wrote the differential equation of equilibrium as follows:

Vertical equilibrium of the arch yields

$$p_v(Cdz) - AV_m + (V_m + dV_m)A - Adz\gamma(g/g_c) = 0 \quad (6.33)$$

where C is bin circumference; p_v, vertical friction force of wall on the layer; V_m, vertical pressure of the mass above the layer; A, bin cross-section area.

Expanding and rearranging the terms of (6.33) yields

$$Cp_v dz + AdV_m - A\gamma(g/g_c)dz = 0$$

$$AdV_m = [A\gamma(g/g_c) - Cp_v]dz$$

$$AdV_m = A[\gamma(g/g_c) - Cp_v/A]dz$$

$$\frac{dV_m}{dz} + \frac{C}{A}p_v = \gamma\left(\frac{g}{g_c}\right)$$

Let $A/C = R$ (hydraulic radius) and $p_v = p_h \tan \mu$ where μ is the friction angle between the wall and the stored material.

From the derivation of the active Rankine state, the relation between horizontal and vertical pressure is

$$p_h = KV$$

where $K = (1 - \sin \phi)/(1 + \sin \phi)$ or $\tan^2(45° - \phi/2)$ then

$$\frac{dV_m}{dz} + \frac{K \tan \mu}{R}V = \gamma\left(\frac{g}{g_c}\right) \quad (6.34)$$

which is a first-order, linear, differential equation. The general solution is

$$V_m = \frac{\int e^{\int [K(\tan \mu)/R]dz}\gamma(g/g_c)\,dz}{e^{\int [K(\tan \mu)/R]dz}} \quad (6.35)$$

If R, K, p, and $\tan \mu$ are not functions of z, then

$$V_m = \frac{RK^{-1}\gamma(g/g_c)(\tan \mu)^{-1} e^{Kz(\tan \mu)/R}}{e^{Kz(\tan \mu)/R}} + \frac{c_1}{e^{Kz(\tan \mu)/R}}$$

at $z = 0$, $V = 0$

172   Load Analysis

$$C_1 = -R\gamma(g/g_c)/K \tan \mu$$

Therefore

$$V_m = \frac{R\gamma(g/g_c)}{K \tan \mu} [1 - e^{-Kz(\tan \mu)/R}]$$

and

$$p_h = KV_m = \frac{R\gamma(g/g_c)}{\tan \mu} [1 - e^{-Kz(\tan \mu)/R}] \tag{6.36}$$

and

$$p_v = p_h \tan \mu = R\gamma(g/g_c) [1 - e^{-Kz(\tan \mu)/R}] \tag{6.37}$$

The equations for p_h and p_v are known as Janssen's equations for vertical and horizontal pressure exerted on the bin walls by the granular mass. The total vertical force in the bin wall per unit length of wall (Σp_v) from $z = 0$ to $z = z$ is given by

$$\Sigma p_v = \int_0^z p_v \, dz = \int_0^z R\gamma \left(\frac{g}{g_c}\right) [1 - e^{-Kz(\tan \mu)/R}] \, dz$$

$$\Sigma p_v = R\gamma \left(\frac{g}{g_c}\right) [z - \frac{R}{K \tan \mu} [1 - e^{-Kz(\tan \mu)/R})]]$$

$$\Sigma p_v = R \left[\gamma \frac{g}{g_c} z - \frac{p_h}{K} \right] \tag{6.38}$$

If the bin is round in cross section the hydraulic radius (R) is

$$R = (\pi D^2/4)/(\pi D) = D/4$$

where D is bin diameter.

If, however, the bin is some other shape, the hydraulic radius is defined as follows. A square with side length s:

$$R = s/4$$

A rectangle with side length l_1 and l_2 where $l_2 > l_1$:

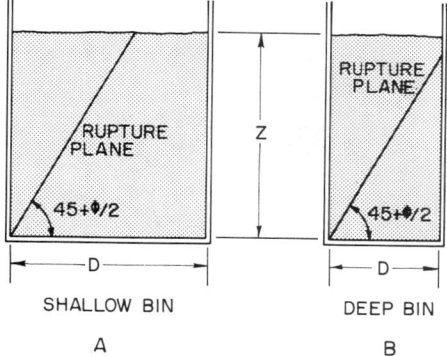

Fig. 6.18. (A) Graphical representation of a shallow bin. (B) Graphical representation of a deep bin.

$$R = l_1 l_2 / 2 \, (l_1 l_2)$$

for the long side:

$$R = l_1/4$$

for the short side.

6.11 DEPTH OF TRANSITION FROM SHALLOW TO DEEP BIN

A number of criteria have been used to determine the transition depth of granular material. At depths less than the transition depth, Rankine's equations predict the pressures (shallow bin) and at depths greater than the transition depth, Janssen's equation predicts the pressures (deep bin). The most used definition is shown graphically in Fig. 6.18. The transition depth is the depth at which the active case rupture plane starting at the wall–floor intersection extends to the grain surface–opposite wall intersection.

Mathematically

$$z < D \tan(45° + \phi/2) \quad \text{shallow bin} \quad (6.39)$$

$$z > D \tan(45° + \phi/2) \quad \text{deep bin} \quad (6.40)$$

Another approach would be to determine the shape of the arch formed in the granular mass and to set the transition depth as the height of the arch. The arch shape is determined by the shape required to support a load without moment. Considering Fig. 6.19 and assuming that the

174 Load Analysis

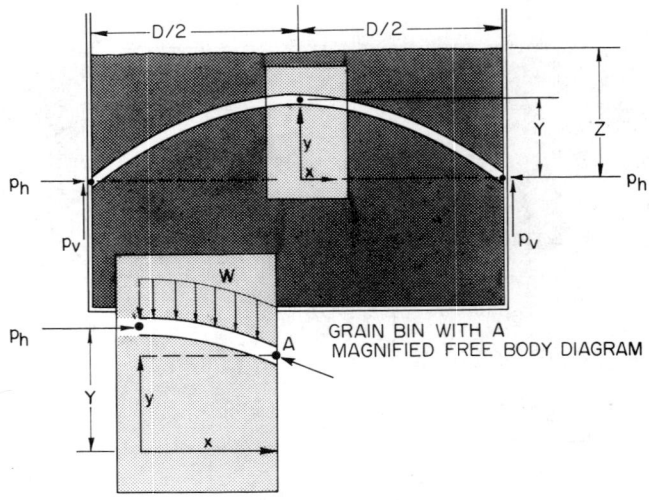

Fig. 6.19. The arch formed in a deep bin of granular material and a free body diagram of a portion of that arch.

arch must support a uniform load of w, then summing moments about point A yields

$$\Sigma M_A = 0 = p_h (Y - y) - wx^2/2$$

Solving for y

$$y = Y - [w/2p_h]x^2 \qquad (6.41)$$

But at $x = r$, $y = 0$, and $Y = wr^2/8p_h$.
  Therefore substituting for p_h

$$Y = \frac{w \tan \mu \; r^2}{8R\rho(g/g_o) [1 - e^{-Kz(\tan \mu)/R}]} \qquad (6.42)$$

Setting $z = Y$ and rearranging

$$Y[1 - e^{-KY(\tan \mu)/R}] = (wr^2 \tan \mu)/8R\rho(g/g_c)$$

If a w is assumed to be the weight of the arch itself and the depth of the arch is unity, then

6.11 Depth of Transition from Shallow to Deep Bin

$$w = \gamma(g/g_c)$$

and

$$Y[1 - e^{-KY(\tan \mu)/R}] = r^2 \tan \mu/8R \qquad (6.43)$$

The height of the arch Y can be determined by trial and error.

Several interesting ideas come from this analysis. If the derivative of y is evaluated at $x = r$, which is the friction angle between the wall and the material, it is found that

$$dy/dx = -wr/2p_h \qquad (6.44)$$

This indicates that the angle of the arch at the wall depends upon the lateral pressure p_h. The maximum angle is μ and as the pressure p_h increases the angle decreases.

Example. Transition Depth Problem. What is the transition depth for a 4-m diameter bin filled with shelled corn? Assume $\tan \mu = 0.4$. Determine the transition depth by both methods shown above.

Solution. From Table 6.11

$$\gamma = (0.77)(1 \text{ Mg m}^{-3}) = 770 \text{ kg m}^{-3}$$
$$\phi = 27° \text{ (emptying angle of repose)}$$

Calculating

$$K = (1 - \sin\phi)/(1 + \sin\phi) = (1 - \sin 27°)/(1 + \sin 27°) = 0.376$$
$$R = D/4 = 4 \text{ m}/4 = 1 \text{ m}$$

The transition depth using the criteria

$$z = D \tan(45° + \phi/2)$$
$$= 4 \text{ m} \tan(45° + 27°/2) = 6.5 \text{ m}$$

Using the height of the arch as the transition depth

$$Y[1 - e^{-KY(\tan \mu)/R}] = r^2 (\tan \mu)/8R$$
$$Y[1 - e^{-(0.376)(Y)(0.4)/1}] = 4^2 \text{ m}(0.4)/8(1 \text{ m})$$
$$Y[1 - e^{-0.150\ Y}] = 0.8 \text{ m}$$

By trial and error $Y \simeq 2.6$ m satisfies the preceding equation.

176   Load Analysis

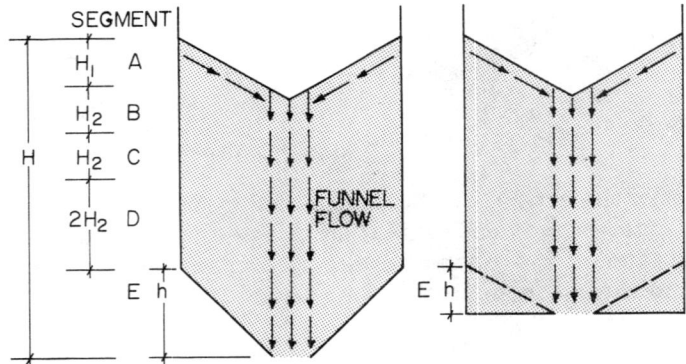

Fig. 6.20. Geometry of funnel flow for calculation of minimum overpressure factors in deep bins.

The preceding problem illustrates the wide variation in transition depth depending on the criteria used. The Rankine solution usually gives higher lateral and vertical pressures, therefore a higher value for the transition depth usually gives the more conservative design.

6.12 DYNAMIC PRESSURES

When the granular material is discharged from the bottom of the bin, two different flow patterns may develop in the granular mass. One flow pattern is for the material on top to flow to an area where a small column of material is moving down to the discharge and out. This is called funnel flow, and if the discharge is in the center of the bin, the column or funnel is formed in the center of the bin and any increase in pressure at the funnel edge is somewhat absorbed by the grain before it reaches the bin walls. The funnel flow pattern is established when the bin walls or hopper do not form a smooth transition to the discharge. Fig. 6.20 illustrates funnel flow during bottom center discharge and the geometry for calculation of overpressure factors.

The other flow pattern exists when the entire granular mass moves down at the same time. This is called mass flow condition. In this situation the pressure caused by the flow can be significant especially where the cross section of the bin changes.

The *Structure and Environmental Handbook* (Midwest Plan Service 1983) makes the following recommendations to account for dynamic pressures:

1. The following apply to *funnel flow* patterns not mass flow.
2. Do not increase lateral loads for dynamic conditions in *shallow* bins.
3. $p_{h_{dynamic}} = p_h(C_d)$ (6.45)
4. $p_{y_{dynamic}} = p_{h_{dynamic}} (\tan \mu)$ (6.46)
5. Minimum overpressure factors, C_d for deep bins, are as follows:

Segment	Material Depth from Surface	$H/D \leq 2$	C_{d_3}	≥ 4
A	0 to $H_1 = D \tan \mu$ or $H_1 = l_1 \tan \mu$ or $H_1 = l_2 \tan \mu$	1.35	1.45	1.50
B	H_1 to $(H_1 + H_2)$	1.45	1.55	1.60
C	$H_1 + H_2$ to $H_1 + 2H_2$	1.55	1.65	1.75
D	$H_1 + 2H_2$ to $H_1 + 4H_2$	1.65	1.75	1.85
E	Hopper depth h	1.65	1.75	1.85

where h is hopper depth or depth of hopper-forming grain (emptying angle of repose sloped hopper); D, bin diameter; l_1, short side of rectangular bin; l_2, long side of rectangular bin; H = total depth of material; $H_2 = (-h - H_1)/4$.

Notes: (1) The pressure can be decreased in the hopper as the hydraulic radius changes. (2) C_d may be linearly interpolated between H/D values shown. (3) If $H_1 < H < 2H_1$, use the C_d from segment B for the total bin. (4) C_d values are for free-flowing noncohesive granular products and funnel flow conditions.

6.12.1 Moist Granular Materials

Silage and soil are examples of granular materials in which the moisture in the granular material can influence the lateral pressure exerted by the material. If the material is not saturated, the moisture affects the mass density, the internal friction, and the friction between the material and the wall. If, however, the material can be considered cohesionless, the pressure pattern is described as previously indicated—i.e., granular materials with or without bridging.

However, when the material becomes saturated and liquid pressure is not reduced by drainage, the horizontal pressure is given as

$$p_h = V_s + \gamma_s (g/g_c)(z - z_s) \qquad (6.47)$$

where p_h is horizontal pressure; V_s, vertical pressure at the point where the material becomes saturated; γ_s, saturated mass density; z, vertical distance from the material surface to the point at which the horizontal pressure is predicted; z_s, vertical distance from the material surface to the point at which the material becomes saturated.

The saturated density of the material can be calculated as follows:

$$\gamma_s = [S_m (1 - M) + M] \gamma_w (1 - r_v) \qquad (6.48)$$

where γ_s is material saturation density; S_m, material specific density; M, water mass/(material mass + water mass); γ_w, water mass density; r_v, void volume/unit volume at saturation.

6.12.2 Silage

Wood (1971) reported that for corn silage the saturation density was achieved at $r_v = 0.1$ and that the specific density of corn silage is 1.60. From these two values, the saturation density from corn silage would be

$$\gamma_s = [1.60(1 - M) + M]1.0 \text{ Mg m}^{-3}(1 - 0.1)$$
$$= 1.44 - 0.54M \text{ Mg m}^{-3}$$

Negri (1983) reported the following empirical expression for the density of corn silage as a function of moisture content, time, and pressure

$$\gamma = \left(\frac{1}{1 - M}\right) [120 + (4440 - 4540M + 3105.3(\ln M) + 14.7(\ln T)(1 - e^{-0.027 V_m})] \qquad (6.49)$$

where γ is silage bulk density; M, as defined in Eq. (6.48); T, time in days; V_m, vertical pressure.

Equation (6.49) applies to whole plant corn silage with moisture contents in the range of 65 to 80%. Negri used a 30-day time period for calculation because silage settlement is essentially complete 30 days after initial fill of the silo.

Equations (6.47), (6.48), and (6.49) along with the equation for vertical pressure in the silage mass,

$$V = \frac{R\gamma(g/g_c)}{k \tan \mu} [1 - e^{-kz(\tan \mu)/R}]$$

will allow solution for the lateral pressure in the unsaturated and the saturated condition. The procedure outlined by Wood (1971) is to divide the silo into a number of finite layers and calculate the vertical pressure and density at the bottom of each layer. These values are then compared to the saturation density and the lateral pressure calculated accordingly.

The National Silo Association recommends an equivalent fluid density of 0.32 Mg m^{-3} (20 lb ft^{-3}) for corn silage at 68–72% moisture content. This equivalent density is determined assuming a maximum $\rho = 1.04$ Mg m^{-3} and $\phi = 32°$. The lateral pressure is given by

$$p_h = 0.32(g/g_c)\, z \text{ Mg m}^{-2}$$

The frictional force on the wall is estimated as follows:

$$V_w = 5.5\, z^{1.08}$$

REFERENCES

American Institute of Steel Construction, Inc. 1980. *Manual of steel construction*. American Institute of Steel Construction, 400 North Michigan Avenue, Chicago, IL 60611.

American National Standards Institute, Inc. 1982. *Minimum design loads for buildings and other structures* (ANSI A58.1-1982). American National Standards Institute, 1430 Broadway, New York, NY 10018.

American Society of Agricultural Engineers. 1978. *1978–1979 Agricultural engineers yearbook*. American Society of Agricultural Engineers, St. Joseph, MI 49085.

———. 1983. *1983–1984 Agricultural engineers yearbook of standards*. American Society of Agricultural Engineers, St. Joseph, MI 49085.

Building Officials and Code Administration International, Inc. 1981. *The BOCA Basic Building Code/1981*. Interstate Printers and Publishers, Inc. Danville, IL.

Davenport, A. G. 1965. The relationship of wind structure to wind loading. In *Wind Effects on Buildings and Structures, Proceedings of the Conference Held at National Physical Laboratory, Teddington, June 1963*, pp. 53–102. H.M.S.O., London.

———. 1968. The dependence of wind loads on meteorological parameters. In *Proceedings of the International Research Seminar, Wind Effects on Buildings and Structures*, Ottawa, Canada, September 1967. Vol. 1. University of Toronto Press, Toronto.

Ghiocel, D. and Lungu, D. 1975. *Wind, snow and temperature effects on structures based on probability.* Abacus Press, Tunbridge Wells, Kent, England.

International Conference of Building Officials. 1976. *Uniform building code.* International Conference of Building Officials, Whittier, CA 90601.

Midwest Plan Service. 1983. *Structures and environmental handbook,* 11th ed. Midwest Plan Service, Iowa State University, Ames, IA 50011.

Negri, S. C., Jofriet, J. C., and Buchanan-Smith, J. 1983. Density-pressure relationship of whole-plant corn silage. Paper No. 83-4001 presented at the Summer Meeting of the American Society of Agricultural Engineers. American Society of Agricultural Engineers, St. Joseph, MI 49085.

Ostavnov, V. A., and Rosenberg, L. S. 1966. Vozmojnosti Snijenia Snegovih Nagruzok na Ploskie Pokritia. In *Promishlenoie Stroitelistvo*, No. 12.

Rusten, A., Slack, R. L., and Molnau, M. 1980. Snow load analysis for structures. In *Journal of the Structural Division ASCE.* Vol. 106, No. STI, January.

Schaerer, P. A. 1970. Variation of ground snow loads in British Columbia. In *Proceedings of the Western Snow Conference.* Victoria, B.C., Canada.

Vellozzi, J., and Cohen, E. 1968. Gust response factors. In *Journal of the Structures Division, ASCE.* June, pp. 1295–1313.

Wood, G. M. 1971. The properties of ensiled crops and why we have to design them. In *Proceedings of the 1971 International Silage Research Conference.* National Silo Association, Des Moines, IA.

PROBLEMS

1. What is the dead load of the following walls per meter (ft.) of wall length?
    a. 203-mm-thick (8-in.) reinforced concrete (stone aggregate) wall 3 m (9.8 ft) high
    b. 203-mm-thick (8-in.) hollow concrete block (lightweight aggregate) wall 3 m (9.8 ft) high

c. a wall consisting of 2 × 4 studs 406 mm (16 in.) o.c., 2 × 6 sill, 2 – 2 × 4's plate, steel siding (28 gauge, 127-mm-thick inside lining), 102-mm-thick fiberglass batt insulation between the studs and 3 m (9.8 ft) high

2. What live loads should be used to design the following:
   a. first floor of a single family dwelling
   b. solid floor in a free stall dairy building that will use a tractor to scrape the floor
   c. floor slats for a swine gestation and breeding building
   d. a 5 in 12 sloped gable roof truss with a 15 m span and a 2 m spacing (specify minimum live load)

3. What is the roof snow load per unit area for a machinery storage building with a 3 in 12 even gable roof? The building is located at the intersection of the South Dakota, Nebraska, and Iowa borders. The building is in a windy area with little shelter. The building is unheated. The roofing is galvanized steel. Building width is 12.2 m (40 ft), length is 36.6 m (120 ft), and eave height is 3.7 m (12 ft).

4. What is the snow load in problem 3, if the building is now insulated at the ceiling level and the attic is well ventilated? In addition, other buildings and a tree windbreak are in the area.

5. What is the maximum uplift reaction due to wind for the truss in example problem 3 if the building is located as described in problem 3 above? The building has one side entirely open. The other side and both ends are closed. The truss spacing is 1.2 m (4 ft).

6. What is the wind design load for the roof purlins in the center of the roof area of the building in problem 5? Purlin spacing is 0.61 m (2 ft).

7. What is the maximum horizontal grain pressure in a corrugated galvanized steel bin 6 m in diameter and 9 m high to the eave? Consider the bin to store shelled corn, soybeans, or wheat. What is the maximum vertical wall force in the bin? What is the maximum floor load?

8. Plot the dynamic horizontal pressure as a function of depth in a 5 m diameter, 9 m tall corrugated galvanized steel bin storing shelled corn. The bin has a center discharge and a flat floor.

9. Compare the horizontal pressure in a 6 m diameter and 18 m high silo full of whole corn silage at 72% moisture content as calculated by the two methods given in the text. Use 2-m-thick layers in the analysis.

NOMENCLATURE FOR CHAPTER 6

A	bin cross-sectional area *or* tributary loaded area, m² (ft²)
C	bin circumference, m (ft)
C_d	dynamic overpressure factor, dimensionless
C_e	exposure factor, dimensionless
C_r	ground to flat roof load factor without drifting, dimensionless
C_s	ground to roof load factor combining C_r, C_α and drift coefficient, dimensionless
C_t	heat loss factor, dimensionless
C_α	roof slope factor, dimensionless
C_1, C_2	constants, dimensionless
D	diameter of bin, tower, chimney, tank, etc., m (ft)
D'	protruding element depth, m (ft)
F	rise per 12 run, m (in.)
G_H	gust factor, dimensionless
GC_p	internal pressure coefficient for components and cladding, dimensionless
GC_{pi}	building internal pressure coefficient, dimensionless
H	elevation above sea level in Eq. 6.2 *or* bin height in Fig. 6.20, m (ft)
H_G	height to the gradient wind velocity, m (ft)
H_1	vertical dimension of bin segment A, m (ft)
H_2	vertical dimension of bin segment B, m (ft)
I	importance factor, dimensionless
K	ratio of horizontal to vertical pressure in a granular mass, dimensionless
M	water mass/(water mass + material mass), dimensionless
R	hydraulic radius, m (ft)
S	design roof snow load, kg m⁻² (lb ft⁻²)
S_g	ground snow load, kg m⁻² (lb ft⁻²)
S_o	ground snow load at a reference elevation, kg m⁻² (lb ft⁻²)
T	lapsed time, days
V	air velocity, m s⁻¹ (ft s⁻¹)
V_G	gradient wind velocity m s⁻¹ (mph)
V_m	vertical pressure in a granular mass, Pa (lbf ft⁻²)
V_{map}	wind velocity from Fig. 6.8 for a specific location at 10m height and a fifty-year recurrence interval, m s⁻¹ (mph)
V_s	vertical pressure in moist granular materials at the point where the material becomes saturated, Pa (lbf ft⁻²)
V_t	total vertical force per unit wall length due to granular mass, N m⁻¹ (lbf ft⁻¹)

Nomenclature for Chapter 6

V_z	design wind velocity at height z, m s^{-1} (mph)
V_w	average winter wind velocity, m s^{-1} (mph)
Y	arch height, m (ft)
a	exponent of z/H_G ratio which is dependent upon ground roughness, dimensionless
c	wind pressure coefficient, dimensionless
d	structure depth in the direction of the wind, ft
g	acceleration of gravity = 9.8 m s^{-2} (32.17 ft s^{-2})
g_c	gravitational constant, 1 m s^{-2} kg N^{-1} (32.17 ft s^{-2} lb lbf^{-1})
h	hopper depth or depth of hopper-forming grain, m (ft)
h_n	structure height, ft
h_o	obstruction height above the roof eave height, m (ft)
l_1	short side length for rectangular bin, m (ft)
l_2	long side length for rectangular bin, m (ft)
p	wind pressure acting on the main frame or components and cladding, Pa (lbf ft^{-2})
p_{fv}	vertical pressure on the floor, Pa (lbf ft^{-2})
p_h	horizontal pressure exerted by the granular mass on the bind wall, Pa (lbf ft^{-2})
p_i	internal pressure, Pa (lbf ft^{-2})
p_n	liquid pressure in any direction at a distance z below the surface, Pa (lbf ft^{-2})
p_v	vertical pressure exerted by the granular mass on the bin wall, Pa (lbf ft^{-2})
q_h	wind pressure calculated at average roof height and exposure C, Pa (lbf ft^{-2})
q_z	basic velocity pressure, Pa (lbf ft^{-2})
r	rise to span ratio for an arch roof, dimensionless
r_v	ratio of void volume to unit volume at saturation, dimensionless
s	square bin side length, m (ft)
S_m	material specific density, kg m^{-3} (lbf ft^{-3})
w	weight of the arch in a granular mass, N m^{-1}
x, y, z	lengths in x, y, and z directions, respectively, m (ft)
z_s	depth of granular material to saturation depth, m (ft)
ΣF_x	summation of forces in the x direction
ΣF_z	summation of forces in the z direction
α	roof slope angle from horizontal, *or* an arbitrary angle used to derive pressure relationships in static granular materials without bridging, *or* the incline of the granular surface from horizontal, degrees

α_f angle between horizontal and failure plane in granular materials without bridging, degrees
τ shear stress on the failure plane, Pa (lbf in^{-2})
μ friction angle between wall and granular material, degrees
γ density, kg m^{-3} (lbf ft^{-3})
γ_e equivalent fluid density, kg m^{-3} (lbf ft^{-3})
γ_s saturated mass density, kg m^{-3} (lbf ft^{-3})
γ_w water mass density, kg m^{-3} (lbf ft^{-3})
σ compression stress perpendicular to the failure plane, Pa (lbf in.$^{-2}$)
σ_1 maximum principal stress, Pa (lbf in.$^{-2}$)
σ_3 minimum principal stress, Pa (lbf in.$^{-2}$)
ϕ angle of internal friction, degrees

7

Fundamentals of Structural Connections

7.1 INTRODUCTION

Connector failures, along with instability failures, account for the vast majority of structural inadequacies. Thus, it is imperative that the designer become familiar with the techniques for analysis of structural joints.

Structural systems consist of combinations of axial, flexural, and torsional members fastened by structural connectors. The connectors may be bolted, glued, or welded joints which must transfer all the axial loads, shear loads, in-plane moments, and out-of-plane moments existing at the ends of the connected members. It is the primary goal of this chapter to learn how groups of connectors are used to transfer various types of loads. Specific details of particular connector-allowable loads (e.g., the allowable load for a single bolt in No. 2 southern pine lumber) will be left to the chapters on timber, steel, and concrete design.

A structural connector has many points of stress concentration, such as bolt holes, corners, and concentrated loads. Consequently, the stress distributions are extremely complex and difficult, if not impossible, to analyze exactly. Thus, greatly simplified procedures are used which yield approximate, but adequate, connector designs. The basic philosophy is to use a simplified, approximate analysis technique to evaluate the loads carried by the individual connector and then to reduce the allowable load per connector to compensate for discrepancies between assumed and actual connector behavior.

The purposes of the chapter are to (1) identify the general classes of connectors; (2) identify the general classes of structural joints; (3) identify how loads are transferred in common types of structural joints;

186 Fundamentals of Structural Connections

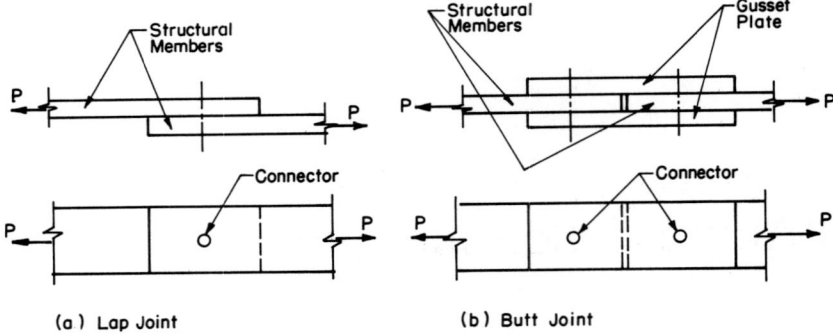

Fig. 7.1. Types of structural joints.

and (4) develop the skills to analyze and design common types of structural joints.

7.2 CLASSES OF STRUCTURAL CONNECTORS

Structural connectors may be classified as being either discrete or continuous. Discrete connectors include rivets, bolts, nails, deformed metal plates, screws, and split rings. Continuous connectors include glue and welds. Of these connectors, the most important are bolts, nails, deformed metal plates, and glue for timber structures and bolts and welds for steel structures.

Discrete and continuous connectors are analyzed within the same basic set of assumptions. Methods for evaluation of some properties of the connector areas do, however, differ.

7.3 CLASSES OF STRUCTURAL JOINTS

Common structural joints can be grouped by their geometry and by the type loads they transmit. Most joints are either lap or butt joints (see Fig. 7.1).

In the lap joint, member loads are transferred directly from one member to another via the connector group. There are no gusset plates in the lap joint. A lap joint may be used to connect two or more members.

In the butt joint, the structural members are butted end-to-end. The load is transferred first to the gusset plates, and then to the other member. There may be one or two gusset plates in the butt joint. If there is one gusset plate, the individual connector is said to be loaded in single shear. With two gusset plates, the connector is loaded in double shear.

Structural joints transfer axial loads, in-plane moments, out-of-plane moments, or any combination of these loads (see Fig. 7.2). The axially loaded lap joint of Fig. 7.2a is typical of upper or lower chord splices in trusses. The load, of course, may be either tensile or compressive. The in-plane moment illustrated in Fig. 7.2b is typical of either the eave or ridge joint in rigid frame structures. The end moments of the vertical legs lie in the plane of the frame and must be transferred by the connectors and gusset plate. In-plane moments are often accompanied by axial and shear loads, thereby creating a combined state of loading in the connectors.

The side bracket in Fig. 7.2c is among the most common out-of-plane connector moment encountered in agricultural structures. The out-of-

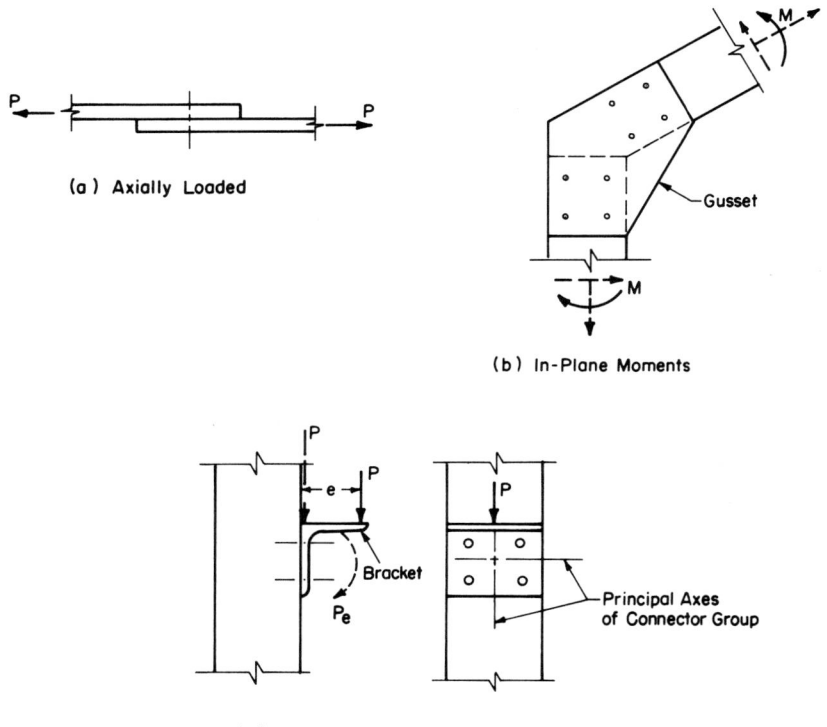

Fig. 7.2. Loads transferred by structural joints.

plane moment is the result of the eccentricity between the line of load application and the shear plane of the connector group. By using couples, the eccentric load is resolved into a shear load acting through the shear plane and an out-of-plane moment, Pe. The connector load is a combination of shearing forces and tensile forces in the upper rows of connectors. It is easily seen that when the load is applied eccentrically with respect to either principal connector group axes in Fig. 7.2c that the connectors are subjected to axial forces, in-plane moments, and out-of-plane moments.

7.4 LOAD AND MOMENT CAPACITY OF JOINTS— AXIALLY LOADED JOINTS

The allowable load per connector, the number of connectors, the connector geometry, and the nature of the loading are the primary factors which determine the load or moment capacity of a structural joint. The load acting on each connector in the joint is obtained by applying principles of statics and strength of materials. The procedures for analyzing several common joints follow. The following discussion assumes that the members are adequate in size, the net section remaining after drilling holes is adequate, and the connectors are adequately spaced with respect to each other and with respect to end and edges of the joint.

The design load capacity, P, of an axially loaded two-member lap joint connected with discrete connectors (see Fig. 7.3a) is simply

$$P = (N)P_{\text{all}}$$

where N = number of connectors and P_{all} = allowable load per connector in single shear. If the discrete connectors are replaced by an adhesive over the entire lap area (Fig. 7.3b), the design load capacity of the connector group becomes

$$P = (b \times d)(q)$$

where $b \times d$ = adhesive area and q = allowable shear strength of the adhesive. If the lap joint is connected by a fillet weld, the design load for the connector group is

$$P = (A_{\text{eff}})(q)$$

7.4 Load and Moment Capacity of Joints   189

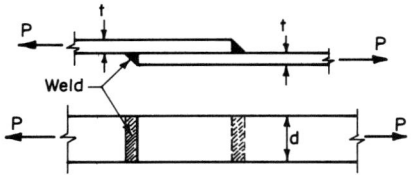

(a) Discrete Connectors   (b) Adhesive Connectors

(c) Welded Connectors

Fig. 7.3. Lap joints.

where A_{eff} is effective weld area, $A_{\text{eff}} = (0.707t)(L)$ (see Fig. 7.4); t, thickness of the connected parts or the weld leg dimension; L, length of the weld; and q, allowable stress of weld material. For the weld group shown in Fig. 7.3c, the design load capacity is

$$P = (0.707t)(2d)(q)$$

The load capacity of the connector group of a three-member discretely connected, axially loaded, lap joint (see Fig. 7.5) is usually greater than

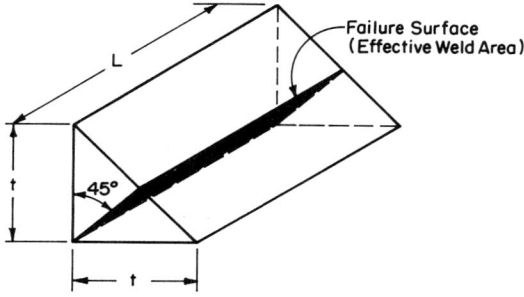

Fig. 7.4. Effective area of fillet welds.

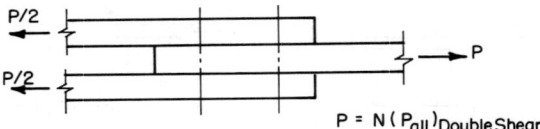

Fig. 7.5. Three-member lap joint.

a two-member joint with the same number of connectors because the connectors are loaded in double, rather than single shear. The allowable load per connector in double shear typically is 33–100% larger than in single shear. The calculation procedure for theoretical load capacities of welded and glue connected three-member lap joints is identical to that for two-member lap joints.

In the butt joint (see Fig. 7.6) the loads need to be transferred twice; first to the gusset plate and then to the other member. The load capacity of the joint is $P = (\frac{1}{2})(N)P_{\text{all}}$. Similarly, the capacity of a glued or welded butt joint is estimated by using only one half of the total glue or effective weld areas.

7.5 JOINTS WITH IN-PLANE MOMENTS

The joint in Fig. 7.7a must transfer moment M from member 1 to member 2. The in-plane moment M is transferred from leg 1 to the lower half of the gusset via the connector group, then to the upper half of the gusset and finally to leg 2 via the connector group in the upper half of the joint. The magnitude of the moment which can be transferred depends largely upon the type of connectors, the connector geometry, and the materials used in the legs and gusset.

The assumptions made in analyzing the moment capacity are (1) that plane sections remain plane, (2) that the moment capacity of the

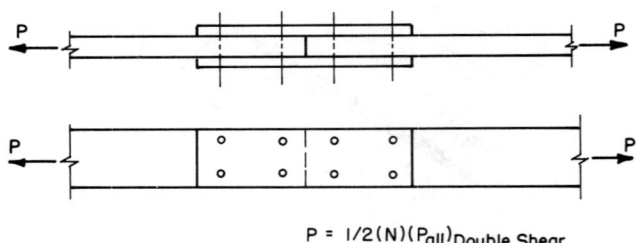

Fig. 7.6. Butt joint.

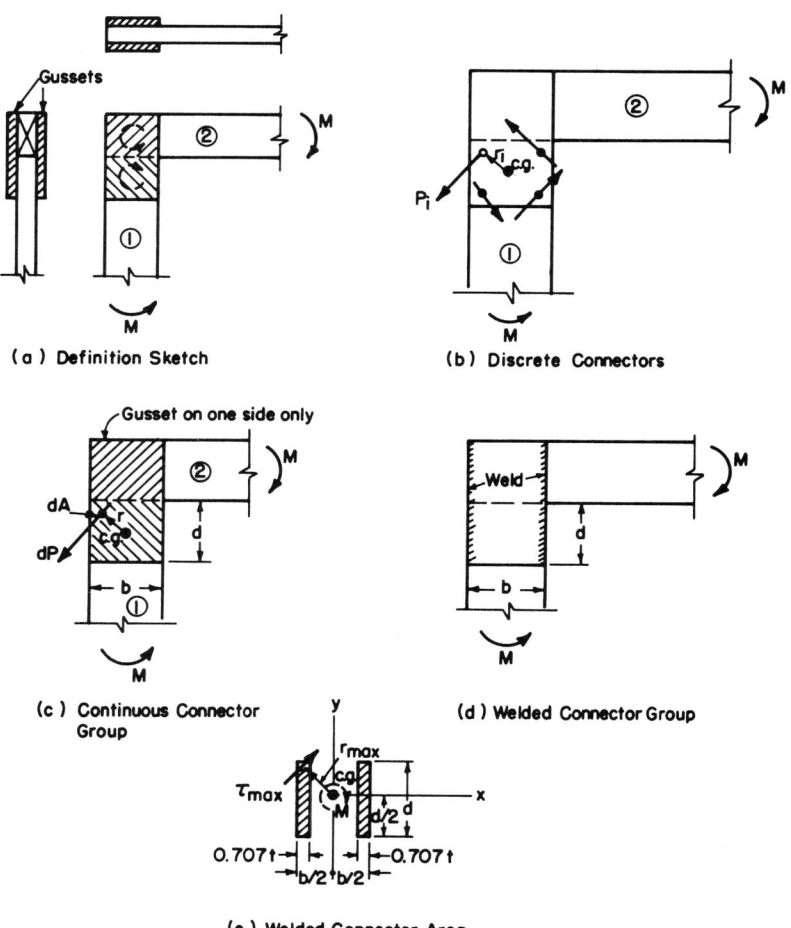

Fig. 7.7. In-plane moment capacity.

joint is reached when any portion of the connector group reaches its allowable load, and (3) that the center of rotation of the connector group coincides with the centroidal axis of the area of the connector group. The consequences of the first and third assumptions are that a connector located at the centroid carries no load due to in-plane moments and connector load varies linearly with distance from the centroidal axis. The consequence of the second assumption (when used in conjunction with assumption 1 and 3) is that the critical, or controlling portion of the connector group, is the portion most distant from the centroidal axis of the connector group.

The in-plane moment capacity of a discrete connector group can be developed by referring to Fig. 7.7b. The joint is first separated into two parts, or into two connector groups. The first group is the portion which transfers the moment from leg 1 to the gusset. The moment of the ith connector is

$$M_i = r_i P_i$$

where M_i is moment capacity of connector i; r_i, perpendicular distance between the c.g. (center of gravity) and the ith connector; and P_i, load carried by connector i. Summing the contribution of all the connectors, the moment capacity of the entire group becomes

$$M = \Sigma r_i P_i$$

Noting that the connectors are loaded in shear and assuming the distribution of shear stress to be uniform over each connector area, the connector load is

$$P_i = \tau_i A_i$$

where τ_i is shear stress in connector i and A_i is cross-sectional area of connector i.

Utilizing the assumption of plane sections remaining plane, the relationship between the shear stress in the most distant connector from the c.g. and the shear stress in the ith connector is

$$\tau_i/\tau_{max} = r_i/r_{max}$$

or

$$\tau_i = r_i \tau_{max}/r_{max}$$

Substituting the expressions for P_i and τ_i into that for M yields

$$M = \Sigma \frac{\tau_{max}}{r_{max}} r_i^2 A_i$$

$$= \frac{\tau_{max}}{r_{max}} \Sigma r_i^2 A_i$$

Noting that $\Sigma r_i^2 A_i$ is the definition of the polar moment of inertia of the connector group with respect to the c.g., we can rewrite the expression for M as

$$M = \frac{\tau_{max}}{r_{max}} J_e$$

where J_e is polar moment of inertia for connector group about c.g. of connector group, $J_e = \Sigma(J_o + A_i r_i^2) \approx \Sigma A_i r_i^2$; J_o is polar moment inertia of individual connector about its centroidal axes.

Rewriting the expression for M in terms of the maximum connector load, P_{max}, and assuming that all connectors are the same size,

$$M = \frac{P_{max}}{A r_{max}} \Sigma A r_i^2 = \frac{P_{max}}{r_{max}} \Sigma r_i^2$$

By replacing P_{max} with P_{all} for the connector, it is possible to evaluate the moment capacity of the group.

In designing a connector for in-plane moments, the magnitudes of M and P_{all} are known and it is necessary to select a satisfactory connector group. A trial and error approach, in which a connector group pattern is selected based on prior design experience or the experience of others, is often used. By calculating the ratio $r_{max}/\Sigma r_i^2$ of the assumed connector group, it is possible to calculate the maximum connector load by $P_{max} = M r_{max}/\Sigma r_i^2$. If $P_{max} \leq P_{all}$, the assumed pattern is adequate. If $P_{max} \geq P_{all}$, a new connector pattern must be tried and checked.

The procedure for evaluating the moment capacity of a continuous connector differs from that for discrete connector groups only in the evaluation of J_e. A brief derivation of the moment capacity for a continuous group follows.

The incremental force and moment capacity of the differential area dA in Fig. 7.7c are

$$dP = \tau \, dA$$

and

$$dM = r_i \, dP = r_i \tau \, dA$$

where τ is stress in the glue line at any distance r from the center of gravity.

194 Fundamentals of Structural Connections

Using arguments similar to those for discrete connectors, the moment capacity for the connector is

$$M = \int_A \left(\frac{\tau}{r}\right) r^2 \, dA = \tau/r \, (J)$$

where $J = \bar{I}_x + \bar{I}_y = \frac{1}{12}(bd^3) + \frac{1}{12}(db^3)$ for the case in Fig. 7.7c.

To obtain the moment capacity of a given connector pattern, set τ_{max} equal to the allowable shear strength of the connector materials and set $r = r_{max}$ in the equation for M. To design a pattern which will carry a given moment, set $\tau_{max} = \tau_{all}$ and rewrite the moment equation as

$$\frac{J_{req'd}}{r_{max}} = \frac{M_{req'd}}{\tau_{all}}$$

By trial and error, a connector pattern can be found in which $J/r_{max} \geq M/\tau_{all}$.

Some connector patterns, such as welds, differ in detail from both the discrete and continuous patterns. However, the only computational difference lies in evaluation of J for the connector pattern. Suppose the gusset in Fig. 7.7d were connected to the vertical leg by two fillet welds of dimension $0.707t$ as shown. A detail of the lower leg-gusset connector group is shown in Fig. 7.7e. The appropriate polar moment in inertia with respect to the connector group center of gravity is

$$J = I_x + I_y = 2\bar{I}_x + 2\bar{I}_y + 2A \, (b/2)^2$$
$$= 2(\tfrac{1}{12})(0.707t)(d^3) + 2(0.707t)(d)(b/2)^2$$

Substitution of J into the equations for M or $J_{req'd}$ yields the moment capacity or the required pattern size.

7.6 COMBINED AXIAL AND IN-PLANE MOMENTS

Combinations of in-plane moments, axial forces, and shear forces may be analyzed by using the principle of superposition. One of the key steps is to determine the critical connector, or portion of a continuous connector pattern, which carries the largest resultant load. The simple joint in Fig. 7.8 will be used to illustrate the procedure.

7.6 Combined Axial and In-Plane Moments

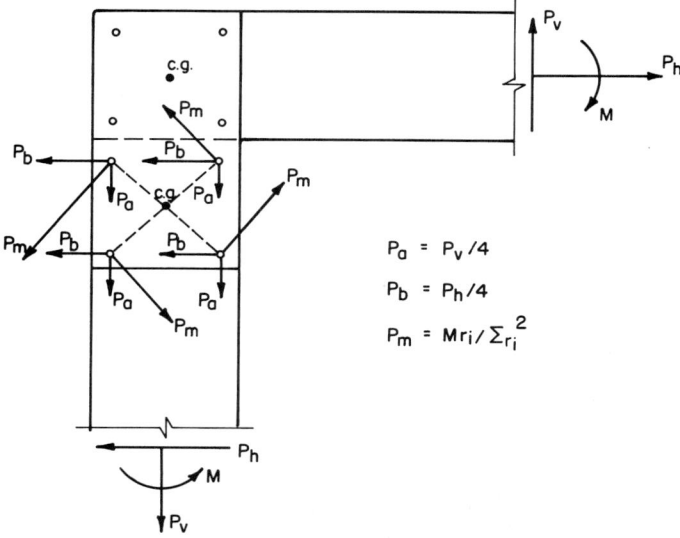

Fig. 7.8. Joint with combined axial loads, shear loads, and in-plane moments.

Considering the connection between the vertical leg and the gusset, each connector carries the following loads

$$P_a = P_v/N = P_v/4$$

$$P_b = P_h/N = P_h/4$$

$$P_m = M\, r_i/\Sigma r_i^2$$

The directions of the individual connector forces are shown in Fig. 7.8. Since the pattern shown is rectangular, the magnitude of P_m is identical in the four connectors. Further inspection of the orthogonal (vertical and horizontal) components of P_m in each connector shows that only in the upper left connector are all the force components additive. Thus, the largest resultant force occurs in the upper left connector and has a magnitude of

$$R_{max} = \sqrt{(P_b + P_{mh})^2 + (P_a + P_{mv})^2}$$

where P_{mh} and P_{mv} are the horizontal and vertical components, respectively, of P_m. The design condition to check for adequacy of the connector is

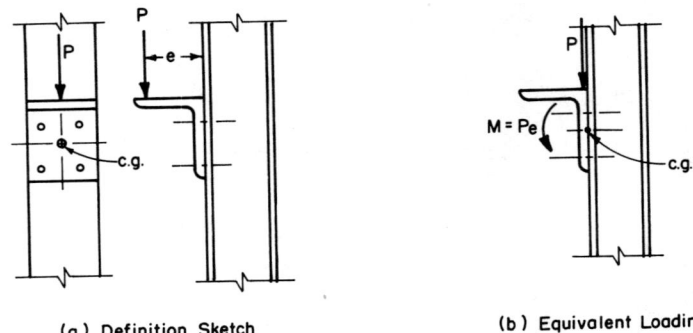

(a) Definition Sketch (b) Equivalent Loading

(c) Forces Acting On Upper Connections

Fig. 7.9. Joint with combined shear loads and out-of-plane moments.

$$R_{max} \leq P_{all}$$

The procedures for analyzing continuous or welded joints with combined in-plane moments and shear forces are identical.

7.7 JOINTS WITH AXIAL LOADS AND OUT-OF-PLANE MOMENTS

Many times the joint connectors are loaded in both shear and tension. The joint holding the bracket in Fig. 7.9 is such a case. After transforming the eccentric load in Fig. 7.9a into an equivalent centric loading as shown in Fig. 7.9b, the shear force in each connector is simply $P_a = P/N = P/4$. Since the couple, Pe, is a free vector, it can be assumed to act about the horizontal centroidal axis of the connector group. Application of the flexure formula from strength of materials provides an estimate of the axial stress, f_t, in the critical connector of the group.

$$f_t = \frac{My_{max}}{I_e} = \frac{(Pe)y_{max}}{I_e}$$

7.7 Joints with Axial Loads and Out-of-Plane Moments

where y_{max} is distance from the horizontal centroidal axis to the outermost connector; I_e is moment of inertia of the connector group about the horizontal centroidal axis.

The adequacy of the connector group is then determined by using a form of the interaction equation. That is,

$$\frac{P_a}{P_{all}} + \frac{f_t}{F_t} \leq 1.0$$

where F_t is allowable tensile stress of the connector; P_{all} is allowable shear load per connector.

If the left side of the equation is less than or equal to 1.0, the connector pattern is adequate. If it is greater than one, a different connector group or size or strength must be tried and rechecked.

In the bracket connection, it is apparent that part of the compressive resultant induced by the moment, Pe, is carried by direct bearing of the bracket on the main member. Because of this, several approaches have been used to evaluate the moment of inertia, I, in the expression for f_t.

The first approach is to assume the bracket to be rigid and that the compressive resultant is carried by the connectors. Then the center of gravity of the connector group is taken as the centroid of the connector group area as shown in Fig. 7.10a. In this case

$$I = I_x = \Sigma (\bar{I}_{xi} + Ay_i^2)$$

or

$$I \approx \Sigma Ay_i^2$$

and if all connectors are of the same size

$$I = A \Sigma y_i^2$$

Another approach that is appropriate for steel members and brackets connected by low strength steel connectors is to assume that the connectors carry the resultant tensile force and the bearing area of the bracket carries the compressive resultant. By assuming elastic behavior, linear variation of the connector and bearing area loads with distance from the neutral axis, and location of the neutral axis at the innermost extent of the bearing area (see Fig. 7.10b), it is possible to evaluate the extent of the compression area.

198 Fundamentals of Structural Connections

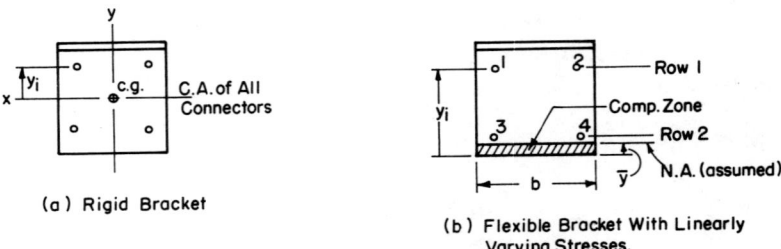

(a) Rigid Bracket

(b) Flexible Bracket With Linearly Varying Stresses.

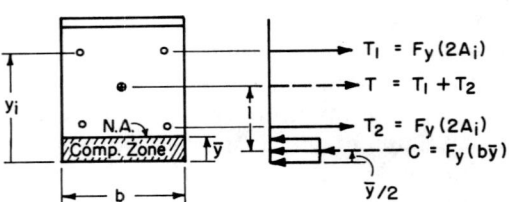

(c) Approach 3 — Plate Bearing and Connector Stresses at Yield.

Fig. 7.10. Effective moment of inertia for tensile stress calculation for out-of-plane moments.

Begin the analysis by assuming a location of the neutral axis (N.A.) noting carefully that the location of the N.A. must be assumed to lie between two rows of connectors. For example, in Fig. 7.10b, the N.A. may fall between the bottom of the bracket and row one, or between row one and row two. Due to the discontinuity of the connector area in the case of discrete connectors, the designer must assume one case to be valid and then check the assumption.

If the N.A. is assumed to fall between the bottom of the bracket and row two in Fig. 7.10b, the N.A. is found by

$$A_{brg}\left(\frac{\bar{y}}{2}\right) = (b\bar{y})\left(\frac{\bar{y}}{2}\right) = \left(\frac{b}{2}\right)\bar{y}^2 = \sum_i A(y_i - \bar{y})$$

or

$$\frac{b}{2}(\bar{y})^2 = A\Sigma(y_i - \bar{y})$$

After solving for $\bar{y}$, the moment of inertia of the connector group is

$$I = \frac{1}{3} b \bar{y}^3 + \sum_i A_i(y_i - \bar{y})^2$$

Note that the last term of the expression for I includes only those connector areas out of the region of direct bearing of the bracket on the member.

A third approach, in which the plate bearing stresses and the effective tensile connector stresses are all assumed to be at yield stress, is sometimes used to locate the N.A. and the moment capacity. This approach is illustrated in Fig. 10c. By summing forces in the horizontal direction

$$\Sigma F_h = 0 : C = T = T_1 + T_2$$

or

$$F_y(b\bar{y}) = \Sigma F_y A_i$$

Once this equation is solved for $\bar{y}$, the centroid of the resultant compressive and tensile forces can be located and the moment capacity calculated by either

$$M = C(l)$$

or

$$M = T(l)$$

7.8 SPECIAL CONSIDERATIONS

Other loading combinations on connectors can be analyzed by superimposing the fundamental methods discussed herein.

The methods presented herein can be used to evaluate the load capacity of connector groups of properly proportioned structural joints. A properly proportioned joint has an adequate net section, adequate end and edge margins, and adequate margins between individual connectors. These requirements are dependent upon the type of structural material and connector being used and will be discussed later in the chapters on timber, steel, and concrete design.

Example 7.1. A structural joint consisting of two groups of 15 discrete connectors is to be constructed as shown in Fig. 7.11. If the al-

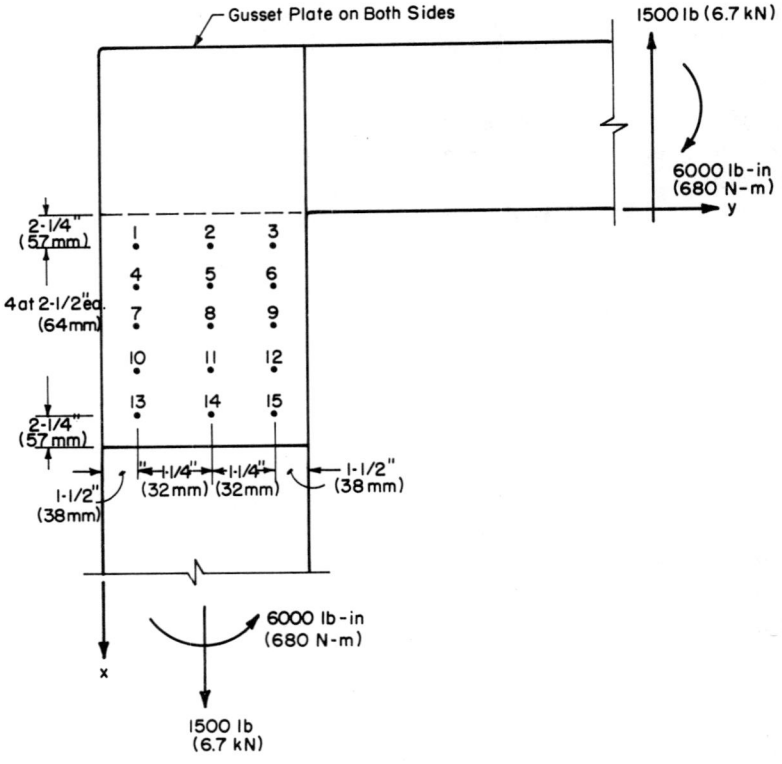

Fig. 7.11. Example 7.1.

lowable load per connector in double shear is 250 lb (1.11 kN), determine if the connector group can transmit the 6000 lb-in. (680 N-m) moment and the 1500 lb (6.7 kN) axial load from the vertical leg to the gusset plates. Assume that the margins, connector spacings, and net sections are adequate and that all connectors are identical.

Solution

Step 1. Locate the center of gravity of the connector group.
    a. Set up a reference axis (see Fig. 7.11).
    b. Locate the center of gravity of the connector group. By inspection in this case

$$\bar{x} = 7.25 \text{ in. } (184 \text{ mm})$$

$$\bar{y} = 2.75 \text{ in. } (69 \text{ mm})$$

7.8 Special Considerations

Step 2. Evaluate the connector shear load due to the axial load

$$P_a = 1500/15 = 100 \text{ lb } (450 \text{ N})$$

Step 3. Evaluate the connector shear load due to the moment

$$P_m = \frac{Mr_{max}}{\Sigma r_i^2}$$

A tabular form for evaluating Σr_i^2 is helpful for complex connector groups. One form follows (units are in inches):

Connector No.	$(x_i - \bar{x})$	$(y_i - \bar{y})$	$(x_i - \bar{x})^2$	$(y_i - \bar{y})^2$	r_i^2
1	−5	−1.25	25	1.6	26.6
2	−5	0	25	0	25.0
3	−5	+1.25	25	1.6	26.6
4	−2.5	−1.25	6.2	1.6	7.8
5	−2.5	0	6.2	0	6.2
6	−2.5	+1.25	6.2	1.6	7.8
7	0	−1.25	0	1.6	1.6
8	0	0	0	0	0
9	0	+1.25	0	1.6	1.6
10	+2.5	−1.25	6.2	1.6	7.8
11	+2.5	0	6.2	0	6.2
12	+2.5	+1.25	6.2	1.6	7.8
13	+5	−1.25	25	1.6	26.6
14	+5	0	25	0	25.0
15	+5	+1.25	25	1.6	26.6
				$\Sigma r_i^2 =$	203 in.² (1310 cm²)

It follows that

$$(P_m)_{1,3,13,15} = \frac{6000 \, (26.6)^{1/2}}{203} = 150 \text{ lb } (670 \text{ N})$$

Step 4. Evaluate the maximum resultant shear force per connector.

From Fig. 7.12 it is apparent that 1 and 13 are the critical connectors.
Using connect 1, resolve P_m into its horizontal and vertical components

$$\theta_i = \tan^{-1}\left(\frac{x_i - \bar{x}}{y_i - \bar{y}}\right) = \tan^{-1}\left(\frac{5}{1.25}\right) = 76°$$

$$P_{mv} = P_{m1} \cos 76° = 36 \text{ lb } \quad (160 \text{ N})$$

$$P_{mh} = P_{m1} \sin 76° = 145 \text{ lb } \quad (645 \text{ N})$$

202 Fundamentals of Structural Connections

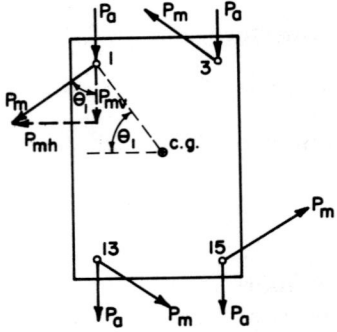

Fig. 7.12. Example 7.1 solution.

Thus, the resultant force in connector 1 is

$$R = \sqrt{(P_a + P_{mv})^2 + P_{mh}^2}$$
$$= \sqrt{(100 + 36)^2 + (145)^2} = 200 \text{ lb } (890 \text{ N}) < 250 \text{ lb } (1.11 \text{ kN})$$

Since the resultant force is less than the allowable connector load, the connector group can easily carry the prescribed loads.

Example 7.2. Check the adequacy of the structural connection of Fig. 7.11 if the connector pattern between the vertical leg and the gusset plate is replaced by a glued area 5.5 in. (140 mm) wide by 6 in. (152 mm) deep (see Fig. 7.13). Assume the allowable shear stress of the adhesive to be 90 psi (620 kPa) and that all margins, spacings, and net sections are adequate.

Step 1: Locate the center of gravity of the connector area

$$\bar{x} = 3 \text{ in. } (76 \text{ mm})$$
$$\bar{y} = 2.75 \text{ in. } (70 \text{ mm})$$

Step 2: Evaluate the connector shear stress due to the axial load

$$\tau_a = \frac{1500}{2(6 \times 5.5)} = 23 \text{ psi } (160 \text{ kPa})$$

Step 3: Evaluate the shear stress due to the in-plane moment. Referring to Fig. 7.14 it is clear that the maximum combined load occurs either at corner a or c.

At point a

$$\tau_{max} = Mr_a/J$$
$$r_a = \sqrt{3^2 + 2.75^2} = 4.07 \text{ in.} \quad (103 \text{ mm})$$
$$J = 2[\tfrac{1}{12}(5.5)(6)^3 + \tfrac{1}{12}(6)(5.5)^3]$$
$$= 364.4 \text{ in.}^4 \ (1.52 \times 10^4 \text{ cm}^4)$$

Thus

$$\tau_{max} = 6000(4.07)/364.4 = 67 \text{ psi } (460 \text{ kPa}).$$

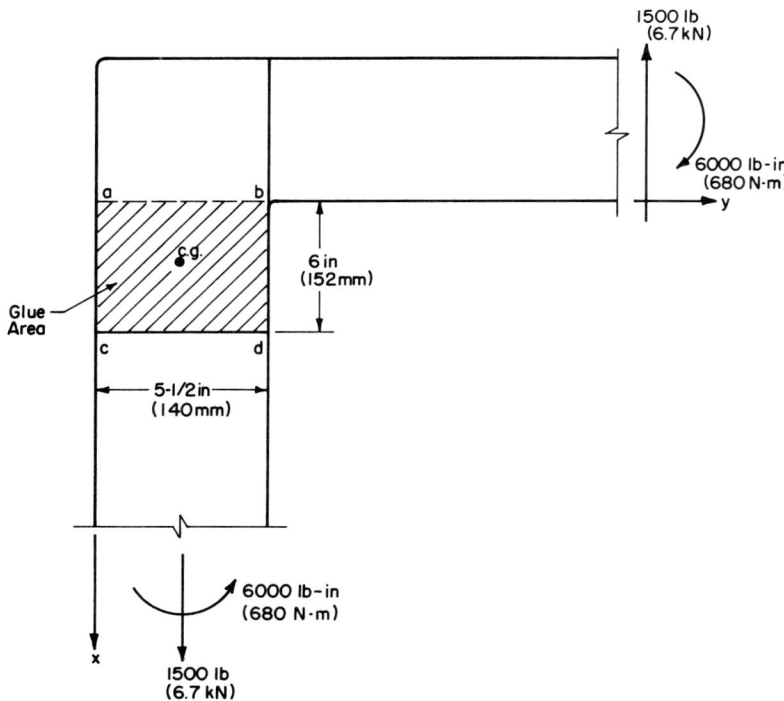

Fig. 7.13. Example 7.2.

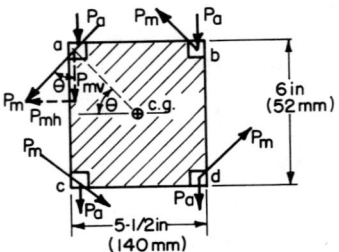

Fig. 7.14. Example 7.2 solution.

Step 4: Evaluate the resultant shear load at point a. (Assume a differential unit area at the corner.)

$$\theta = \tan^{-1}(3/2.75) = 47.5°$$
$$P_{mh} = (\tau\ dA)\sin 47.5 = 49.4\ lb\ dA \quad (220\ N)$$
$$P_{mv} = (\tau_{dA})\cos 47.5 = 45.3\ dA\ lb \quad (201\ N)$$
$$P_a = (\tau_a\ dA) = 23\ dA\ lb \quad (102\ N)$$

The resultant load at the corner a is

$$\tau dA = \sqrt{(P_{mv} + P_a)^2 + (P_{mh})^2}$$
$$= \sqrt{(45.3 + 23)^2 + 49.4^2}\ dA$$

and

$$\tau = 84.3\ psi \quad (580\ kPa)$$

The allowable unit stress at a is 90 psi (620 kPa). Since $\tau < \tau_{all}$, the connector group is adequate to transmit the load to the gusset.

PROBLEMS

7.1. Determine the moment capacity of the structural joint of Fig. 7.15 with the discrete connectors (neglecting the welds) if each connector is of the same size and has an allowable load of 100 lb (440 N) and if $P_h = 240$ lb (1.07 kN) and $P_v = 300$ lb (1.33 kN).

7.2. Repeat Problem 7.1 if $P_h = P_v = 0$.

7.3. Evaluate the moment capacity of the discretely connected structural joint of Fig. 7.15 (neglecting the welds) if connectors 1 and 2 have an area of 0.6 in.^{2} (390 mm^2) and an allowable load of 200 lb (890 N), and if connectors 3, 4, 5, and 6 have an area of 0.3 in.^{2} (190 mm^2) and an allowable load of 150 lb (670 N).

7.4. Evaluate the load capacity of the joint in Problem 7.1 if $M = P_h = 0$.

7.5. Evaluate the load capacity of the joint in Problem 7.3 if $M = P_h = 0$.

7.6. Evaluate the moment capacity of the joint in Fig. 7.15 if the discrete connectors and welds are replaced with an adhesive on both gussets with an allowable connector shear stress of 120 psi (827 kPa), $P_h = 1000$ lb (4.45 kN), and $P_v = 1500$ lb (6.67 kN).

7.7. Repeat problem 7.6 if $P_v = 0$. If $P_h = 0$. If $P_v = P_h = 0$.

7.8. Evaluate the moment capacity of the joint in Fig. 7.15 if it is connected with fillet welds (neglecting the discrete connectors) along the lines shown, if the allowable stress for the welds is 10 ksi (70 MPa), if $P_h = P_v = 0$, if the thickness of the connected members equals 0.25 in. (6.4 mm), and if gussets are welded to both sides.

7.9. Repeat Problem 7.8 if $P_v = 10,000$ lb (44 kN).

7.10. Repeat Problem 7.8 if $P_v = 10,000$ lb (44 kN) and $P_h = 15,000$ lb (67 kN).

7.11. Evaluate the allowable load, P_v, which can be carried by the welded connection of Problem 7.8 if $P_h = M = 0$.

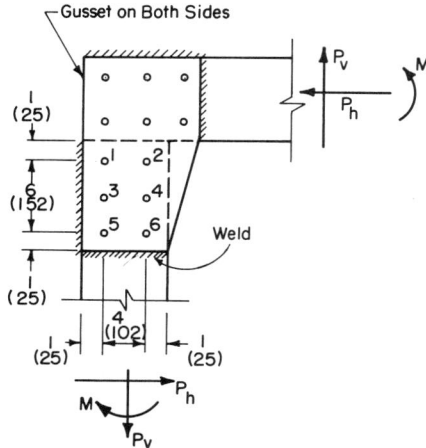

Fig. 7.15. Problem 7.1 All dimensions are in in. (mm).

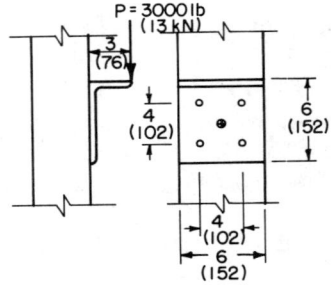

Fig. 7.16. Problem 7.12 All dimensions are in in. (mm).

7.12. Determine if the connector group for the bracket in Fig. 7.16 is adequate to carry the load. Each connector has a cross-sectional area of 0.4 in.² (260 mm²), an allowable shear load of 1200 lb (5.3 kN), and an allowable axial stress of 20 ksi (140 MPa).

NOMENCLATURE FOR CHAPTER 7

A	Cross-sectional area of individual connector or area of a continuous connector grouping, in.² (mm²)
A_{brg}	Bearing area, in.² (mm²)
A_{eff}	Effective weld area, in.² (mm²)
F_t	Allowable axial stress, psi (Pa)
I_x	Second moment of area about the x-axis, in.⁴ (mm⁴)
I_y	Second moment of area about the y-axis, in.⁴ (mm⁴)
$\bar{I}$	Second moment of area about centroidal axis, in.⁴ (mm⁴)
J	Polar moment of inertia of continuous connector area about the c.g. of connector area, in.⁴ (mm⁴)
J_e	Polar moment of inertia of discrete connector group area about the c.g. of the connector group, in.⁴ (mm⁴)
J_o	Polar moment of inertia of a discrete connector area about its centroidal axis, in.⁴ (mm⁴)
L	Length of weld, in. (mm)
M	In-plane moment, lb-in (N·−m)
N	Number of connectors
P	Applied concentrated load, lb (N)
P_{all}	Allowable load per connector in single shear, lb (N)
R	Resultant force, lb (N)
b	Width of glue area, in. (mm)
d	Depth of glue or weld length area, in. (mm)
e	Eccentricity of load, in. (mm)

f_t	Axial stress, psi (Pa)
i	Subscript denoting the ith connector or element in a connector grouping
q	Allowable shear strength of glue or weld material, psi (Pa)
r_i	Perpendicular distance between the c.g. of a connector group and the ith connector
t	Thickness of connected welded parts, in. (mm)
y	Distance from centroidal axis of cross section, in. (mm)
τ	Connector shear stress, psi (Pa)

8

Structural Steel Design

8.1 INTRODUCTION

Structural steel is classified as either hot rolled or cold formed. The primary differences between the two classes are the method of forming the structural shapes, the thickness of the elements of the structural sections, and the availability of standard shapes.

Hot-rolled structural steel shapes are either rolled while the steel is hot or they are built up by welding together two or more hot-rolled plates. Cold-formed shapes are developed by one of three processes: (1) cold rolling thin steel sheets—as with corrugated roofing and siding, (2) press brake operations, or (3) bend brake operations. Typical element thicknesses of cold-formed shapes range from 0.0149 in. (0.4 mm) to 0.25 in. (6.4 mm). Elements to 0.75 in. (19.0 mm) thick can be cold-formed, but are not common. Hot-rolled thicknesses range from 0.1875 in. (4.8 mm) to 4.9375 in. (125.4 mm). Hot-rolled sections used in agricultural applications are usually standard wide flange, I, channel, or tubular shapes. Cold-formed sections are often nonstandard shapes.

The thickness of the elements—webs, flanges, and corners—of structural shapes is important because of the role it plays in local buckling. The thinner the element, the more susceptible it is to buckling locally before the entire section reaches its full elastic or plastic* moment capacity. It follows that cold-formed shapes, having thinner elements, are more susceptible to local buckling than thicker hot-rolled sections. Although there are other factors, such as cold working, which cause

---

*The plastic moment of a steel section is the moment capacity when the entire section has yielded. That is, the tensile and compressive flexural stresses equal the yield stress everywhere on their respective sides of the neutral axes.

hot- and cold-formed steel specifications to differ, the primary differences relate to element thickness and considerations of local buckling.

There are two primary specifications for steel building design in the United States. The cold-formed specification, prepared by the American Iron and Steel Institute (AISI), is entitled *Specification for the Design of Cold-Formed Steel Structural Members*. The specification for hot-rolled steel is the American Institute of Steel Construction (AISC) publication, *Specification for the Design, Fabrication, and Erection of Structural Steel Buildings*.

In this chapter the discussion will be limited to the design procedures for hot-rolled steel tension, compression, and flexural members. The design procedures conform to the AISC design specification as published in the *Manual of Steel Construction* (AISC, 8th ed.).

The objectives of the chapter are to introduce the engineer to the concepts that make steel design unique from wood or concrete structural members, to show how steel structural member analysis techniques differ from the elementary theory of strength of materials, and to show how the steel design procedures compensate for these differences in behavior. Most agricultural engineering applications of hot-rolled steel will involve standard shapes for which there are many design and selection aids. An objective of the chapter is to outline the rationale behind the design of steel beams and columns so that the engineer can use the design aids correctly and confidently.

8.2 WORKING STRENGTH DESIGN PHILOSOPHY

Two basic philosophies of structural steel design are the ultimate and the working strength methods. While the ultimate strength method is gaining acceptance, the traditional working strength method is still the most widely used design approach and is usually a more familiar approach for the beginning student. This chapter will consider only working strength concepts.

The working strength, or working stress, concept assumes that the steel behaves as a perfect linearly elastic material up to yield. Then, to account for discrepancies between the theoretical and actual behavior of steel sections, the allowable stresses are reduced to some level less than the yield stress. Thus, the stress levels in the structural elements are predicted by simple strength of materials equations such as $f_b = Mc/I$, while some appropriate design specification establishes the amount by which the allowable stress must be reduced for a variety of service conditions.

To illustrate the concept consider a flexural member. The actual stress in the beam is assumed to be $f_b = Mc/I$, while the allowable stress, F_b, equals kF_y; where k is a reduction factor and F_y is the yield stress. The design condition which must be satisfied is that $f_b \leq F_b$, or $Mc/I \leq kF_y$. The value of k is obtained from the design specification. Among the factors influencing the value of F_b are the steel alloy, heat treatment, rate of load application, temperature, cold working, repeated stresses, type loading (tension, compression, flexure, etc.), and stability (buckling) of the section elements. The nomenclature used in this chapter is listed at the end of the chapter.

8.3 STABILITY OF STEEL SECTIONS

Instability (buckling) type failures play a major role in steel design because the elements of sections are relatively slender. Thus, there is potential for local buckling of elements as well as gross buckling of the entire structural member. The four most common types of buckling failure are local buckling of the webs, local buckling of the flanges, flexural buckling of the member, and lateral torsional buckling of the member.

Local buckling of the webs is illustrated in Fig. 8.1a. The factors which influence the load at which local buckling of the web occurs include the depth-to-thickness ratio of the web, d/t, and the element type (stiffened or unstiffened).

Local buckling of the flanges, which is illustrated in Fig. 8.1b, is a concern on both flexural and compression members. The primary factors influencing the stress levels at which flanges will buckle locally include the compression element width to thickness ratio, b/t, and the element type.

Flexural buckling of the member is illustrated in Fig. 8.1c. Columns and beam-columns are both susceptible to this type failure. Flexural buckling can occur before the member average compressive stress reaches the material yield stress (elastic buckling), or it can occur after the member compressive stress reaches the yield stress (inelastic buckling). Factors which influence the load at which gross buckling occurs include the slenderness ratio, L/r (the ratio of the unsupported column length to the radius of gyration of the cross section), the degree of restraint (free, fixed, or pinned) at the ends of the member, and the steel stiffness (modulus of elasticity).

Lateral torsional buckling is an instability-type failure common to beams and beam-columns which have long unbraced compression flanges.

212 Structural Steel Design

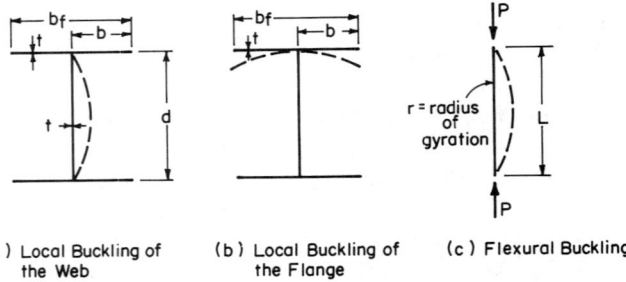

(a) Local Buckling of the Web
(b) Local Buckling of the Flange
(c) Flexural Buckling

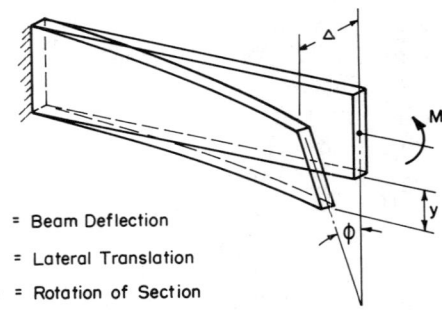

y = Beam Deflection
Δ = Lateral Translation
φ = Rotation of Section

(d.) Lateral Torsional Buckling of Beams

Fig. 8.1. Modes of instability failures in structural members.

The flexural stress at which the instability occurs is dependent upon the unbraced length of the compression flange, the beam depth, and the area of the compression flange. Lateral torsional buckling is a gross buckling phenomenon in which a beam not only deflects in the direction of an applied flexural load, but it also translates in the plane normal to and rotates about the longitudinal axis. That is, the cross section twists as illustrated in Fig. 8.1d.

The allowable stress level of a structural steel member must be lower than the stress levels (with appropriate factors of safety) which will cause any type of instability failure or failure by yielding. The designer must investigate carefully the member geometry, bracing, and loading to ascertain which mode of failure controls the design before establishing an allowable stress for the structural member.

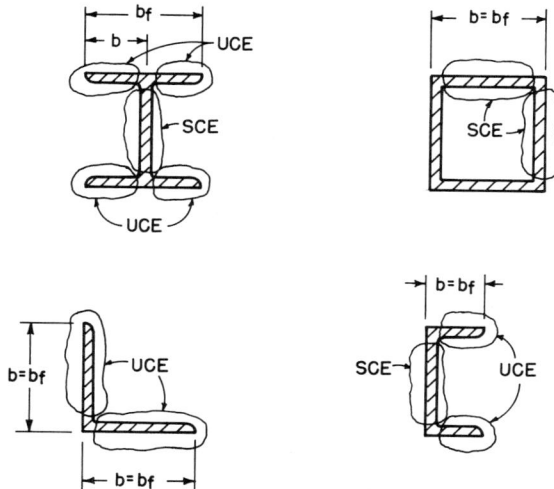

Fig. 8.2. Stiffened and unstiffened compression elements of common structural shapes.

8.4 STIFFENED AND UNSTIFFENED COMPRESSION ELEMENTS

The elements of a structural cross section are classified as either stiffened compression elements (SCE) or unstiffened compression elements (UCE). The classification indicates the degree of end restraint of the individual section element. If an element has one end restrained and one end free, it is an unstiffened element. If both ends of the element are restrained, the element is stiffened. Stiffened and unstiffened elements are both illustrated in Fig. 8.2.

8.5 MECHANICAL PROPERTIES OF STRUCTURAL STEELS

The mechanical properties of structural steels are affected by a wide variety of factors, including chemical composition, heat treatment, strain history, and rate of loading. Structural steels are grouped into four categories. *Low carbon steels,* of which A36 is the most common in

214   Structural Steel Design

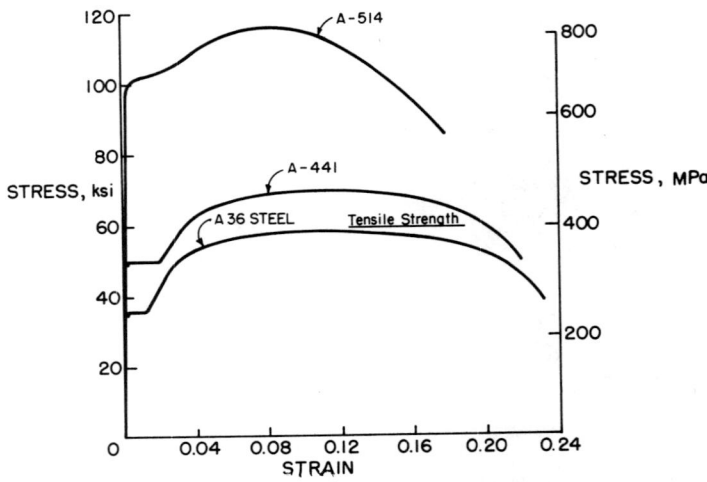

Fig. 8.3. Typical stress-strain curves for structural steels with specified minimum tensile properties. Reproduced with permission from *Steel Design Manual*, 1980 ed., U. S. Steel Corporation, Pittsburgh, PA.

agricultural applications, have yield stresses in the 30 to 40 ksi (210 to 280 MPa) range. These steels are used where stresses are low and rigidity is a primary design criterion. *High strength–low alloy steels* with yield strength above 40 ksi (280 MPa) achieve their strength through the hot-rolling process rather than by heat treatment. These steels are readily available and are corrosion resistant. *Heat-treated high-strength carbon steels* achieve yield strengths in a range of 46 to 80 ksi (320 to 550 MPa) by heat treatment. *Heat-treated construction alloy steels* have yield strengths of 90 to 100 ksi (620 to 690 MPa) and are used primarily in bridge construction.

The stress–strain behavior of most structural steels is similar to those of Figs. 8.3 and 8.4. Figure 8.3 represents the behavior up to failure. Since failure strains are in the range of 0.18 to 0.24 (18 to 24%) and since yield strains are in the order of 1.4×10^{-3}, the elastic behavior and initial yield behavior are illustrated in a partial stress–strain diagram similar to that given in Fig. 8.4. Most structural steels are elastoplastic. That is, they are linearly elastic to yield. Then they experience a range of strain many times greater than the yield strain in which the stress level does not increase. After reaching a strain level five to ten times the yield strain, the stress begins to increase again as the steel enters the strain hardening range.

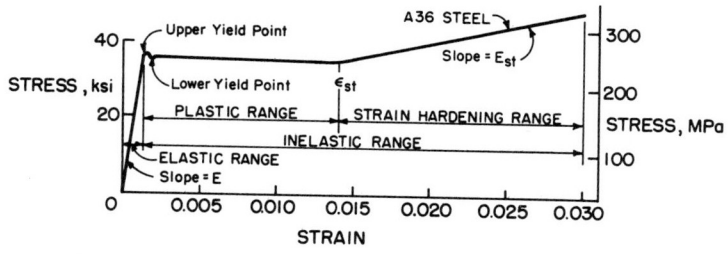

Fig. 8.4. Typical initial stress-strain curve for A-36 steel. Reproduced with permission from *Steel Design Manual*, 1980 ed., U. S. Steel Corporation, Pittsburgh, PA.

8.6 TENSION MEMBERS

Tension members are designed by assuring that the tensile stress, f_t, is less than $F_t = 0.60F_y$ on the net section of the member and less than $0.45F_y$ at pin holes in eye bolts, pin-connected plates, and built-up members. That is,

$$f_t = \frac{P}{A_{net}} \leq F_t = (0.45 \text{ or } 0.6)F_y \tag{8.1}$$

The net area of a section depends on the type, number, size, and placement of connectors. In welded connections, A_{net} is the gross section area. In bolted connections, A_{net} depends upon the member thickness (t), member width (w), connector diameter (d), connector spacing (s), and connector row spacing (g). For example, if a tensile member is fastened by two rows of bolts as shown in Fig. 8.5, the net area is the smaller of

$$A_{net} = (t)(w - 2d + s^2/4g) \leq 0.85(w \times t) \tag{8.2}$$

or

$$A_{net} = (t)(w - d) \leq 0.85(w \times t) \tag{8.3}$$

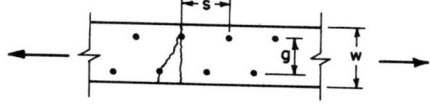

Fig. 8.5. Tensile member.

The term $s^2/4g$ accounts for normal and shear forces on the diagonal area between bolts.

Tension member slenderness ratios are limited to $L/r \leq 240$ in main members and $L/r \leq 300$ in secondary members. This requirement precludes the use of very slender members which are likely to vibrate excessively in service.

8.7 COMPRESSION MEMBERS

8.7.1 Case of No Local Buckling

8.7.1.1 Design Rationale. If the width–thickness ratios, b/t, of the compression elements of a section meet the following requirements, they will yield before buckling locally.

1. Unstiffened compression elements (UCE)
   - Single and double angle struts
     $b/t < 76/\sqrt{F_y}$*
   - Double angle struts in contact; angles or plates projecting from girders, columns, or other compression members; and compression flanges of beams
     $b/t < 95/\sqrt{F_y}$
   - Stems of tees
     $b/t < 127/\sqrt{F_y}$
2. Stiffened compression elements (SCE)
   - Square or rectangular box sections
     $b/t < 238/\sqrt{F_y}$
   - Unsupported width of cover plates
     $b/t < 317/\sqrt{F_y}$
   - Other uniformly compressed elements
     $b/t < 253/\sqrt{F_y}$
3. Circular tubular elements $D/t < 3300/F_y$ (F_y in ksi)

Most standard structural steel sections satisfy these requirements. In this section and in the sections on flexural members and beam-columns, it is assumed that the standard shapes meet these requirements.

*In this chapter, whenever the yield stress, F_y, appears under the radical (i.e., $\sqrt{F_y}$), F_y is restricted to units of ksi.

8.7 Compression Members 217

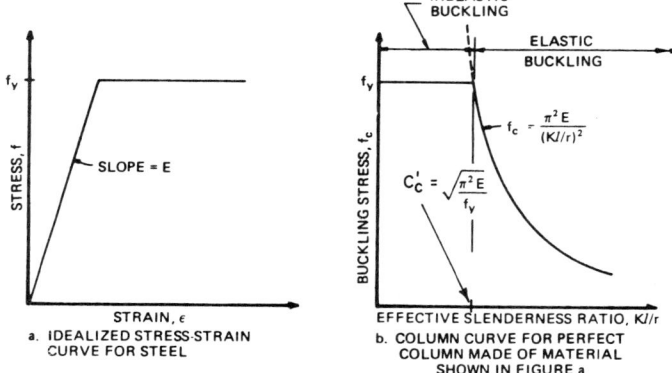

Fig. 8.6. Buckling stress for a perfect column for a perfect elastoplastic material. Reproduced with permission from *Steel Design Manual*, 1980 ed., U. S. Steel Corporation, Pittsburgh, PA.

A perfect column which has no eccentricity, crookedness, or residual stresses will have material stress–strain curves and compressive strength curves similar to those in Figs. 8.6 and 8.7. In a perfect column fabricated from a perfect elastoplastic material, elastic buckling will occur whenever the Euler buckling stress is less than the yield stress. The magnitude of the slenderness ratio at which elastic buckling commences $(KL/r)_{\text{lim}}$ is obtained by equating the yield stress to the Euler

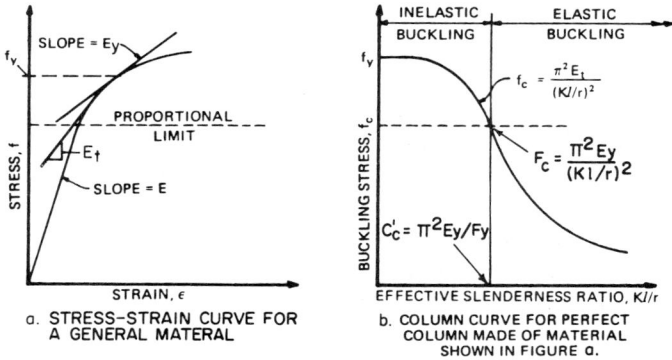

Fig. 8.7. Buckling stress for a perfect column for a material with a nonperfect elastic-plastic stress-strain behavior. Reproduced with permission from *Steel Design Manual*, 1980 ed., U. S. Steel Corporation, Pittsburgh, PA.

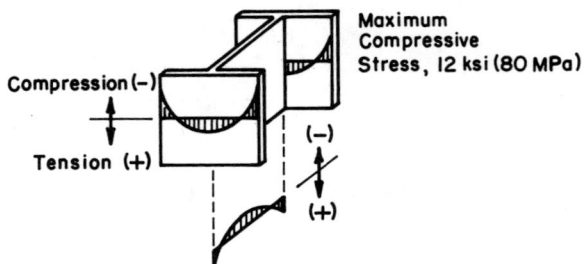

Fig. 8.8. Typical residual stress pattern on rolled shapes. Reproduced with permission from *Steel Design Manual*, 1980 ed., U. S. Steel Corporation, Pittsburgh, PA.

buckling stress. That is, $F_y = \pi^2 E/(KL/r)^2$. Solving for (KL/r), $(KL/r)_{\lim} = \sqrt{\pi^2 E/F_y} = C'_c$. If KL/r exceeds C'_c the column undergoes elastic buckling; and if KL/r is less than C'_c the column yields before buckling and experiences inelastic buckling.

If the perfect column is fabricated from a material similar to that in Fig. 8.7a, the limiting slenderness ratio becomes $(KL/r)_{\lim} = \sqrt{\pi^2 E_y/F_y}$. Elastic buckling loads are still predicted by Euler's equation, $P_{cr}/A = \pi^2 E/(KL/r)^2$ and inelastic buckling loads are predicted by $P_{cr}/A = \pi^2 E_t/(KL/r)^2$, where E_t is the tangent modulus of elasticity at stress level P_{cr}/A.

Hot-rolled steel columns do not behave like perfect columns because of factors such as initial curvature, accidental eccentricities of loading, and residual stresses. The most critical discrepancy between real and perfect columns is the effect of residual stresses.

Residual stresses are caused by differential rates of cooling in formed cross sections. Differential cooling induces large compressive stresses in parts of the section before any external load is applied. The maximum residual stresses, which average 12 ksi (80 MPa) and are as high as 20 ksi (140 MPa), occur near the edges of the flanges and near the center of the webs (see Fig. 8.8). The stress–strain curve and cross sections in Fig. 8.9 illustrate the influence of residual stresses upon the stress–strain behavior of the section. Since the buckling strength of a column is dependent upon the stiffness of the material (the slope of the stress–strain curve) and since the stress–strain curve shape is significantly altered by residual stresses, it follows that residual stresses significantly alter the compressive strength of hot rolled steel columns. The AISC specification alters the Euler and tangent modulus equations for perfect columns to account for the discrepancy.

The Column Research Council (CRC) of AISC investigated the effect

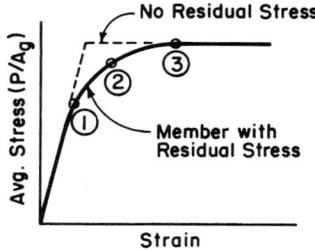

(a.) Stress-Strain Curve for a Real Column

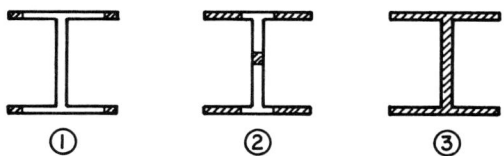

(b.) Cross-Sections of Column in (a.) Shaded Portions Have Yielded at Indicated Stress Levels.

Fig. 8.9. Compressive stress-strain behavior of a hot-rolled section with residual stresses. Reproduced with permission from *Steel Design Manual*, 1980 ed., U. S. Steel Corporation, Pittsburgh, PA.

of residual stresses on steel column strength. The results of their studies on W sections* (other standard structural sections behave similarly) are summarized in Fig. 8.10. They observed, in laboratory tests, columns to have buckling strengths indicated by the curves labeled "strong axis" and "weak axis." The curve labeled the "basic strength curve" is CRC's best fit strength curve for actual columns. Note that it is slightly conservative for strong axis buckling, but nonconservative for weak axis buckling in the inelastic range. The nonconservative estimate for weak axis buckling is more than compensated for by the factors of safety applied in specification requirements. It is important to note that the CRC strength curve is empirical and adequately defines the column strength in the inelastic range. The equation for the CRC curve is $F_c = F_y [1 - (F_y/4\pi^2 E)(KL/r)^2]$.

8.7.1.2 Design Specification. In the elastic buckling range the column strength is still predicted by the Euler equation. The curves of Fig. 8.10 show a smooth transition between the CRC and Euler equa-

---

*A W section is shorthand for a wide flange section.

220 Structural Steel Design

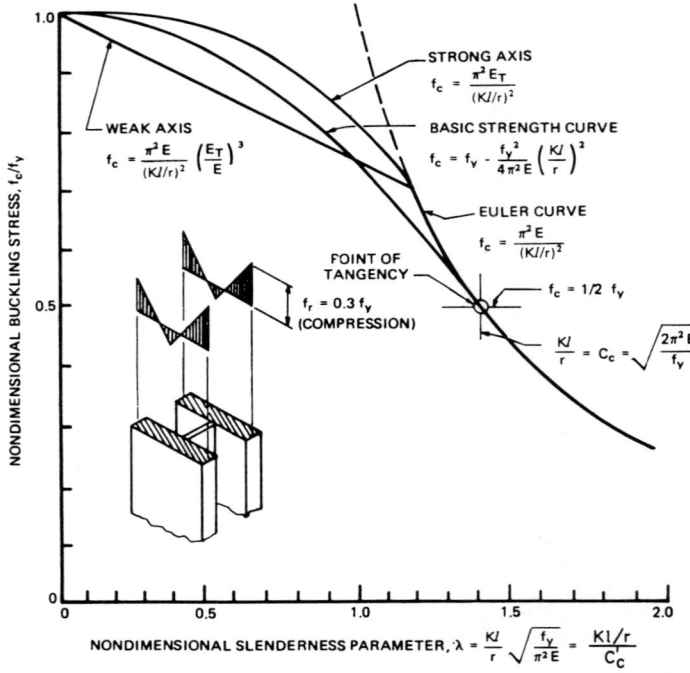

Fig. 8.10. Effect of residual stress on column strength. Reproduced with permission from *Steel Design Manual,* 1980 ed., U. S. Steel Corporation, Pittsburgh, PA.

tions. This smooth transition occurs because the CRC arbitrarily forced the CRC equation to be tangent to the Euler curve at a buckling strength of $\frac{1}{2}$ the yield stress; i.e., when $F_c = P/A_g = 0.5F_y$. The value of the limiting slenderness ratio for elastic buckling is obtained by equating the CRC buckling strength equation to $0.5F_y$, or $F_c = 0.5F_y$. Solving for KL/r, $(KL/r)_{\text{lim}} = \sqrt{2\pi^2 E/F_y} = C_c$.

The limiting slenderness ratio is defined as C_c. If $KL/r \geq C_c$ the column will buckle elastically and Euler's equation governs the design. Conversely, if $KL/r < C_c$, the column will buckle inelastically and the CRC equation predicts the column strength.

The AISC specification for columns consists of the CRC and Euler equations with appropriate factors of safety. Specifically, for main structural members, and secondary members with $KL/r < 120$,

$$F_a = \frac{[1 - (KL/r)^2/2C_c^2]F_y}{5/3 + 3(KL/r)/8C_c - (KL/r)^3/8C_c^3} \qquad \text{if } \frac{KL}{r} < C_c \quad (8.4)$$

and

$$F_a = \frac{12}{23} \frac{\pi^2 E}{(KL/r)^2} \quad \text{if} \quad \frac{KL}{r} \geq C_c \tag{8.5}$$

and

$$F_a = 0 \quad \text{if} \quad \frac{KL}{r} > 200 \tag{8.6}$$

For secondary members, such as those used for bracing, with $120 < KL/r < 300$,

$$F_{a_s} = \frac{F_a}{1.6 - \dfrac{KL}{200r}} \tag{8.7}$$

where F_a is computed as for main members and K equals 1.0.

The factors of safety for short columns increase from 1.67 to 1.92 as KL/r increases from 0 to C_c. Above $KL/r = C_c$, the factor of safety is constant at 1.92.

8.7.2 Effective Slenderness Ratios

The buckling strength of columns is influenced by the degree of restraint provided at the ends. For example, if all other factors are equal, a long column with two fixed ends will buckle at a theoretical load four times as large as will a long column with two pinned ends. The influence of end restraint on column strength is incorporated into the column specification by the effective length factor K. The theoretical and recommended design values of K are summarized in Fig. 8.11 for several ideal columns.

Effective length factors cannot always be estimated by the ideal conditions given in Fig. 8.11. For example, in a single bay gable rigid frame, the connection between the vertical column and the inclined flexural member is rigid, but rotation and translation can occur. The degree of restraint against rotation is dependent upon the relative stiffness of the column to the stiffness of the other members tying into the joint. K values for columns in frames with and without sidesway may be evaluated by using the charts in Fig. 8.12.

In Fig. 8.12, G is defined as the ratio of the relative rigidity of all the columns to the relative rigidity of all the girders or other restraining members tying into the joint. In equation form

$$G = \frac{\Sigma I_c/L_c}{\Sigma I_g/L_g} \tag{8.8}$$

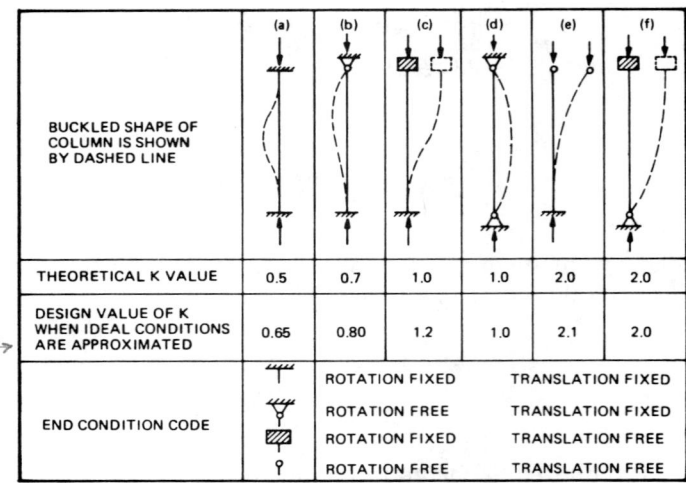

Fig. 8.11. Effective length factors for columns. Reproduced with permission from *Steel Construction Manual,* 8th ed., 1980, AISC, Chicago, IL.

where I and L are, respectively, the moment of inertia and length of the members connecting at the joint; and subscripts c and g refer to the columns and girders, respectively. The subscripts A and B refer to end A and end B of the column. Two special cases are important. If one end of the column is supported but not rigidly connected to a footing, the ideal end condition is a pin and $G = \infty$, but practical design procedures use $G = 10$. If one end of a column is rigidity attached to a footing, the ideal end condition is fixed and $G = 0$. However, use $G = 1.0$ for design purposes. To determine K, calculate G for both ends of the column, select the appropriate nomograph, draw a straight line between G_A and G_B, and read K from the center scale.

8.7.3 Design Aids for A36 Steel Columns

The allowable axial compressive loads for selected wide flange, round pipe, and square tubing shapes are given in Tables 8.1–8.4. These tables are applicable to columns made of A36 steel and are based upon weak axis buckling. That is, the allowable load tabulated for an effective length KL is based upon a slenderness ratio KL/r_y. Thus, the allowable load for an A36 column can be obtained directly from the table if $(KL)_y > (KL)_x$. When $(KL)_x > (KL)_y$, then buckling may occur about the strong axis and the tables cannot be used directly. The tables can be used for strong axis buckling by entering with an effective length $(KL)' = (KL)_x/(r_x/r_y)$. The allowable load for the column is then based

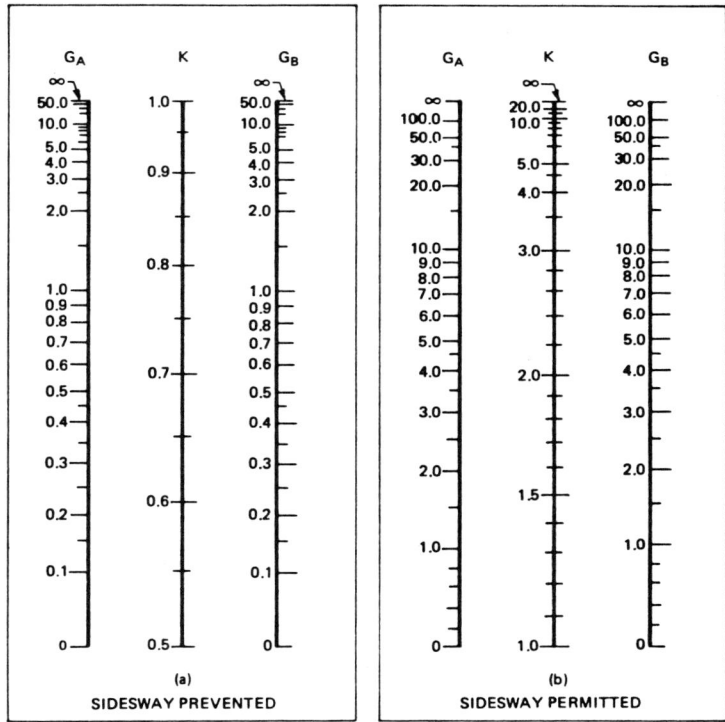

Fig. 8.12. Charts for effective length of columns in continuous frames. Reproduced with permission from *Steel Construction Manual,* 8th ed., 1980, AISC, Chicago, IL.

upon the larger of $(KL)'$ or $(KL)_y$. The tabulated loads are in kips. Load values above the solid line are for columns with $KL/r < 120$. Loads are omitted whenever $KL/r > 200$.

Table 8.5 lists the allowable compressive stress for A36 steel columns as a function of slenderness ratio. The stresses are all in ksi and are listed separately for main and secondary members.

8.7.4 Steel Column Illustrations

Example 8.1. Steel Column Selection

   a. An A36 steel, W 8 × 35* (W 203 × 52) section is used as a 24 ft (7.32 m) long column. Use the allowable load tables to de-

---

*The designation W $X \times Y$ identifies a wide flange section with a nominal depth X and a dead load per unit length, Y. X has units of in. (mm) and Y has units of lb/ft (kg/m).

Table 8.1. Allowable Column Loads (in kip) for A36 W and M Shapes When Buckling Occurs about the Weak Axis.

Name	W8						W6						W5		W4	M8		M6		M5	M4
Width	8	8	6½	6½	5¼	5¼	6	6	6	4	4	4	5	5	4	8	8	6	6	5	4
lb/ft	35	31	28	24	20	17	25	20	15.5	16	12	8.5	18.5	16	12	34.3	32.6	22.5	20	18.9	13
2[a]	—	—	172	148	122	103	153	123	95	96	72	51	112	97	78	212	201	137	122	114	77
3	—	—	168	144	118	100	150	120	93	92	68	48	109	94	74	208	198	134	119	111	74
4	—	—	164	141	114	96	146	117	90	87	64	45	105	91	71	204	194	130	116	106	70
5	—	178	160	137	109	92	142	113	87	81	60	42	101	87	66	200	190	125	112	102	65
6	201	174	155	133	104	88	137	109	84	75	55	38	97	83	62	195	185	120	107	96	60
7	197	169	150	128	98	83	132	105	81	69	50	34	92	79	57	190	180	115	103	91	54
8	191	164	144	123	93	78	126	100	77	62	44	30	86	74	51	184	175	109	98	85	48
9	186	159	138	118	86	72	120	96	73	54	38	25	81	69	45	178	170	103	92	78	42
10	180	154	132	113	79	66	114	91	69	46	31	21	75	64	39	172	164	96	87	71	35
11	174	148	125	107	72	60	108	85	65	38	26	17	68	58	32	165	158	89	81	64	29
12	168	142	118	101	65	53	101	80	60	32	21	14	62	52	27	158	151	82	74	56	24
13	162	136	111	94	56	46	94	74	55	27	18	12	54	46	23	151	144	74	68	48	21
14	155	130	103	88	49	39	86	68	50	23	16	10	47	39	20	143	137	66	61	42	18
15	148	123	95	81	42	34	78	61	45	20	14		41	34	17	135	130	57	53	36	15
16	141	117	86	73	37	30	70	54	39	18			36	30	15	127	122	50	47	32	
17	133	110	78	66	33	27	62	48	35				32	27		118	114	45	41	28	
18	125	102	69	59	29	24	55	43	31				28	24		109	106	40	37	25	
19	117	95	62	53	26	21	49	39	28				26	21		100	98	36	33	23	
20	109	79	56	47	24	19	45	35	25				23	19		91	89	32	30		
22	91	76	46	39			37	29	21							75	73	27	25		
24	76	66	39	33			31	24	17							63	62				
26	65	57	33	28												54	52				
28	56	49														46	45				
30	49	42														40	39				
32	43	37																			

224

Properties

Area																	
A, in.^{2}	10.3	9.12	8.23	7.06	5.89	5.01	7.35	5.88	4.56	4.72	3.54	2.51	5.43	4.70	3.82	10.1	9.58	6.62	5.89	5.55	3.81
I_x, in.^{4}	126.0	110.0	97.8	82.5	69.4	56.6	53.3	41.5	30.1	31.7	21.7	14.8	25.4	21.3	11.3	116.0	114.0	41.2	39.0	39.0	10.5
I_y, in.^{4}	42.5	37.0	21.6	18.2	9.22	7.44	17.1	13.3	9.67	4.42	2.98	1.98	8.89	7.51	3.76	34.9	34.1	12.4	11.6	11.6	3.36
Ratio																	
r_x/r_y	1.72	1.73	2.13	2.12	2.74	2.75	1.76	1.76	1.76	2.68	2.70	2.73	1.69	1.69	1.74	1.83	1.82	1.82	1.84	1.84	1.77
r_y, in.	2.03	2.01	1.62	1.61	1.25	1.22	1.53	1.51	1.46	0.97	0.92	0.89	1.28	1.26	0.99	1.86	1.89	1.37	1.40	1.40	0.94
B_x, in.$^{-1}$	0.332	0.333	0.339	0.340	0.347	0.356	0.441	0.439	0.456	0.463	0.489	0.495	0.547	0.551	0.701	0.348	0.338	0.484	0.454	0.577	0.728
B_y, in.$^{-1}$	0.972	0.988	1.246	1.259	1.683	1.771	1.308	1.328	1.412	2.156	2.376	2.486	1.534	1.567	2.065	1.157	1.117	1.623	1.511	1.768	2.229
$a_x{}^b$	18.8	16.4	14.60	12.30	10.32	8.43	7.92	6.20	4.49	4.72	3.24	2.21	3.77	3.18	1.68	17.40	16.89	6.12	5.80	3.58	1.56
$a_y{}^b$	6.3	5.5	3.22	2.73	1.37	1.11	2.56	2.00	1.45	0.66	0.44	0.30	1.33	1.11	0.56	5.21	5.10	1.85	1.72	1.17	1.50

Reproduced with permission from *Steel Construction Manual*, 7th ed. 1973. AISC, Chicago, IL.
a. This column gives effective length, KL, in feet with respect to least radius of gyration r_y.
b. Multiply values by 10^6

Table 8.2. Allowable Column Load (in kip) for A36 Extra-Strong Pipe.

Diameter (in.)	6	5	4	$3\frac{1}{2}$	3
Thickness (in.)	0.432	0.375	0.337	0.318	0.300
lb/ft	28.5	20.7	14.9	12.5	10.2
6^a	166	118	81	66	52
7	162	114	78	63	48
8	159	111	75	59	45
9	155	107	71	55	41
10	151	103	67	51	37
11	146	99	63	47	<u>33</u>
12	142	95	59	43	28
13	137	91	54	<u>38</u>	24
14	132	86	<u>49</u>	33	21
15	127	81	44	29	18
16	122	76	39	25	16
17	116	71	34	23	14
18	111	<u>65</u>	31	20	12
19	105	59	28	18	11
20	<u>99</u>	54	25	16	
22	86	44	21		
24	73	37	17		
26	62	32			
28	54	27			
30	47	24			
Properties					
Area A, in.^{2}	8.40	6.11	4.41	3.68	3.02
I, in.^{4}	40.5	20.7	9.61	6.28	3.89
r, in.	2.19	1.84	1.48	1.31	1.14
B, in.$^{-1}$	0.688	0.822	1.03	1.17	1.36
a^b	6.00	3.08	1.44	0.941	0.585

Reproduced with permission from *Steel Construction Manual*, 7th ed. 1973. AISC, Chicago, IL.
a. This column gives effective length, KL, in feet, with respect to radius of gyration.
b. Multiply values by 10^6.

termine the allowable compressive load the section can carry if both ends are pinned with respect to both the strong and weak axis and if the column is braced at midspan with respect to the weak axis.

  i. Obtain the effective lengths.

$$(KL)_x = 1.0\,(24) = 24 \text{ ft} \quad (7.32 \text{ m})$$

$$(KL)_y = 1.0\,(24/2) = 12 \text{ ft} \quad (3.66 \text{ m})$$

  ii. Check for the controlling axis for buckling of the section.

Table 8.3. Allowable Column Load (in kip) for A36 Standard Pipe.

Diameter (in.)	6	5	4	$3\frac{1}{2}$	3
Thickness (in.)	0.280	0.258	0.237	0.226	0.216
lb/ft	18.9	14.6	10.7	9.1	7.5
6[a]	110	83	59	48	38
7	108	81	57	46	36
8	106	78	54	44	34
9	103	76	52	41	31
10	101	73	49	38	28
11	98	71	46	35	$\underline{25}$
12	95	68	43	32	22
13	92	65	40	$\underline{29}$	19
14	89	61	36	25	16
15	86	58	$\underline{33}$	22	14
16	82	55	29	19	12
17	79	51	26	17	11
18	75	$\underline{47}$	23	15	10
19	71	43	21	14	9
20	67	39	19	12	
22	$\underline{59}$	32	15	10	
24	51	27	13		
26	43	23			
28	37	20			
30	32	17			
Properties					
Area A, in.^{2}	5.58	4.30	3.17	2.68	2.23
I, in.^{4}	28.1	15.2	7.23	4.79	3.02
r, in.	2.25	1.88	1.51	1.34	1.16
B, in.$^{-1}$	0.657	0.789	0.987	1.12	1.29
a^b	4.21	2.26	1.08	0.717	0.447

Reproduced with permission from *Steel Construction Manual*, 7th ed. 1973. AISC, Chicago, IL.
a. This column gives effective length, KL, in feet with respect to radius of gyration.
b. Multiply values by 10^6.

   Since $r_x/r_y = 1.72 < KL_x/KL_y = 2$, $(KL/r)_x > (KL/r)_y$ and the strong axis controls.
   iii. Enter Table 8.1 with $KL = 24/1.72 = 13.95$ or 14 ft (4.27 m), $P_{all} = 155k$ (689 kN)
b. What is the magnitude of the allowable stress for the steel column in part a?
   i. Evaluate the slenderness ratio.

$$\left(\frac{KL}{r}\right)_{max} = \left(\frac{KL}{r}\right)_x = \frac{24 \times 12}{1.72 \times 2.03} = 82.5$$

   ii. Enter Table 8.5 and interpolate.

Table 8.4 Axial Column Load (in Kip) for A36 Square Structural Tubing.

Size (in.)	6 × 6			5 × 5			4 × 4		
Thickness (in.)	3/8	5/16	1/4	3/8	5/16	1/4	3/8	5/16	1/4
lb/ft	27.0	23.0	18.8	21.9	18.7	15.4	16.8	14.5	12.0
6[a]	157	134	110	124	107	88	91	79	66
7	154	132	108	121	104	86	87	76	63
8	151	129	106	117	101	83	83	72	60
9	147	126	103	113	98	81	79	69	57
10	144	123	101	109	94	78	74	65	54
11	140	120	98	105	91	75	69	61	51
12	136	116	95	101	87	72	64	57	48
13	131	113	93	96	83	69	59	52	44
14	127	109	90	91	79	66	53	47	40
15	122	105	87	86	75	62	47	42	36
16	118	101	83	81	70	59	42	37	32
17	113	97	80	75	66	55	37	33	29
18	108	93	77	69	61	51	33	30	25
19	102	88	73	63	56	47	29	27	23
20	97	84	69	57	51	43	27	24	21
22	85	74	62	47	42	35	22	20	17
24	73	64	54	40	35	30	18	17	14
26	62	54	46	34	30	25			
28	54	47	39	29	26	22			
30	47	41	34	25	22	19			
32	41	36	30						
34	36	32	27						
36	32	28	24						
38		25	21						
Properties									
Area A, in.²	7.95	6.77	5.54	6.45	5.52	4.54	4.95	4.27	3.54
I, in.⁴	40.5	35.5	29.9	22.0	19.5	16.6	10.2	9.23	8.00
r, in.	2.26	2.29	2.32	1.85	1.88	1.91	1.44	1.47	1.50
B, in^{-1}	0.589	0.572	0.556	0.773	0.708	0.684	0.971	0.925	0.885
a^b	6.05	5.29	4.44	3.29	2.91	2.47	1.53	1.37	1.19

Reproduced with permission from *Steel Construction Manual.* 7th ed. 1973. AISC, Chicago, IL.
a. This column gives effective length, KL, in feet with respect to radius of gyration.
b. Multiply values by 10^6.

Table 8.5. Allowable Compressive Stress for A36 Steel Columns.

\multicolumn{4}{c}{Main and Secondary Members, KL/r not over 120}				\multicolumn{4}{c}{Main Members, KL/r 121 to 200}				\multicolumn{4}{c}{Secondary members,[a] L/r 121 to 200}			
KL/r	F_a (ksi)	KL/r	F_a (ksi)	KL/r	F_a (ksi)	KL/r	F_a (ksi)	L/r	F_{as} (ksi)	L/r	F_{as} (ksi)
5	21.39	45	18.78	85	14.79	125	9.55	165	5.49	125	9.80	165	7.08
10	21.16	50	18.35	90	14.20	130	8.84	170	5.17	130	9.30	170	6.89
15	20.89	55	17.90	95	13.60	135	8.19	175	4.88	135	8.86	175	6.73
20	20.60	60	17.43	100	12.98	140	7.62	180	4.61	140	8.47	180	6.58
25	20.28	65	16.94	105	12.33	145	7.10	185	4.36	145	8.12	185	6.46
30	19.94	70	16.43	110	11.67	150	6.64	190	4.14	150	7.81	190	6.36
35	19.58	75	15.90	115	10.99	155	6.22	195	3.93	155	7.53	195	6.28
40	19.19	80	15.36	120	10.28	160	5.83	200	3.73	160	7.29	200	6.22

Reproduced with permission from *Steel Construction Manual*, 8th ed. 1980. AISC, Chicago IL.
a. K taken as 1.0 for secondary members.

230 Structural Steel Design

$$F_a = 15.08 \text{ ksi} \quad (104 \text{ MPa})$$

    iii. The allowable stress agrees with the result from part a. That is, $F_c = P_{all}/A = 155/10.3 = 15.05$ ksi (104 MPa)

c. What is the allowable stress and load for the column in part a if steel with a yield stress, F_y, equal to 50 ksi (340 MPa) is used in place of A36 steel?

    i. The specification equations must be used since design aids are not given for this steel.

    ii. Classify the column

$$C_c = \sqrt{2\pi^2 E/F_y} = \sqrt{2\pi^2 (29 \times 10^3)/50} = 107$$

Since $KL/r < C_c$ the column is intermediate.

    iii. Evaluate F_a with Eq. (8.4)

$$F_a = \frac{F_y[1 - (KL/r)^2/2C_c^2]}{\frac{5}{3} + 3(KL/r)/8C_c - (KL/r)^3/8C_c^3}$$

$$= \frac{50[1 - 82.5^2/2(107)^2]}{\frac{5}{3} + \frac{3}{8}(82.5/107) - \frac{1}{8}(82.5/107)^3}$$

$$F_a = 18.48 \text{ ksi} \quad (127 \text{ MPa})$$

d. What size square structural A36 tubing is required to carry a compressive load of 60 k (267 kN) if the effective lengths are $KL_x = 24$ ft (7.32 m) and $KL_y = 12$ ft (3.66 m)?

    i. Since $r_x/r_y = 1.0$, KL_x is the controlling slenderness ratio.

    ii. Enter Table 8.4 with $KL = 24$ ft (7.32 m). Use 6 × 6 × 5/16 in. (152 × 152 × 7.9 mm).

e. What size A36 round pipe is required to carry a compressive load of 80 k (356 kN) if $KL_x = 22$ ft (7.32 m) and $KL_y = 12$ ft (3.66 m)?

    i. Enter Table 8.3 with $KL = 22$ ft (7.32 m). No standard pipe qualifies.

    ii. Enter Table 8.2 with $KL = 22$ ft (7.32 m). Use 0.432 in. × 6 in. diameter (11.0 × 152 mm) extra strong pipe.

f. A steel column, 16 ft (4.88 m) long, must carry a compressive load of 80 k (356 kN). If both the top and bottom are fixed with

respect to both the strong and weak axis, select the lightest A36, W section which can carry the load.
i. Evaluate the effective lengths and controlling axis

$$KL_x = KL_y = 0.65(16) = 10.40 \text{ ft} \quad (3.17 \text{ m})$$

since $r_y < r_x$, KL_y controls.
ii. Enter tables with $KL = 10.4$ ft (3.17 m) and select a W 6 × 20 (W 152 × 30).

g. How would the solution to problem f change if the weak axis were supported laterally at midspan?
i. Evaluate the effective lengths:

$$KL_x = 10.4 \text{ ft} \quad (3.17 \text{ m})$$
$$KL_y = 0.8\,(8) = 6.4 \text{ ft} \quad (1.95 \text{ m})$$

ii. Since $KL_x > KL_y$, compare KL_x/KL_y to r_x/r_y for the section.

$$\frac{KL_x}{KL_y} = \frac{10.4}{6.4} = 1.63$$

Since $r_x/r_y > 1.63$ for most sections (Table 8.1), it is likely that $(KL/r)_y > (KL/r)_x$ and the weak axis will control the allowable load.

iii. Selecting the column size based on KL_y, enter table 8.1 with $KL = 6.4$ ft (1.95 m), the allowable load for a W 5 × 16 (W 127 × 24) equals 81 k (360 kN) and the allowable load for a W 6 × 15.5 (W 152 × 23) equals 83 k (369 kN). Thus, use a W 6 × 15.5 (W 152 × 23) section.

iv. Check the slenderness ratio and allowable load for the section selected. Since $(KL/r)_y = 52.6$ and $(KL/r)_x = 48.6$, the weak axis does control. From the stress table, $F_a = 18.13$ psi (120 MPa) and the allowable load equals 82.67 k (368 kN). Thus, the solution checks.

Example 8.2 Column Design. Select the lightest weight A36 W or M section that can be used to carry an axial compressive load of 100 k (440 kN) if the column is 10 ft (3.04 m) long, fixed at both the top and bottom ends with respect to the weak axis and fixed at the bottom

232 Structural Steel Design

and pinned at the top with respect to the strong, and braced at the third points to prevent bending about the weak axis of the column.

Solution

a. The effective lengths about the weak axis are
$(KL)_y = 0.8 \,(0.33 \times 10) = 2.67$ ft (0.81 m)
for the two end sections and
$(KL)_y = 0.33(10) = 3.33$ ft (1.01 m)
for the middle section. The strong axis effective length equals
$(KL)_x = 0.8(10) = 8.0$ ft (2.48 m)

b. Whether weak or strong axis bending controls the design depends upon the relative KL/r values. Since $(KL)_x/(KL)_y = 2.4$ and r_x/r_y is less than 2.4 for many of the W and M sections in Table 8.1, it appears that strong axis buckling will control.

c. Taking an average value for r_x/r_y of 2.0, enter the column load tables with $KL = (KL)_x/(r_x/r_y) = 4$ ft (1.22 m) and $P = 100$ kip (440 kN). A W 5×18.5 (W 127×28) appears to be adequate, but $r_x/r_y = 1.69 < 2.0$ and must be checked.

d. To check the section, enter the load tables with $KL = 8/1.69 = 4.73$ ft (1.44 m) and $P = 100$ kip (440 kN). Since the selected section can carry 101 kip (450 kN) when KL = 5.0 ft (1.52 m), the section is adequate.

e. Check the allowable load for the section using stress tables. Since

$(KL/r)_y = 3.33 \times \dfrac{12}{1.28} = 31.3$ and $(KL/r)_x =$

$\dfrac{(8 \times 12)}{(1.69 \times 1.28)} = 44.38,$

the strong axis controls. From the table of allowable stresses, $F_a = 18.83$ ksi (130 MPa)
and
$P_{all} = F_a A = 18.83 \,(5.43) = 102$ kip (450 kN)
> 100 kip (440 kN).

f. Check solution using the AISC equations.

Since $C_c = \sqrt{2\pi^2 E/F_y} = \sqrt{2\pi^2(29 \times 10^6)/36} = 126.1,$

and $(KL/r)_{max} = 44.4$, inelastic buckling controls the load capacity and

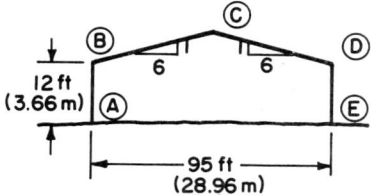

Fig. 8.13. Example 8.3.

$$F_a = \frac{[1 - \frac{(44.4)^2}{2(126.1)^2}]36}{\frac{5}{3} + \frac{3(44.4)}{8(126.1)} - \frac{(44.4)^3}{8(126.1)^3}} = 18.80 \text{ ksi} \quad (130 \text{ MPa})$$

Example 8.3. Column Design. The A36 steel rigid frame given in Fig. 8.13 is used in a commercial fertilizer storage facility. The frames are spaced 25 ft (7.62 m) o.c. (on center) and carry a design $DL + SL$ of 25 psf (1.2 kPa). The vertical members are W sections oriented such that strong axis bending takes place in the plane of the frame. If the bottom end of column AB is pinned with respect to both the strong and weak axis, if the top of AB is completely fixed with respect to weak axis bending, if the top of AB is rigidly attached to flexural member BC with respect to strong axis bending, and if there are no intermediate bracing members between A and B, then

a. Determine the pure compressive load (neglecting flexural loads) which the column can carry if BC is a W 5 × 16 (W 127 × 24) section and AB is a W 6 × 20 (W 152 × 30) section.
b. Determine if the column section is large enough to carry the pure compressive loads in the absence of flexure.
c. Determine the lightest weight W section from those in Table 8.1 which can be used to carry the compressive loads in column AB.

Solution

a. The allowable compressive load in AB.
   i. The weak axis effective length, $KL_y = 0.8(12) = 9.6$ ft (2.93

m) and $(KL/r)_y = 76.3$. Bending about the strong axis of AB is characterized as a continuous frame with sidesway permitted. Entering Fig. 8.12 with $G_A = \infty$ (use 10 for the pinned end) and $G_B = I_c L_g / I_g L_c = 8.12$, $K = 2.9$. Thus, $KL_x = 34.8$ ft (10.6 m) and $(KL/r)_x = 157$ and the strong axis controls the allowable load. Entering Table 8.1 with $KL = (KL_x)/(r_x/r_y) = 19.8$ ft (6.04 m), the allowable load equals 35 k (156 kN).

b. The column AB must carry a compressive load equal to $25(47.5)(25) = 29.7$ k (132 kN). Thus the W 6 × 20 (W 152 × 30) is adequate. Note that it may not be adequate to carry the combined compression and flexural loads in AB. Combined loading will be treated in a later section.

c. The lightest W section to carry the compressive load in AB is found as follows:

   i. Since KL_x depends upon relative values of I_g and I_c, as a first estimate assume the weak axis controls. Entering Table 8.1 with $KL = 9.6$ ft (2.93 m) and $P = 29.7$ k (132 kN), there are eight possible sections. One of the more likely candidates appears to be the W 6 × 15.5 (W 152 × 23) section.

   ii. To check the W 6 × 15.5 (W 152 × 23) section for strong axis buckling, evaluate $G_A = 10$ (pin) and $G_B = (30.1/12)/(21.3/50) = 5.9$. Entering Fig. 8.12, $K = 2.65$ and $KL_x = 31.8$ ft (9.69 m). Entering Table 8.1 with $KL = KL_x/(r_x/r_y) = 18.1$ ft (55.2 m), the allowable load equals 30.7 k (136 kN). Thus the section is adequate.

   iii. It is easily shown that the load capacity of the W 6 × 12 (W 152 × 18), W 4 × 12 (W 102 × 18), and M 4 × 13 (M 102 × 19) sections are all inadequate. Verify this as a student exercise.

   iv. Thus, a W 6 × 15.5 (W 152 × 23) is the lightest adequate section for member AB.

8.8 FLEXURAL MEMBER DESIGN

The governing equation for flexural design of steel beams is $M_{all} \leq F_b S$. The allowable bending stress is dependent upon the ultimate yield strengths of the material and the buckling characteristics of the beam cross section. The allowable stress is influenced primarily by the type section (compact or noncompact), local buckling, lateral bracing, and the shape factor of the cross section.

There are two basic classifications of steel beam cross sections—compact and noncompact sections. A compact section is proportioned such that the cross section reaches its full plastic moment M_p before either local buckling of the elements or lateral torsional buckling occurs. Conversely, a noncompact cross section will experience either local buckling or lateral torsional buckling before the plastic moment is developed.

The discussion herein is restricted to steels with yield stresses in the 36 and 50 ksi (250 and 340 MPa) range. The allowable stresses discussed are not valid for hybrid beams. Unless otherwise specified all flexural loads are applied through the shear center of the beam cross section.

8.8.1 Allowable Flexural Stresses*

8.8.1.1 Compact Sections. The allowable flexural stress for a compact section, symmetric about and loaded in the plane of its minor axis, is two-thirds the yield stress—$F_b = 0.66F_y$. The requirements for compactness are moderately severe. Most standard sections of A36 steel, however, do meet the compact section requirements if the beam is adequately braced to prevent lateral torsional buckling.

A section is compact if the following are true:

1. The flanges are continuously connected to the webs.
2. The flange width-to-thickness ratio, b/t, of UCEs (unstiffened compression elements) is less than or equal to $65/\sqrt{F_y}$.
3. The flange width-to-thickness ratio, b/t, of SCEs (stiffened compression elements) is less than or equal to $190/\sqrt{F_y}$.
4. The depth-to-thickness ratio, d/t, of the webs is less than $(640)/\sqrt{F_y})(1 - 3.74 f_a/F_y)$ if $f_a/F_y \leq 0.16$ or $257/\sqrt{F_y}$ if $f_a/F_y > 0.16$.
5. The compression flange of the beam is laterally supported at intervals, L_b, less than both $L_c = (76b_f/\sqrt{F_y})$ and $20,000/[(d/A_f)(F_y)]$. In the case of box beams whose depth is less than six times the width, and whose flange thickness is less than twice the web thickness, the compression flange is braced at intervals less than $(1950 + 1200M_1/M_2)(b/F_y)$.

*Except in the case of dimensionless equations or ratios, the equations in this section are restricted to English (customary) units of in., in.², and ksi.

6. The diameter-to-thickness ratio of hollow circular sections is less than $3300/F_y$.

Condition 1 assures continuity between the elements of the section. The second and third conditions assure that the flanges will not buckle locally before the plastic moment is reached, while condition 4 proportions the section such that the plastic moment is reached before the webs buckle locally. Requirement 5 assures that the beam is braced sufficiently so the plastic moment is developed before the beam buckles laterally.

8.8.1.2 Nearly Compact Sections. If a cross section is compact *except* that b/t of unstiffened flanges is between $65/\sqrt{F_y}$ and $95/\sqrt{F_y}$, then $F_b = 0.79 - 0.002(b/t)\sqrt{F_y}$. Using this equation, F_b decreases from $0.66F_y$ to $0.60F_y$ as the UCE becomes more slender. This special case allows the designer to avoid an abrupt reduction in allowable stress when b/t exceeds $65/\sqrt{F_y}$.

8.8.1.3 Minor Axis Bending and Solid Sections. Solid rectangular sections bent about their minor axis, doubly symmetric I and H sections bent about their minor axis, and solid round or square cross sections have large shape factors. That is, the ratio of their plastic-to-elastic moment carrying capacity, M_p/M_y, is large—within the range of 1.5 to 1.7. Such sections are also very stable and the possibility of lateral buckling is remote. Thus, if the first two criteria for compactness are met, these sections have an allowable stress, $F_b = 0.75F_y$. If all the criteria for compactness of doubly symmetric I and H sections are met except b/t lies between $65/\sqrt{F_y}$ and $95/\sqrt{F_y}$, then $F_b = F_y \times [1.075 - 0.0005 (b/t)\sqrt{F_y}]$. Depending upon the slenderness of the UCEs, F_b varies between $0.60F_y$ and $0.75F_y$ for these sections.

8.8.1.4 Box-Type Sections. Box sections are relatively stable sections constructed entirely of stiffened elements and are less subject to local buckling and lateral buckling than most other structural shapes. The allowable flexural stress of these sections when bent about the minor axis equals $0.66 F_y$ provided conditions 1, 3, and 4 for compactness are met. If $190/\sqrt{F_y} \leq b/t \leq 238/\sqrt{F_y}$ (condition 4 for compactness), $\leq d_w/t_w \leq 253/\sqrt{F_y}$, and if the depth is less than six times the section width, F_b equals $0.60 F_y$.

8.8.1.5 Noncompact Sections. If a noncompact section has adequate bracing ($L_b \leq L_c = 76b/\sqrt{F_y}$), a width–thickness ratio of UCEs less than $95.0/\sqrt{F_y}$ and a width–thickness ratio of SCEs less than $253/\sqrt{F_y}$, then $F_b = 0.60F_y$. These conditions ensure that the beam

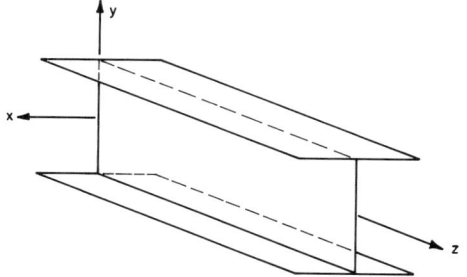

Fig. 8.14. Axis orientation for a W section.

will neither buckle locally nor laterally before the beam moment reaches the yield moment.

If a section is noncompact due to inadequate bracing, L_b is greater than L_c. In this case the beam buckles laterally before reaching the plastic moment, and F_b is always less than or equal to $0.6F_y$. The basic procedure for determining the allowable stress for these sections is to (1) evaluate F_b for pure Euler-type buckling of the compression flange; (2) evaluate F_b for pure St. Venant-type buckling (torsional); and (3) use the larger F_b.

Euler-type buckling (warping) consists of a translation of the compression flange in the x direction and twisting about the y axis (see Fig. 8.14). Resistance to warping is provided by the material stiffness and slenderness ratio of the compression flange plus one third of the compression web area with respect to the x axis (see Fig. 8.15). Torsional buckling consists of twisting about the z axis (St. Venant buckling). Resistance to St. Venant buckling is dependent upon the torsional rigidity and polar moment of inertia of the section.

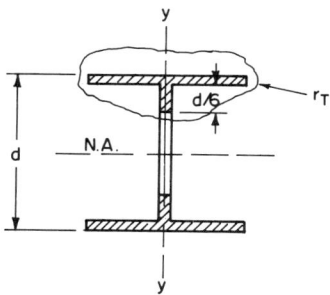

Fig. 8.15. Definition sketch for r_T. Table 8.7 gives r_T, for the "circled" portion of compression flange.

The allowable flexural stress level for inelastic Euler-type buckling of the compression flange is

$$F_b = \left[\frac{2}{3} - \frac{F_y(L_b/r_T)^2}{1530 \times 10^3 C_b}\right](F_y) \qquad (8.9)$$

and is used if

$$\left(\frac{102 \times 10^3 C_b}{F_y}\right)^{1/2} \leq \frac{L_b}{r_T} \leq \left(\frac{510 \times 10^3 C_b}{F_y}\right)^{1/2}$$

where L_b/r_T is the slenderness ratio of the compression flange plus one third of the area of the compression web about the x axis (see Fig. 8.15). L_b is the length between lateral supports along the compression flange.

The allowable flexural stress level for elastic Euler-type buckling of the compression flange is

$$F_b = \left[\frac{170 \times 10^3 C_b}{(L_b/r_T)^2}\right] \qquad (8.10a)$$

and is used if

$$\frac{L_b}{r_T} > \left(\frac{510 \times 10^3 C_b}{F_y}\right)^{1/2}$$

If the compression area of a section is approximately rectangular and if the compression areas are greater than or equal to the tension areas, the allowable flexural stress level for torsional buckling of the compression flange is

$$F_b = \frac{12 \times 10^3 C_b}{(d/A_f)L_b} \qquad (8.10b)$$

The allowable stress in bending for noncompact, inadequately braced sections is to be taken as the larger of the values from Eqs. (8.9), (8.10a), or (8.10b), but not larger than $0.60F_y$. Equations (8.9), (8.10a), and (8.10b) may be used for any section having an axis of symmetry in and loaded in the plane of the web. For channels the allowable stress is computed by Eq. (8.10b) only.

Table 8.6. Summary of Allowable Flexural Stresses for Hot Rolled Beams.

Type Member	F_b	Remarks
Special compact	$0.75F_y$	Sections with large shape factors ($M_p = 1.5$ to $1.7M_y$); solid rectangular sections; minor axis bending
Compact	$0.66F_y$	Sections that develop full plastic moment before buckling locally or laterally; $L_b \leq L_c$
Nearly compact and adequately braced	$0.66F_y$ to $0.60F_y$	Flanges may buckle locally before developing full plastic moment, but not before developing yield moment; braced such that $L_b \leq L_c$ so plastic moment developed before buckling laterally
Compact but only partially braced	$0.60F_y$	Braced such that $L_c < L_b \leq C_b L_u$; section can develop full plastic moment before buckling locally; section can develop full yield moment before buckling laterally, but buckles laterally before developing full plastic moment
Noncompact and inadequately braced	$<0.60F_y$ (Eq. 8.9, 8.10a, or 8.10b)	Section buckles laterally before developing full yield moment; braced such that $L_b > L_c$ or L_u

8.8.1.6 Commentary. The simplified summary of the variation of F_b for common structural shapes in Table 8.6 illustrates that F_b varies from $0.75F_y$ to less than $0.60F_y$. The highest allowable stresses are for sections which are very stable locally and torsionally and thus can achieve their plastic moment before becoming unstable. The lowest allowable stresses are for slender, relatively unstable sections which are unbraced and will buckle locally or laterally before reaching their yield moment.

The definition of and relationship between L_b, L_c, and L_u are extremely important to flexural design of steel members. To clarify the relationship note that (1) L_b is the actual length between lateral supports; (2) L_c is the maximum allowable distance between lateral supports for which the section can develop its full plastic moment without buckling laterally and equals the smaller of $76b_f/\sqrt{F_y}$ and $20,000/[(d/A_f)(F_y)]$; and (3) L_u is the maximum allowable distance between lateral supports for which the section can develop its full elastic moment without buckling laterally and equals the larger value of L_b evaluated from Eqs. (8.9) and (8.10b) for $F_b = 0.60F_y$ and $C_b = 1.0$. Thus, if $L_b \leq L_c$ and the section meets the other compact criteria, the

240 Structural Steel Design

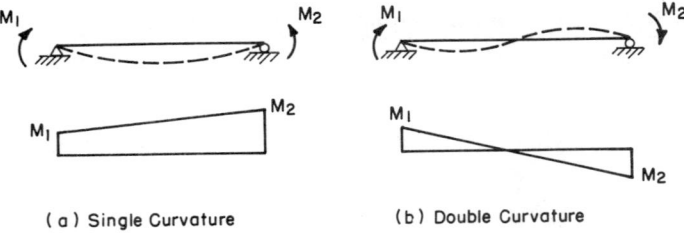

Fig. 8.16. Definition sketch for C_b.

section can develop M_p without buckling laterally, and $F_b = 0.66F_y$; if $L_c < L_b \leq C_b L_u$, the section can develop M_y without buckling laterally and $F_b = 0.60F_y$; and if $L_b > C_b L_u$, the section buckles laterally before developing M_e and $F_b < 0.60F_y$.

The tendency for a beam to buckle laterally is somewhat dependent upon the variation of moment between bracing points. For example, the flange of an unbraced length of a beam subjected to a constant moment with no intermediate loads has a constant compressive stress along its entire length. In contrast, the flange of an unbraced length of a beam subjected to moments as in Fig. 8.16b is compressed to the maximum stress level at only one point along its length. Qualitatively, it is apparent that, since compressive stresses are the forces which induce lateral buckling, the constant moment situation is the more severe one.

The term C_b in the specification equations allows the designer to compensate for the relative magnitudes of moment variations in lateral buckling considerations. C_b is defined by the relationship

$$C_b = 1.75 + 1.05(M_1/M_2) + 0.3(M_1/M_2)^2 \qquad (8.11)$$

where M_1 and M_2 are bending moments at the ends of an unbraced beam section, $|M_1| < |M_2|$; M_1/M_2 is positive when the section is bent in double curvature; and M_1/M_2 is negative when the section is bent in single curvature.

Limitations on C_b are that it shall never exceed 2.3 and that C_b shall be taken as 1.0 if the absolute value of an interior moment between the ends of the braced segment is greater than the absolute value of both M_1 and M_2.

The sketches in Fig. 8.16 define the differences between single curvature and double curvature. When an unbraced section is bent in double curvature, the section is said to have undergone a moment

8.8 Flexural Member Design

reversal. The moment diagrams in Fig. 8.16 also demonstrate that in single curvature a flange will be subjected to large compressive stresses over a longer length of beam than when bent in double curvature.

8.8.2 Shear Stresses

The allowable shear stress for a flexural member without web stiffeners between bearing points is $F_v = 0.4F_y$. The actual shear stress in the beam is evaluated by use of the basic strength of materials relationship $f_v = VQ/(It)$. However, for standard structural steel shapes, the shear stress can be approximated with adequate accuracy by the average shear stress over the web area. Thus $f_v \approx V/(d_w t_w)$ and the shear design criterion is $f_v \leq F_v$.

Web stiffeners are required in beams which have very slender webs. They assure that a slender web may be stressed to $0.4F_y$ before buckling locally. If the depth–thickness ratio, d_w/t_w, of the web is lower than 62.6 and 53.7 for steels with yield stresses of 36 and 50 ksi, respectively, the web does not require stiffeners. Most applications in agricultural structures will meet this requirement.

8.8.3 Bearing Stresses

Web crippling, localized yielding, or buckling due to transverse compression may occur at the location of end reactions or concentrated loads unless either web stiffeners or properly sized bearing areas are provided. In agricultural applications the magnitudes of the loads and size beams normally encountered do not require web stiffeners even at concentrated loads. The only requirement for most applications is that the bearing plate dimensions satisfy the following relationships: (1) at interior concentrated loads $R/[t_w(N + 2k)] \leq 0.75F_y$; and (2) at end reactions $R/[t_w(N + k)] \leq 0.75F_y$. The term k is the distance from the outer face of the flange of the web toe of the fillet and is defined in Fig. 8.17.

8.8.4 Deflection

Ordinary methods of elastic analysis are acceptable for evaluating deflections in steel beams. Allowable deflections may be found in the appropriate building or structural codes governing the design.

Example 8.4. Allowable Flexural Stresses in Beams. Determine the allowable bending stress for an A36 W 8 × 28 (W 203 × 42) simply

242 Structural Steel Design

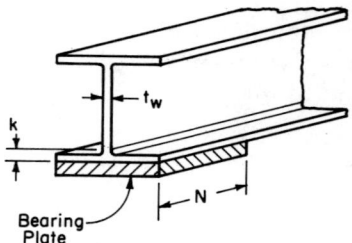

Fig. 8.17. Definition sketch for k and N in bearing stress equation.

supported steel beam if it is unsupported laterally and has a span L_b of (a) 6 ft (1.83 m), (b) 10 ft (3.05 m), (c) 15 ft (4.57 m), (d) 20 ft (6.10 m), and (e) 30 ft (9.14 m). The section is bent about its major axis and the loads are applied through the shear center. Let $C_b = 1.0$.

Solution. The section properties for the beam section are found in Table 8.7. These and some calculated properties are

$$b_f = 6.54 \text{ in. (166 mm)}$$

$$b = b_f/2 = 3.27 \text{ in. (83 mm)}$$

$$r_t = 1.80 \text{ in. (46 mm)}$$

$$\frac{d}{A_f} = 2.66 \text{ in.}^{-1} \; (0.10 \text{ mm}^{-1})$$

$$\frac{b}{t} = \frac{b_f}{2t_f} = 7.06 < \left(\frac{b}{t}\right)_{\lim} = \frac{65}{\sqrt{F_y}} = 10.83$$

$$\frac{d_w}{t_w} = 28.3 < \left(\frac{d}{t}\right)_{\lim} = \frac{640}{\sqrt{F_y}} = 106.7$$

$$L_c = \frac{76 b_f}{\sqrt{F_y}} = \frac{76(6.54)}{\sqrt{36}} = 83.1 \text{ in. } (2.11 \text{ m})$$

or

$$L_c = \frac{20{,}000}{(d/A_f)(F_y)} = \frac{20{,}000}{2.66(36)} = 208 \text{ in. } (5.28 \text{ m})$$

From Eqs. (8.9) and (8.10b),

$$L_u = \frac{12{,}000C_b}{(d/A_f)(0.6F_y)} = 208 \text{ in.} \quad (5.28 \text{ m})$$

$$\text{or } L_u = r_T\sqrt{107 \times 10^3 C_b/F_y} = 98 \text{ in.} \quad (2.49 \text{ m})$$

Using the smaller L_c value, and the larger L_u value, $L_c = 6.9$ ft (2.11 m) and $L_u = 17.4$ ft (5.28 m). If $0 < L_b < 6.9$ ft (2.11 m), the section satisfies all the criteria for compact sections.

a. If $L_b = 6$ ft (1.83 m) $< L_c$, then $F_b = 0.66\, F_y = 24$ ksi (170 MPa).
b. If $L_b = 10$ ft (3.05 m), then $L_c < L_b < L_u$ and $F_b = 0.60\, F_y = 22$ ksi (150 MPa).
c. If $L_b = 15$ ft (4.57 m), then $L_c < L_b < L_u$ and $F_b = 22$ ksi (150 MPa).
d. If $L_b = 20$ ft (6.10 m), $L_b > L_u$, and $F_b \leq 0.6\, F_y$. Using Eqs. (8.9), (8.10a), and (8.10b), $L_b/r_t = (20 \times 12)/1.8 = 133.3$, and $C_b = 1.0$, the limiting parameters are

$$\sqrt{102 \times 10^3 C_b/F_y}$$
$$= 53\sqrt{C_b} = 53 \quad \text{and} \quad \sqrt{510 \times 10^3 C_b/F_y} = 119.$$

Therefore,

$$F_b = \frac{170 \times 10^3 C_b}{(L_b/r_t)^2} = \frac{170 \times 10^3}{(133.3)^2} = 9.6 \text{ ksi} \quad (70 \text{ MPa})$$

or

$$F_b = \frac{12 \times 10^3}{240(2.66)} = 18.8 \text{ ksi} \quad (130 \text{ MPa})$$

The allowable stress is the greater of these, or 18.8 ksi (130 MPa).

e. If $L_b = 30$ ft (9.14 m) $> L_u$, then $F_b < 0.6\, F_y$ and $L_b/r_t = 30 \times 12/1.80 = 200 > 119$. Therefore, $F_b = 170{,}000/(200)^2 = 4.25$ ksi (30 MPa) or $F_b = 12{,}000/240(2.66) = 12.53$ ksi (90 MPa). The allowable stress is the greater of these, or 12.53 ksi (90 MPa).

8.8.5 Design Aids

Numerous design aids are available for structural steel. Steel fabricators are a source of design aids as is the AISC. The AISC *Manual of*

Table 8.7. Section Properties for Designing.

Designation	Area A (in.²)	Depth d (in.)	Flange Width b_f (in.)	Flange Thickness t_f (in.)	Web Thickness t_w (in.)	k (in.)	Axis X-X I (in.⁴)	Axis X-X S (in.³)	Axis X-X r (in.)	Axis Y-Y I (in.⁴)	Axis Y-Y S (in.³)	Axis Y-Y r (in.)	r_T (in.)	$\dfrac{d}{A_f}$	Compact $\dfrac{b}{2t_f}$	Compact $\dfrac{d}{t_w}$
W Shapes																
W 12 × 36	10.6	12.24	6.565	0.540	0.305	1.06	281	46.0	5.15	25.5	7.77	1.55	1.77	3.45	6.08	40.1
12 × 31	9.13	12.09	6.525	0.465	0.265	1.00	239	39.5	5.12	21.6	6.61	1.54	1.75	3.98	7.02	45.6
12 × 27	7.95	11.96	6.497	0.400	0.237	0.93	204	34.2	5.07	18.3	5.63	1.52	1.74	4.60	8.12	50.5
W 12 × 22	6.47	12.31	4.030	0.424	0.260	0.93	156	25.3	4.91	4.64	2.31	0.847	1.03	7.20	4.75	47.3
12 × 19	5.59	12.16	4.007	0.349	0.237	0.87	130	21.3	4.82	3.76	1.88	0.820	1.01	8.70	5.74	51.3
12 × 16.5	4.87	12.00	4.000	0.269	0.230	0.81	105	17.6	4.65	2.88	1.44	0.770	0.975	11.2	7.43	52.2
12 × 14	4.12	11.91	3.968	0.224	0.198	0.75	88.0	14.8	4.62	2.34	1.18	0.754	0.957	13.4	8.86	60.2
W 10 × 45	13.2	10.12	8.022	0.618	0.350	1.18	249	49.1	4.33	53.2	13.3	2.00	2.21	2.04	6.49	28.9
10 × 39	11.5	9.94	7.990	0.528	0.318	1.12	210	42.2	4.27	44.9	11.2	1.98	2.19	2.36	7.57	31.3
10 × 33	9.71	9.75	7.964	0.433	0.292	1.00	171	35.0	4.20	36.5	9.16	1.94	2.16	2.83	9.20	33.4
W 10 × 29	8.54	10.22	5.799	0.500	0.289	1.06	158	30.8	4.30	16.3	5.61	1.38	1.57	3.52	5.80	35.4
10 × 25	7.36	10.08	5.762	0.430	0.252	1.00	133	26.5	4.26	13.7	4.76	1.37	1.56	4.07	6.70	40.0
10 × 21	6.20	9.90	5.750	0.340	0.240	0.87	107	21.5	4.15	10.8	3.75	1.32	1.53	5.06	8.46	41.3
W 10 × 19	5.61	10.25	4.020	0.394	0.250	0.93	96.3	18.8	4.14	4.28	2.13	0.874	1.05	6.47	5.10	41.0
10 × 17	4.99	10.12	4.010	0.329	0.240	0.87	81.9	16.2	4.05	3.55	1.77	0.844	1.03	7.67	6.09	42.2
10 × 15	4.41	10.00	4.000	0.269	0.230	0.81	68.9	13.8	3.95	2.88	1.44	0.809	1.00	9.29	7.43	43.5
10 × 11.5	3.39	9.87	3.950	0.204	0.180	0.75	52.0	10.5	3.92	2.10	1.06	0.787	0.975	12.2	9.68	54.8
W 8 × 35	10.3	8.12	8.027	0.493	0.315	1.00	126	31.1	3.50	42.5	10.6	2.03	2.22	2.05	8.14	25.8
8 × 31	9.12	8.00	8.000	0.433	0.288	0.93	110	27.4	3.47	37.0	9.24	2.01	2.21	2.31	9.24	27.8
8 × 28	8.23	8.06	6.540	0.463	0.285	0.93	97.8	24.3	3.45	21.6	6.61	1.62	1.80	2.66	7.06	28.3
8 × 24	7.06	7.93	6.500	0.398	0.245	0.87	82.5	20.8	3.42	18.2	5.61	1.61	1.78	3.07	8.17	32.4

W	8 × 20	5.89	8.14	5.268	0.378	0.248	0.87	69.4	17.0	3.43	9.22	3.50	1.25	1.42	4.09	6.97	32.8
	8 × 17	5.01	8.00	5.250	0.308	0.230	0.81	56.6	14.1	3.36	7.44	2.83	1.22	1.40	4.95	8.52	34.8
W	8 × 15	4.43	8.12	4.015	0.314	0.245	0.81	48.1	11.8	3.29	3.40	1.69	0.876	1.04	6.44	6.39	33.1
	8 × 13	3.83	8.00	4.000	0.254	0.230	0.75	39.6	9.90	3.21	2.72	1.36	0.842	1.02	7.87	7.87	34.8
	8 × 10	2.96	7.90	3.940	0.204	0.170	0.68	30.8	7.80	3.23	2.08	1.06	0.839	1.00	9.83	9.66	46.5
W	6 × 25	7.35	6.37	6.080	0.456	0.320	0.93	53.3	16.7	2.69	17.1	5.62	1.53	1.69	2.30	6.67	19.9
	6 × 20	5.88	6.20	6.018	0.367	0.258	0.87	41.5	13.4	2.66	13.3	4.43	1.51	1.66	2.81	8.20	24.0
	6 × 15.5	4.56	6.00	5.995	0.269	0.235	0.75	30.1	10.0	2.57	9.67	3.23	1.46	1.53	3.72	11.1	25.5
W	6 × 16	4.72	6.25	4.030	0.404	0.260	0.87	31.7	10.2	2.59	4.42	2.19	0.967	1.10	3.84	4.99	24.0
	6 × 12	3.54	6.00	4.000	0.279	0.230	0.75	21.7	7.25	2.48	2.98	1.49	0.918	1.07	5.38	7.17	26.1
	6 × 8.5	2.51	5.83	3.940	0.194	0.170	0.68	14.8	5.08	2.43	1.98	1.01	0.889	1.04	7.63	10.2	34.3
W	5 × 18.5	5.43	5.12	5.025	0.420	0.265	0.81	25.4	9.94	2.16	8.89	3.54	1.28	1.40	2.43	5.98	19.3
	5 × 16	4.70	5.00	5.000	0.360	0.240	0.75	21.3	8.53	2.13	7.51	3.00	1.26	1.39	2.78	6.94	20.8
W	4 × 13	3.82	4.16	4.060	0.345	0.280	0.81	11.3	5.45	1.72	3.76	1.85	0.991	1.11	2.97	5.88	14.9
M Shapes																	
M	14 × 17.2	5.05	14.00	4.000	0.272	0.210	0.62	147	21.1	5.40	2.65	1.33	0.725	0.925	12.8	7.34	66.7
M	12 × 11.8	3.47	12.00	3.065	0.225	0.177	0.56	71.9	12.0	4.55	0.980	0.639	0.532	0.690	17.4	6.81	67.8
M	10 × 29.1	8.56	9.88	5.937	0.389	0.427	0.87	131	26.6	3.92	11.2	3.76	1.14	1.40	4.28	7.63	23.1
M	10 × 22.9	6.73	9.88	5.752	0.389	0.242	0.87	117	23.6	4.16	10.0	3.48	1.22	1.40	4.42	7.39	40.8
M	10 × 9	2.65	10.00	2.690	0.206	0.157	0.50	38.8	7.76	3.83	0.609	0.453	0.480	0.616	18.0	6.53	63.7
M	8 × 34.3	10.1	8.00	8.003	0.459	0.378	1.06	116	29.1	3.40	34.9	8.73	1.86	2.08	2.18	8.72	21.2
M	8 × 32.6	9.58	8.00	7.940	0.459	0.315	1.06	114	28.4	3.44	34.1	8.58	1.89	2.08	2.20	8.65	25.4
	8 × 6.5	1.92	8.00	2.281	0.189	0.135	0.50	18.5	4.62	3.10	0.343	0.301	0.423	0.535	18.6	6.03	59.3
M	7 × 5.5	1.62	7.00	2.080	0.180	0.128	0.43	12.0	3.44	2.73	0.249	0.239	0.392	0.493	18.7	5.78	54.7
M	6 × 22.5	6.62	6.00	6.060	0.379	0.372	0.81	41.2	13.7	2.49	12.4	4.08	1.37	1.55	2.61	7.98	16.1
	6 × 20	5.89	6.00	5.938	0.379	0.250	0.81	39.0	13.0	2.57	11.6	3.90	1.40	1.54	2.66	7.82	24.0
	6 × 4.4	1.29	6.00	1.844	0.171	0.114	0.37	7.20	2.40	2.36	0.165	0.179	0.358	0.444	19.0	5.39	52.6
M	5 × 18.9	5.55	5.00	5.003	0.416	0.316	0.87	24.1	9.63	2.08	7.86	3.14	1.19	1.32	2.40	6.01	15.8
M	4 × 13	3.81	4.00	3.940	0.371	0.254	0.81	10.5	5.24	1.66	3.36	1.71	0.939	1.04	2.73	5.30	15.7

Reproduced with permission from *Steel Construction Manual*, 7th ed. 1973. AISC, Chicago, IL.

Table 8.8. Total Uniformly Distributed Beam Loads in Kips, W & M Shapes (A36 Steel).

Name				W12				
Width	6 5/8	6 1/2	6 1/2	4	4	4	4	
lb/ft	36	31	27	22	19	16.5	14	
L_c, ft	6.9	6.9	6.9	4.3	4.2	4.1	3.5	Defl.
L_u, ft	13.4	11.6	10.1	6.4	5.3	4.3	4.2	in.
Span 8	92	79	68	50	42	35	29	0.13
9	81	70	60	45	37	31	26	0.17
10	73	63	54	40	34	28	23	0.21
11	66	57	49	36	31	25	21	0.25
12	61	52	45	33	28	23	19	0.30
13	56	48	42	31	26	21	18	0.35
14	52	45	39	28	24	20	16	0.41
15	49	42	36	27	22	18	15	0.47
16	46	39	34	25	21	17	14	0.53
17	43	37	32	23	20	16	13	0.60
18	40	35	30	22	18	15	13	0.67
19	38	33	28	21	17	14	12	0.75
20	36	31	27	20	17	14	11	0.83
21	35	30	26	19	16	13	11	0.91
22	33	28	24	18	15	12	10	1.00
23	32	27	23	17	14	12	10	1.09
24	30	26	22	16	14	11	9	1.19
25	29	25	21	16	13	11	9	1.29
			Properties and Reaction Values					
S, in.^{3}	46.0	39.5	34.2	25.3	21.3	17.6	14.8	
V, kip	54.1	46.5	41.1	46.4	41.8	40.0	34.2	
R, kip	37.6	32.2	28.4	31.2	28.0	26.8	22.7	
R_i, kip	8.2	7.2	6.4	7.0	6.4	6.2	5.3	
N_0, in.	5.5	5.5	5.5	5.7	5.7	5.6	5.6	

Steel Construction contains section properties, allowable load tables, allowable moment tables, and other helpful data to assist the steel designer. Design aids for structural sizes commonly used in agriculture are included herein.

8.8.5.1 Allowable Loads in Beams. The total uniformly distributed load, including the beam weight plus the applied load, which can be carried by selected W- and M-shaped A36 steel beam sections are listed in Table 8.8. The values in Table 8.8 are excerpted from more extensive tables in the 7th edition of the AISC Manual. The allowable loads in the table assume the beam to be simply supported and are based upon $M_{\max} = wL^2/8$. The tabular values are the total allowable load. Thus the tabulated load W_{tab} equals wL. The tabulated loads also assume

Table 8.8. Continued.

Name						W10					
Width	8	8	8	5 3/4	5 3/4	5 3/4	4	4	4	4	
lb/ft	45	39	33	29	25	21	19	17	15	11.5	
L_c, ft	8.5	8.4	8.4	6.1	6.1	6.1	4.2	4.2	4.2	3.8	Defl.
L_u, ft	22.7	19.6	16.4	13.2	11.4	9.1	7.2	6.0	5.0	4.3	in.
8^a	98	84	69	61	53	43	37	32	27	20.5	0.16
9	87	75	61	54	47	38	33	28	24	18.2	0.20
10	79	68	55	49	42	34	30	25	22	16.4	0.25
11	71	61	50	44	38	31	27	23	20	14.9	0.30
12	65	56	46	41	35	28	25	21	18	13.7	0.36
13	60	52	42	37	32	26	23	19	17	12.6	0.42
14	56	48	39	35	30	24	21	18	15	11.7	0.49
15	52	45	37	32	28	22	20	17	14	10.9	0.56
16	49	42	34	30	26	21	18	16	13	10.3	0.64
17	46	40	32	29	24	20	17	15	13	9.7	0.72
18	44	38	31	27	23	19	16	14	12	9.1	0.80
19	41	36	29	25	22	18	15	13	11	8.6	0.90
20	39	34	28	24	21	17	15	13	11	8.2	0.99
21	37	32	26	23	20	16	14	12	10	7.8	1.09
22	36	31	25	22	19	15	13	11	10	7.5	1.20
				Properties and Reaction Values							
S, in^3	49.1	42.2	35.0	30.8	26.5	21.5	18.8	16.2	13.8	10.5	
V, kip	51	46	41	42.8	36.8	34.5	37.2	35.2	33.4	25.8	
R, kip	44	40	35	35.6	30.6	28.4	30.0	28.4	26.8	20.7	
R_i, kip	9.5	8.6	7.9	7.8	6.8	6.5	6.8	6.5	6.2	4.9	
N_0, in.	4.2	4.2	4.2	4.4	4.4	4.4	4.6	4.6	4.6	4.6	

(*continued*)

that the beam is braced adequately to prevent lateral buckling and are valid only when $L_b \leq L_c$ and $F_b = 0.66F_y = 24$ ksi (170 MPa). The magnitude of L_c is given for each section. If $L_c < L_b < L_u$, then $F_b = 22$ ksi (150 MPa) and the allowable load equals (22/24) × (tabulated load). If $L_b > L_u$, then the load table cannot be used and the load capacity must be evaluated by applications of Eqs. (8.9), (8.10a), and (8.10b) or by using moment capacity charts.

The allowable load table is adaptable to laterally supported simple beams which carry equally spaced concentrated loads (see Table 8.9). The maximum bending moment in a simply supported beam carrying a concentrated load at midspan equals $PL/4$. A simple beam carrying a uniform load has a maximum moment of $wL^2/8 = (wL)(L/8) = W_{tab}(L/8)$. Thus, a given section can carry a midspan concentrated load of $\frac{1}{2}$ the total uniformly distributed load. The equivalent load of $2P$ in Table 8.9 indicates that a total uniform load of $2P$ will yield the same maximum moment as a concentrated midspan load of P. To obtain the

Table 8.8. Continued.

Name	W8								W6			
Width	6 1/2	6 1/2	5 1/4	5 1/4	4	4	4		4	4	4	
lb/ft	28	24	20	17	15	13	10		16	12	8.5	
L_c, ft	6.9	6.9	5.6	5.5	4.2	4.2	4.2	Defl.	4.3	4.2	4.2	Defl.
L_u, ft	17.4	15.1	11.3	9.4	7.2	5.9	4.7	in.	12.1	8.6	6.1	in.
8^a	48	41	34	28	23	19.8	15.3	0.20	20	14.5	9.9	0.26
9	43	37	30	25	21	17.6	13.6	0.25	18	12.9	8.8	0.34
10	38	33	27	22	18	15.8	12.2	0.31	16	11.6	7.9	0.41
11	35	30	24	20	17	14.4	11.1	0.38	14	10.5	7.2	0.50
12	32	27	22	18	15	13.2	10.2	0.45	13	9.7	6.6	0.60
13	29	25	20	17	14	12.2	9.4	0.52	12	8.9	6.1	0.70
14	27	23	19	16	13	11.3	8.7	0.61				
15	25	22	18	15	12	10.6	8.1	0.70				
16	24	20	17	14	11	9.9	7.6	0.79				
17	22	19	16	13	11	9.3	7.2	0.90				
Properties and Reaction Values												
S, in.	24.3	20.8	17.0	14.1	11.8	9.9	7.8		10.2	7.3	5.1	
V, kip	33.3	28.2	29.3	26.7	28.8	26.7	19.5		23.6	20.0	14.4	
R, kip	34.1	28.9	29.3	26.8	28.5	26.4	19.2		30.7	26.4	19.2	
R_i, kip	7.7	6.6	6.7	6.2	6.6	6.2	4.6		7.0	6.2	4.6	
N_0, in.	3.4	3.4	3.5	3.5	3.5	3.5	3.6		2.5	2.5	2.4	

allowable concentrated load a beam can carry at midspan, simply divide the tabulated load in Table 8.8 by 2.0. Deflections for concentrated loads equal the listed deflection times the coefficient in Table 8.9.

Several important properties of the beam section are summarized at the bottom of the table. Among them are the strong axis section modulus S_x; the allowable shear force V; the maximum allowable end reaction R for a $3^1/_2$ in. (89 mm) bearing length; the increase in the allowable end reaction for each additional inch of bearing length R_i; and the length of bearing required to develop V, N_0.

Example 8.5. Beam Selection Using Beam Load Tables. Using A36 steel, select a 10 in. (254 mm) deep beam to span 20 ft (6.10 m) and support three equal concentrated loads of 7 kips (31 kN) located at the quarter points of the span.

Solution. Refer to Table 8.9 and note that the equivalent uniform load W_{eq} equals 4.0 P, the deflection coefficient is 0.95 and $C_b = 1.0$ since internal moments are greater than either end moment.

Table 8.8. Continued.

Name	M14	M12	M10			M8	M7	M6
Width	4	3 1/8	5 7/8	5 3/4	2 3/4	2 1/4	2 1/8	1 7/8
lb/ft	17.2	11.8	29.1	22.9	9	6.5	5.5	4.4
L_c, ft	3.6	2.7	6.3	6.1	2.6	2.4	2.2	1.9
L_u, ft	4.1	3.1	10.8	10.5	2.7	2.5	2.5	2.4
8[a]	42	24.0	53	47	15.5	9.2	6.9	4.8
9	37	21.3	47	42	13.8	8.2	6.1	4.3
10	33	19.2	42	37	12.4	7.4	5.5	3.8
11	30	17.5	38	34	11.3	6.7	5.0	3.5
12	28	16.0	35	31	10.3	6.2	4.6	3.2
13	26	14.8	32	29	9.6	5.7	4.2	3.0
14	24	13.7	30	27	8.9	5.3	3.9	
15	22	12.8	28	25	8.3	4.9	3.7	
16	21	12.0	26	23	7.8	4.6		
17	19	11.3	25	22	7.3	4.3		
18	18	10.7	23	21	6.9			
19	17	10.1	22	19	6.5			
20	16	9.6	21	18	6.2			
21	16	9.1	20	18	5.9			
22	15	8.7						
23	14	8.3						
24	14	8.0						
25	13	7.7						
26	13							
27	12							
28	12							
29	11							
30	11							
Properties and Reaction Values								
S, in.	21.1	12.0	26.6	23.6	7.8	4.6	3.4	2.4
V, kip	42.6	30.8	61.2	34.7	22.8	15.7	13.0	9.9
R, kip	23.4	19.4	50.4	28.6	17.0	14.6	13.6	11.9
R_i, kip	5.7	4.8	11.5	6.5	4.2	3.6	3.5	3.1
N_0, in.	6.9	5.9	4.4	4.4	4.9	3.8	3.3	2.8

Reproduced with permission from *Steel Construction Manual*, 7th ed. 1973. AISC, Chicago, IL.
[a]This column gives span in feet.

    a. The equivalent uniform load is $W_{eq} = 4.0 \times 7 = 28$ kip (124 kN).
    b. Enter beam load tables for a W 10 section and span of 20 ft (6.10 m). A W 10 × 33 (W 254 × 49) section has a tabulated load of 28 kip (124 kN).
    c. Check the unsupported lengths of the section.

$$L_b = 20 \text{ ft } (6.10 \text{ m}), L_c = 8.4 \text{ ft } (2.56 \text{ m}), L_u = 16.4 \text{ ft } (5.00 \text{ m})$$

Table 8.9. Equivalent Uniform Loads.

Type of Loading: Equal Loads, Equal Spaces	Equivalent Uniform Load ($W = wL$)	Deflection Coefficient
P at L/2, L/2	2.00P	0.80
P, P at L/3, L/3, L/3	2.67P	1.02
P, P, P at L/4, L/4, L/4, L/4	4.00P	0.95
P, P, P, P at L/5, L/5, L/5, L/5, L/5	4.80P	1.01

Reproduced with permission from *Steel Construction Manual*, 7th ed. 1973. AISC, Chicago, IL.

Since $L_b > L_u$, $F_b < 0.6F_y$ and the tables are not applicable. However, when $F_b = 0.6F_y$, $W_{\text{all}} = {}^{22}/_{24} (W_{\text{tab}}) = 25.7$ k (122 kN) < 28 k (124 kN).

d. Next try a W 10 × 39 (W 254 × 58) with $L_c = 8.4$ ft (2.56 m) and $L_u = 19.6$ ft (5.97 m). Since $L_b \approx L_u$, $F_b \approx 0.6F_y$ and $W_{\text{all}} \approx {}^{22}/_{24} (34) = 31.2$ kip (139 kN). The section is adequate since $W_{\text{all}} > W_{\text{eq}}$.

e. Evaluate the beam deflection $\Delta_{\max} = 0.95 \times 0.99 = 0.94$ in. (24 mm)

8.8.5.2 Allowable Moments in Beams. The moment charts of Fig. 8.18 give the allowable moment capacity of selected M and W sections fabricated of A36 steel. The allowable moment capacity is plotted as a function of the unbraced length of the beam (not necessarily the span) and for single curvature ($C_b = 1.0$). The moment capacity is the total moment capacity. Thus, if the allowable live load capacity is required, the dead load moment must be subtracted from the plotted capacity. Or, to select a beam, enter the chart with the dead plus the live load moment.

For each section, when $L_b < L_c$, the moment capacity is $M = (0.66F_y)(S_x)$. When $L_c < L_b < L_u$, the capacity is $M = (0.60F_y)(S_x)$; and when $L_b > L_u$, $M = F_b S_x$ where F_b is less than $0.6F_y$ and is evaluated from Eqs. (8.9), (8.10a or 8.11b). The solid dot represents the maximum unbraced length for which $F_b = 0.66F_y$. The open dot represents the maximum unbraced length for which $F_b = 0.60F_y$.

If $C_b \neq 1.0$, the moment capacity cannot be obtained directly from the charts. Instead F_b must be evaluated using Eqs. (8.9), (8.10a), or (8.10b) and $M = F_b S_x$.

The charts are very useful for selecting the lightest weight W or M section for a given application. Once the required moment capacity and unbraced length are plotted on the chart, the lightest adequate section is represented by the first solid line directly above and to the right of the plotted point. The dashed lines are for adequate beams which are heavier than the most economical section.

The dashed lines terminate at the point where $F_b = 11$ ksi (76 MPa). Further extension of the lines is not practical for normal design.

8.8.6 Illustrations—Flexural Members

Example 8.6. Beam Selection Using Moment Charts. Using A36 steel, determine the size of the lightest weight W or M girder with a span of 35 ft (10.67 m) which supports two equal concentrated loads located 10 ft (3.05 m) from its left and right reaction points (Fig. 8.19).The compression flange is laterally supported at the concentrated load points only. The loads produce a maximum calculated moment of 65 kip-ft (88 kN-m) in the section between the loads.

Solution

a. The center section is critical, because $C_b = 1.0$ for all portions, it is the location of $M_{\max}$, and it has the greatest unbraced length.
b. Assume a beam dead load of 40 lb/ft (580 N/m). Then $M_{\max} = 65 + w_D L^2/8$ or 71 k-ft (96 kN-m).
c. Select size using moment chart. Entering Fig. 8.18 with an unbraced length equal to 15 ft (4.57 m) on the bottom scale (abscissa), proceed upward to meet the horizontal line corresponding to a moment equal to 71 kip-ft (96 kN-m) on the left-hand scale (ordinate). Any beam listed above and to the right of the point so located satisfies the allowable bending stress requirement. In this case, the lightest section satisfying this criterion is a W 12 × 36 (W 305 × 54) for which the total allowable moment with an unbraced length of 15 ft (4.57 m) is 73.5 k-ft (100 kN-m). (*Note:* $L_b > L_u$ and this design could not be accomplished with the load tables.)

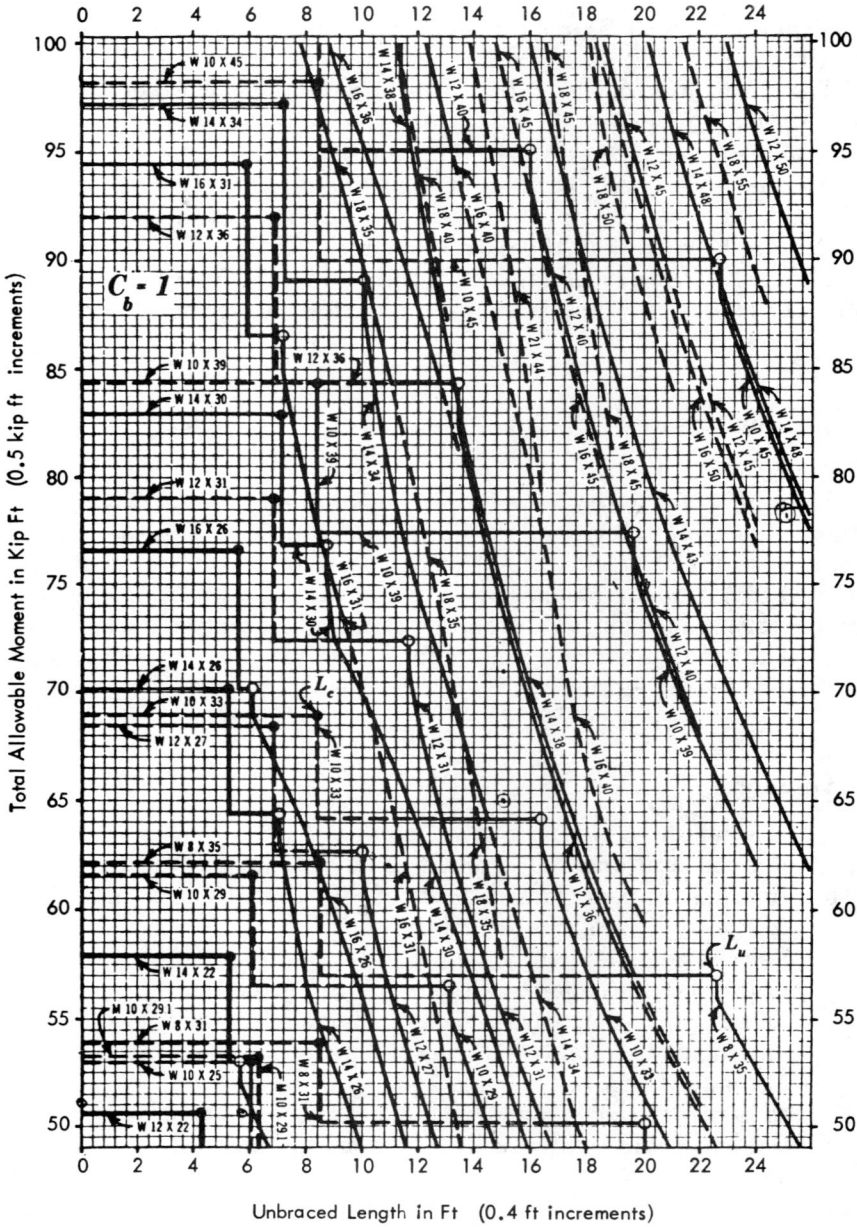

Fig. 8.18. Allowable moments in beams, $F_y = 36$ ksi. Reproduced with permission from *Steel Construction Manual,* 7th ed., 1973, AISC, Chicago, IL.

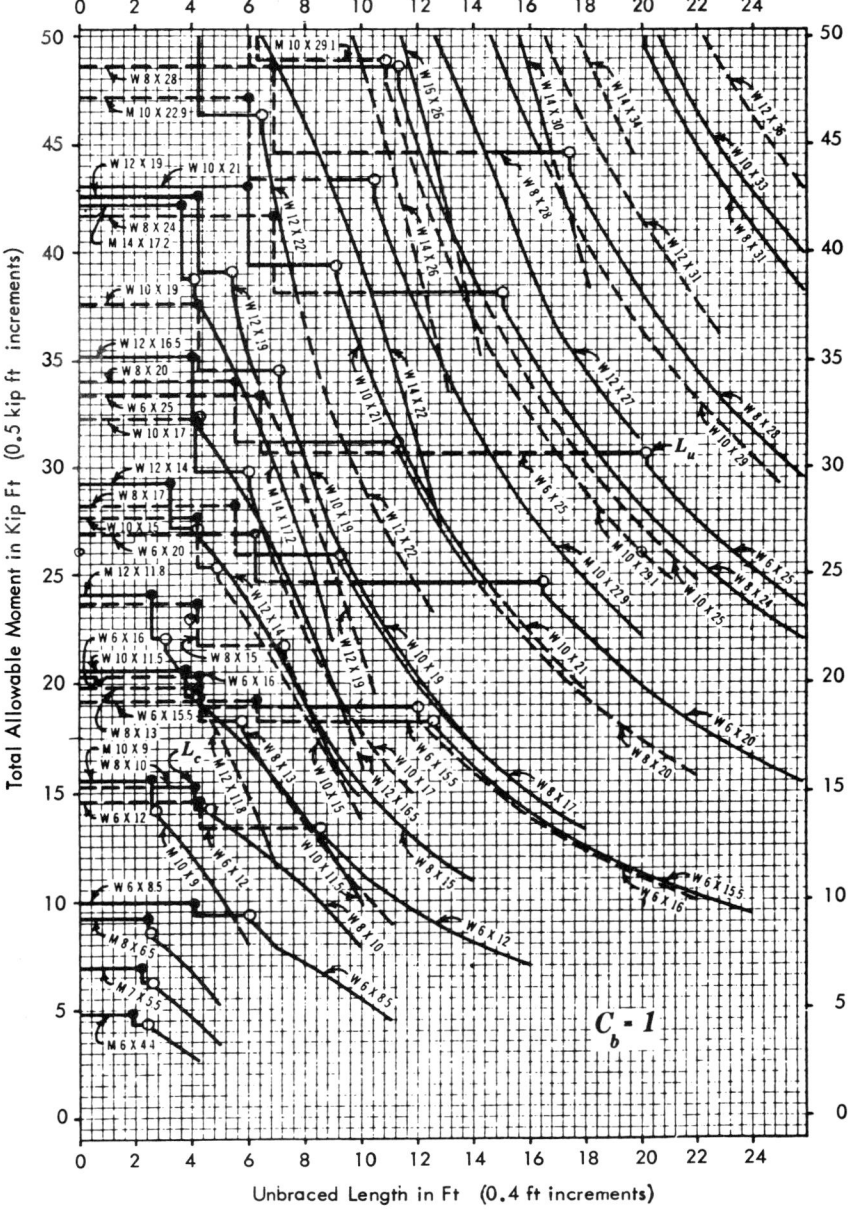

Fig. 8.18. Continued.

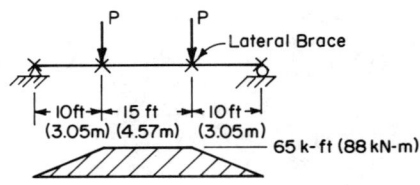

Fig. 8.19. Example 8.6.

Example 8.7. *Steel Beam Selection Using Design Aids*

a. What is the allowable bending stress for a simply supported A36, W 10 × 33 (W 254 × 49) steel beam if it spans 6 ft (1.83 m) and has no lateral bracing between supports? From Table 8.8, L_b = 6 ft (1.83 m) < L_c = 8.4 ft (2.56 m). Thus, F_b = 24 ksi (170 MPa).

b. What is the allowable bending stress for a simply supported W 8 × 17 (W 203 × 25) A36 steel beam if it spans 8 ft (2.44 m) and has no lateral bracing between supports? From Table 8.8, L_b = 8 ft (2.44 m) < L_u = 9.4 ft (2.56 m). Thus, F_b = 22 ksi (150 MPa).

c. What is the allowable bending stress for the section in part b if the simple beam spans 14 ft (4.27 m) with no intermediate support?

   i. From Table 8.8, L_b = 14 ft (4.27 m) > $C_b L_u$. Thus, F_b < $0.6F_y$.

   ii. Evaluate F_b using Eqs. (8.9), (8.10a), or (8.10b).

      1. Euler buckling (C_b = 1.0). Evaluate the limiting conditions for Eqs. (8.9) and (8.10a).

$$\left(\frac{510 \times 10^3 C_b}{F_y}\right)^{1/2} = \left[\frac{510 \times 10^3(1.0)}{36}\right]^{1/2} = 119$$

$$\left(\frac{102 \times 10^3 C_b}{F_y}\right)^{1/2} = \left[\frac{102 \times 10^3(1.0)}{36}\right]^{1/2} = 53$$

Then evaluate L_b/r_t = (14 × 12)/1.40 = 120 > 119. Therefore

$$F_b = \frac{170 \times 10^3 C_b}{(L_b/r_t)^2} = \frac{170 \times 10^3}{(120)^2} = 11.81 \text{ ksi} \quad (81 \text{ MPa})$$

2. St. Venant buckling. Evaluate the allowable stress with Eq. (8.10b).

$$F_b = \frac{12 \times 10^3 C_b}{L_b d/A_f} = \frac{12 \times 10^3 (1.0)}{(4.95)(14 \times 12)}$$
$$= 14.43 \text{ ksi} \quad (100 \text{ MPa})$$

3. The allowable stress is the larger, or 14.43 ksi (100 MPa).

d. What is the allowable uniformly distributed load that a W 12 × 31 (W 305 × 46) A36 steel beam can carry over an unbraced simple span of 10 ft (3.05 m)?
   i. Entering Table 8.8, L_c = 6.9 ft (2.10 m) and $C_b L_u$ = 11.6 ft (3.54 m). Thus F_b = $0.6 F_y$ = 22 ksi (150 MPa).
   ii. Adjusting the tabulated load for the reduction in allowable stress,

$$W = {}^{22}/_{24} W_{tab} = {}^{22}/_{24} (63) = 57.8 \text{ k} (257 \text{ kN})$$

   iii. The allowable distributed load, including the beam weight, is equal to 57.8/10, or 5.78 k/ft (84.3 kN/m).

e. What is the allowable concentrated load that a W 10 × 45 (W 254 × 67) A36 steel beam can carry at the center of an unbraced simple span of 20 ft (6.10 m)?
   i. Entering Table 8.8, L_c = 8.5 ft (2.59 m) and $C_b L_u$ = 22.7 ft (6.92 m). Thus F_b equals $0.6 F_y$ = 22 ksi (150 MPa). It follows that the allowable distributed load $({}^{22}/_{24})(39)$ = 35.8 k (159 kN).
   ii. Entering Table 8.9, the allowable concentrated load equals 35.8/2 = 17.9 k (80 kN).

f. A simply supported W 8 × 17 (W 203 × 25) A36 steel beam spans 12 ft (3.66 m), carries a uniformly distributed load, and is unbraced between the supports. Evaluate the magnitude of the allowable load the section can carry.
   i. Entering Table 8.8, $C_b L_u$ = 9.4 ft (2.86 m). Since $L_b > C_b L_u$, $F_b < 0.60 F_y$ and the moment charts must be used.
   ii. Entering Fig. 8.18 with L_b = 12 ft (3.66 m), M_{all} = 20 k-ft (27 k N-m). The maximum moment equals $M_{max} = wL^2/8 + 0.017 L^2/8$. Equating M_{max} and M_{all} and solving for w yields an allowable distributed load of 1.09 k/ft (15.9 kN/m).

256 Structural Steel Design

g. Select the lightest A36 steel W section which can be used to carry a uniformly distributed load (including DL) of 1500 lb/ft (21.9 kN/m) over an unbraced simple span of 20 ft (6.10 m).
  i. The required moment capacity is $M_{max} = wL^2/8 = 1.5(20)^2/8 = 75$ k-ft (102 kN-m).
  ii. Enter Fig. 8.18 with $L_b = 20$ ft (6.10 m) and $M = 75$ k-ft (102 kN-m). The lightest member is a W 12 × 40 (W 305 × 60).

Example 8.8. Laterally Supported Beam Design. Determine the lightest A36 W section to carry a uniformly distributed load of 500 lb/ft (7.3 kN/m) over a simply supported beam, 20 ft (6.10 m) long. Assume the beam is *laterally supported*.

Solution A. Flexure Equation ($S = M/F_b$)

a. The maximum moment equals $wL^2/8 = (0.5)(20^2)/8 = 25.0$ k-ft (34 kN-m).
b. $C_b = 1.0$ since interior moments are greater than both end moments.
c. Assume a compact section. Then $F_b = 0.66\ F_y = 24$ ksi (170 MPa).
d. $S_x = M/F_b = 25.0(12)/24 = 12.5$ in.^{3} (204.8 cm^3).
e. From bottom of load Table 8.8, a W 12 × 14 (W 305 × 21) has a section modulus of $S_x = 14.8$ in.^{3} (242.5 cm^3). It is the lightest section with adequate S_x. If adequately braced, all the sections in Table 8.8 are compact. Thus, a W 12 × 14 (W 305 × 21) is adequate and the lightest section if lateral bracing is provided every 3.5 ft (1.07 m).

Solution B. Beam Loads, Table 8.8

a. The total load equals $0.5(20) = 10$ kip (44 kN).
b. Directly from the beam load table, Table 8.8, a W 12 × 14 (W 305 × 21) is the lightest section with an allowable load of 11 kip (49 kN).

Solution C. Beam Allowable Moment Charts, Fig. 8.18

a. Assume beam dead load of 20 lb/ft (290 N/m).
b. Since adequately braced, L_b must be less than 3.5 ft (1.07 m).

8.8 Flexural Member Design

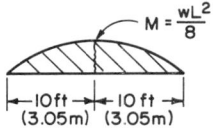

Fig. 8.20. Example 8.9—Moment diagram.

c. The required moment $M = 25 + 0.02(20)^2/8 = 26$ k-ft (34 kN-m).
d. From the figure the smallest size is W 12 × 14 (W 305 × 21) with a moment capacity of 29.3 k-ft (40 kN-m).
e. Note from the moment charts that if the beam were unbraced, $L_b = 20$ ft (6.10 m), and a W 8 × 24 (W 205 × 36) section is required.

Example 8.9. *Laterally Braced and Unbraced Beam Design.* A simply supported beam carries a uniformly distributed load of 1 k/ft (15 kN/m) over a span of 20 ft (6.10 m). What is the lightest weight A36 W10 or smaller section which will carry the loads if $L_b = 0$ ft (0 m)? 10 ft (3.05 m)? 20 ft (6.10 m)? See Fig. 8.20.

Solution

a. The beam must carry the following moments:

$$M_L = wL^2/8 = (1)(20)^2/8 = 50 \text{ k-ft} \quad (68 \text{ kN-m})$$

b. If $L_b = 0$, assume a compact section with $F_b = 0.66 F_y$.
   i. The required section modulus, $S = M/F_b = 50(12)/24 = 25$ in.^{3} (409.7 cm^3).
   ii. From the bottom of the beam load tables, a W 10 × 25 (W 254 × 37) is lightest with $S = 26.5$ in.^{3} (434.3 cm^3). The section is compact if bracing provided at $L_b = L_c = 6.1$ ft (1.86 m).
   iii. Check the actual beam stresses for the section with the dead load included. The calculated stress, $f_b = (1 + 0.025)(20)^2(12)/26.5 = 23.2$ ksi (160 MPa) $< 0.66 F_y$. Therefore, the section is adequate.
   iv. An alternate solution is obtained directly from the load tables. Assuming a beam weight of 25 lb/ft (360 N/m), the

total load is 1.025 × 20 = 20.5 kip (91 kN). For a span of 20 ft (6.10 m) a W 10 × 25 (W 254 × 37) is the best choice.

c. If L_b = 10 ft (3.05 m), C_b = 1.00 (M_1/M_2 = 0 for both left and right sections, but the absolute value of internal moments between the ends of both braced sections is greater than one of the end moments).
   i. Enter Fig. 8-18 with L_b = 10 ft (3.06 m) and M = 50 k-ft (68 kN-m). Select a W 16 × 26 (W 406 × 38) with L_u = 5.6 ft (1.70 m).

d. If $L = L_b$ = 20 ft (6.10 m), C_b = 1.0 since internal moments are greater than end moments. Using Fig. 8.18 with L_b = 20 ft (6.10 m) and M = 50 k-ft (68 kN-m), select a W 10 × 33 (W 254 × 49).

8.9 COMBINED AXIAL COMPRESSION AND FLEXURAL STRESS—BEAM COLUMNS

To determine the adequacy of a steel member subject to combined flexure and axial stress, the designer may use a form of the interaction equation. In conceptual form the equation is

$$\frac{\text{Actual axial stress}}{\text{Allowable axial stress}} + \frac{\text{Actual flexural stress}}{\text{Allowable flexural stress}} \leq 1.0 \quad (8.12)$$

The AISC specification separates beam-columns into those for which f_a/F_a is greater than or less than 0.15. When $f_a/F_a > 0.15$ the moment magnification due to P-Δ effects becomes critical and is included in the interaction equation. If $f_a/F_a < 0.15$ these effects are neglected and the beam-column must satisfy the stability requirement in

$$\frac{f_a}{F_a} + \frac{f_{bx}}{F_{bx}} + \frac{f_{by}}{F_{by}} \leq 1.0 \quad (8.13)$$

If $f_a/F_a > 0.15$, the beam column must satisfy both the stability requirement of Eq. (8.14) and the strength requirement of Eq. (8.15).

$$\frac{f_a}{F_a} + \frac{C_{mx}f_{bx}}{F_{bx}(1 - f_a/F'_{ex})} + \frac{C_{my}f_{by}}{F_{by}(1 - f_a/F'_{ey})} \leq 1.0 \quad (8.14)$$

8.9 Combined Axial Compression and Flexural Stress—Beam Columns

$$\frac{f_a}{0.6F_y} + \frac{f_{bx}}{F_{bx}} + \frac{f_{by}}{F_{by}} \leq 1.0 \qquad (8.15)$$

The definition of each term in the three equations is: f_a, f_{bx}, f_{by} are the actual axial and bending stress levels; F_a, the allowable axial stress for pure compression based upon $(KL/r)_{\max}$; F_{bx}, F_{by} are the allowable flexure stresses about the x and y axis, respectively, for pure flexure with $C_b = 1.0$; F'_{ex}, F'_{ey} are Euler buckling stresses based upon the slenderness ratio about the axis of bending, $12\pi^2 E/[(23)(KL/r)_b^2]$; $C_m = 0.85$ for members in frames subject to sidesway or joint translation; $C_m = 0.6 - 0.4 M_1/M_2 \geq 0.4$ for members in frames restrained from sidesway or joint translation and not subject to transverse loads between supports in the plane of bending; M_1/M_2, the ratio of the smaller-to-larger moments at ends of unbraced sections, is positive for reverse curvature bending and negative for single curvature bending; $C_m = 0.85$ for compression members with restrained ends in frames with transverse loads between ends in the plane of bending, but braced against sidesway or joint translation; and $C_m = 1.00$ for compression members with unrestrained ends in frames with transverse loads between ends, but braced against sidesway or joint translation.

To analyze an existing beam-column simply substitute the section properties and allowable stress values into the appropriate interaction equations to assure that the sum is less than 1.0. Design of beam-columns is a more complex procedure since the allowable stress levels and section properties are unknown until the member is designed. A trial and error procedure, in which an initial estimate is based on either pure compression or pure flexure and then refined, is required. This procedure can become tedious and time consuming if the lightest adequate section is desired.

Beam-column design can be simplified considerably by using the column load tables and an equivalent load technique. The technique consists of transforming an axial load and bending moment into an equivalent axial load and selecting the section which can carry the equivalent load for the column effective length. The equivalent load method is approximate in many cases, so the final design should be checked by substitution into the correct interaction equations.

The equivalent load method for bending about only one axis, say the ith axis, follows. Interaction equation (8.13) can be rewritten into the form

$$\frac{P}{AF_a} + \frac{M_i}{S_i F_{bi}} \leq 1.0$$

260 Structural Steel Design

Multiplying by AF_a yields

$$P + \frac{A}{S_i}\frac{F_a}{F_{bi}} M_i \leq F_a A = P_{\text{equivalent}}$$

Thus the equivalent axial load equals the axial load P plus an axial load P' or

$$P_{\text{equivalent}} = P + P' = P + B_i (F_a/F_{bi}) M_i \qquad (8.16)$$

where B_i = cross-section area/section modulus = A/S_i.

Similarly, for bending about only the ith axis, interaction equations (8.14) and (8.15) can be transformed, respectively, into

$$P_{\text{equivalent}} = P + B_i M_i C_{mi} \left(\frac{F_a}{F_{bi}}\right)\left(\frac{a_i}{a_i - P(KL)^2}\right) \qquad (8.17)$$

and

$$P + P'_i = P\left(\frac{F_a}{0.6F_y}\right) + B_i M_i \left(\frac{F_a}{F_{bi}}\right) \qquad (8.18)$$

where KL is effective length in the plane of bending, in. (mm) and

$$a_i = 5.15 EA r_i^2, \text{ lb-in}^2 \quad (\text{kN} \cdot \text{m}^2)$$

Tables 8.1 and 8.7 include values for most of the parameters. It is advisable to start with the simplest modified interaction equation (Eq. (8.16)) and a column a little larger than needed for axial load $F_a/F_{bi} = 1.0$, and an average value for B_i. Then refine the solution until an economical section is found.

Example 8-10. Beam-Column Design. A W- or M-shaped A36 steel column is to be selected for a pole building. The bottom end of the column is pinned; the top can translate, but not rotate. There is no lateral support between the ends of the column. The column is 10 ft (3.05 m) long and supports a 16-kip (71-kN) axial compressive load and a 5 kip-ft (6.8 kN-m) moment acting in the plane of the strong axis. Select the lightest adequate section.

Solution

a. For a rough estimate use $P_{\text{req'd}} = P + B_x M_x (F_a/F_{bi})$. Using an average value for B_i of 0.30 and $F_a/F_{bx} = 1.0$ (see Tables 8.1, 8.7, and 8.8 for section properties),

8.9 Combined Axial Compression and Flexural Stress—Beam Columns

$P_{req'd} = 16 + 0.3(5 \times 12)(1.0) = 34.0$ kip (151 kN)

Since the effective length is the same in both planes $(KL)_x = (KL)_y = 2.0(10) = 20$ ft (6.10 m) and weak axis buckling controls. Entering Table 8.1 with $KL = 20$ ft (6.10 m) and $P_{req'd} = 34.0$ kip (151 kN), a W 6 × 20 (W 152 × 30) is the lightest section.

b. Check the W 6 × 20 (W 152 × 30) by substituting into the interaction equations. Since $(KL/r)_y = (20 \times 12)/1.51 = 158.9$, $F_a = 5.92$ ksi (41 MPa) (from Table 8.5), and the ratio $f_a/F_a = (16/5.88)/5.92 = 0.46 > 0.15$. Thus Eqs. (8.17) and (8.18) must be checked.

$$P_{req'd} = P + B_x M_x C_{mx}\left(\frac{F_a}{F_b}\right)\left(\frac{a_x}{a_x - P(KL)_x^2}\right)$$

$$= 16 + 0.439(5 \cdot 12)(0.85)\left(\frac{5.92}{22.0}\right)$$

$$\left(\frac{6.20 \times 10^6}{6.20 \times 10^6 - 16(20 \cdot 12)^2}\right)$$

$$= 23.05 \text{ kip} \quad (102 \text{ kN})$$

and

$$P_{req'd} = P(F_a/0.6F_y) + B_x M_x (F_a/F_{bx})$$

$$= 16(5.92/22) + 0.439(5 \times 12)(5.92/22)$$

$$= 11.39 \text{ kip} \quad (51 \text{ kN})$$

Since $P_{req'd} < P_{tabulated}$, a W 6 × 20 is adequate.

c. Check for possibility of a lighter section. From Table 8.1, a W 6 × 15.5 (W 152 × 23) has an allowable load of 25 k (111 kN) and section properties similar to the W 6 × 20 (W 152 × 30). It is the only other lighter candidate and will be checked. For this section, $(KL/r)_y = 240/1.46 = 164.4$. From Table 8.5, $F_a = 5.53$ ksi (38 MPa) and from Fig. 8.18, $F_b = 22$ ksi (150 MPa). Substituting the strength and section properties into Eq. (8.17)

$$P_{req'd} = P + B_x M_x C_m \left(\frac{F_a}{F_{bx}}\right)\left(\frac{a_x}{a_x - P(KL_x)^2}\right)$$

$$= 16 + 0.456(5 \times 12)(0.85)\left(\frac{5.53}{22}\right)$$

$$\left(\frac{4.49 \times 10^6}{4.49 \times 10^6 - 16(240)^2}\right)$$

$$= 16 + 7.36 = 23.36 \text{ kip} \quad (104 \text{ kN})$$

Substituting the section properties into Eq. (8.18),

$$P_{req'd} = P(F_a/0.6F_y) + B_x M_x (F_a/F_{bx})$$

$$= 16(5.53/22) + 0.456(5 \times 12)(5.53/22)$$

$$= 10.90 \text{ kip } (48 \text{ kN})$$

Since the required load is less than the tabulated load the new section appears adequate. However, since the equivalent load equations are approximate, the section will be checked by direct substitution into Eqs. (8.14) and (8.15).

$$\frac{f_a}{F_a} + \frac{f_{bx} C_{mx}}{F_{bx}(1 - f_a/F'_{ex})} \leq 1.00$$

$$\frac{3.51}{5.53} + \frac{(5 \times 12/10)(0.85)}{22(1 - 3.51/13.79)} \leq 1.00$$

$$0.63 + 0.31 = 0.94 < 1.00$$

and

$$\frac{f_a}{0.6F_y} + \frac{f_{bx}}{F_{bx}} \leq 1.0$$

$$\frac{3.51}{22} + \frac{(5 \times 12)/10}{22} = 0.43 < 1.00$$

Since the left sides of both interaction equations are less than 1.0 and since the first interaction equation yields a value near 1.0, the design solution checks.

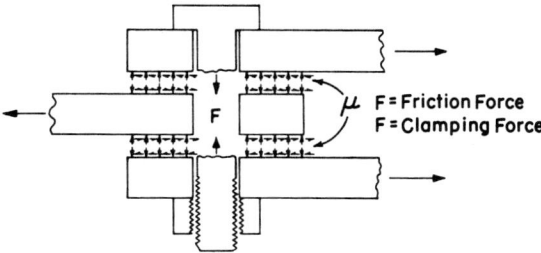

Fig. 8.21. Friction-type bolted connection in double shear. Reproduced with permission from *Steel Construction Manual*, 8th ed., 1980, AISC, Chicago, IL.

8.10 BOLTED CONNECTIONS*

There are two primary classes of bolted connections—friction type and bearing type. The main difference between the two is the factor of safety with respect to slip. The friction type, having a higher factor of safety against slip, has higher fatigue life and is recommended when cyclical loading is encountered.

The friction-type connection is clamped together with sufficient tensile force that the connected parts cannot move relative to one another. That is, the friction between the connected parts is great enough to carry the design loads. Bearing failure of the connected parts is not a concern in the friction-type connection. Only an "apparent" shear stress of the bolt is considered in design of the connector. If the "apparent" shear stress is exceeded, the connected parts will slip and the connection is assumed to have failed. High strength bolts, such as A325, A490, or A449 with hardened washers and A325 nuts, must be used if friction-type connectors are to be adequately clamped. Figure 8.21 illustrates how loads are transferred in a friction-type connector.

Bearing-type connections are not clamped sufficiently to prevent slippage between the connected members. Since these connections are allowed to slip, the modes of failure are by excessive shear stresses in the bolt and excessive bearing stresses in the connected parts. Figure 8.22 illustrates how loads are transferred in bearing-type connectors.

Allowable stresses in both shear and tension are given in Table 8.10 in kip per square inch of nominal bolt area for both friction- and bearing-type connections. Note that allowable stresses are given for connectors when threads are included in the shear plane and when threads

---

*The student is referred to Chapter 7 on Fundamentals of Structural Connections for a general discussion of load transfer in and analysis of structural joints.

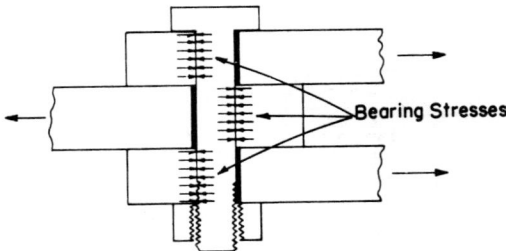

Fig. 8.22. Bearing-type bolted connection in double shear. Reproduced with permission from *Steel Construction Manual,* 8th ed., 1980, AISC, Chicago, IL.

are not included in the shear plane. There are no differences in the allowable shear stresses for friction-type connectors with and without threads in the shear plane because they do not fail by bolt shear.

The allowable bearing stress on the projected area of bolts in bearing-type connections is $1.5F_u$, where F_u is the specified minimum tensile strength of the connected part. Bearing stress is not restricted in friction-type connectors.

In any bolted connection the net area of the connected parts must be sufficient to assure that $f_t = P/A_{net} \leq 0.45\, F_y$.

To develop the allowable stresses in Table 8.10, bolts must be placed and spaced properly. The required center-to-center distance between bolts is the larger of $2\frac{2}{3}$ times the nominal bolt diameter or $2P/(F_u t) + d/2$, where P is force transmitted to one connector; F_u, specified minimum tensile stress of the critical connected part; t, thickness of the critical connected part; and d, nominal bolt diameter. Edge distances must be

Table 8.10. Allowable Stresses for Bolts (in ksi).

Bolt Type by ASTM Designation	Allowable Tension (F_t)	Allowable Shear (F_v)	
		Friction Type[a] (ksi)	Bearing Type (ksi)
A325, threads in the shear plane	44.0	17.5	21.0
A325, threads not in the shear plane	44.0	17.5	30.0
A490, threads in the shear plane	54.0	22.0	28.0
A490, threads not in the shear plane	54.0	22.0	40.0

Reproduced with permission from *Steel Construction Manual,* 8th ed. 1980. AISC, Chicago IL.
[a]Assumes a standard size hole. Hole diameter tolerances of $+\frac{1}{16}$ in. (1.6 mm).

Table 8.11. Minimum Edge Distances[a] for Bolts (in.).

Nominal Bolt Diameter (in.)	At Sheared Edges	At Rolled or Gas-Cut Edges	At Sheared Edges
$1/2$	$7/8$	$3/4$	$7/8$
$5/8$	$1 1/8$	$7/8$	$1 1/8$
$3/4$	$1 1/4$	1	$1 1/4$
$7/8$	$1 1/2$	$1 1/8$	$1 1/2$
1	$1 3/4$	$1 1/4$	$1 3/4$
$1 1/8$	2	$1 1/2$	2
$1 1/4$	$2 1/4$	$1 5/8$	$2 1/4$
over $1 1/4$	$1 3/4 \times$ diam.	$1 1/4 \times$ diam.	$1 3/4 \times$ diam.

Reproduced with permission from *Steel Construction Manual*, 8th ed. 1980. AISC, Chicago, IL.
[a]Distance between center of hole and edge of connected part.

greater than the values in Table 8.11. Also, along a line of transmitted force, and in the direction of the force, the distance between the center of the hole and the edge of the connected part must be greater than $2P/(F_u t)$; and the distance between the edge and bolt hole center line is not to exceed the smaller of 6 in. (152 mm) or 12 times the thickness of the connected parts.

8.11 WELDED CONNECTIONS

8.11.1 General

The basic types of structural welds are illustrated in Fig. 8.23. The strength of each weld depends upon the allowable stress, the effective thickness, and the length of the weld.

8.11.2 Allowable Stresses in Welds

The allowable stresses for groove, fillet, plug, and slot welds are given in Table 8.12 for a variety of loading conditions and base materials. Allowable base metal strengths are the allowable tensile or compressive stresses defined in prior sections of this chapter. Tensile, strengths of the weld materials are 60, 70, 80, 90, 100, and 110 ksi (410, 480, 550, 620, 690, and 760 MPa) and depend upon the electrode used. A typical electrode designation is E60XX and indicates the weld metal has a nominal tensile strength of 60 ksi (410 MPa).

8.11.3 Effective Weld Areas

The effective areas of welds depend upon the type weld. This section defines, in accordance with the AISC specifications, the effective areas for several weld types.

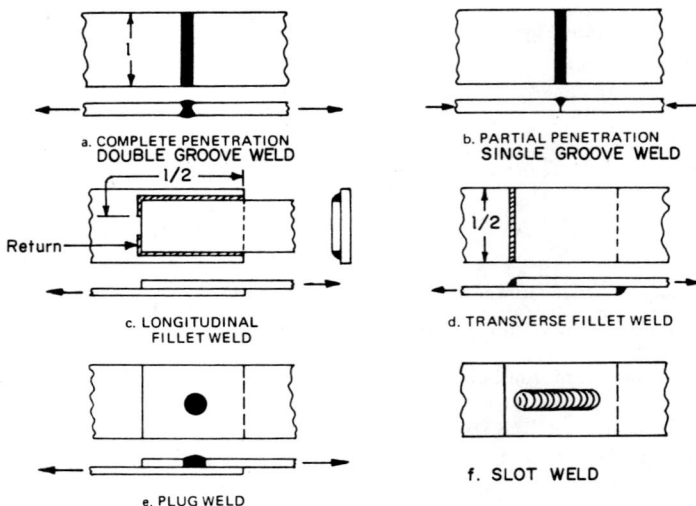

Fig. 8.23. Principal types of structural welds (l = total weld length).

The effective area of groove and fillet welds shall be considered as the effective length of the weld times the effective throat thickness. The effective throat thickness of a fully penetrating groove weld is the thickness of the thinnest part joined. The effective throat thickness of a partially penetrating groove weld is given in Table 8.13. The effective throat thickness of a fillet weld is usually the shortest distance from the root of the weld to the face. This is illustrated in Fig. 7.4. The effective area of fillet welds in holes and slots shall be computed using for effective length, the length of center line of the weld through the center of the plane through the throat. However, in the case of overlapping fillets, the effective area shall not exceed the nominal cross-sectional area of the hole or slot in the plane of the faying surface; i.e., the nominal area of the shearing surface. The effective length of a groove weld is the width of the parts jointed. The effective length of a fillet weld is the overall length of the weld including returns (see Fig. 8.23). The effective area of plug and slot welds is simply the nominal cross-sectional area of the hole or slot.

8.11.4 Size Limitations

Section 1.17 of the AISC specification imposes several size and spacing limitations for welds. They are as follows:

Table 8.12. Allowable Stress on Welds.

Type of Weld and Stress	Allowable Stress	Required Weld Strength Level[a]
Complete-Penetration Groove Welds		
Tension normal to effective area	Same as base metal	"Matching" weld metal must be used
Compression normal to effective area	Same as base metal	Weld metal with a strength level equal to or less than "matching" weld metal may be used
Tension or compression parallel to axis of weld	Same as base metal	
Shear on effective area	0.30 × nominal tensile strength of weld metal (ksi), except shear stress on base metal shall not exceed 0.40 × yield stress of base metal	
Partial-Penetration Groove Welds		
Compression normal to effective area	Same as base metal	Weld metal with a strength level equal to or less than "matching" weld metal may be used
Tension or compression parallel to axis of weld[b]	Same as base metal	
Shear parallel to axis of weld	0.30 × nominal tensile strength of weld metal (ksi), except shear stress on base metal shall not exceed 0.40 × yield stress of base metal	
Tension normal to effective area	0.30 × nominal tensile strength of weld metal (ksi), except tensile stress on base metal shall not exceed 0.60 × yield stress of base metal	
Fillet Welds		
Shear on effective area	0.30 × nominal tensile strength of weld metal (ksi), except shear stress on base metal shall not exceed 0.40 × yield stress of base metal	Weld metal with a strength level equal to or less than "matching" weld metal may be used
Tension or compression parallel to axis of weld[b]	Same as base metal	
Plug and Slot Welds		
Shear parallel to faying surfaces (on effective area)	0.30 × nominal tensile strength of weld metal (ksi), except shear stress on base metal shall not exceed 0.40 × yield stress of base metal	Weld metal with a strength level equal to or less than "matching" weld metal may be used

Reproduced with permission from *Steel Construction Manual,* 8th ed. 1980. AISC, Chicago, IL.
[a]Weld metal one strength level stronger than "matching" weld metal will be permitted.
[b]Fillet welds and partial-penetration groove welds joining the component elements of built-up members, such as flange-to-web connections, may be designed without regard to the tensile or compressive stress in these elements parallel to the axis of the welds.

Table 8.13. Effective Throat Thickness of Partial-Penetration Groove Welds.

Welding Process	Welding Position	Included Angle at Root of Groove	Effective Throat Thickness
Shielded metal arc or submerged arc	All	$<60°$ but $\geqslant 45°$	Depth of chamfer minus $\tfrac{1}{8}$-inch
		$\geqslant 60°$	Depth of chamfer
Gas metal arc or flux cored arc	All	$\geqslant 60°$	Depth of chamfer
	Horizontal or flat	$<60°$ but $\geqslant 45°$	Depth of chamfer
	Vertical or overhead	$<60°$ but $\geqslant 45°$	Depth of chamfer minus $\tfrac{1}{8}$-inch
Electrogas	All	$\geqslant 60°$	Depth of chamfer

Reproduced with permission from *Steel Construction Manual,* 8th ed. 1980. AISC, Chicago, IL.

1. Fillet and partially penetrating groove welds should conform to minimum sizes listed in Table 8.14.
2. The maximum effective size fillet weld for along edges of material $\tfrac{1}{4}$ in. (7 mm) thick, or less, is the thickness of the material.
3. The maximum effective size for materials thicker than $\tfrac{1}{4}$ in. (7 mm) is $1/16$ in. (2 mm) less than the material thickness.
4. The minimum effective length of a fillet weld designed on the basis of strength shall not be less than four times the nominal weld size to develop its full strength. Shorter welds shall have a nominal size equal to one fourth their length.
5. If longitudinal fillet welds are used along in end connections of flat bar tension members, the length of each fillet shall not be less than the perpendicular distance between them. The transverse distance between such welds shall not exceed 8 in. (203 mm).
6. The minimum amount of lap on lap joints shall be five times the thickness of the thinner part but not less than 1 in. (25 mm).
7. Side or end fillet welds terminating at ends or sides, respectively, of parts or members shall, wherever practical, be returned con-

Table 8.14. Minimum Size Fillet and Partial-Penetration Groove Welds.

Material Thickness of Thicker Part Joined (in.)	Minimum Size of Fillet or Partial-Penetration Groove Weld (in.)	Material Thickness of Thicker Part Joined (in.)	Minimum Size of Fillet or Partial-Penetration Groove Welds (in.)
To $1/4$ inclusive	$1/8$	Over $1\tfrac{1}{2}$ to $2\tfrac{1}{4}$	$3/8$
Over $1/4$ to $1/2$	$3/16$	Over $2\tfrac{1}{4}$ to 6	$1/2$
Over $1/2$ to $3/4$	$1/4$	Over 6	$5/8$
Over $3/4$ to $1\tfrac{1}{2}$	$5/16$		

Reproduced with permission from *Steel Construction Manual,* 8th ed. 1980. AISC, Chicago, IL.

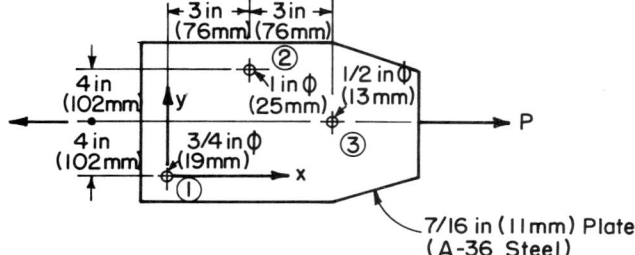

Fig. 8.24. Example 8.11.

tinuously around the corners for a distance not less than twice the nominal weld size.

8. The diameter of the holes for a plug weld shall be not less than the thickness of the part containing it plus $5/16$ in. (8 mm), rounded to the next greater odd $1/16$ in (2 mm), nor greater than $2^{1}/_{4}$ times the thickness of the weld metal.
9. The minimum center-to-center spacing of plug welds shall be four times the diameter of the hole.
10. The length of slot for a slot weld shall not exceed ten times the thickness of the weld. The width of the slot shall be not less than the thickness of the part containing it, plus $5/16$ in. (8 mm), rounded to the next greater odd $1/16$ in. (2 mm), nor shall it be greater than $2^{1}/_{4}$ times the thickness of the weld. The ends of the slot shall be semicircular or shall have the corners rounded to a radius not less than the thickness of the part containing it, except those ends that extend to the edge of the part.
11. The minimum spacing of lines of slot welds in a direction transverse to their length shall be four times the width of the slot. The minimum center-to-center spacing in a longitudinal direction on any line shall be two times the length of the slot.
12. The thickness of plug or slot welds in material $5/8$ in. (16 mm) or less in thickness shall be equal to the thickness of the material. In material over $5/8$ in. (16 mm) in thickness, it shall be at least one half the thickness of the material but not less than $5/8$ in. (16 mm).

Example 8.11. Bolted Connection. The connection shown in Fig. 8.24 carries a centric load P in single shear. If the bolts are A-325 in a bearing-type connection with threads excluded from the shearing plane, evaluate the allowable load P which can be carried.

270 Structural Steel Design

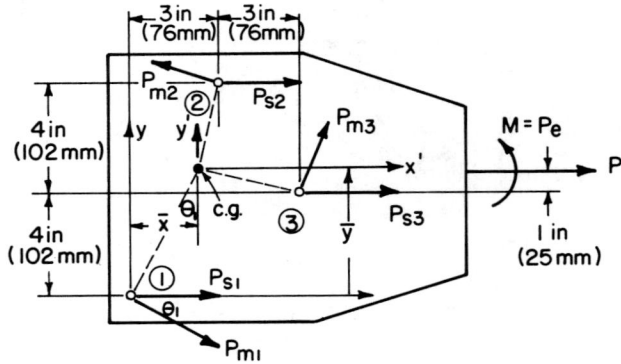

Fig. 8.25. Example 8.11 solution.

Solution

a. Assume the center of rotation and center of gravity of the bolt group coincide. Setting the origin of the reference axis at bolt 1, the center of gravity of the connector group can be determined by the first moment of the bolt areas as follows:

$$(A_1 + A_2 + A_3)\bar{x} = A_1(0) + A_2(3) + A_3(6)$$

$$1.43\bar{x} = 0.79(3) + 0.20(6)$$

$$\bar{x} = 2.50 \text{ in. } (64 \text{ mm})$$

$$1.43\bar{y} = A_1(0) + 0.79(8) + 0.20(4)$$

$$\bar{y} = 4.98 \text{ in. } (126 \text{ mm}) \approx 5 \text{ in. } (127 \text{ mm})$$

b. The load P can now be replaced by a load P, acting through the center of gravity and a couple Pe (Fig. 8.25).

c. Referring to Fig. 8.25 the resultant load carried by each bolt is

$$R_i = P_{si} \mathrel{+\!\!\!+} P_{mi}$$

where $P_{si} = PA_i/\Sigma A_i$ = the shear force in connector i due to the load P and $P_{mi} = A_i(Pe)r_i/\Sigma(A_i r_i^2) = (Pe)A_i r_i/\Sigma A_i(x_i^2 + y_i^2)$ is the shear force in connector i due to the moment Pe. The student is referred to the chapter on connectors (Chapter 7) for analysis details.

d. The connector group properties are evaluated with the following table:

Bolt	A (in.^{2})	x' $(x-\bar{x})$ (in.)	y' $(y-\bar{y})$ (in.)	$(x')^2$ (in.^{2})	$(y')^2$ (in.^{2})	$(x')^2 + (y')^2$ (in.^{2})	Ar_i^2 (in.^{3})
1	0.44	-2.5	-5	6.25	25	31.25	13.25
2	0.79	$+0.5$	$+3$	0.25	9	9.25	7.31
3	0.20	$+3.5$	-1	12.25	1	13.25	2.45
						$\Sigma(A_i r_i^2) =$	$\overline{23.51}$ in.^{3}
							(385.3 cm^3)

e. Bolt 1 is most distant from the center of gravity; however, bolt 3 has a smaller area than bolt 1, and bolt 2 obviously carries the greater shear due to P only. Thus any one of the bolts may control the allowable connector group load. In bolt 1

$$P_{s1} = P(0.44)/1.43 = 0.31P$$

$$P_{m1} = \frac{P(1)(0.44)(31.25)^{1/2}}{23.51} = 0.10P$$

$$R_1 = \sqrt{(P_{m1}\sin\theta_1)^2 + (P_{m1}\cos\theta_1 + P_{s1})^2}$$

where

$$\theta_1 = \tan^{-1}(x_i'/y_i') = \tan^{-1}(-2.5/-5.0) = 26.6°$$
$$R_1 = \sqrt{(0.10 \times 0.45)^2 + (0.10 \times 0.89 + 0.31)^2}P$$
$$R_1 = 0.40P$$

272 Structural Steel Design

The resultant loads in connector 3 are

$$P_{s3} = P(0.20)/1.43 = 0.14P$$

$$P_{m3} = \frac{P(1)(0.2)(13.25)^{1/2}}{23.51} = 0.03P$$

$$R_3 = \sqrt{(P_{m3} \sin \theta_3)^2 + (P_{m3} \cos \theta_3 + P_{s3})^2}$$

$$\theta_3 = \tan^{-1}[3.5/(-1)] = -74°$$

$$R_3 = \sqrt{(0.03 \times 0.96)^2 + (0.03 \times 0.28 + 0.14)^2} P$$

$$= 0.15P$$

The resultant loads in connector 2 are

$$P_{s2} = P(0.79)/1.43 = 0.55P$$

$$P_{m2} = P(1)(0.79)(9.25)^{1/2}/23.51 = 0.10P$$

$$\theta_2 = \tan^{-1}[0.5/3] = 9.5°$$

$$R_2 = \sqrt{(0.10 \times 0.17)^2 + (0.1 \times 0.99 + 0.55)^2} P$$

$$= 0.65P$$

f. Allowable connector load based on bolt shear from Table 8.10. The allowable resultant shear for bolt 1 is $(R_1)_{\text{all}} = 30$ ksi $(0.44) = 13.2$ k (59 kN). Thus, the allowable load, P, for bolt 1 equals $13.2/0.4 = 33$ k (147 kN).

The allowable resultant shear load for bolt 3 is $(R_3)_{\text{all}} = 30(0.20) = 6$ k (27 kN) and the allowable load, P, for bolt 3 equals $6/0.15 = 40$ k (178 kN).

The allowable load for bolt 2 is $(R_2)_{\text{all}} = 30(0.79) = 23.7$ k (105 kN) and the allowable load, P, for bolt 2 equals $23.7/0.65 = 36.5$ k (162 kN). Thus, based on bolt shear, the allowable connector load equals 33 k (147 kN).

g. The bearing stress on the connector plate must also be checked in the bearing-type connector. The allowable bearing stress is $F_{\text{brg}} = 1.5F_y = 1.5(36) = 54$ ksi (370 MPa). The bearing stress at bolt 1 when $P = 33$ k (147 kN) is

$$F_{\text{brg}} = \frac{R_3}{D_1 t} = \frac{13.2}{0.44 \times 0.75} = 40.0 \text{ ksi (280 MPa)} < F_{\text{brg}}$$

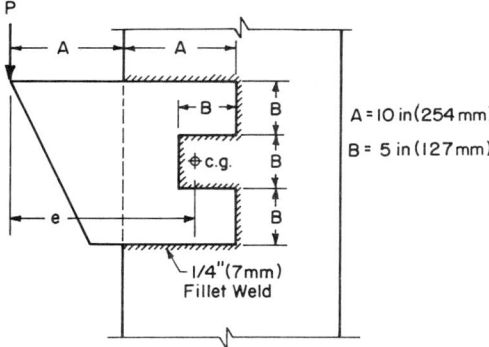

Fig. 8.26. Example 8.12.

The bearing stress at bolt 2 when $P = 33$ k (147 kN) is

$$f_{brg} = \frac{R_2}{(D_2 xt)} = \frac{0.65 \times 33}{1.0 \times 0.44} = 48.8 \text{ ksi (340 MPa)} < F_{brg}$$

The bearing stress at bolt 3 when $P = 33$ k (147 kN) is

$$f_{brg} = \frac{R_3}{(D_3 xt)} = \frac{0.15 \times 33}{0.5 \times 0.44} = 22.5 \text{ ksi (150 MPa)} < F_{brg}$$

Thus the shear in bolt 1 is the controlling factor for the connector and $P_{all} = 33$ kip (147 kN).

Example 8.12. Welded Connections. A bracket is welded to a structural member with a $1/4$ in. (7 mm) fillet weld with E70 electrodes. If the bracket is made of A36 steel and if the load is applied as shown in Fig. 8.26, evaluate the allowable load P.

Solution

a. Since the weld is subjected to shear, the allowable stress τ from Table 10.12 is the smaller of $0.3(70) = 21$ ksi (140 MPa) in the weld material or $0.4(36) = 14.4$ ksi (100 MPa) on the base metal.
b. Effective throat dimension t_e of the weld = $0.707(0.25) = 0.18$ in. (4.5 mm).
c. Locate the center of gravity of the weld group, with A_t = effec-

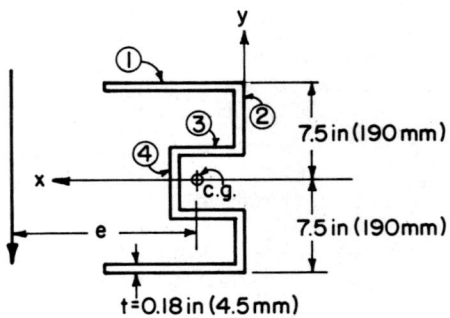

Fig. 8.27. Solution c to example 8.12.

tive weld area $= t_e \times \Sigma L_i$, where $\Sigma L_i =$ length of the weld (Fig. 8.27).

$$\bar{y} = 0$$
$$A_t \bar{x} = 2A_1 \bar{x}_1 + 2A_2 \bar{x}_2 + 2A_3 \bar{x}_3 + A_4 \bar{x}_4$$
$$(45t_e)\bar{x} = (20t_e)(5) + (10t_e)(2.5) + (5t_e)(5)$$
$$\bar{x} = 3.33 \text{ in. } (85 \text{ mm})$$

d. The applied load can be transformed to an equivalent load P acting through the center of gravity and a counterclockwise couple equal to Pe (Fig. 8.28).

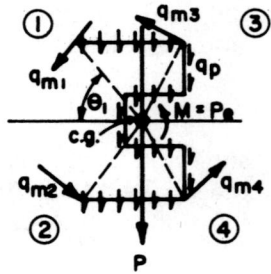

Fig. 8.28. Solution to Example 8.12 (part d).

e. The allowable load per inch of weld, or the shear flow,

$$q_{\text{all}} = (\tau)(A_{\text{eff}}) = 21(t_e \times 1) = 3.78 \text{ k/in.} \quad (660 \text{ kN/m})$$

for the weld material and

$$q_{\text{all}} = (\tau)(A) = 14.4(t \times 1) = 14.4(0.25 \times 1)$$
$$= 3.60 \text{ k/in.} \quad (630 \text{ kN/m})$$

for the base material.

f. The shear stress due to the shear load P is equal to $P/A_t = P/(t_e \Sigma L_i)$ and the subsequent shear flow q_p equals $q_p = (P/A_t)(t_e \times 1) = P/\Sigma L_i$. Thus $q_p = P/45 = 0.022P$.

g. The shear stress due to M is $\tau_m = Mr_i/J_e$ (see Chapter 7 on connectors for details of this analysis) and the subsequent shear flow at any point is $q_m = (\tau_m)(t_e \times 1) = Mr/J_e\,(0.707t)$. From the sketch it is apparent that points 1 and 2 are the most distant from the center of gravity and are the critical points. At point l

$$q_{m1} = \left[\frac{16.67P\,(6.67^2 + 7.5^2)^{1/2}}{J_e}\right](0.707t)$$

$J_e = \Sigma(I_{xi} + I_{yi}) = I_x + I_y$, where i represents the ith element of the weld group.

$$I_x = 2(A_1 d_1^2) + 2(\bar{I}_2 + A_2 d_2^2) + 2(A_3 d_3^2) + \bar{I}_4$$
$$= 2(10 \times 0.707t \times 7.5^2) + 2[(^1/_{12})(0.707t)(5)^3$$
$$+ 5(0.707t)(5)^2] + 2(5 \times 0.707t)(2.5)^2$$
$$+ \,^1/_{12}(0.707t)(5)^3$$
$$= 1468(0.707t)$$

$$I_y = 2(\bar{I}_1 + A_1 d_1^2) + 2A_2 d_2^2 + 2(\bar{I}_3 + A_3 d_3^2) + A_4 d_4^2$$
$$= 2[^1/_{12}(0.707t)(10^3) + 10(0.707t)(1.67)^2]$$
$$+ 2(0.707t)(5)(3.33)^2 + 2\,[^1/_{12}(0.707t)(5)^3$$
$$+ (0.707t)(0.83)^2] + (0.707t)(5)(1.67)^2$$
$$= 369(0.707t)$$

and

$$J_e = 1837(0.707t) = 330.7 \text{ in}^4 \ (5419.2 \text{ cm}^4)$$

Thus, upon substituting values

$$q_{m1} = 16.67P(6.67^2 + 7.5^2)^{1/2}/1837 = 0.091P$$

h. Resolving q_{m1} into x and y components and evaluating the resultant shear flow, q_1

$$(q_{m1})_x = q_{m1} \sin \theta_1, \quad \theta_1 = \tan^{-1} \frac{y_1}{x_1} = \tan^{-1} \frac{7.5}{6.67} = 48.3°$$

$$(q_{m1})_x = 0.091P(\sin 48.3) = 0.068\,P,$$

$$(q_{ml})_y = 0.019P(\cos 48.3) = 0.060P$$

$$q_1 = \sqrt{(q_{m1})_x^2 + [(q_{m1})_y + q_p]^2}$$
$$= \sqrt{(0.068P)^2 + [(0.060P + 0.022P)]^2}$$
$$= 0.106P$$

i. Since $q_1 = q_{\max}$ and $q_{\max}$ must be less than q_{all},

$$q_1 = 0.106P \leq q_{\text{all}} = 3.60 \text{ k/in.} \quad (630 \text{ kN/m})$$

Thus $P_{\text{all}} = 34 \text{ k } (150 \text{ kN})$.

PROBLEMS

8.1. Determine the allowable compressive load for a 20 ft (6.10 m) long M 6 × 20 (M 152 × 30) A36 steel column pinned at both ends with respect to bending about both the weak and strong axes.

8.2. Repeat Problem 8.1 if both ends are fixed with respect to each axis.

8.3. Repeat Problem 8.1 if both ends are fixed with respect to weak axis bending.

8.4. Repeat Problem 8.3 if the column is braced laterally with respect to the strong axis at midlength.

8.5. Select the lightest A36 W or M steel section to carry a kN) centric compressive load if the column is 16 ft (4.88 m) long and if the ends are both fixed with respect to both axes.

8.6. Repeat Problem 8.5 if the ends are both fixed with respect to the weak axis.

8.7. Determine the lightest A36 steel W section to carry a uniformly distributed flexural load of 2 k/ft over an unbraced span of (a) 6 ft (1.83 m); (b) 8 ft (2.44 m); (c) 12 ft (3.66 m); (d) 16 ft (4.88 m); (e) 20 ft (6.10 m); (f) 24 ft (7.32 m).

8.8. Determine the allowable flexural stress for a W 8 × 15 (W 203 × 22), A36 steel beam if it carries a load of 1 k/ft (14.6 kN/m) over an unbraced simply supported span of (a) 4 ft (1.22 m); (b) 5 ft (1.52 m); (c) 7 ft (2.13 m); (d) 10 ft (3.05 m); (e) 15 ft (4.57 m).

8.9. Repeat Problem 8.8 if the A36 steel is replaced with steel with a yield stress of 50 ksi (340 MPa).

8.10. Select the lightest A36 steel W section to carry five equally spaced truss reactive forces of 1240 lb (55.2 kN) over a simply supported span of 24 ft (7.32 m) if the beam is unbraced over its entire span. Repeat the problem if the beam is braced every 4 ft (1.22 m) at each truss reactive point.

8.11. Prove that Eq. (8.14) can be transformed into Eq. (8.17) for flexure about one axis.

8.12. Prove that Eq. (8.15) can be transformed into Eq. (8.18) for flexure about one axis.

8.13. Develop the form of Eqs. (8.16), (8.17), and (8.18) for flexure about both the x and y axes.

8.14. Repeat Example 8.10 if the bottom of the pole is fixed.

8.15. Repeat Example 8.10 if the column is 15 ft (4.57 m) long.

8.16. Repeat Example 8.10 if the axial load equals 25 kip (111 kN) and the moment equals 8 k-ft (10.8 kN-m).

8.17. Determine the moment capacity M of the connector groups shown in Figure 8.29 if $P = 0$.

8.18. Determine the load capacity of the connector groups shown in Figure 8.29 if $M = 0$.

8.19. Determine the allowable load P of the connector group in Figure 8.29 if $P = 50$ k (222 kN).

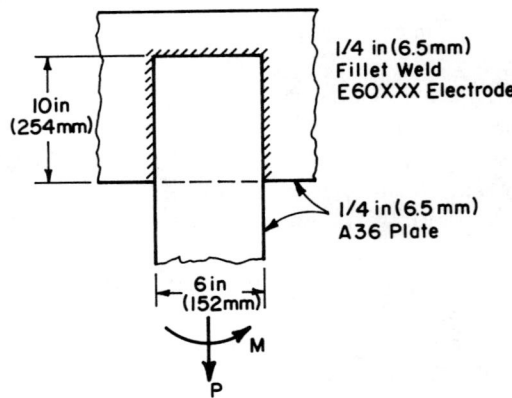

Fig. 8.29. Problems 8.17–8.19.

8.20. Determine the moment capacity of the single shear bearing-type connector group in Fig. 8.30 if the dimension is $A = 6$ in. (152 mm), the connectors are $\frac{1}{2}$-in. (13 mm) diameter A325 bolts with threads excluded from the shear plane, and the connected parts are $\frac{1}{4}$ in. (6.5 mm) thick.

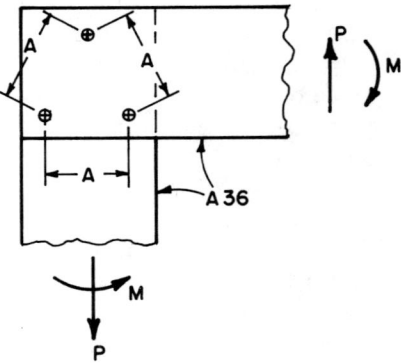

Fig. 8.30. Problems 8.20–8.21.

8.21. Repeat Problem 8.20 for a friction-type connector group.

NOMENCLATURE FOR CHAPTER 8

A	Gross cross-sectional area of a load carrying member, in.² (mm²)
A_{net}	Net cross-sectional area of a member after removal of area for holes, etc., in.² (mm²)
A_w	Area of girder web, in.² (mm²)
C_b	Bending coefficient dependent upon moment gradient; equal to $1.75 + 1.05(M_1/M_2) + 0.3(M_1/M_2)^2$
C_c	Column slenderness ratio dividing elastic and inelastic buckling; equal to $(\pi^2 E/F_y)^{1/2}$
C_m	Coefficient applied to bending term in interaction formula and dependent upon column curvature caused by applied moments
D	Diameter of a circular tube section, in. (mm)
E	Modulus of elasticity of steel, 29,000 ksi (200 GPa)
F_a	Axial stress permitted in the absence of bending moment, ksi (MPa)
F_{as}	Axial compressive stress, permitted in the absence of bending moment, for bracing and other secondary members, ksi (MPa)
F_b	Bending stress permitted in the absence of axial force, ksi (MPa)
F'_e	Euler stress for a prismatic member divided by factor of safety; $(12\pi^2 E)/[23(KL/r)_b^2]$, ksi (MPa)
F_t	Allowable tensile stress, ksi (MPa)
F_v	Allowable shear stress, ksi (MPa)
F_y	Specified minimum yield stress of the type of steel being used (kip per square inch). In this chapter, "yield stress" denotes either the specified minimum yield point (for those steels that have a yield point) or specified minimum yield strength (for those steels that do not have a yield point), ksi (MPa)
F'_y	Theoretical yield stress above which the section is not compact due to flange buckling, ksi (MPa)
F''_y	Theoretical yield stress above which the section is not compact due to web crippling, ksi (MPa)
F'''_y	Theoretical yield stress above which the section is not compact under combined axial and flexural loading, ksi (MPa)
I	Second moment of the area of a cross section
K	Effective length factor for columns

L	Beam span length, ft (m); column actual unbraced length, ft (m)
L_b	For beams, the distance between cross sections braced against twist or lateral displacement of the compression flange, ft (m); for columns or beam columns, the unbraced length in the plane of bending, ft (m)
L_c	The allowable unbraced length of the compression flange of a beam for which the plastic moment is reached before the section buckles laterally, ft (m)
L_u	The allowable unbraced length of the compression flange of a beam for which the elastic moment is reached before the section buckles laterally, ft (m)
M	Moment, k-ft (N-m)
M_{all}	Allowable moment of a section, k-ft (N-m)
M_1	Smaller moment at end of unbraced length of beam-column, k-ft (N-m)
M_2	Larger moment at end of unbraced length of beam-column, k-ft (N-m)
M_D	Moment produced by dead load, k-ft (N-m)
M_L	Moment produced by live load, k-ft (N-m)
M_p	Plastic moment, k-ft (N-m)
N	Length of bearing of applied load, in. (mm)
P	Applied concentrated load, k (N)
R	Reaction or concentrated transverse load applied to beam or girder, k (N)
S	Section modulus of a cross section, in.^{3} (cm^3)
b	Actual width of stiffened and unstiffened compression elements, in. (mm)
b_e	Effective width of stiffened compression element, in. (mm)
b_f	Flange width of rolled beam or plate girder, in. (mm)
c	Distance from neutral axis to extreme fiber of beam, in. (mm)
d	Depth of beam or girder, in. (mm)
d_e	Column web depth clear of fillets, in. (mm)
f_a	Computed axial stress, ksi (MPa)
f_t	Computed tensile stress, ksi (MPa)
f_v	Computed shear stress, ksi (MPa)
g	Transverse spacing between fastener gauge lines, in. (mm)
r	Governing radius of gyration, in. (mm)
r_b	Radius of gyration about axis of concurrent bending in beam-columns, in. (mm)

r_t	Radius of gyration of a section comprising the compression flange plus one third of the compression web area taken about an axis in the plane of the web, in. (mm)
r_y	Radius of gyration about the weak (y) axis, in. (mm)
s	Spacing (pitch) between successive holes in line of stress, in. (mm)
t	Girder, beam, or column web thickness, in. (mm)
t_f	Flange thickness, in. (mm)
t_w	Web thickness, in. (mm)
x	Subscript relating symbol to strong axis bending
y	Subscript relating symbol to weak axis bending
ν	Poisson's ratio, may be taken as 0.3 for steel

9

Cold-Formed Steel Design

9.1 INTRODUCTION

9.1.1 General

The applications of light gauge cold-formed steel in agricultural and other light structures, equipment, and machinery are probably more numerous and varied than those of hot-rolled steel. Cold-formed sections find use as purlins, wall studs, siding, roofing, grain bin walls, components of feeding and watering equipment, components of livestock housing equipment (such as cage laying systems, coverings, and panels), and light secondary framing for machinery. Cold-formed steel is an economical material choice where loads are light to moderate and where structural strength is required along with a unique shape which can be fabricated by a press brake or cold-forming process.

Cold-formed steel design procedures are similar to the working strength methods for hot-rolled steel design. Elastic behavior of the material is assumed; e.g., bending stresses are assumed to be equal to M/S and then the allowable stress F_b is reduced accordingly to compensate for discrepancies between actual and assumed behavior and to simplify some complex and tedious theoretical procedures.

9.1.2 List of Symbols

The symbols used throughout the chapter are defined in the list at the end of the chapter. In dimensional equations, the customary English units should be used. In dimensionless equations, either the customary or SI units may be used. In specification equations which are dimensional, only the customary units will be given. The final result, of course, may then be converted to SI units. When allowable stress equations or limiting values of parameters such as w/t are evaluated, it is suggested that customary units always be used. In numerous places in the chapter a stress term appears under the radical; e.g., $\sqrt{F_y}$, $\sqrt{F}$, $\sqrt{f}$. Whenever this form appears the units of stress are ksi.

9.2 DIFFERENCES BETWEEN COLD-FORMED AND HOT-ROLLED STEEL

9.2.1 Element Thickness

The primary difference between the two types of steel sections is the thickness of the elements. Cold-formed sections may have element thickness as great as 1 in. (25.4 mm). Typical sections, however, are fabricated from sheet stock less than $\frac{1}{8}$ in. (3.2 mm) thick. Local buckling of the thin elements, at stresses *below the elastic strength of the element,* is a common mode of failure. The cold-formed steel specification considers and accounts for allowable stress reductions due to local buckling by reducing allowable stresses in UCEs (unstiffened compression elements) and reducing effective element widths of SCEs (stiffened compression elements). These procedures will be discussed herein.

In hot-rolled steel many sections are able to reach plastic moments before buckling locally. In cold-formed shapes most sections are unserviceable long before developing the plastic moment. Thus, the cold-formed specification has no provisions for plastic design.

Many light-gauge cold-formed members are also open, nonsymmetric shapes. These shapes, especially when made of thin elements, have a low torsional rigidity. Thus, they can fail by buckling torsionally or by flexural–torsional buckling at stresses below the yield stress. That is, they can fail by suddenly twisting about the shear center of the section either before flexural buckling or simultaneously with flexural buckling.

Thin, cold-formed sections can fail in any of the following modes: yielding, flexural buckling (gross), local buckling of elements, lateral–torsional buckling, torsional buckling, or flexural–torsional buckling. The designer must establish the mode of failure which commences first along with the associated stress level.

9.2.2 Section Geometry

Since cold-formed sections are often fabricated by press brake processes, a wide variety of section shapes are possible. Actually, the only limitations on section shapes are the needs and imagination of the designer. This is in contrast to the limited number of standard sections, such as wide flange, channel, angle, and flat, available in hot-rolled steel. Figure 9.1 illustrates a few cold-formed shapes. Probably the sections which are most common among standard shapes are the channel, hat, Z, and angle. Properties of some of the standard shapes are given in Tables 9.1 to 9.9. The section properties of the many nonstandard cold-formed shapes are not available in tables and must be evaluated by the designer. Procedures for calculating the properties are the topic of a later section.

9.2.3 Strain Hardening

There are no residual thermal stresses in cold-formed steel sections. Instead, residual stresses due to cold-forming processes develop in corners of the sections. Cold forming induces strain hardening and strain aging of the material, which increases the material yield stress and reduces the material ductility in the vicinity of the corners.

9.2.4 Stiffened and Unstiffened Compression Elements

The definition of a cold-formed UCE is identical to that for hot-rolled sections. Stiffened compression elements in cold-formed steel are similar to those in hot-rolled sections except that edge-stiffened elements (see Figs. 9.1a–c, 9.1f) are considered to be SCEs if the formed edge is sufficiently large. For example, if the edge stiffeners are sufficiently large in Fig. 9.1c, the unstiffened legs of the angle can be treated as stiffened elements.

Both SCEs and UCEs (with one edge stiffened in the case of an UCE) deform in a two-dimensional pattern when loaded to the point of local, elastic buckling by in-plane compressive forces. Figure 9.2a illustrates local buckling of a compression flange of a hat section with both edges stiffened. The deformation pattern for a plate element with two edges stiffened (SCE) will be similar to that of Fig. 9.2b at loads which produce local elastic buckling. This deformation pattern is sometimes called "postbuckling action," because it occurs only if the plate element is loaded beyond the point of incipient buckling. The two-dimensional

(text continues on page 297)

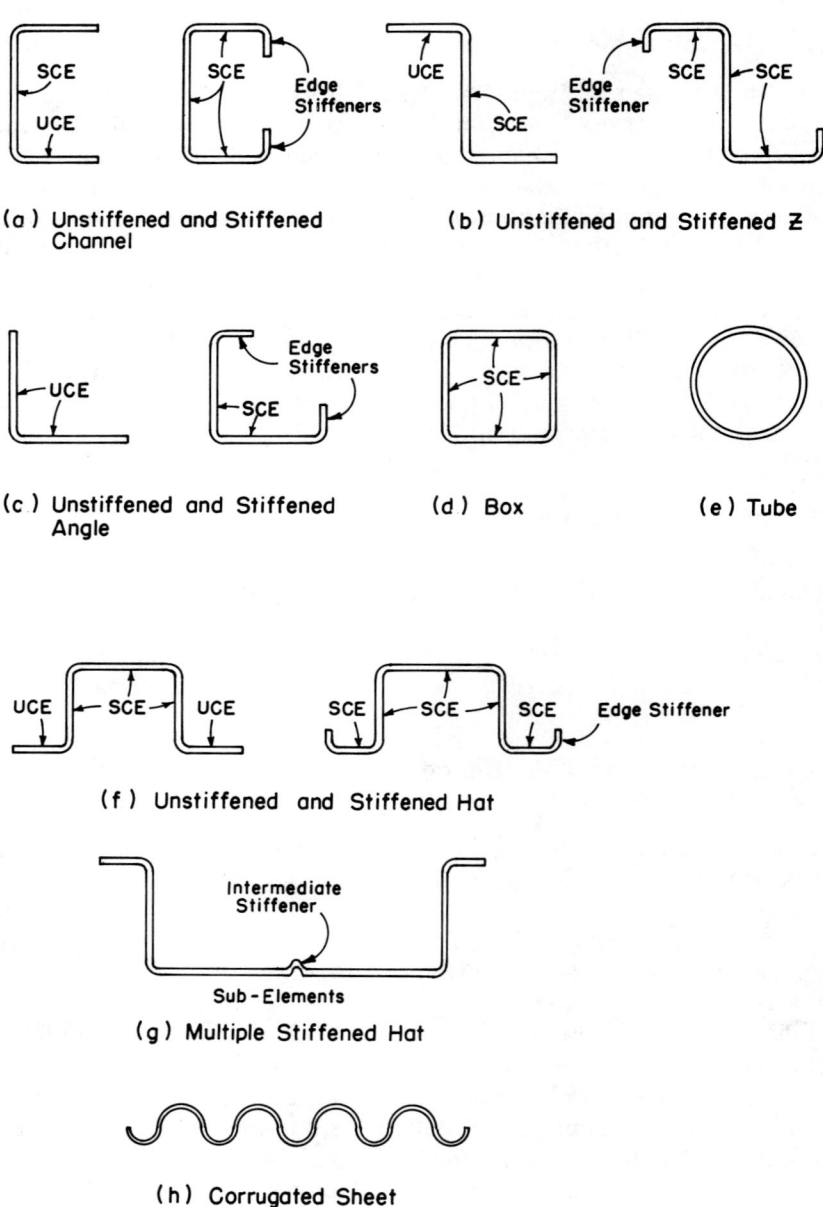

Fig. 9.1. Typical cold-formed steel sections (all edge stiffeners assumed adequate in size).

9.2 Differences Between Cold-Formed and Hot-Rolled Steel

Table 9.1. Channel with Stiffened Flanges.

Effective section for beam strength about the x-x axis where w/t of compression flange exceeds 38.2 for F=20 ksi and 31.2 for F=30 ksi

Size			t	d	R	Area	Wgt per Foot	Beam Strength Effective S_x		Axis x-x				Axis y-y								
D	B							$F=20$ ksi	$F=30$ ksi	I_x	S_x	r_x		I_y	S_y	r_y	x	m	J	C_w	j	r_0	x_0
In.	In.		In.	In.	In.	In.²	Lb	In.³	In.³	In.⁴	In.³	In.		In.⁴	In.³	In.	In.	In.	In.⁴	In.⁶	In.	In.	In.
12.0	3.50		.135	1.00	3/16	2.70	9.42	9.37	9.37	56.2	9.37	4.56		4.02	1.55	1.22	.910	1.48	.0164	117.	6.79	5.26	−2.33
			.105	0.90	3/16	2.10	7.31	7.31	7.31	43.8	7.31	4.57		3.09	1.18	1.21	.883	1.46	.00770	88.6	6.84	5.26	−2.29
10.0	3.50		.135	1.00	3/16	2.43	8.48	7.30	7.30	36.5	7.30	3.87		3.81	1.52	1.25	1.00	1.58	.0148	78.0	6.60	4.78	−2.51
			.105	0.90	3/16	1.89	6.57	5.70	5.70	28.5	5.70	3.89		2.93	1.16	1.25	.975	1.55	.00693	59.1	5.64	4.77	−2.48
			.075	0.70	3/32	1.34	4.67	4.00	4.00	20.5	4.10	3.91		2.02	.783	1.23	.926	1.46	.00251	38.8	5.59	4.72	−2.35
9.0	3.25		.135	1.00	3/16	2.23	7.77	6.03	6.03	27.2	6.03	3.49		3.07	1.34	1.17	.965	1.51	.0135	52.2	5.03	4.40	−2.40
			.105	0.80	3/16	1.71	5.95	4.66	4.66	21.0	4.66	3.50		2.26	.964	1.15	.910	1.44	.00628	36.7	5.12	4.34	−2.30
			.075	0.70	3/32	1.23	4.28	3.39	3.21	16.3	3.40	3.53		1.63	.689	1.15	.888	1.39	.00230	25.7	5.03	4.34	−2.24
8.0	3.00		.135	0.90	3/16	2.00	6.97	4.82	4.82	19.3	4.82	3.11		2.32	1.11	1.08	.901	1.40	.0122	31.2	4.53	3.97	−2.23
			.105	0.80	3/16	1.55	5.40	3.78	3.78	15.1	3.78	3.12		1.71	.839	1.07	.873	1.37	.00570	23.5	4.57	3.96	−2.19
			.075	0.70	3/32	1.12	3.89	2.76	2.67	11.0	2.76	3.15		1.29	.600	1.08	.851	1.32	.00209	16.3	4.48	3.95	−2.14
			.060	0.60	3/32	.885	3.08	2.11	1.99	8.79	2.20	3.15		.997	.458	1.06	.822	1.29	.00106	12.4	4.51	3.92	−2.08
7.0	2.75		.135	0.80	3/16	1.77	6.17	3.75	3.75	13.1	3.75	2.72		1.70	.891	.981	.837	1.29	.0108	17.5	4.02	3.55	−2.06
			.105	0.80	3/16	1.45	5.06	2.98	2.98	10.4	2.98	2.74		1.38	.722	.996	.837	1.30	.00512	14.3	4.02	3.58	−2.09
			.075	0.70	3/32	1.00	3.50	2.19	2.17	7.66	2.19	2.76		1.09	.517	.999	.815	1.25	.00188	9.94	3.95	3.57	−2.03
			.060	0.60	3/32	.796	2.77	1.72	1.62	6.10	1.74	2.77		.772	.393	.986	.786	1.22	.000954	7.47	3.98	3.54	−1.97
6.0	2.50		.135	0.70	3/16	1.54	5.37	2.80	2.80	8.41	2.80	2.34		1.21	.698	.884	.774	1.18	.00937	9.10	3.53	3.13	−1.88
			.105	0.70	3/16	1.21	4.23	2.24	2.24	6.72	2.24	2.35		.982	.569	.899	.774	1.19	.00446	7.49	3.53	3.16	−1.91
			.075	0.60	3/32	.891	3.10	1.68	1.68	5.03	1.68	2.38		.755	.439	.921	.780	1.18	.00167	5.71	3.43	3.19	−1.93
			.060	0.60	3/32	.705	2.46	1.34	1.28	4.01	1.34	2.39		.583	.333	.909	.752	1.15	.000854	4.26	3.46	3.16	−1.87
5.0	2.50		.135	0.70	3/16	1.27	4.43	1.87	1.87	4.68	1.87	1.92		.648	.478	.714	.644	1.18	.00773	3.61	2.89	2.58	−1.56
			.105	0.70	3/16	1.00	3.50	1.50	1.50	3.76	1.50	1.93		.533	.393	.728	.643	1.00	.00369	3.01	2.90	2.61	−1.59
			.075	0.60	3/32	.726	2.53	1.12	1.12	2.80	1.12	1.96		.389	.283	.733	.622	1.19	.00136	2.06	2.83	2.60	−1.53
			.060	0.60	3/32	.613	2.10	.891	.891	2.23	.891	1.97		.298	.212	.721	.594	1.18	.000687	1.51	2.86	2.57	−1.48
			.048	0.50	3/32	.461	1.61	.722	.697	1.80	.722	1.98		.244	.173	.709	.594	1.15	.000354	1.25	2.86	2.58	−1.49
4.0	2.00		.135	0.70	3/16	1.14	3.96	1.38	1.38	2.75	1.38	1.56		.598	.464	.725	.712	1.05	.00691	2.31	2.52	2.41	−1.70
			.105	0.70	3/16	.900	3.13	1.11	1.11	2.22	1.11	1.57		.492	.382	.740	.712	1.07	.00331	1.93	2.53	2.45	−1.73
			.075	0.60	3/32	.651	2.27	.832	.832	1.66	.832	1.60		.361	.275	.745	.689	1.01	.00122	1.31	2.48	2.42	−1.66
			.060	0.60	3/32	.613	2.00	.665	.665	1.33	.665	1.61		.277	.206	.734	.660	.972	.000615	.943	2.49	2.39	−1.60
			.048	0.50	3/32	.413	1.44	.539	.539	1.08	.539	1.62		.226	.169	.740	.660	.979	.000317	.777	2.49	2.40	−1.61
3.5	2.00		.135	0.70	3/16	1.07	3.73	1.14	1.14	2.00	1.14	1.37		.568	.456	.729	.753	1.09	.00650	1.79	2.39	2.36	−1.78
			.105	0.70	3/16	.847	2.95	.926	.926	1.62	.926	1.38		.468	.376	.743	.753	1.11	.00311	1.50	2.40	2.39	−1.81
			.075	0.60	3/32	.613	2.14	.698	.698	1.22	.698	1.41		.344	.271	.749	.729	1.04	.00115	1.01	2.35	2.36	−1.73
			.060	0.50	3/32	.483	1.68	.559	.559	.979	.559	1.42		.264	.203	.749	.699	1.00	.000679	.717	2.36	2.32	−1.67
			.048	0.50	3/32	.389	1.36	.454	.454	.795	.454	1.43		.216	.166	.746	.699	1.01	.000299	.592	2.33	2.33	−1.69
3.0	1.75		.105	0.70	3/16	.742	2.59	.678	.678	1.02	.678	1.17		.318	.299	.654	.689	1.02	.00273	.835	2.10	2.13	−1.65
			.075	0.50	3/32	.528	1.82	.509	.509	.763	.509	1.19		.219	.196	.647	.635	.906	.000981	.460	2.04	2.04	−1.50
			.060	0.50	3/32	.423	1.47	.416	.416	.624	.416	1.21		.181	.162	.654	.635	.912	.000507	.385	2.05	2.05	−1.52
			.048	0.40	3/32	.351	1.16	.332	.332	.497	.332	1.23		.136	.119	.622	.604	.872	.000256	.267	2.05	2.01	−1.45

Reproduced with permission from *Cold Formed Steel Design Manual: Part V—Charts and Tables*. 1977 ed. American Iron and Steel Institute, Washington, DC.

288 Cold-Formed Steel Design

Table 9.2. Channel with Unstiffened Flanges.

Size		t	R	Area	Wgt per Foot	Axis x-x			Axis y-y				Properties of Full Section							Allowable Beam Stress F_b	
D	B					I_x	S_x	r_x	I_y	S_y	r_y	x	m	J	C_w	j	r_0	x_0	$F_y=20$ ksi	$F_y=30$ ksi
In.	In.	In.	In.	In.²	Lb.	In.⁴	In.³	In.	In.⁴	In.³	In.	In.	In.	In.⁴	In.⁶	In.	In.	In.	ksi	ksi
8.	2.00	.135	3/16	1.55	5.42	13.1	3.27	2.90	.485	.302	.559	.393	.555	.00944	5.12	4.27	3.08	-.881	19.1	26.8
		.105	3/16	1.22	4.24	10.3	2.58	2.91	.386	.238	.563	.381	.569	.00447	4.18	4.41	3.10	-.897	17.2	23.2
		.075	3/32	.880	3.07	7.60	1.90	2.94	.283	.173	.567	.366	.574	.00165	3.13	4.57	3.13	-.903	13.1	13.4
		.060	3/32	.706	2.46	6.13	1.53	2.95	.229	.140	.569	.360	.581	.000848	2.57	4.64	3.13	-.911	11.2	11.2
7.	1.50	.135	3/16	1.28	4.47	7.84	2.24	2.47	.204	.168	.399	.284	.378	.00780	1.63	3.88	2.57	-.595	19.8	30.0
		.105	3/16	1.01	3.51	6.22	1.78	2.49	.164	.133	.404	.272	.391	.00370	1.35	4.06	2.59	-.610	19.6	27.6
		.075	3/32	.730	2.55	4.60	1.32	2.51	.121	.0975	.407	.257	.397	.00137	1.03	4.26	2.62	-.617	16.4	21.8
		.060	3/32	.586	2.04	3.72	1.06	2.52	.0984	.0788	.410	.251	.403	.000704	.846	4.33	2.63	-.624	14.1	15.9
6.	1.50	.135	3/16	1.15	4.00	5.33	1.78	2.15	.197	.166	.414	.310	.406	.00698	1.13	3.03	2.29	-.648	19.8	30.0
		.105	3/16	.901	3.14	4.24	1.41	2.17	.158	.132	.419	.298	.419	.00331	.937	3.17	2.31	-.664	19.6	27.6
		.075	3/32	.656	2.28	3.15	1.05	2.19	.117	.0962	.423	.283	.424	.00123	.715	3.34	2.33	-.669	16.4	21.8
		.060	3/32	.526	1.83	2.55	.849	2.20	.0961	.0777	.425	.277	.431	.000632	.590	3.41	2.34	-.677	14.1	15.9
		.048	3/32	.423	1.47	2.06	.685	2.21	.0770	.0627	.427	.272	.436	.000325	.484	3.46	2.35	-.684	11.9	11.9
5.	1.25	.105	3/16	.744	2.59	2.40	.960	1.80	.0695	.0900	.347	.256	.344	.00273	.360	2.56	1.91	-.547	19.8	29.8
		.075	3/32	.543	1.89	1.80	.719	1.82	.0667	.0661	.350	.241	.349	.00102	.278	2.72	1.93	-.553	18.1	24.9
		.060	3/32	.436	1.52	1.46	.583	1.83	.0543	.0535	.353	.235	.355	.000524	.231	2.79	1.94	-.561	16.2	21.3
		.048	3/32	.351	1.22	1.18	.471	1.83	.0441	.0432	.355	.230	.361	.000269	.190	2.84	1.96	-.567	13.8	15.0
4.	1.125	.105	3/16	.613	2.13	1.29	.643	1.45	.0623	.0713	.319	.251	.322	.00225	.156	1.89	1.57	-.521	19.8	30.0
		.075	3/32	.449	1.56	.973	.486	1.47	.0467	.0525	.323	.235	.327	.000842	.122	2.08	1.60	-.525	18.9	26.4
		.060	3/32	.361	1.26	.791	.396	1.48	.0382	.0426	.325	.229	.334	.000434	.102	2.09	1.61	-.534	17.2	23.2
		.048	3/32	.291	1.01	.640	.320	1.48	.0310	.0345	.327	.225	.340	.000223	.0841	2.14	1.61	-.540	15.1	19.1
3.	1.125	.105	3/16	.508	1.77	.636	.424	1.12	.0573	.0688	.336	.292	.360	.00187	.0782	1.35	1.31	-.599	19.8	30.0
		.075	3/32	.374	1.30	.487	.324	1.14	.0432	.0508	.340	.275	.363	.000701	.0615	1.44	1.33	-.601	18.9	26.4
		.060	3/32	.301	1.05	.397	.265	1.15	.0353	.0412	.342	.269	.371	.000362	.0615	1.49	1.34	-.610	17.2	23.2
		.048	3/32	.243	.846	.322	.215	1.15	.0287	.0334	.344	.264	.376	.000185	.0437	1.52	1.35	-.617	15.1	19.1
2.	1.125	.105	3/16	.403	1.40	.241	.241	.773	.0497	.0646	.351	.355	.407	.00148	.0294	1.01	1.11	-.710	19.8	30.0
		.075	3/32	.299	1.04	.188	.188	.792	.0379	.0480	.355	.335	.408	.000561	.0234	1.07	1.12	-.706	18.9	26.4
		.060	3/32	.241	.841	.154	.154	.798	.0310	.0389	.358	.329	.416	.000290	.0197	1.11	1.13	-.716	17.2	23.2
		.048	3/32	.195	.678	.126	.126	.804	.0253	.0315	.360	.324	.422	.000149	.0164	1.13	1.14	-.722	15.1	19.1

Reproduced with permission from *Cold Formed Steel Design Manual: Part V—Charts and Tables*, 1977 ed. American Iron and Steel Institute, Washington, DC.

9.2 Differences Between Cold-Formed and Hot-Rolled Steel

Table 9.3. Z Section with Stiffened Flanges.

Effective section for beam strength about the x-x axis where w/t of compression flange exceeds 38.2 for F = 20 ksi and 31.2 for F = 30 ksi.

Size		t	d	R	Area	Wgt per Foot	Effective S_x		Beam Strength Properties of Full Section										$90° - \theta$
							F = 20 ksi	F = 30 ksi	L_x		Axis x-x			Axis y-y				Axis x_c-x_c	
D	B									I_x	S_x	r_x	L_y	I_y	S_y	r_y	r_{min}	
In.	In.	In.	In.	In.	In.²	Lb	In.³	In.³	In.⁴	In.⁴	In.³	In.	In.⁴	In.⁴	In.³	In.	In.	Deg.
12.0	3.50	.135	1.00	3/16	2.70	9.42	9.37	9.37	13.1	56.2	9.37	4.56	5.94	6.94	1.73	1.48	1.00	0.999	13.8
		.105	0.90	3/16	2.10	7.31	7.31	7.31	10.1	43.8	7.31	4.57	4.54	4.54	1.32	1.47	0.999		13.6
10.0	3.50	.135	1.00	3/16	2.43	8.48	7.30	7.30	10.8	36.5	7.30	3.87	5.94	5.94	1.73	1.56	1.01		17.7
		.105	0.90	3/16	1.89	6.57	5.70	5.70	8.35	28.5	5.70	3.89	4.54	4.54	1.32	1.55	1.01		17.4
		.075	0.70	3/32	1.34	4.67	4.00	3.78	5.80	20.5	4.10	3.91	3.07	3.07	0.888	1.51	0.992		16.8
9.0	3.25	.135	1.00	3/16	2.23	7.77	6.03	6.03	8.50	27.2	6.03	3.49	4.87	4.87	1.53	1.48	0.946		18.7
		.105	0.80	3/16	1.71	5.95	4.66	4.66	6.32	21.0	4.66	3.50	3.51	3.51	1.10	1.43	0.924		18.0
		.075	0.70	3/32	1.23	4.28	3.39	3.21	4.56	15.3	3.40	3.53	2.52	2.52	0.783	1.43	0.927		17.8
8.0	3.00	.135	0.90	3/16	2.00	6.97	4.82	4.82	6.28	19.3	4.82	3.11	3.71	3.71	1.27	1.36	0.864		19.4
		.105	0.80	3/16	1.55	5.40	3.78	3.78	4.85	15.1	3.78	3.12	2.83	2.83	0.959	1.35	0.859		19.2
		.075	0.70	3/32	1.12	3.89	2.67	2.76	3.50	11.0	2.76	3.15	2.03	2.03	0.685	1.35	0.862		18.9
		.060	0.60	3/32	0.886	3.08	2.11	1.99	2.73	8.79	2.20	3.15	1.55	1.55	0.622	1.32	0.851		18.5
7.0	2.75	.135	0.80	3/16	1.77	6.17	3.75	3.75	4.49	13.1	3.75	2.72	2.76	2.76	1.03	1.25	0.781		20.4
		.105	0.70	3/16	1.39	4.85	2.98	2.98	3.61	10.4	2.98	2.74	2.24	2.24	0.829	1.27	0.793		20.7
		.075	0.60	3/32	1.00	3.50	2.19	2.17	2.61	7.66	2.19	2.76	1.61	1.61	0.592	1.27	0.796		20.4
		.060	0.60	3/32	0.795	2.77	1.72	1.62	2.03	6.10	1.74	2.77	1.23	1.23	0.451	1.24	0.785		19.9
6.0	2.50	.135	0.70	3/16	1.54	5.37	2.80	2.80	3.06	8.41	2.80	2.34	1.97	1.97	0.812	1.13	0.698		21.8
		.105	0.70	3/16	1.21	4.23	2.24	2.24	2.47	6.72	2.24	2.35	1.61	1.61	0.659	1.15	0.709		22.1
		.075	0.60	3/32	0.891	3.10	1.68	1.68	1.88	5.03	1.68	2.38	1.25	1.25	0.506	1.18	0.728		22.4
		.060	0.60	3/32	0.706	2.46	1.34	1.28	1.46	4.01	1.34	2.39	0.960	0.960	0.384	1.16	0.718		21.8
5.0	2.00	.135	0.70	3/16	1.27	4.43	1.87	1.87	1.68	4.68	1.87	1.92	1.07	1.07	0.554	0.917	0.568		21.5
		.105	0.70	3/16	1.00	3.50	1.50	1.50	1.37	3.76	1.50	1.93	0.884	0.884	0.454	0.938	0.579		21.8
		.075	0.60	3/32	0.726	2.53	1.12	1.12	0.998	2.80	1.12	1.96	0.637	0.637	0.325	0.937	0.583		21.4
		.060	0.50	3/32	0.573	2.00	0.891	0.891	0.771	2.23	0.891	1.97	0.480	0.480	0.244	0.915	0.573		20.7
		.048	0.50	3/32	0.461	1.61	0.722	0.697	0.629	1.80	0.722	1.98	0.393	0.393	0.199	0.924	0.577		20.9
4.0	2.00	.135	0.70	3/16	1.14	3.96	1.38	1.38	1.31	2.75	1.38	1.56	1.07	1.07	0.554	0.970	0.568		28.7
		.105	0.70	3/16	0.900	3.13	1.11	1.11	1.07	2.22	1.11	1.57	0.884	0.884	0.454	0.991	0.567		29.0
		.075	0.60	8/32	0.651	2.27	0.832	0.832	0.786	1.66	0.832	1.60	0.637	0.637	0.325	0.990	0.570		28.4
		.060	0.50	8/32	0.513	1.79	0.665	0.665	0.610	1.33	0.665	1.61	0.480	0.480	0.244	0.967	0.561		27.6
		.048	0.50	8/32	0.413	1.44	0.539	0.520	0.497	1.08	0.539	1.62	0.393	0.393	0.199	0.976	0.565		27.7
3.5	2.00	.135	0.70	3/16	1.07	3.73	1.14	1.14	1.13	2.00	1.14	1.37	1.07	1.07	0.554	1.00	0.542		33.8
		.105	0.60	3/16	0.847	2.96	0.926	0.926	0.923	1.62	0.926	1.38	0.884	0.884	0.454	1.02	0.552		34.1
		.075	0.60	3/32	0.613	2.14	0.698	0.698	0.680	1.22	0.698	1.41	0.637	0.637	0.325	1.02	0.556		33.4
		.060	0.50	3/32	0.483	1.68	0.559	0.559	0.529	0.979	0.559	1.42	0.480	0.480	0.244	0.997	0.546		32.4
		.048	0.50	3/32	0.389	1.36	0.454	0.437	0.432	0.796	0.454	1.43	0.393	0.393	0.199	1.01	0.551		32.5
3.0	1.75	.105	0.70	3/16	0.742	2.59	0.678	0.678	0.610	1.02	0.678	1.17	0.618	0.618	0.364	0.913	0.487		35.9
		.075	0.50	3/32	0.623	1.82	0.509	0.509	0.430	0.763	0.509	1.21	0.406	0.405	0.237	0.880	0.476		35.7
		.060	0.50	3/32	0.423	1.47	0.416	0.416	0.354	0.624	0.416	1.21	0.335	0.335	0.196	0.891	0.481		33.9
		.048	0.40	3/32	0.331	1.16	0.332	0.332	0.272	0.497	0.382	1.23	0.248	0.248	0.144	0.865	0.471		32.7

Reproduced with permission from *Cold Formed Steel Design Manual: Part V – Charts and Tables*, 1977 ed. American Iron and Steel Institute, Washington, DC.

Table 9.4. Z Section with Unstiffened Flanges.

Size		t	R	Area	Weight per Foot	I_{xy}	Axis x-x			Axis y-y			Axis x_1-x_1		90°-θ	Allowable Beam Stress F_a	
D	B						I_x	S_x	r_x	I_y	S_y	r_y	r_{min}			F=20	F=30
In.	In.	In.	In.	In.²	Lb	In.⁴	In.⁴	In.³	In.	In.⁴	In.³	In.	In.		Deg.	ksi	ksi
8.	2.00	.135	3/16	1.55	5.42	1.99	13.1	3.27	2.90	.650	.336	.647	.468		8.87	19.1	26.8
		.105	3/16	1.22	4.24	1.58	10.3	2.58	2.91	.517	.266	.652	.472		8.90	17.2	23.2
		.075	3/32	.880	3.07	1.15	7.60	1.90	2.94	.378	.193	.655	.477		8.80	13.1	13.4
		.060	3/32	.706	2.46	.925	6.13	1.53	2.95	.306	.155	.658	.480		8.81	11.2	11.2
7.	1.50	.135	3/16	1.28	4.47	.955	7.84	2.24	2.47	.265	.185	.454	.337		7.07	19.8	30.0
		.105	3/16	1.01	3.51	.761	6.22	1.78	2.49	.212	.147	.459	.342		7.11	19.6	27.6
		.075	3/32	.730	2.55	.556	4.60	1.32	2.51	.156	.107	.463	.347		7.02	16.4	21.8
		.060	3/32	.586	2.04	.450	3.72	1.06	2.52	.127	.0864	.466	.349		7.04	14.1	15.9
6.	1.50	.135	3/16	1.15	4.00	.816	5.33	1.78	2.15	.265	.185	.480	.345		8.92	19.8	30.0
		.105	3/16	.901	3.14	.651	4.24	1.41	2.17	.212	.147	.485	.349		8.95	19.6	27.6
		.075	3/32	.655	2.28	.476	3.15	1.05	2.19	.156	.107	.489	.355		8.82	16.4	21.8
		.060	3/32	.526	1.83	.385	2.55	.849	2.20	.127	.0864	.491	.357		8.84	14.1	16.9
		.048	3/32	.423	1.47	.311	2.05	.685	2.21	.103	.0697	.494	.359		8.85	11.9	11.9
5.	1.25	.106	3/16	.744	2.59	.370	2.40	.960	1.80	.120	.100	.402	.288		9.01	29.8	29.8
		.075	3/32	.543	1.89	.272	1.80	.719	1.82	.0891	.0735	.405	.294		8.83	18.1	24.9
		.060	3/32	.436	1.52	.221	1.46	.583	1.83	.0726	.0595	.408	.296		8.85	16.2	21.3
		.048	3/32	.351	1.22	.179	1.18	.471	1.83	.0590	.0481	.410	.298		8.87	13.8	15.0
4.	1.125	.105	3/16	.613	2.13	.237	1.29	.643	1.45	.0864	.0806	.376	.260		10.8	19.8	30.0
		.075	3/32	.449	1.56	.174	.973	.486	1.47	.0643	.0691	.378	.267		10.5	18.1	26.4
		.060	3/32	.361	1.26	.142	.791	.395	1.48	.0525	.0480	.381	.269		10.5	18.6	23.2
		.048	3/32	.291	1.01	.115	.640	.350	1.48	.0427	.0388	.383	.271		10.5	15.1	19.1
3.	1.125	.106	3/16	.508	1.77	.176	.636	.424	1.12	.0864	.0806	.413	.263		16.3	19.8	30.0
		.075	3/32	.374	1.30	.130	.487	.324	1.14	.0643	.0691	.415	.271		15.8	18.9	26.4
		.060	3/32	.301	1.05	.106	.397	.265	1.15	.0525	.0480	.417	.274		15.8	17.2	23.2
		.048	3/32	.243	.845	.086	.322	.215	1.15	.0427	.0388	.420	.276		15.8	15.1	19.1
2.	1.125	.105	3/16	.403	1.40	.115	.241	.241	.773	.0864	.0806	.463	.249		28.1	19.8	30.0
		.075	3/32	.299	1.04	.0855	.188	.188	.792	.0643	.0691	.464	.262		27.1	18.9	26.4
		.060	3/32	.241	.841	.0699	.154	.154	.798	.0525	.0480	.466	.264		27.0	17.2	23.2
		.048	3/32	.195	.678	.0669	.126	.126	.804	.0427	.0388	.469	.266		27.0	15.1	19.1

Reproduced with permission from *Cold Formed Steel Design Manual: Part V—Charts and Tables*, 1977 ed. American Iron and Steel Institute, Washington, DC.

Table 9.5. Two Channels with Stiffened Flanges Back-to-Back.

Effective section for beam strength about the x-x axis where w/t of compression flange exceeds 38.2 for F=20 ksi and 31.2 for F=30 ksi

Size			t	d	R	Area	Weight per Foot	Beam Strength			Properties of Full Section						
								S_x Effective		S_x	Axis x-x				Axis y-y		
D	B							F=20 ksi	F=30 ksi	F=20 & 30 ksi	I_x	S_x	r_x	I_y	S_y	r_y	I_t
In.	In.		In.	In.	In.	In.²	Lb	In.³	In.³	In.³	In.⁴	In.³	In.	In.⁴	In.³	In.	In.⁴
12.0	7.0		.135	1.00	3/16	5.41	18.8	18.7	18.7	18.7	112.	18.7	4.56	12.5	3.58	1.52	44.3
			.105	0.90	3/16	4.19	14.6	14.6	14.6	14.6	87.7	14.6	4.57	9.46	2.70	1.50	34.9
10.0	7.0		.135	1.00	3/16	4.87	17.0	17.0	14.6	17.0	73.0	14.6	3.87	12.5	3.57	1.60	37.9
			.105	0.90	3/16	3.77	13.1	11.4	11.4	11.4	57.0	11.4	3.89	9.45	2.70	1.58	29.9
			.075	0.70	3/32	2.68	9.34	8.01	7.56	8.01	41.0	8.19	3.91	6.33	1.81	1.54	21.8
9.0	6.5		.135	1.00	3/16	4.46	15.5	12.1	12.1	12.1	54.3	12.1	3.49	10.3	3.17	1.52	29.4
			.105	0.80	3/16	3.42	11.9	9.32	9.32	9.32	41.9	9.32	3.50	7.34	2.26	1.47	22.2
			.075	0.70	3/32	2.46	8.56	6.78	6.42	6.78	30.6	6.79	3.53	5.19	1.60	1.45	17.0
8.0	6.0		.135	0.90	3/16	4.00	13.9	9.65	9.65	9.65	38.6	9.65	3.11	7.89	2.63	1.40	22.3
			.105	0.80	3/16	3.10	10.8	7.55	7.55	7.55	30.2	7.55	3.12	5.93	1.98	1.38	17.6
			.075	0.70	3/32	2.23	7.78	5.52	5.35	5.52	22.1	5.52	3.15	4.20	1.40	1.37	12.9
			.060	0.60	3/32	1.77	6.17	4.22	3.98	4.22	17.6	4.40	3.15	3.19	1.06	1.34	10.4
7.0	5.5		.135	0.80	3/16	3.54	12.3	7.49	7.49	7.49	26.2	7.49	2.72	5.89	2.14	1.29	16.4
			.105	0.80	3/16	2.79	9.71	5.96	5.96	5.96	20.9	5.96	2.74	4.71	1.71	1.30	13.0
			.075	0.70	3/32	2.01	6.99	4.38	4.35	4.38	15.3	4.38	2.76	3.33	1.21	1.29	9.51
			.060	0.60	3/32	1.59	5.54	3.44	3.24	3.44	12.2	3.48	2.77	2.53	0.919	1.26	7.68
6.0	5.0		.135	0.80	3/16	3.08	10.7	5.61	5.61	5.61	16.8	5.61	2.34	4.26	1.70	1.17	11.6
			.105	0.70	3/16	2.43	8.47	4.48	4.48	4.48	13.4	4.48	2.35	3.42	1.37	1.19	9.20
			.075	0.60	3/32	1.78	6.21	3.36	3.36	3.36	10.1	3.36	2.38	2.60	1.04	1.21	6.78
			.060	0.60	3/32	1.41	4.91	2.67	2.57	2.67	8.02	2.67	2.39	1.96	0.785	1.18	5.48
5.0	4.0		.135	0.70	3/16	2.54	8.87	3.75	3.75	3.75	9.37	3.75	1.92	2.35	1.18	0.961	5.98
			.105	0.70	3/32	2.01	7.00	3.01	3.01	3.01	7.52	3.01	1.93	1.90	0.949	0.972	4.76
			.075	0.60	3/32	1.46	5.06	2.24	2.24	2.24	5.59	2.24	1.96	1.34	0.670	0.961	3.53
			.060	0.50	3/32	1.15	3.99	1.78	1.78	1.78	4.45	1.78	1.97	0.994	0.500	0.934	2.86
			.048	0.50	3/32	0.922	3.21	1.44	1.39	1.44	3.61	1.44	1.98	0.812	0.406	0.939	2.31
4.0	4.0		.135	0.70	3/16	2.27	7.92	2.75	2.75	2.75	5.50	2.75	1.56	2.35	1.17	1.02	4.97
			.105	0.60	3/16	1.80	6.27	2.22	2.22	2.22	4.44	2.22	1.60	1.90	0.949	1.03	3.97
			.075	0.60	3/32	1.30	4.53	1.66	1.66	1.66	3.33	1.66	1.60	1.34	0.670	1.01	2.96
			.060	0.50	3/32	1.03	3.57	1.33	1.33	1.33	2.66	1.33	1.61	0.999	0.500	0.987	2.40
			.048	0.50	3/32	0.826	2.88	1.08	1.04	1.08	2.16	1.08	1.62	0.812	0.406	0.992	1.94
4.0	4.0		.135	0.70	3/16	2.14	7.45	2.29	2.29	2.29	4.01	2.29	1.37	2.35	1.17	1.05	4.46
			.105	0.60	3/32	1.69	5.90	1.85	1.85	1.85	3.24	1.85	1.38	1.90	0.948	1.05	3.57
			.075	0.50	3/32	1.23	4.27	1.40	1.40	1.40	2.44	1.40	1.41	1.34	0.670	1.05	2.67
			.060	0.50	3/32	0.966	3.37	1.12	1.12	1.12	1.96	1.12	1.42	0.999	0.500	1.02	2.18
			.048	0.50	3/32	0.778	2.71	0.909	0.875	0.909	1.59	0.909	1.43	0.812	0.406	1.02	1.75
3.5	3.5		.135	0.70	3/16	1.48	5.17	1.36	1.36	1.36	2.03	1.38	1.17	1.34	0.766	0.950	2.31
			.105	0.60	3/32	1.06	3.55	1.02	1.02	1.02	1.53	1.02	1.20	0.859	0.491	0.906	1.74
			.075	0.50	3/32	0.846	2.95	0.832	0.832	0.832	1.25	0.832	1.21	0.702	0.401	0.911	1.41
3.0			.048	0.40	3/32	0.663	2.31	0.663	0.663	0.663	0.995	0.663	1.23	0.515	0.294	0.881	1.14

Reproduced with permission from *Cold Formed Steel Design Manual: Part V—Charts and Tables*, 1977 ed. American Iron and Steel Institute, Washington, DC.

Table 9.6. Two Channels with Unstiffened Flanges Back-to-Back.

Size		t	R	Area	Weight per Foot	Properties of Full Section						Allowable Beam Stress F_b	
						Axis x-x			Axis y-y				
D	B					I_x	S_x	r_x	I_y	S_y	r_y	$F=20$ ksi	$F=30$ ksi
In.	In.	In.	In.	In.²	Lb	In.⁴	In.³	In.	In.⁴	In.³	In.		
8.	4.00	.135	3/16	3.11	10.8	26.2	6.54	2.90	1.46	0.726	.683	19.1	26.8
		.105	3/16	2.43	8.48	20.7	5.17	2.91	1.12	0.562	.680	17.2	23.2
		.075	3/32	1.76	6.14	15.2	3.80	2.94	0.802	0.401	.676	13.1	13.4
		.060	3/32	1.41	4.92	12.3	3.06	2.95	0.641	0.320	.674	11.2	11.2
7.	3.00	.135	3/16	2.57	8.96	15.7	4.48	2.47	0.616	0.411	.490	19.8	80.0
		.105	3/16	2.01	7.01	12.4	3.55	2.49	0.477	0.318	.487	19.6	27.6
		.075	3/32	1.46	5.09	9.21	2.63	2.51	0.339	0.226	.482	16.4	21.8
		.060	3/32	1.17	4.09	7.44	2.12	2.52	0.271	0.181	.481	14.1	16.9
6.	3.00	.135	3/16	2.30	8.01	10.7	3.56	2.15	0.615	0.410	.517	19.8	30.0
		.105	3/16	1.80	6.28	8.48	2.83	2.17	0.476	0.317	.514	19.6	27.6
		.075	3/32	1.31	4.57	6.30	2.10	2.19	0.339	0.226	.508	16.4	21.8
		.060	3/32	1.05	3.67	5.09	1.70	2.20	0.271	0.180	.507	14.1	16.9
		.048	3/32	0.845	2.95	4.11	1.37	2.21	0.216	0.144	.506	11.9	11.9
5.	2.50	.105	3/32	1.49	5.18	4.80	1.92	1.80	0.276	0.221	.431	19.8	29.8
		.075	3/32	1.09	3.78	3.59	1.44	1.82	0.196	0.157	.425	18.1	24.9
		.060	3/32	0.873	3.04	2.91	1.17	1.83	0.157	0.126	.424	16.2	21.3
		.048	3/32	0.701	2.44	2.35	0.942	1.83	0.125	0.100	.423	13.8	15.0
4.	2.25	.105	3/32	1.23	4.27	2.57	1.29	1.45	0.202	0.179	.406	19.8	30.0
		.075	3/32	0.898	3.13	1.95	0.973	1.47	0.143	0.127	.399	18.9	26.4
		.060	3/32	0.723	2.52	1.58	0.790	1.48	0.114	0.102	.398	17.2	23.2
		.048	3/32	0.581	2.03	1.28	0.640	1.48	0.0914	0.0812	.397	15.1	19.1
3.	2.25	.105	3/16	1.02	3.54	1.27	0.843	1.12	0.201	0.179	.445	19.8	30.0
		.075	3/32	0.748	2.61	0.973	0.649	1.14	0.143	0.127	.437	18.9	26.4
		.060	3/32	0.603	2.10	0.794	0.529	1.15	0.114	0.102	.435	17.2	23.2
		.048	3/32	0.485	1.69	0.645	0.430	1.15	0.0913	0.0812	.434	15.1	19.1
2.	2.25	.105	3/16	0.805	2.81	0.482	0.481	0.773	0.201	0.178	.499	19.8	30.0
		.075	3/32	0.598	2.08	0.375	0.375	0.792	0.143	0.127	.489	18.9	26.4
		.060	3/32	0.483	1.68	0.308	0.308	0.798	0.114	0.101	.486	17.2	23.2
		.048	3/32	0.389	1.36	0.251	0.251	0.804	0.0913	0.0811	.484	15.1	19.1

Reproduced with permission from *Cold Formed Steel Design Manual: Part V–Charts and Tables*, 1977 ed. American Iron and Steel Institute, Washington, DC.

Table 9.7. Equal Leg Angle with Stiffened Legs.

Size			t	d	R	Area	Wgt per Foot	Axis x-x and Axis y-y					Properties of Full Section						
D	B							I	S	r	x=y	r_{r_1}	J	C_w	r_o	I_{r_2}	r_{r_2}	J	x_o
In.	In.		In.	In.	In.	In.²	Lb.	In.⁴	In.³	In.	In.	In.	In.⁴	In.⁶	In.	In.⁴	In.	In.	In.
4.0	4.0		.135	1.10	3/16	1.28	4.45	2.59	.957	1.42	1.29	1.92	.00776	1.17	2.80	1.24	.987	3.11	−1.94
3.0	3.0		.135	0.90	3/16	0.954	3.32	1.07	.532	1.06	0.991	1.45	.00580	0.341	2.10	0.526	.742	2.32	−1.47
			.105	0.80	3/16	0.732	2.55	0.824	.404	1.06	0.961	1.43	.00269	0.198	2.06	0.391	.731	2.32	−1.41
2.5	2.5		.135	0.70	3/16	0.765	2.67	0.571	.339	0.864	0.815	1.19	.00465	0.105	1.68	0.273	.598	1.90	−1.15
			.105	0.70	3/16	0.606	2.11	0.467	.276	0.877	0.812	1.20	.00223	0.0889	1.71	0.224	.608	1.93	−1.18
2.0	2.0		.135	0.70	3/16	0.630	2.20	0.300	.230	0.690	0.693	0.978	.00383	0.0654	1.41	0.157	.499	1.53	−1.01
			.105	0.70	3/16	0.501	1.75	0.248	.189	0.703	0.689	0.985	.00184	0.0557	1.44	0.130	.509	1.55	−1.04
			.075	0.60	3/32	0.360	1.26	0.184	.137	0.714	0.659	0.972	.000676	0.0259	1.43	0.0907	.502	1.56	−1.01
			.060	0.50	3/32	0.280	0.974	0.141	.103	0.710	0.631	0.950	.000336	0.0125	1.37	0.0658	.485	1.55	−0.94

Reproduced with permission from *Cold Formed Steel Design Manual: Part V—Charts and Tables*, 1977 ed. American Iron and Steel Institute, Washington, DC.

294 Cold-Formed Steel Design

Table 9.8. Equal Leg Angle with Unstiffened Legs.

Size		t	R	Area	Wgt. per Foot	Axis x-x and Axis y-y						Properties of Full Section			Axis y				Beam Strength								
						I	S	r	x=y	r_{z_1}	J	r_o	I_{z_2}	r_{z_2}	j	x_o		$F = 20$ ksi				$F = 30$ ksi					
																		F_c	M max.				F_c	M max.			
																			Comp. ⊥ Tension	Comp. ⊥ Tension	Comp. ⊤ Tension	Comp. ⊤ Tension		Comp. ⊥ Tension	Comp. ⊥ Tension	Comp. ⊤ Tension	Comp. ⊤ Tension
D	B																		In.-kip	In.-kip	In.-kip	In.-kip		In.-kip	In.-kip	In.-kip	In.-kip
In.	In.	In.	In.	In.²	Lb	In.⁴	In.³	In.	In.	In.	In.⁴	In.	In.⁴	In.	In.	In.		ksi					ksi				
4.0	4.0	.135	3/16	1.05	3.65	1.69	.577	1.27	1.07	1.66	.00636	2.22	.654	.790	2.76	−1.31		12.2	7.03	11.4	7.03		12.2	7.03	17.3	7.03	17.3
3.0	3.0	.135	3/16	0.777	2.71	0.700	.321	0.949	0.815	1.25	.00472	1.65	.267	.586	2.06	−0.952		15.4	4.93	6.35	4.93		19.8	6.36	9.62	6.36	9.62
		.105	3/16	0.608	2.12	0.554	.252	0.954	0.803	1.25	.00224	1.66	.211	.590	2.07	−0.962		12.6	3.17	4.99	3.17		12.6	3.17	7.56	3.17	7.56
2.5	2.5	.135	3/16	0.642	2.24	0.399	.220	0.788	0.691	1.05	.00390	1.36	.150	.483	1.70	−0.776		17.2	3.80	4.36	3.80		23.3	5.13	6.61	5.13	6.61
		.105	3/16	0.503	1.75	0.316	.174	0.793	0.678	1.04	.00185	1.37	.120	.487	1.71	−0.786		14.8	2.57	3.44	2.57		18.1	3.14	5.21	3.14	5.21
2.0	2.0	.135	3/16	0.507	1.77	0.199	.139	0.626	0.566	0.844	.00308	1.07	.0737	.381	1.35	−0.599		19.1	2.65	2.75	2.65		26.8	3.71	4.16	3.71	4.16
		.105	3/16	0.398	1.39	0.159	.110	0.631	0.554	0.840	.00146	1.08	.0591	.385	1.36	−0.609		17.2	1.88	2.17	1.88		23.2	2.54	3.29	2.54	3.29
		.075	3/32	0.290	1.01	0.117	.0799	0.635	0.536	0.831	.000544	1.11	.0451	.394	1.38	−0.650		13.1	1.05	1.58	1.05		13.4	1.07	2.40	1.07	2.40
		.060	3/32	0.233	0.813	0.0947	.0644	0.637	0.530	0.829	.000280	1.11	.0366	.396	1.38	−0.655		11.2	0.721	1.28	0.721		11.2	0.721	1.93	0.721	1.93

Reproduced with permission from *Cold Formed Steel Design Manual: Part V–Charts and Tables*, 1977 ed. American Iron and Steel Institute, Washington, DC.

9.2 Differences Between Cold-Formed and Hot-Rolled Steel

Table 9.9. Hat Sections.

Reproduced with permission from *Cold Formed Steel Design Manual: Part V—Charts and Tables*, 1977 ed. American Iron and Steel Institute, Washington, DC.

296 Cold-Formed Steel Design

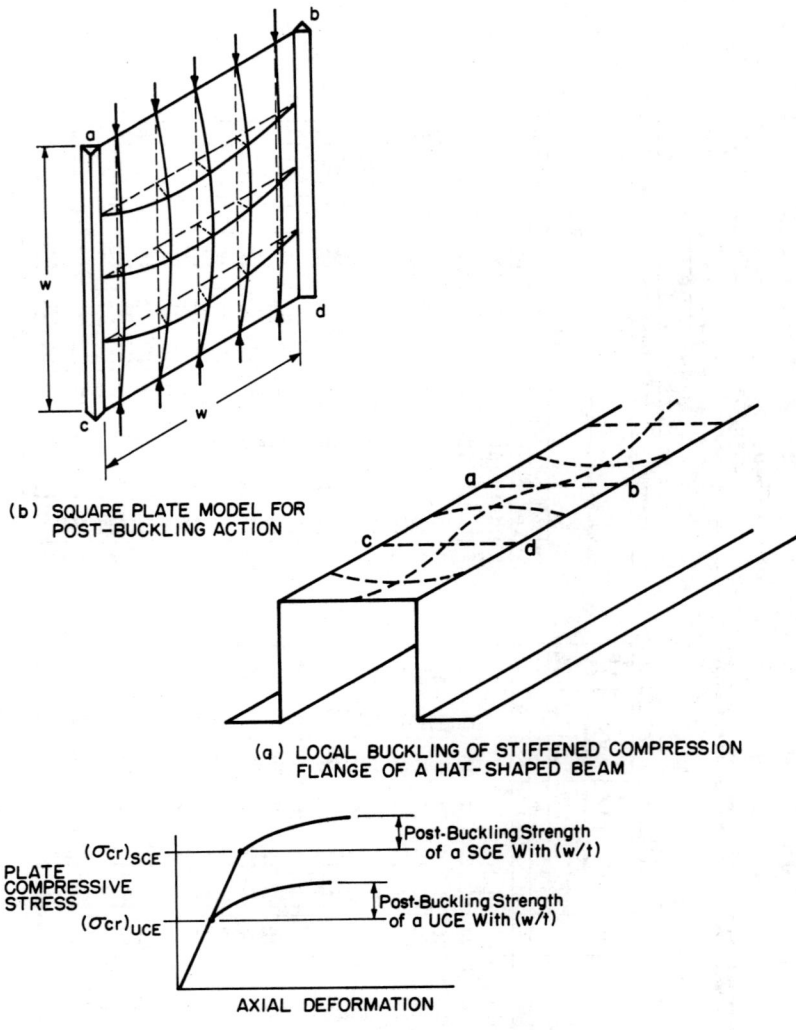

Fig. 9.2. Buckling and postbuckling strength of compression elements. Reproduced with permission from *Cold Formed Steel Design Manual: Part II—Commentary,* 1982 ed., American Iron and Steel Institute, Washington, DC.

pattern of deformation enables the SCE to carry additional load beyond the incipient buckling load because deformation is resisted by bending stresses in two planes, rather than just one as in the case of a thin plate compression element with no stiffened edges.

For an element with only one edge stiffened (UCE), the buckled mode of deformation will be comparable to that of Fig. 9.2b, but only one edge will be restrained against deformation normal to the plane of the element. Consequently, the postbuckling strength will be less for an UCE (one edge unstiffened) as illustrated in Fig. 9.2c. For an element with no stiffened edges, deformation will be one dimensional, as in the case of an ordinary column.

Since the postbuckling strength of a SCE is considerably greater than that of an UCE, it is advantageous to make as many compression elements behave as SCEs as possible. Edge and intermediate stiffeners provide this opportunity by providing resistance to buckling in a plane normal to the plane of the compression element. A compression element may be considered a SCE (see Fig. 9.3a) if each edge is stiffened by a web or lip having a moment of inertia about its own centroidal axis parallel to the longitudinal axis of the stiffened element at least as great as

$$I_{min} = 1.83t^4 \sqrt{w/t^2 - 4000/F_y} \qquad (9.1)$$

but $\not< 9.2t^4$

where I is stiffener moment of inertia about its centroidal axis, in.^{4}; t, element thickness, in.; w, element flat width, in.; and F_y, yield stress, ksi. If the edge stiffener is a simple lip, as in Fig. 3b, an additional requirement is that the stiffener depth d_{min} (in.), be at least as great as both

$$d_{min} = 2.8t \sqrt[6]{(w/t)^2 - 4000/F_y} \quad \text{and} \quad 4.8\,t \qquad (9.2)$$

Finally a simple lip is not a satisfactory stiffener if the element to be stiffened has a $w/t > 60$.

An intermediate stiffener (see Fig. 9.1g) must stiffen a compression element on either side. Thus, an intermediate stiffener must have a minimum moment of inertia twice as large as that for an edge stiffener to be effective. Additionally, the following limitations, in which f has the units of ksi, are imposed on multiple stiffener effectiveness.

1. For strength (load) considerations:
    a. If w/t of subelements between 2 webs $> 177/\sqrt{f}$, only two stiffeners are effective.

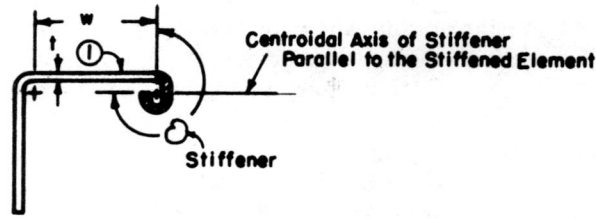

(a) Edge Stiffener

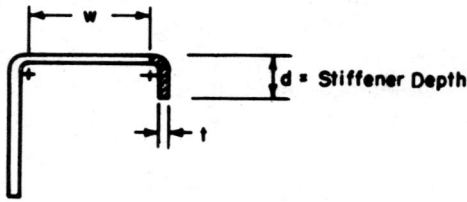

(b) Single Lip Stiffener

Fig. 9.3. Edge stiffeners. In (a) note that (*1*) w is flat width of element 1; (2) hashed area is stiffener for element 1; (3) w/t = flat width ratio.

    b. If w/t of subelements between a web and an edge stiffener $> 171/\sqrt{f}$, only one intermediate stiffener is effective.

2. For deflection considerations, the same limitations are imposed if $w/t > 221/\sqrt{f}$.

If intermediate stiffeners are spaced such that $w/t <$ limits set in the preceding, all stiffeners are effective. In this event, the multiple stiffened element may be replaced by a rectangular element with an effective thickness, t_s, for evaluating the flat width ratio (w/t) of such multiple stiffened elements.

$$t_s = \sqrt[3]{12\, I_s/w_s} \tag{9.3}$$

where w_s is entire width between webs or between web and edge stiffener, in.; I_s is moment of inertia of the full area of the multiple stiffened element about its centroidal axis, in.^{4}. An example of an element which meets these requirements is corrugated sheet metal siding.

9.2.5 Steels and Strength Properties

Twelve ASTM designation steels are included in the AISI Specification for cold-formed steel. They are steels conforming to ASTM A446-76, A570-79, A606-75, A607-75, A611-72, A715-75, A36-77a, A242-79, A441-79, A572-79, A588-80, and A529-79. Properties of several of these steels are given in Table 9.10. Note that the steel sheets and strips may be either hot or cold rolled. However, the structural sections are cold formed from the sheets.

Yield strengths of the steels range from 25 to 80 ksi (170 to 550 MPa). The accepted modulus of elasticity for all the steels is 29.5×10^6 psi (200 GPa) and ductility ranges from 12 to 27% in 2 in. (51 mm). The ductility of steels used in cold-formed sections should be greater than either 10% in 2 in. (51 mm) or 7% in 8 in. (203 mm). Furthermore, the ratio of tensile strength (F_u) to yield point stress (F_y) should be greater than 1.08. Some special steels, such as A446-76 (Grade E) and 611-72 (Grade E), which do not meet the aforementioned ductility requirements, may also be used provided the design stress is less than both $0.45F_y$ and 36 ksi (250 MPa).

The stress–strain curve for cold-formed steels are of two types. One exhibits a sharp yield point, whereas the other is characterized by a gradual yielding (see Fig. 9.4). In the case of a gradual yielding steel, the yield strength is defined as the divergence of the stress–strain curve from linear or by the offset yield point method as in Fig. 9.4b. The offset strain ε_o is usually taken as 0.2%.

When the corners of the cold-formed steels are bent to develop a section, the corners are cold worked. That is, the material in the corners is stressed beyond the yield point. Figure 9.5 illustrates the change in steel properties during cold working. The 0.2% yield stress of the virgin steel before cold forming is F_y. When the section is bent the steel in the vicinity of the corners yields and reaches a state of stress (1) in Fig. 9.5. After bending, the corner material unloads along path 1–2 parallel to the original linear stress–strain curve. If the material is immediately reloaded, the stress–strain behavior returns linearly to point 1 and then follows the path of the virgin steel to point 4. The material, when it follows this path, is said to be strain hardened; the result of which is an increase in the yield strength to F'_y and a decrease in ductility from D_e to D'_e. If, as is the usual case, the corner steel is

Table 9.10. Summary of Typical Steels Used in Cold-Formed Sections.

ASTM Designation	Grade	F_y, ksi (min)	F_u, ksi (min)	Percent elongation in 2 in. (min)
A 446-76	A	33	45	20
	B	37	52	18
	C	40	55	16
	D	50	65	12
	E	80	82	—
A 570-79	A	25	45	23 to 27
	B	30	49	21 to 25
	C	33	52	18 to 23
	D	40	55	15 to 21
	E	42	58	13 to 19
A 606-75	Hot rolled—as rolled cut lengths	50	70	22
	Hot rolled—as rolled coils	45	65	22
	Hot rolled—annealed or normalized	45	65	22
	Cold rolled	45	65	22
A 607-75	45	45	60	Hot-rolled 25 / Cold-rolled 22
	50	50	65	Hot-rolled 22 / Cold-rolled 20
	55	55	70	Hot-rolled 20 / Cold-rolled 18
	60	60	75	Hot-rolled 18 / Cold-rolled 16
	65	65	80	Hot-rolled 16 / Cold-rolled 15
	70	70	85	14
A 611-72	A	25	42	26
	B	30	45	24
	C	33	48	22
	D	40	52	20

Reproduced with permission from *Cold Formed Steel Design Manual: Part III—Supplementary Information,* 1977 ed. American Iron and Steel Institute, Washington, DC.

kept unloaded for a period of time before again being stressed, the stress–strain behavior is characterized as strain aging and follows path 2–3. The result of strain aging is a further increase in yield strength to F_y'' with a subsequent reduction in ductility to D_e''.

The AISI specification allows the designer to utilize the increased strength due to cold working. Evaluation of the increase will be taken up in the section on allowable stresses.

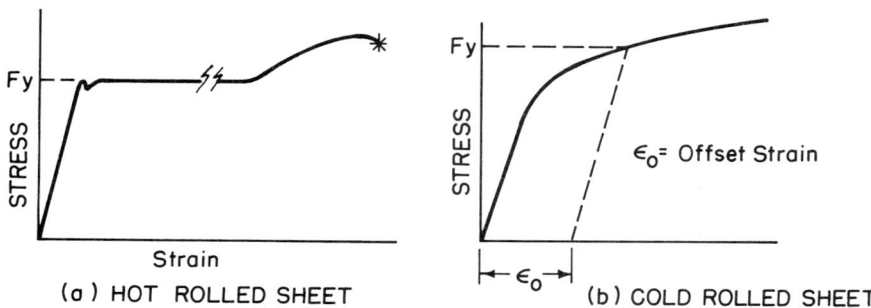

Fig. 9.4. Stress–strain behavior of steels.

9.2.6 Summary

Cold-formed steel structural sections have numerous unique features which cause their behavior to be different from that of hot-rolled sections. Among the more important differences relative to structural strength are the following:

1. Cross-section elements are usually thinner in cold-formed sections.

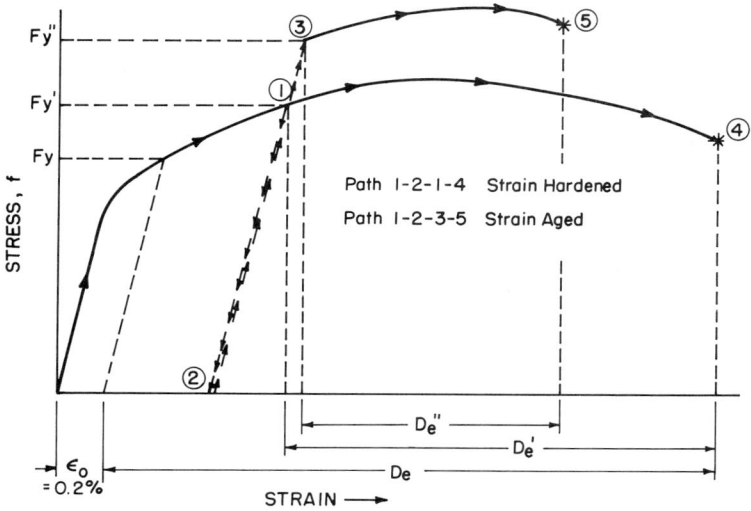

Fig. 9.5. Effects of cold working steel.

2. Local buckling is more common in cold-formed sections and often occurs at stress levels below the yield stress.
3. Flexural–torsional and torsional buckling are a more common mode of failure in cold-formed sections.
4. Edge stiffeners are commonly used in cold-formed sections and significantly strengthen compression elements.
5. Plastic design procedures are not used in cold-formed design.
6. Web crippling is more critical in cold-formed sections since the sections depth-to-element thickness ratio is large in typical sections and web stiffeners are often not practical.
7. In bolted connections, bearing stresses in materials fastened are more critical in cold-formed design since the bolt diameter-to-material-thickness ratio is large.
8. Yield stress of cold-formed sections is higher than virgin steel due to strain hardening and strain aging at bent corners.

9.3 ALLOWABLE STRESSES

9.3.1 Basic Design Stresses

The basic allowable stress for tension members and tension and compression on the extreme fibers of flexural members is defined by F

$$F = 0.60 F_y \tag{9.4}$$

The maximum allowable shear stress in cold-formed sections is defined by F_v

$$F_v = 0.40 F_y \tag{9.5}$$

These stresses are the allowable stresses if yielding occurs before any other mode of failure. When cold working is considered the basic stress for an axially loaded compression member and for the flanges of flexural members may be increased to

$$F = 0.60 F_{ya} \tag{9.6}$$

where F_{ya} is average yield point of the entire section. The average yield point may be used since the flats undergo little, if any, cold work when the corners are bent. When the section is compact and the section yields

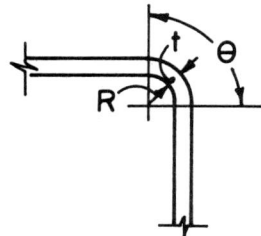

Fig. 9.6. Typical corner.

before buckling locally ($Q = 1.0$), F_{ya} is obtained from Eq. (9.7) or from full section tensile tests or stub column tests.*

$$F_{ya} = CF_{yc} + (1 - C)F_{yf} \tag{9.7}$$

where C is the ratio of corner area to total section area in compression members, or the ratio of corner area to the full cross-sectional area of the controlling flange; F_{yc}, yield point of corners, ksi (MPa),

$$F_{yc} = \frac{B_c F_y}{(R/t)^m} \quad \text{if} \quad \frac{F_u}{F_y} \geq 1.2, \quad \frac{R}{t} \leq 7, \quad \text{and} \quad \theta \leq 120°$$

(See Fig. 9.6.) F_{yf} is weighted average yield point of flats, ksi (MPa); $B_c = 3.69(F_u/F_y) - 0.819(F_u/F_y)^2 - 1.79$; $m = 0.192 (F_u/F_y) - 0.068$; R is inside bend radius, in. (mm); t, element thickness, in. (mm); and F_u, ultimate tensile strength of the virgin steel, ksi (MPa).

When sections are noncompact ($Q < 1.0$) then F_{ya} may be taken as the tensile yield point of the virgin steel specified by ASTM or the weighted average of the tensile yield points of the flats as determined by testing.

9.3.2 Allowable Stresses in Thin Compression Elements

9.3.2.1 Unstiffened Compression Elements. The allowable compressive stress F_c in an UCE is dependent upon its flat width ratio. If $w/t \leq 63.3/\sqrt{F_y}$, the element yields before buckling locally and $F_c = 0.60F_y$. If $63.3/\sqrt{F_y} < w/t < 144/\sqrt{F_y}$, the element will buckle inelastically before yielding and

---

*Article 6 of the 1980 edition of the AISI Specification for Cold-Formed Steel outlines the appropriate test procedure.

304 Cold-Formed Steel Design

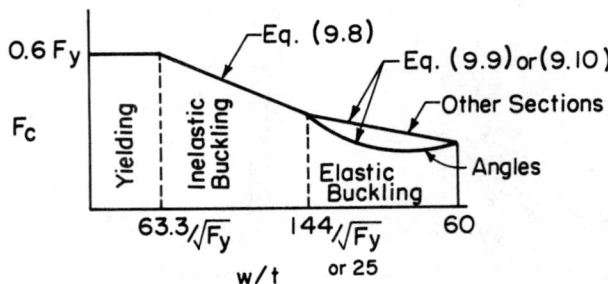

Fig. 9.7. Allowable stresses in unstiffened compression elements. Reproduced with permission from *Cold Formed Steel Design Manual: Part II—Commentary*, 1982 ed., American Iron and Steel Institute, Washington, DC.

$$F_c = F_y [0.767 - (2.64/10^3)(w/t)\sqrt{F_y}] \quad (9.8)$$

If $25 < w/t < 60$, elastic buckling considerations control the allowable stress and

$$F_c = 8000/(w/t)^2 \quad \text{for angle struts} \quad (9.9)$$

and

$$F_c = 19.8 - 0.28(w/t) \text{ for other sections} \quad (9.10)$$

In some steels with $F_c > 33$ ksi (230 MPa) elastic buckling controls at $w/t < 25$. For these steels, when $144/\sqrt{F_y} \leq w/t < 25$,

$$F_c = 8000/(w/t)^2 \quad (9.11)$$

Figure 9.7 illustrates the general relationship between F_c for UCEs and the w/t ratio of the element. Note the analogy between w/t for plates and L/r for flexural buckling of columns, and the decrease in allowable compressive stress as the compression elements become more slender.

9.3.2.2 Stiffened Compression Elements. In a section with all SCEs, such as a box section used as a centrically loaded column, the element design stress F is $0.60F_y$ or $0.6 F_{ya}$. In a flexural member the design stress for SCEs in flanges or webs is also either $0.60F_y$ or $0.6F_{ya}$. In sections used as columns with both SCEs and UCEs, the design stress is the smaller of $0.60F_y$ ($0.6F_{ya}$ may be used for cold working considerations) or F_c for the UCEs.

If any local buckling occurs in SCEs before yielding, the AISI procedure is not to reduce the allowable stress. Instead the flat width w of the SCE is reduced to an effective width b. In Fig. 9.8 the effective width of the flats of the compression elements is shown by the unhashed areas. Section properties such as I, S, and A, are then based on effective widths rather than full widths.

9.4 EFFECTIVE WIDTHS OF SCEs

In order to account for local buckling and postbuckling strength of SCEs used in flexural or compression members, the following effective widths, b, are used to evaluate the section properties. For elements without intermediate stiffeners and for load capacity determinations (strength) and for all sections except square or rectangular tubes,

$$b = w \quad \text{if} \quad w/t \leq 171/\sqrt{f} \tag{9.12}$$

and

$$\frac{b}{t} = \frac{253}{\sqrt{f}}\left[1 - \frac{55.3}{(w/t)\sqrt{f}}\right] \quad \text{if} \quad \frac{w}{t} > \frac{171}{\sqrt{f}} \tag{9.13}$$

For elements without intermediate stiffeners and for deflection determinations and for all sections except square or rectangular tubes,

$$b = w \quad \text{if} \quad w/t \leq 221/\sqrt{f} \tag{9.14}$$

and

$$\frac{b}{t} = \left(\frac{326}{\sqrt{f}}\right)\left[1 - \frac{71.3}{(w/t)\sqrt{f}}\right] \quad \text{if} \quad \frac{w}{t} > \frac{221}{\sqrt{f}} \tag{9.15}$$

For square or rectangular tubes without intermediate stiffeners, the effective widths for load capacity determinations are

$$b = w \quad \text{if} \quad w/t \leq 184/\sqrt{f} \tag{9.16}$$

and

$$\frac{b}{t} = \left(\frac{253}{\sqrt{f}}\right)\left[1 - \frac{50.3}{(w/t)\sqrt{f}}\right] \quad \text{if} \quad \frac{w}{t} > \frac{184}{\sqrt{f}} \tag{9.17}$$

306 Cold-Formed Steel Design

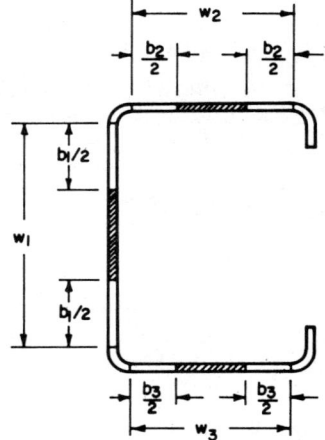

(a) All Elements Stiffened and Subject To Compressive Stresses

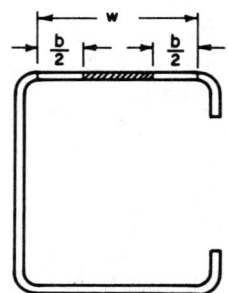

(b) All Elements Stiffened and Section Subject to a Positive Moment (Top Fiber in Compression, Bottom in Tension)

Fig. 9.8. Effective width concept for stiffened compression elements.

For deflection determinations in square or rectangular tubes,

$$b = w \quad \text{if} \quad w/t \leq 237/\sqrt{f} \tag{9.18}$$

and

$$\frac{b}{t} = \left(\frac{326}{\sqrt{f}}\right)\left[1 - \frac{64.9}{(w/t)\sqrt{f}}\right] \quad \text{if} \quad \frac{w}{t} > \frac{237}{\sqrt{f}} \tag{9.19}$$

If multiple stiffened elements, or stiffened elements with only one edge connected to a web have subelement $w/t < 60$, then the effective width of subelements is determined with Eqs. (9.12)–(9.15). If $w/t > 60$ for these subelements, then

$$\frac{b_e}{t} = \frac{b}{t} - 0.10\left(\frac{w}{t} - 60\right) \qquad (9.20)$$

where b is subelement effective width for $w/t < 60$, and b_e is reduced subelement effective width for $w/t > 60$. When subelements of multiple stiffened elements have $w/t > 60$ and b_e is used in place of b, the stiffener area is reduced, when evaluating cross-sectional area, to:

$$A_{ef} = \alpha A_{\text{stiffener}} \qquad \text{if} \quad w/t = 60\text{–}90 \qquad (9.21)$$

where

$$\alpha = \left(3 - \frac{2b_e}{w}\right) - \frac{1}{30}\left(1 - \frac{b_e}{w}\right)\left(\frac{w}{t}\right)$$

and

$$A_{ef} = \frac{b_e}{w} A_{\text{stiffener}} \qquad \text{if} \quad \frac{w}{t} > 90 \qquad (9.22)$$

The centroid and the moment of inertia of the stiffener, however, are still based on the full stiffener area.

Whenever a portion of the flat width is removed to arrive at the effective width, it is removed symmetrically about the center line of the element. This concept is illustrated in Fig. 9.8.

9.5 MAXIMUM FLAT-WIDTH RATIOS

The maximum allowable flat-width ratios for section elements are summarized in Table 9.11. These w/t ratios are based on total flat width disregarding any intermediate stiffeners.

Table 9.11. Maximum Allowable Flat Width Ratios for Compression Elements.

Element Type	Maximum w/t
SCE with one edge at a web and one edge stiffened with simple lip	60
SCE with one edge at a web and one edge stiffened by any adequate stiffener except a simple lip	90
SCE with both edges connected to other SCEs	500
UCE	60

9.6 PROPERTIES OF SECTIONS WITH THIN ELEMENTS

Tables 9.1–9.9 summarize the section properties of a few standard stiffened and unstiffened channel, hat, angle, and built-up I cold-formed shapes. Since most cold-formed sections are unique in shape, section properties are often not tabulated and need to be calculated.

The three most commonly required section properties are the area, moment of inertia, and section modulus. These properties can be evaluated by standard relationships, including the first and second moments of areas. Another method, which is useful and quite accurate for sections with thin elements, is the linear, or midline method.

In the midline method all the element material is assumed concentrated along the midline of the element. Then the area is replaced by the element length and without regard to thickness. Since the section properties are linearly related to element thickness, the section properties are easily calculated by multiplying midline properties by t.

The properties of several line elements are given in Fig. 9.9. Example 9.1 demonstrates the midline method for a section with all UCEs. In this case, all elements are fully effective and $b_i = w_i$.

Example 9.1. Midline Method for Section Properties Where All Elements Are Fully Effective. Find the area, moment of inertia about the x axis, and the section modulus about the x axis for the unstiffened Z section in Figure 9.10a.

Solution

1. Replace the area elements with line elements (Fig. 9.10b). The line elements are located at the centroidal axis of each area.
2. Evaluate line lengths, L, and distance between centroids of the line elements and the x axis, d.

9.6 Properties of Sections with Thin Elements   309

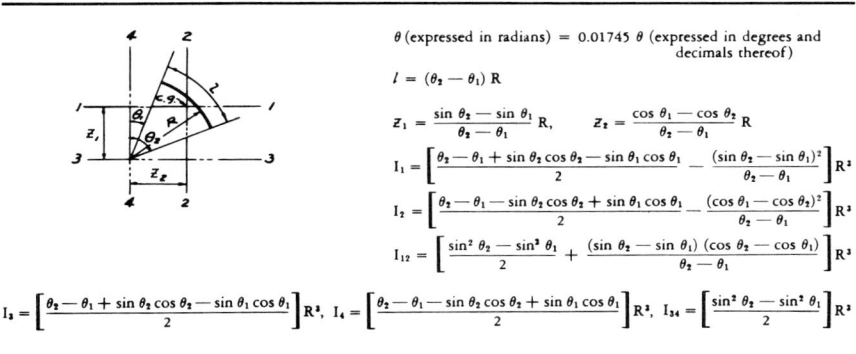

$I_1 = \dfrac{l^3}{12}$, $I_2 = 0$ $I_1 = 0$, $I_2 = \dfrac{l^3}{12}$ $I_{12} = \left[\dfrac{\sin\theta\cos\theta}{12}\right]l^3 = \dfrac{lmn}{12}$

$I_3 = la^2 + \dfrac{l^3}{12} = l\left(a^2 + \dfrac{l^2}{12}\right)$ $I_2 = la^2$ $I_3 = la^2 + \dfrac{ln^2}{12} = l\left(a^2 + \dfrac{n^2}{12}\right)$

$I_1 = \left[\dfrac{\cos^2\theta}{12}\right]l^3 = \dfrac{ln^2}{12}$

$I_2 = \left[\dfrac{\sin^2\theta}{12}\right]l^3 = \dfrac{lm^2}{12}$

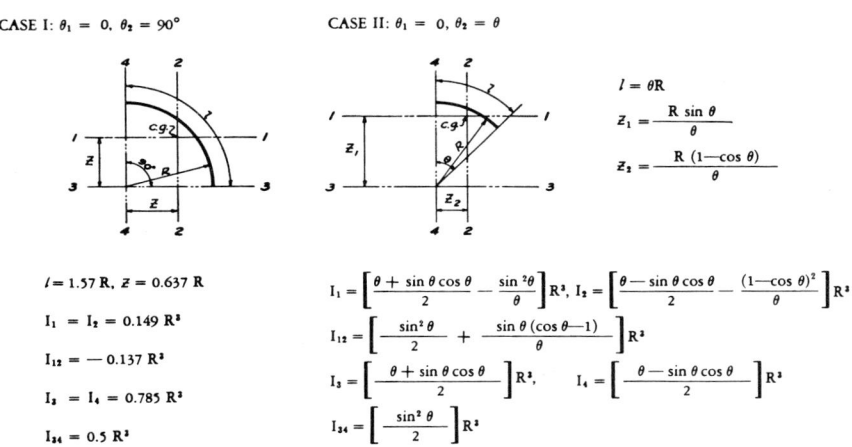

θ (expressed in radians) = 0.01745 θ (expressed in degrees and decimals thereof)

$l = (\theta_2 - \theta_1) R$

$\bar{z}_1 = \dfrac{\sin\theta_2 - \sin\theta_1}{\theta_2 - \theta_1} R$, $\bar{z}_2 = \dfrac{\cos\theta_1 - \cos\theta_2}{\theta_2 - \theta_1} R$

$I_1 = \left[\dfrac{\theta_2 - \theta_1 + \sin\theta_2\cos\theta_2 - \sin\theta_1\cos\theta_1}{2} - \dfrac{(\sin\theta_2 - \sin\theta_1)^2}{\theta_2 - \theta_1}\right] R^3$

$I_2 = \left[\dfrac{\theta_2 - \theta_1 - \sin\theta_2\cos\theta_2 + \sin\theta_1\cos\theta_1}{2} - \dfrac{(\cos\theta_1 - \cos\theta_2)^2}{\theta_2 - \theta_1}\right] R^3$

$I_{12} = \left[\dfrac{\sin^2\theta_2 - \sin^2\theta_1}{2} + \dfrac{(\sin\theta_2 - \sin\theta_1)(\cos\theta_2 - \cos\theta_1)}{\theta_2 - \theta_1}\right] R^3$

$I_3 = \left[\dfrac{\theta_2 - \theta_1 + \sin\theta_2\cos\theta_2 - \sin\theta_1\cos\theta_1}{2}\right] R^3$, $I_4 = \left[\dfrac{\theta_2 - \theta_1 - \sin\theta_2\cos\theta_2 + \sin\theta_1\cos\theta_1}{2}\right] R^3$, $I_{34} = \left[\dfrac{\sin^2\theta_2 - \sin^2\theta_1}{2}\right] R^3$

CASE I: $\theta_1 = 0$, $\theta_2 = 90°$ CASE II: $\theta_1 = 0$, $\theta_2 = \theta$

$l = 1.57 R$, $\bar{z} = 0.637 R$

$I_1 = I_2 = 0.149 R^3$

$I_{12} = -0.137 R^3$

$I_3 = I_4 = 0.785 R^3$

$I_{34} = 0.5 R^3$

$l = \theta R$

$\bar{z}_1 = \dfrac{R\sin\theta}{\theta}$

$\bar{z}_2 = \dfrac{R(1 - \cos\theta)}{\theta}$

$I_1 = \left[\dfrac{\theta + \sin\theta\cos\theta}{2} - \dfrac{\sin^2\theta}{\theta}\right] R^3$, $I_2 = \left[\dfrac{\theta - \sin\theta\cos\theta}{2} - \dfrac{(1 - \cos\theta)^2}{\theta}\right] R^3$

$I_{12} = \left[\dfrac{\sin^2\theta}{2} + \dfrac{\sin\theta(\cos\theta - 1)}{\theta}\right] R^3$

$I_3 = \left[\dfrac{\theta + \sin\theta\cos\theta}{2}\right] R^3$, $I_4 = \left[\dfrac{\theta - \sin\theta\cos\theta}{2}\right] R^3$

$I_{34} = \left[\dfrac{\sin^2\theta}{2}\right] R^3$

Fig. 9.9. Properties of line elements. Reproduced with permission from *Cold Formed Steel Design Manual: Part III—Supplementary Information,* 1977 ed., American Iron and Steel Institute, Washington, DC.

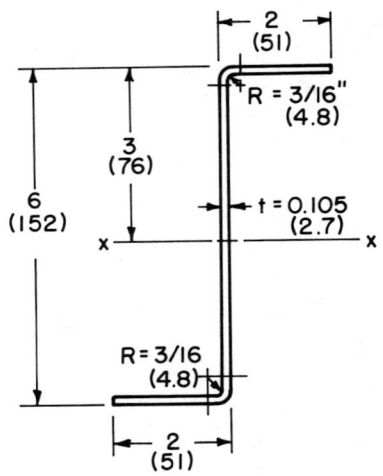 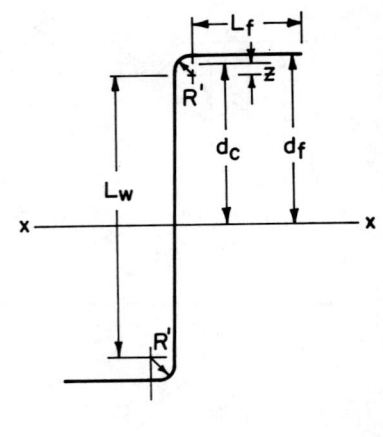

(a) Section Dimensions   (b) Midline Dimensions

Fig. 9.10. Example 9.1. All dimensions are in in. (mm).

a. For the flanges

$$L_f = 2 - (3/16 + 0.105) = 1.708 \text{ in.} \quad (43.4 \text{ mm})$$
$$d_f = 3 - 0.105/2 = 2.948 \text{ in.} \quad (74.9 \text{ mm})$$

b. For the webs

$$L_w = 6 - 2(3/16 + 0.105) = 5.415 \text{ in.} \quad (137.5 \text{ mm})$$
$$d_w = 0 \text{ in.} \quad (0 \text{ mm})$$

c. For the corners

$$R' = 3/16 + 0.105/2 = 0.240 \text{ in.} \quad (6.1 \text{ mm})$$
$$L_c = 1.57 R' = 0.377 \text{ in.} \quad (9.6 \text{ mm}) \quad (\text{see Fig. 9.9})$$
$$z = 0.637 R' = 0.153 \text{ in.} \quad (3.9 \text{ mm}) \quad (\text{see Fig. 9.9})$$
$$d_c = 5.415/2 + 0.153 = 2.861 \text{ in.} \quad (72.7 \text{ mm})$$

3. Evaluate the section area,

$$A = \sum L_i t$$
$$= (2L_f + 2L_c + L_w)t$$
$$= [2(1.708) + 2(0.377) + (5.415)](0.105)$$
$$= 1.006 \text{ in.}^2 \ (650 \text{ mm}^2)$$

4. Evaluate I'_x (moment of inertia of the line elements)

$$I'_x = 2(\bar{I}'_f + L_f d_f^2) + \bar{I}'_w + 2(\bar{I}'_c + L_c d_c^2)$$

where $\bar{I}'_f = 0$ (approximately), $\bar{I}'_w = L_w^3/12$, and $\bar{I}'_c = 0.149(R')^3$

$$I'_x = 2[1.708 \ (2.948)^2] + \frac{(5.415)^3}{12}$$
$$+ 2 \ [0.149(0.240)^3 + 0.377(2.861)^2]$$
$$= 49.096 \text{ in.}^3 \ (804.5 \text{ cm}^3)$$

5. Evaluate I_x and S_x

$$I_x = I'_x \ (t) = 49.096 \ (0.105) = 5.16 \text{ in.}^4 \ (215 \text{ cm}^4)$$

and

$$S_x = \frac{I_x}{c} = \frac{5.16}{3} = 1.72 \text{ in.}^3 \ (28.2 \text{ cm}^3)$$

6. Commentary on the section properties for the section of Example 9.1: The assumption of fully effective elements is valid for this section when used as a flexural member bent about the x axis since the compression elements (flanges) are UCEs. If the section were used as a column, then the web becomes a SCE and may not be fully effective. The flat-width ratio, w/t, would have to be checked to see if $b = w$ or if $b < w$. If $b < w$ then a portion of L_w would be removed symmetrically from the web

about the x axis and the section area would be reduced by the amount $(L_w - b)(t)$. To illustrate the procedure for column loading, assume A36 steel with $F_y = 36$ ksi (250 MPa) were used in the section.

a. Evaluate w/t of the UCE (flange)

$$(w/t)_f = L_f/t = 1.708/0.105 = 16.27$$

b. Evaluate limiting w/t of UCE for $F_c = 0.6F_y$

$$(w/t)_{\text{lim}} = 63.3/\sqrt{F_y} = 63.3/\sqrt{36} = 10.55$$

c. Evaluate w/t limit for inelastic local buckling

$$(w/t)_{\text{lim}} = 144/\sqrt{F_y} = 144/6 = 24$$

Thus, the UCE experiences inelastic buckling before yielding and $F_c < 0.60\ F_y$.

d. The allowable compressive stress in the UCE from Eq. (9.8)

$$F_c = F_y [0.767 - (2.64/10^3)(w/t)\sqrt{F_y}]$$
$$= 36 [0.767 - (2.64/10^3)(16.27)(6)]$$
$$= 18.33 \text{ ksi} \quad (130 \text{ MPa})$$

e. Since the section is uniformly stressed, since this stress level is less than $0.6\ F_y$, and since effective widths of SCEs depend on the actual stress level f, the effective width of the webs will be based on a maximum stress of 18.33 ksi (130 MPa).

f. Evaluate w/t for the SCE (web)

$$w/t = 5.415/0.105 = 51.57$$

g. Evaluate the limiting w/t for $b = w$ for strength calculations

$$(w/t)_{\text{lim}} = 171/\sqrt{f} = 171/\sqrt{18.33} = 39.94$$

Thus, $w/t > (w/t)_{\text{lim}}$

9.6 Properties of Sections with Thin Elements

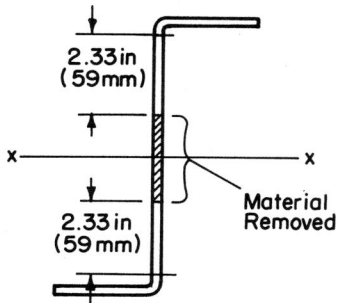

Fig. 9.11. Effective width of web for Example 9.1 (part i).

h. Evaluating the effective width, b, for strength and allowable load determination from Eq. (9.13),

$$\frac{b}{t} = \frac{253}{\sqrt{f}} \left[1 - \frac{55.3}{(w/t)\sqrt{f}} \right]$$

$$b = \frac{253}{\sqrt{18.33}} \left[1 - \frac{55.3}{51.57\sqrt{18.33}} \right] (0.105)$$

$$= 4.65 \text{ in.} \quad (118 \text{ mm})$$

i. Remove material from the SCE (web) and evaluate the effective section area. Figure 9.11 illustrates how the area is to be removed from the web.

$$A = (2L_f + 2L_c + b)t$$

$$= [2(1.708) + 2(0.377) + 4.65](0.105)$$

$$= 0.926 \text{ in.}^2 \quad (600 \text{ mm}^2)$$

Example 9.1 shows how to use the midline method for fully effective sections and also shows how to find properties of sections without fully effective elements. It is important to note that the effective width of a SCE depends upon the stress level in the compression element. In a flexural member which has SCEs for flanges, the stress level in the flanges is determined by the location of the neutral axis (N.A.). If the N.A. is closer to the tension flange than the compression flange, the flexural stress in the compressive flange is greater than that in the tension flange. The stress level in the compression flange at design loads is thus $F = 0.60F_y$ and b is evaluated by using $f = F = 0.60F_y$.

Table 9.12. Error Introduced by Midline Method.

Channel Section[a]	Thickness of Material (in.)	Expected Error in I_x (%)
A	0.50	3.3
	0.25	0.7
	0.10	0.1
B	0.50	0.6
	0.25	0.15
	0.10	0.02

Reproduced with permission from Yu, Wei-Wen. *Cold-Formed Steel Structures,* 1st ed, 1972. McGraw-Hill, New York.
a. Section B deeper (8 in.) than section A (3 in.).

If the N.A. is closer to the compression flange than the tension flange, at design loads the tension flange is stressed to $0.60F_y$ and the stress in the compression flange is less than $0.6F_y$. A problem arises because the stress level in the flange f and the effective width of the flange b are interdependent. One way to resolve the problem is to assume a location of the N.A., evaluate f and b, and compare the N.A. location for f and b to that assumed. Several trials usually yield a converging solution. The section on flexural members treats those situations in more detail.

The midline method is approximate and does introduce some error into section properties. The variation, however, is usually small. A summary of expected errors for channel sections is in Table 9.12.

9.7 FLEXURAL MEMBERS

9.7.1 Beams with Adequate Lateral Bracing

If a cold-formed steel flexural member is adequately braced it may fail by any of the following modes: yielding, local buckling of webs or flanges, or web crippling at points of application of concentrated loads. It does not fail by buckling laterally. Thus, the allowable flexural stress is either $F_b = 0.6F_y$ or F_c as defined for UCEs with large w/t ratios.

There are many sections which are laterally stable and will not buckle laterally. Some of these are (1) sections bent about their minor axis, and (2) closed box beams with an unbraced length-to-section-width ratio less than 75. Channel, I, Z, and hat shapes are laterally stable if sufficiently braced. The bracing requirements for lateral stability of these sections are as follows:

1. I sections which are symmetrically shaped about an axis in the

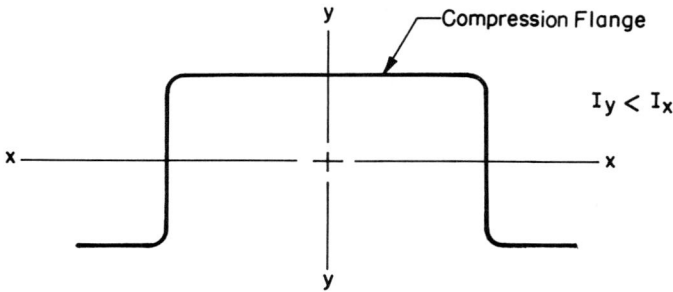

Fig. 9.12. Definition sketch for an adequately braced hat section.

plane of the web and bent about the axis normal to the web, or symmetrically shaped channel sections which are bent about a centroidal axis perpendicular to the web are adequately braced if the parameter $L_b^2 S_{xc}/(dI_{yc}) \leq 0.36\pi^2 EC_b/F_y$.

2. Point symmetric Z-shaped sections bent about the centroidal axis perpendicular to the web are adequately braced if the parameter $L_b^2 S_{xc}/(dI_{yc}) \leq 0.18\pi^2 EC_b/F_y$.
3. Hat sections bent about the x axis (see Fig. 9.12), and with $I_y < I_x$, are adequately braced if the allowable stress $F_b = 151{,}900/(L_b/r_y)^2$ [in ksi] is greater than $0.6F_y$, where L_b is distance between lateral bracing, in. (mm); d, depth of section, in. (mm); C_b, $1.75 + 1.05(M_1/M_2) + 0.3(M_1/M_2)^2$ (see Section 8.8.1.6) in the text for a discussion of this parameter); I_{yc}, moment of inertia of the compressive portion of the section about the centroidal axis of the *entire* section parallel to the web, in.^{4} (cm^4); and S_{xc} is compression section modulus of *entire* section about the major axis. (I_x divided by distance from centroidal axis (C.A.) to outside fiber of the compression flange) in.^{3} (cm^3)

The design procedure for adequately braced beams is to do the following:

1. Satisfy the flexure equation to assure that flanges do not fail.

$$M/S \leq F_b$$

where $F_b = 0.60F_y$ or the allowable compressive strength F_c of any UCEs used as compression flanges and which have w/t ratios larger than limits set for no local buckling; M, required moment capacity of the section; S, section modulus of the entire beam shape about the axis of bending if compression flanges

are UCEs, or the section modulus of the effective area of the beam shape if the compression flanges are SCEs and w/t exceeds limits for no local buckling.
2. Satisfy the following requirements to prevent web failure due to excessive shear stress f_v.
   a. If $h/t \leq 237\sqrt{k_v/F_y}$, then $F_v = 65.7\sqrt{k_v F_y}/(h/t)$. (9.23)
   b. If $h/t > 237\sqrt{k_v/F_y}$, then $F_v = 15{,}600\, k_v/(h/t)^2$. (9.24)
   c. But in no case may $F_v > 0.4 F_y$ where f_v = average shear stress, ksi; $f_v = V/A_w$; A_w, area of webs, in.^{2}; F_v, allowable shear stress, ksi; h, clear distance between flanges measured along the web, in.; k_v, shear buckling coefficient, C (5.34 for unreinforced webs—webs without transverse stiffeners).
3. Satisfy the following requirements to prevent the web from failure due to excessive flexural stresses.
   a. In beams with SCEs for flanges the maximum compressive flexural stress in the flat of the web must be less than F_{bw} from Eq. (9.25). The actual flexural stress, $f_{bw} = M_{yw}/I$, is based on the full web area and the effective flange area.

$$F_{bw} = [1.21 - 0.00034(h/t)\sqrt{F_y}]\,(0.60 F_y) \leq 0.6 F_y \quad (9.25)$$

   b. In beams with UCEs for flanges, the maximum compressive flexural stress in the flat of the web must be less than F_{bw} from Eq. (9.26). The actual web flexural stress is based on the full web area and a reduced flange area equal to $(F_c/0.60 F_y)$ (gross compression flange area).

$$F_{bw} = [1.26 - 0.00051(h/t)\sqrt{F_y}]\,(0.6 F_y) \leq 0.6 F_y \quad (9.26)$$

   In Eqs. (9.25) and (9.26), F_{bw} and F_y both have the units of ksi.
4. Satisfy the following interaction equation requirement to assure that the unreinforced (unstiffened) webs do not fail due to excessive combined shear and flexure stresses in the web.

$$\left(\frac{f_{bw}}{F_{bw}}\right)^2 + \left(\frac{f_v}{F_v}\right)^2 \leq 1.0 \quad (9.27)$$

   In Eq. (9.27), the allowable web flexural stress limitation of $0.6 F_y$ does not apply.
5. Satisfy the requirements summarized in Tables 9.13 and 9.14 to assure that the webs do not fail (cripple) due to transverse compressive stresses in the web at concentrated end reactions

Table 9.13. Allowable Concentrated Loads (in kip) or End Reactions to Prevent Web Crippling in Beam Sections Having Single Webs and $h/t < 200$.[a]

At locations of one concentrated load or reaction acting either on top or bottom flange, when clear distance between bearing edges of this and adjacent opposite concentrated loads or reactions is greater than $1.5h$	For end reactions of beams or concentrated loads on end of cantilevers when distance from edge of bearing to end of the beam is less than $1.5h$	Stiffened flanges $t^2k\ C_3C_4C_\theta[179 - 0.33(h/t)]$ $[1 + 0.01\ (N/t)]$ Unstiffened flanges $t^2k\ C_3C_4C_\theta[117 - 0.15(h/t)]$ $[1 + 0.01\ (N/t)]^b$
	For reactions and concentrated loads when distance from edge of bearing to end of the beam is equal to or larger than $1.5h$	Stiffened and unstiffened flanges $t^2k\ C_1C_2C_\theta[291 - 0.41(h/t)]$ $[1 + 0.007\ (N/t)]^c$
At locations of two opposite concentrated loads or of a concentrated load and an opposite reaction acting simultaneously on the top and bottom flanges, when the clear distance between their adjacent bearing edges is equal to or less than $1.5h$	For end reactions of beams or concentrated loads on end of cantilevers when distance from edge of bearing to end of the beam is less than $1.5h$	Stiffened and unstiffened flanges $t^2k\ C_3C_4C_\theta[132 - 031(h/t)]$ $[1 + 0.01\ (N/t)]$
	For reactions and concentrated loads when distance from edge of bearing to end of beam is equal to or larger than $1.5h$	Stiffened and unstiffened flanges $t^2k\ C_1C_2C_\theta[417 - 1.22(h/t)]$ $[1 + 0.0013\ (N/t)]$

Reproduced with permission from *Cold Formed Steel Design Manual: Part 1—Specification for the Design of Cold Formed Steel Structural Members*, 1980 ed. American Iron and Steel Institute, Washington, DC.
[a]Valid if $R/t \leq 6$ in beams, $R/t \leq 7$ in decks, $N/t \leq 210$, and $N/h \leq 3.5$.
[b]When $N/t > 60$, the factor $[1 + 0.01(N/t)]$ may be increased to $[0.71 + 0.015(N/t)]$.
[c]When $N/t > 60$, the factor $[1 + 0.007(N/t)]$ may be increased to $[0.75 + 0.011(N/t)]$.
N is actual length in bearing, in. (mm); k is $F_y/33$; C_1 is $1.22 - 0.22\ k$; C_2 is $(1.06 - 0.06\ R/t) \leq 1.0$; C_3 is $1.33 - 0.33\ k$; C_4 is $(1.15 - 0.15\ R/t) \leq 1.0$ but not less than 0.5; C_θ is $0.7 + 0.3(\theta/90)^2$; θ is angle between plane of the web and the plane of the bearing surface $\geq 45°$ but $\leq 90°$.

Table 9.14. Allowable Concentrated Loads (in kip) or End Reactions to Prevent Web Crippling in I Beams Made of Two Channels Connected Back to Back.[a]

At locations of one concentrated load or reaction acting either on top or bottom flange, when clear distance between bearing edges of this and adjacent opposite concentrated loads or reactions is greater than $1.5h$	For end reactions of beams or concentrated loads on end of cantilevers when distance from edge of bearing to end of beam is less than $1.5h$	Stiffened and unstiffened flanges $t^2 F_y C_7 (5.0 + 0.63\sqrt{N/t})$
	For reactions and concentrated loads when distance from edge of bearing to end of the beam is equal to or larger than $1.5h$	Stiffened and unstiffened flanges $t^2 F_y C_5 C_6 (7.50 + 1.63\sqrt{N/t})$
At locations of two opposite concentrated loads or of a concentrated load and opposite reaction acting simultaneously on top and bottom flanges, when clear distance between their adjacent bearing edges is equal to or less than $1.5h$	For end reactions of beams or concentrated loads on end of cantilevers when distance from edge of bearing to end of beam is less than $1.5h$	Stiffened and unstiffened flanges $t^2 F_y C_{10} C_{11} (5.0 + 0.63\sqrt{N/t})$
	For reactions and concentrated loads when distance from edge of bearing to end of the beam is equal to or larger than $1.5h$	Stiffened and unstiffened flanges $t^2 F_y C_8 C_9 (7.50 + 1.63\sqrt{N/t})$

Reproduced with permission from *Cold Formed Steel Design Manual: Part 1—Specification for the Design of Cold Formed Steel Structural Members*, 1980 ed. American Iron and Steel Institute, Washington, DC.

[a] For similar sections that provide a high degree of restraint against rotation of the web, such as I Sections made by welding two angles to a channel and $h/t < 200$. Valid if $R/t \leq 6$ in beams, $R/t \leq 7$ in decks, $N/t \leq 210$, and $N/t \leq 3.5$.
C_5 is $(1.49 - 0.53k) \leq 0.6$; C_6 is $0.88 - 0.12 m$; C_7 is $1 + (h/t)/750$ when $h/t \leq 150$; otherwise C_7 is 1.20; C_8 is $1/k$ when $h/t \leq 66.5$; otherwise C_8 is $[1.10 - (h/t)/665](1/k)$; C_9 is $0.82 + 0.15 m$; C_{10} is $[0.98 - (h/t)/865](1/k)$; C_{11} is $0.64 + 0.31 m$; m is $t/0.075$. All other coefficients are defined in Table 9.13.

or at points of application of transverse concentrated loads. Web crippling will not occur in webs having a flat width ratio $h/t <$ 200 if the concentrated loads or reactive forces are less than those in Tables 9.13 and 9.14.

6. Satisfy the conditions of interaction equations (9.28) and (9.29) to assure that the webs do not fail due to combined flexural and transverse compressive stresses. For unreinforced (unstiffened) webs of shapes with single webs

$$1.2(P/P_{\text{all}}) + (M/M_{\text{all}}) \leq 1.5 \qquad (9.28)$$

For unreinforced (unstiffened) webs of shapes made by welding two channels back-to-back

$$1.1(P/P_{\text{all}}) + (M/M_{\text{all}}) \leq 1.5 \qquad (9.29)$$

7. Satisfy the deflection requirements.

$$\Delta \leq \Delta_{\text{all}} \qquad (9.30)$$

where Δ is calculated by the elastic analysis of beams and is based on the effective width of SCEs for deflection considerations.

Adequately braced flexural members may have (1) flanges with UCE with no local buckling; (2) flanges with UCE and local buckling; (3) flanges with SCE and no local buckling; (4) flanges with SCE and local buckling.

The following examples illustrate procedures for handling the second and fourth cases. Cases 1 and 3 are special simpler cases of these two in which $b = w$ for SCEs and $F_b = 0.60 F_y$ or $0.6 F_{ya}$ for UCEs.

Example 9.2: UCEs with Local Buckling. Evaluate the moment capacity of an I section fabricated from two 7 × 1.5 × 0.06 in. (two 178 × 38 × 1.5 mm) channels with unstiffened flanges if $F_y = 50$ ksi (340 MPa) and the flanges are adequately braced (Fig. 9.13). Neglect any increases in yield strength due to cold working.

Solution

a. Since all compression flanges are UCEs, the entire section is effective. Thus, from Table 9.6, $S_x = 2(1.06) = 2.12$ in.^{3} (34.7 cm^3).

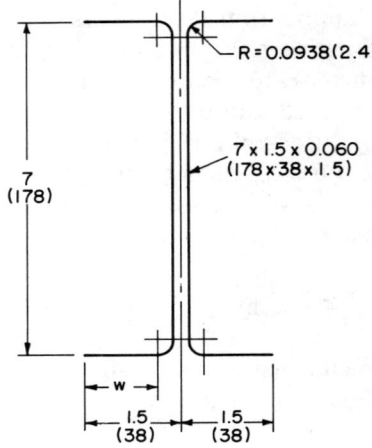

Fig. 9.13. Unstiffened I section for Example 9.2. All dimensions are in in. (mm).

b. Evaluate w/t and $(w/t)_{\text{lim}}$ for the compression flange

$$w = 1.5 - R - t = 1.5 - 0.0938 - 0.06$$
$$= 1.35 \text{ in.} \quad (34.3 \text{ mm})$$

$$\frac{w}{t} = \frac{1.35}{0.06} = 22.52$$

$$\left(\frac{w}{t}\right)_{\text{lim}} = \frac{63.3}{\sqrt{F_y}} = 8.95 < \frac{w}{t}$$

$$\left(\frac{w}{t}\right)_{\text{lim}} = \frac{144}{\sqrt{F_y}} = 20.36$$

c. Since w/t is greater than both limiting ratios and is less than 25, from Eq. (9.11) $F_c = \dfrac{8000}{(w/t)^2} = \dfrac{8000}{(22.52)^2} = 15.77$ ksi (108 MPa) and

$$M_{\text{all}} = F_c S_x = 15.77 \times 2.12 = 33.44 \text{ k-in.} \quad (3.78 \text{ kN-m})$$

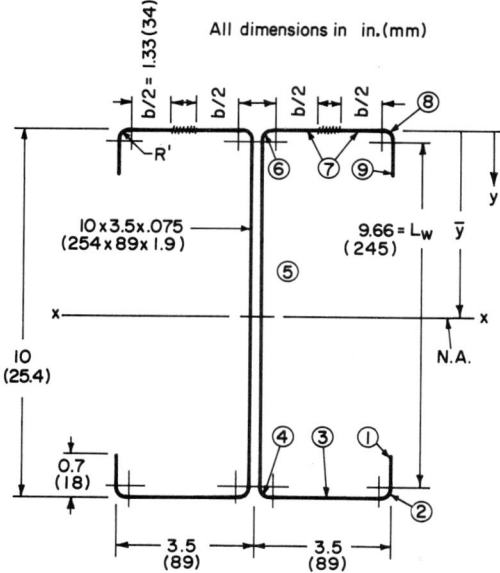

Fig. 9.14. Stiffened I section for Example 9.3.

d. Note that the design aid in Table 9.6 gives $F_c = 15.9$ ksi (110 MPa) directly for the I shape and $F = 0.6(50) = 30$ ksi (210 MPa).

Example 9.3. SCE with NA Closer to Tension Flange. Evaluate the allowable moment capacity for an I-section made of $2 - 10 \times 3.5 \times 0.075$ in. ($254 \times 89 \times 1.9$ mm) channels with edge-stiffened flanges if $F_y = 50$ ksi (340 MPa) and the flanges are adequately braced (Fig. 9.14). Neglect cold working effects.

Solution

a. Sketch the section and note that $R = 3/32$ in. (2.4 mm).
b. Since the edges are stiffened, the N.A. will be at the C.A. or will shift away from the compression flanges because of the potential reduction in effective width of the compression flange. Thus, the compressive stress in the top flange will control the section capacity and $F_b = 0.60 F_y = 30$ ksi (210 MPa).
c. Evaluate the effective width of the compression flange

322 Cold-Formed Steel Design

$$w_f = 3.5 - 2(R + t) = 3.16 \text{ in.} \quad (80.3 \text{ mm})$$

$$(w/t)_f = 3.16/.075 = 42.1$$

$$(w/t)_{\lim} = 171/\sqrt{f} = 171/\sqrt{30} = 31.22$$

Since $(w/t)_f > (w/t)_{\lim}$

$$b = \frac{253}{\sqrt{f}} \left[1 - \frac{55.3}{(w/t)\sqrt{f}} \right] t = \frac{253}{\sqrt{30}} \left[1 - \frac{55.3}{(42.1)\sqrt{30}} \right] (0.075)$$

$$= 2.63 \text{ in.} \quad (66.8 \text{ mm})$$

d. Evaluate the location of the N.A. and I_{NA} of the reduced section using the element numbers in Fig. 9.14 and the following tabular format.

Element	L_i	$\bar{y}_i$	$L_i\bar{y}_i$	I'	d_i	$L_i d_i^2$
1	0.530	9.560	5.067	0.01	4.407	10.290
2	0.206	9.920	2.044	0	4.760	4.670
3	3.160	9.960	31.474	0	4.810	73.110
4	0.206	9.920	2.044	0	4.760	4.670
5	9.660	5.000	48.300	75.1	0.153	0.226
6	0.206	0.042	0.009	0	5.070	5.300
7	2.630	0.040	0.105	0	5.116	68.840
8	0.206	0.042	0.009	0	5.070	5.300
9	0.530	0.435	0.231	0.01	4.720	11.810
	$\Sigma L_i = 17.334$ in.		$\Sigma L_i \bar{y}_i = 89.283$ in.^{2}			$\Sigma L_i d_i^2 = 184.216$ in.^{3}
	(440.1 mm)		(576.0 cm^2)			(3019 cm^3)

$$\bar{y} = \frac{\Sigma L_i \bar{y}_i}{\Sigma L_i} = \frac{89.283}{17.33} = 5.152 \text{ in.} \quad (130.9 \text{ mm})$$

$$I_{NA}' = (\Sigma I' + \Sigma L_i d_i^2)(2)$$
$$= (75.1 + 184.22)(2) = 518.64 \text{ in.}^3 \quad (8498.9 \text{ cm}^3)$$
$$I_{NA} = I_{NA}' \times t = 518.64 \times 0.075$$
$$= 38.90 \text{ in.}^4 \quad (1619 \text{ cm}^4)$$
$$S_x = \frac{38.90}{5.153} = 7.55 \text{ in.}^3 \quad (123.7 \text{ cm}^3)$$
$$M_{\text{all}} = 7.55 (30) = 226 \text{ k-in.} \quad (25.5 \text{ kN-m})$$

e. The section properties could also be obtained from Table 9.5. Note that $I_x = 41.0$ in.^{4} (1706 cm^4). However, that property is based on the full section. The section modulus, which is based

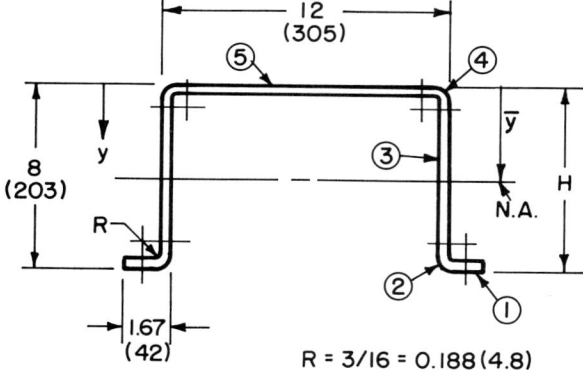

Fig. 9.15. Hat section for Example 9.4. All dimensions are in in. (mm).

on effective area, is listed as 7.56 in.³ (123.9 cm³) for $F = 0.6(50) = 30$ ksi (210 MPa).

Example 9.4. SCE with N.A. Closer to the Compression Flange Evaluate the moment capacity of an 8 × 12 × 0.135 in. (203 × 305 × 3.4 mm) hat section if $F_y = 50$ ksi (340 MPa) and the compression flange is adequately braced (see Fig. 9.15). Neglect effects of cold working.

Solution

a. Sketch section—see Fig. 9.15.
b. It is apparent that the N.A. for the fully effective section is closer to the top compression flange than to the bottom flange. Thus, the tensile stress in the bottom flange controls. That is, $F_t = 0.60F_y = 30$ ksi (210 MPa) tension on the bottom fibers and $F_c = F < 0.60F_y$ compression on the top fibers.
c. Since the effective width of the compression flange depends on the stress level in the top flange, which in turn depends on the N.A. location, it is necessary to assume something.
d. Trial 1—Assume $f = 15$ ksi (100 MPa)

$$w = 12 - 2(0.323) = 11.36 \text{ in.} \quad (288.5 \text{ mm})$$
$$w/t = 11.36/0.135 = 84.12$$
$$(w/t)_{\text{lim}} = 171/\sqrt{f} = 171/\sqrt{15} = 44.15$$

324 Cold-Formed Steel Design

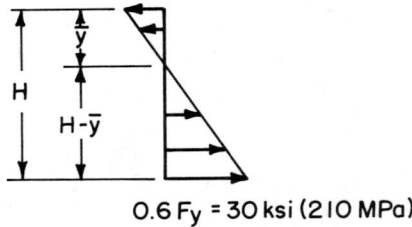

0.6 F_y = 30 ksi (210 MPa)

Fig. 9.16. Stress distribution for solution of Example 9.4.

and

$$b = \frac{253}{\sqrt{15}}\left[1 - \frac{55.3}{(84.12)\sqrt{15}}\right](0.135) = 7.32 \text{ in.} \quad (185.9 \text{ mm})$$

Evaluate the section properties.

Element	L_i	$\bar{y}_i$	$L_i\bar{y}_i$
1	1.35 × 2	7.93	21.42
2	0.40 × 2	7.84	6.29
3	7.36 × 2	4.00	58.88
4	0.40 × 2	0.16	0.13
5	7.32 × 1	0.068	0.50
	$\Sigma L_i = 26.34$ in.		$\Sigma L_i\bar{y}_i = 87.21$ in.^{2}
	(669.0 mm)		(562.6 cm^2)

$$\bar{y} = \frac{\Sigma L_i \bar{y}_i}{\Sigma L_i} = \frac{87.21}{26.34} = 3.31 \text{ in.} \quad (84.1 \text{ mm})$$

Since the stress distribution is linear and the bottom fiber stress equals 30 ksi, use proportionality to check f (see Fig. 9.16)

$$f = \frac{\bar{y}}{H - \bar{y}}(0.6F_y) = \frac{3.31}{8 - 3.31}(30) = 21.17 \text{ ksi} \quad (150 \text{ MPa})$$

$$\neq \text{ assumed } f$$

Thus, the assumed f is not correct and another trial is necessary.

e. Trial 2—Assume $f = 22.7$ ksi (160 MPa).

$$b = \frac{253}{4.76}\left[1 - \frac{55.3}{84.12(4.76)}\right](0.135) = 6.19 \text{ in.} \quad (157.2 \text{ mm}).$$

Element	L_i	$\bar{y}_i$	$L_i\bar{y}_i$	$\bar{I}'$	d_i	$L_i d_i^2$
1	2.70	7.93	21.38	0	−4.48	54.19
2	0.80	7.84	6.29	0	−4.39	15.42
3	14.72	4.00	58.88	66.2	−0.55	4.45
4	0.80	0.16	0.13	0	3.29	8.66
5	6.19	0.068	0.42	0	3.38	70.72
	$\Sigma L_i = 25.21$ in. (640.3 mm)		$\Sigma L_i\bar{y}_i = 87.10$ in.² (561.9 mm²)	$\Sigma = 66.2$ in.³ (1084.8 cm³)		$\Sigma = 153.44$ in.³ (2514.4 cm³)

$$\bar{y} = 87.10/25.21 = 3.45 \text{ in. } (87.6 \text{ mm})$$

Calculating f

$$f = \frac{3.45}{8 - 3.45}(30) = 22.78 \text{ ksi } (160 \text{ MPa})$$

Thus, $F_c = f \approx 22.74$ ksi (160 MPa).

f. Evaluate I_x and M_{all}

$$d_i = \bar{y} - \bar{y}_i$$

$$I_x' = 66.2 + 153.44 = 219.64 \text{ in.}^3 \text{ (3599.3 cm}^3\text{)}$$

$$I_x = I_x'(t) = 29.65 \text{ in.}^4 \text{ (1234 cm}^4\text{)}$$

$$M_{\text{all}} = (F_c)(I_x/c_b) = 30\,[29.65/(8 - 3.45)]$$

$$M_{\text{all}} = 195.5 \text{ k-in. } (22.1 \text{ kN-m})$$

9.7.2 Laterally Unsupported Beams

Unstable beam sections which are not adequately braced may buckle laterally when flange stresses are below either $0.6F_y$ or F_c. The criteria for bracing and the design stresses for laterally unstable beams are defined in Eqs. (9.31)–(9.37). Any consistent set of units may be used in each of the equations except Eq. 9.37, in which F_b has the units of ksi.

1. For symmetrical I or channel shapes bent about a centroidal axis perpendicular to the web: If

326 Cold-Formed Steel Design

$$\frac{L_b^2 S_{xc}}{dI_{yc}} \leq \frac{0.36\pi^2 E C_b}{F_y}$$

$$F_b = 0.60 F_y \qquad (9.31)$$

If

$$\frac{0.36\pi^3 E C_b}{F_y} < \frac{L_b^2 S_{xc}}{dI_{yc}} \leq \frac{1.8\pi^2 E C_b}{F_y}$$

$$F_b = \frac{2}{3} F_y - \left[\frac{F_y^2}{5.4\pi^2 E C_b}\right]\left(\frac{L_b^2 S_{xc}}{dI_{yc}}\right) \qquad (9.32)$$

If

$$\frac{L_b^2 S_{xc}}{dI_{yc}} \geq \frac{1.8\pi^2 E C_b}{F_y},$$

$$F_b = 0.6\pi^2 E C_b \left(\frac{dI_{yc}}{L_b^2 S_{xc}}\right) \qquad (9.33)$$

In the first condition, the section yields before buckling laterally. If the second condition exists, inelastic lateral buckling commences before yielding and if the third condition exists, elastic lateral buckling commences before yielding.

2. For point symmetric Z sections bent about an axis perpendicular to the web: If

$$\frac{L_b^2 S_{xc}}{dI_{yc}} \leq \frac{0.18\pi^2 E C_b}{F_y}$$

$$F_b = 0.6 F_y \qquad (9.34)$$

If

$$\frac{0.18\pi^2 E C_b}{F_y} < \frac{L_b^2 S_{xc}}{dI_{yc}} < \frac{0.90\pi^2 E C_b}{F_y}$$

$$F_b = \frac{2}{3} F_y - \left(\frac{F_y^2}{2.7\pi^2 E C_b}\right)\left(\frac{L_b^2 S_{xc}}{dI_{yc}}\right) \qquad (9.35)$$

If

$$\frac{L_b^2 S_{xc}}{dI_{yc}} > \frac{0.9\pi^2 E C_b}{F_y}$$

$$F_b = 0.3\pi^2 E C_b \left(\frac{dI_{yc}}{L_b^2 S_{xc}}\right) \qquad (9.36)$$

Equation (9.34) is a yielding criterion, Eq. (9.35) is a criterion for inelastic lateral buckling, and Eq. 9.36 is an elastic buckling criterion.

3. For hat sections with $I_x > I_y$,

$$F_b = \frac{151{,}900}{(L_b/r_y)^2} \quad \text{but} \quad \leq 0.6 F_y \qquad (9.37)$$

Laterally unbraced beams might buckle locally before buckling laterally. Thus, use the smaller of the stresses at which lateral buckling or local buckling of UCEs commences for the allowable flexure stress. Also, if local buckling is not a factor when SCEs are encountered, use full flat widths for designing. Conversely, use effective widths for SCEs when local buckling is a factor.

Example 9.5 Unbraced Beam. Evaluate the allowable uniformly distributed load for an I section composed of two 6 × 1.5 × 0.135 in. (152 × 38 × 3.4 mm) channels with unstiffened flanges and bent about the major axis. The beam is simply supported over a span of 6 ft (1.83 m) with no intermediate lateral support of the compression flange (Fig. 9.17). Neglect the effect of cold work and assume $F_y = 50$ ksi (340 MPa).

Solution

a. Sketch the system—See Fig. 9.17.
b. Check if the flanges buckle locally

$$w_f = 1.5 - (R + t) = 1.18 \text{ in.} \quad (30.0 \text{ mm})$$

$$(w/t)_f = 1.18/0.135 = 8.74$$

$$(w/t)_{\text{lim}} = 63.3/\sqrt{F_y} = 8.91$$

Since $(w/t)_f < (w/t)_{\text{lim}}$ no local buckling occurs in the flanges, and $F_c = 0.6 F_y$.

328 Cold-Formed Steel Design

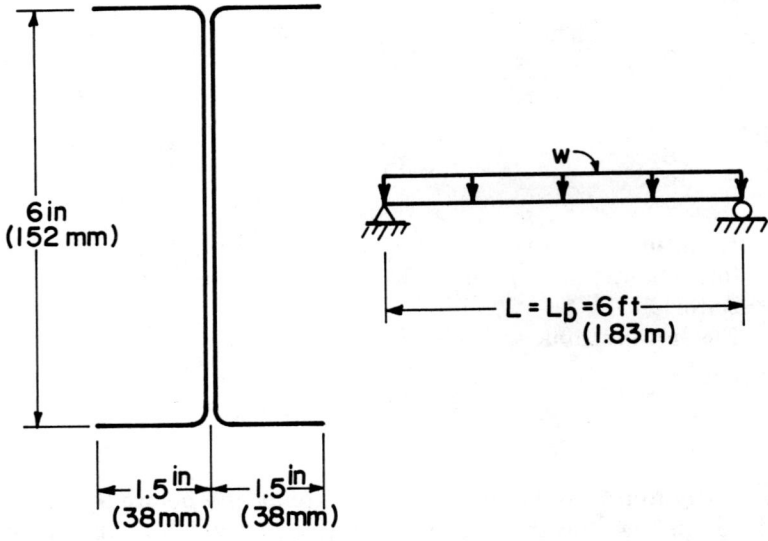

Fig. 9.17. Unbraced beam for Example 9.5.

c. Check for lateral buckling

$$L_b = 6 \text{ ft} \times 12 = 72 \text{ in.} \quad (1.83 \text{ m})$$

$$C_b = 1.0$$

since the moment between braced ends is greater than either end moment

$$E = 29 \times 10^3 \text{ ksi} \quad (200 \text{ GPa})$$

i. Evaluate the limits on the beam bracing parameter:

$$\left(\frac{L_b^2 S_{xc}}{dI_{yc}} \right)_{\text{lim}} = \frac{0.36 \pi^2 E C_b}{F_y} = 2060$$

and

$$\left(\frac{L_b^2 S_{xc}}{dI_{yc}} \right)_{\text{lim}} = \frac{1.8 \pi^2 E C_b}{F_y} = 10{,}640$$

ii. Evaluate the bracing parameter for the beam: From Table 9.6

$$d = 6 \text{ in.} \quad (152 \text{ mm})$$

$$I_{yc} = I_y/2 = 0.615/2 = 0.308 \text{ in.}^4 \quad (13 \text{ cm}^4)$$

$$S_{xc} = S_x = 3.56 \text{ in.}^3 \quad (58.3 \text{ cm}^3)$$

$$\frac{L_b^2 S_{xc}}{dI_{yc}} = \frac{(72)^2(3.56)}{6(0.308)} = 9986$$

iii. Thus the controlling criterion for F_b is Eq. (9.32) (inelastic lateral buckling). Evaluate the allowable stress:

$$F_b = \frac{2}{3} F_y - \frac{F_y^2}{5.4\pi^2 EC_b} \left(\frac{L_b^2 S_{xc}}{dI_{yc}}\right)$$

$$= 0.67(50) - \frac{(50)^2}{5.4\pi^2 (29 \times 10^3)(1.0)} (9986)$$

$$= 17.35 \text{ ksi} \quad (120 \text{ MPa})$$

$$< 30 \text{ ksi} \quad (210 \text{ MPa})$$

Thus, the allowable flexural stress equals 17.35 ksi (120 MPa).

d. Evaluate the moment capacity of the beam: Since all flanges are UCEs, the full section is effective and

$$M_{\text{all}} = F_b S_x = 17.35 \, (3.56)$$

$$M_{\text{all}} = 61.8 \text{ k-in.} \quad (6.9 \text{ kN-m})$$

e. Evaluate the allowable load including the beam dead weight: For a SS beam, $M = wL^2/8$. Thus,

$$w_{\text{all}} = \frac{8M_{\text{all}}}{L^2} = \frac{8(61.8) \times 12}{(72)^2}$$

$$= 1.2 \text{ k/ft} \quad (1.6 \text{ kN/m})$$

330 Cold-Formed Steel Design

f. Check the shear stress in the webs: The shear force and average shear stress in the web are

$$V_{max} = wL/2 = 1.2(6)/2 = 3.6 \text{ k} \quad (16.0 \text{ kN})$$

$$f_v = V/A_{web} = \frac{3.6}{[6.00 - 2(0.188 + 0.135)](0.270)}$$

$$= 2.5 \text{ ksi} \quad (17 \text{ MPa})$$

Evaluate the limiting value of h/t from Eq. (9.23) and (9.24).

$$(h/t)_{lim} = 237\sqrt{k_v/F_y} = 237\sqrt{5.34/50} = 77.45$$

Evaluate h/t:

$$h/t = (6 - 0.270)/0.135 = 42.44$$

Thus, from Eq. (9.23),

$$F_v = \frac{65.7\sqrt{k_v/F_y}}{(h/t)} = \frac{65.7\sqrt{5.34(50)}}{42.44} = 25.2 \text{ ksi}$$

and

$$F_v = 0.4(50) = 20 \text{ ksi}$$

Since $f_v < F_v$, the section can carry V_{max}.

g. Check the maximum flexural stresses in the web. Since the flanges are unstiffened, the allowable flexural stress in the webs equals

$$F_{bw} = [1.26 - 0.00051(h/t)\sqrt{F_y}](0.6F_y) \leq 0.6 F_y$$

$$= [1.26 - 0.00051(42.44)\sqrt{50}](0.6 \times 50)$$

$$= 33.21 \text{ ksi } (229 \text{ MPa}) > 30 \text{ ksi } (210 \text{ MPa})$$

Thus, use $F_{bw} = 30$ ksi (210 MPa). The maximum flexural stress in the web is based on the full web area and $(F_c/0.6F_y)$(gross flange area)

$$F_{bw} = M_{max} y_w/I_c$$

Since $F_c/0.6F_y = 1.0$, $I_c = 10.7$ in.^{4} (450 cm^4) and

$$f_{bw} = 61.8\,(3 - 0.135)/10.7 = 16.55 \text{ ksi} \quad (110 \text{ MPa})$$

Since $f_{bw} < F_{bw}$, and since the combined stress criteria, Eq. (9.27), is also satisfied (when the shear is maximum the moment equals zero and vice versa), the web is adequate to carry the flexure stresses and transverse shear.

h. Web crippling criteria at the end reactions would have to be checked using the allowable loads of Table 9.14 and deflections would need to be checked with Eq. (9.30) to completely analyze the adequacy of the beam.

9.8 AXIALLY LOADED COMPRESSION MEMBERS

9.8.1 Introduction

Axially loaded columns of cold-formed steel sections may fail by yielding, local buckling of section elements, flexural buckling about their axis with largest KL/r ratio, or simultaneous torsional–flexural buckling. Torsional–flexural buckling is characterized by simultaneous twisting of the section about its longitudinal axis as the member undergoes flexural buckling. The mode of failure is dictated in large part by the shape of the cross section, the slenderness of the entire column, and the degree of bracing against both lateral movement of the column and twisting.

The degree of symmetry (whether it is singly, doubly, or point symmetric) of a cross section is an important factor for determining the mode of failure in a column. The box and tube sections of Figs. 9.1d and 9.1e are doubly symmetric; the channels, angles, and hat sections of Figs. 9.1a, 9.1c, 9.1f, and 9.1g are singly symmetric; and the Z section of Fig. 9.1d is point symmetric. The unequal leg angle sections of Table 9.7 are unsymmetric. Whether a section is open or closed is also of importance in column behavior. In Fig. 9.1, only the box and tube sections are closed sections.

Cold-formed steel column shapes are either torsionally stable or torsionally unstable. A torsionally stable axially loaded column is defined as one which fails by yielding, local buckling, or flexural buckling but not by torsional (twisting) buckling or simultaneous flexural–torsional buckling. Typical torsionally stable columns are those whose cross sections are either doubly symmetric, closed, cylindrical, solid, point symmetric such that the centroid and shear center coincide, or a section which is braced against twisting by adequate diaphragms or intermittent bracing.

Torsionally unstable columns include sections which are open, singly symmetric, or nonsymmetric. Possible failure modes for these sections include all those for torsionally stable columns plus torsional buckling (rarely a factor) and flexural–torsional buckling.

9.8.2 Torsionally Stable Columns

Torsionally stable columns are classified as being long, short, or intermediate in length. The slenderness ratio, KL/r, is used to classify the column. The 1980 AISI specification for the allowable compressive stress F_{a1} in any consistent set of units of torsionally stable sections is given by Eqs. (9.38), (9.39), (9.40), and (9.41). If $KL/r < C_c/\sqrt{Q}$, inelastic buckling is the failure mode and

$$F_{a1} = \frac{12}{23}(QF_y) - \frac{3(QF_y)^2}{23\pi^2 E}\left(\frac{KL}{r}\right)^2 \qquad (9.38)$$

$$= 0.522(QF_y) - \frac{1}{75.7E}[(QF_y)(KL/r)]^2 \qquad (9.39)$$

If $KL/r \geq C_c/\sqrt{Q}$, the column is long and elastic buckling is the failure mode. Then

$$F_{a1} = \frac{12\pi^2 E}{23(KL/r)^2} = 5.15\frac{E}{(KL/r)^2} \qquad (9.40)$$

Since the limiting KL/r for cold-formed columns is 200, if $KL/r > 200$

$$F_{a1} = 0 \qquad (9.41)$$

Of course, for very short columns, it is possible that yielding occurs before inelastic buckling. In this case $F_{a1} > F$ and the allowable compressive stress is $F_{a1} = 0.60F_y$. In Eqs. (9.38)–(9.40), $C_c = \sqrt{2\pi^2 E/F_y}$; P, total loads, kip (kN); A, full unreduced cross-sectional area, in.^{2} (mm^2); F_{a1}, allowable average axial compressive stress, ksi (MPa); E, modulus of elasticity (MOE) = 29.5×10^3 ksi (200 GPa); K, effective length factor; L, unbraced length of column, in. (mm); r, radius of gyration of full unreduced section, in. (mm); Q, form factor for local buckling.

If the column elements do not buckle locally before yielding (w/t is small), the form factor $Q = 1.0$. Conversely, if either the UCEs or SCEs of the section buckle locally before yielding, then $Q < 1.0$. Local buckling considerations are thus handled simply by basing flexural buckling loads on a reduced effective yield stress (QF_y).

If a section is composed of only SCEs, Q is defined as the ratio of the effective area to the full area of the cross section. The effective area is based upon the effective width of the flats using the basic design stress in the SCE ($f = 0.60F_y$). Thus, for a section with all SCEs,

$$Q = Q_a = \frac{A_{\text{effective}}}{A_{\text{total}}} \quad (9.42)$$

If the section is composed of all UCEs, Q is the ratio of the allowable compressive stress, F_c, for the weakest UCE to the basic design stress. That is,

$$Q = Q_s = \frac{F_c}{F} = \frac{F_c}{0.6F_y} \quad (9.43)$$

If the section is composed of both SCEs and UCEs, Q is the product of Q_a and Q_s. That is,

$$Q = Q_a Q_s = \left(\frac{A_{\text{eff}}}{A_{\text{tot}}}\right)\left(\frac{F_c}{0.6F_y}\right) \quad (9.44)$$

where Q_a is based upon the allowable stress F_c for UCEs.

For the special cases where $Q = 1.0$, element thickness greater than 0.09 in. (2.3 mm), and $KL/r < C_c$, the allowable column compressive stress, in any consistent set of units, is identical to that for intermediate length hot-rolled sections and is defined in Eq. (9.45).

$$F_{a1} = \frac{[1 - \tfrac{1}{2}\left(\frac{KL/r}{C_c}\right)^2]F_y}{\tfrac{5}{3} + \tfrac{3}{8}\left(\frac{KL/r}{C_c}\right) - \tfrac{1}{8}\left(\frac{KL/r}{C_c}\right)^3} \quad (9.45)$$

Example 9.6 Doubly Symmetric Column. Evaluate the allowable axial compressive load for a $4 \times 2\tfrac{1}{4} \times 0.075$ in. ($102 \times 58 \times 1.9$ mm) I section (see Fig. 9.18) built up with two A36 cold-formed channels if $KL = 4$ ft (1.22 m); if $KL = 6$ ft (1.83 m).

334 Cold-Formed Steel Design

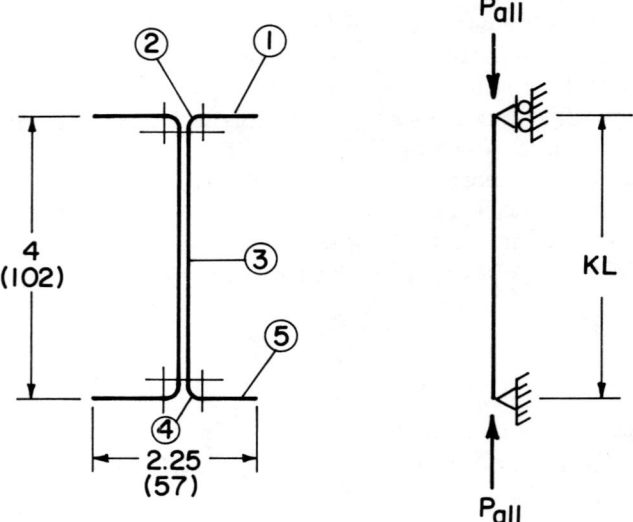

Fig. 9.18. Doubly symmetric column section. All dimensions are in in. (mm).

Solution

a. The section properties from Table 9.6 are

$$A = 0.898 \text{ in}^2. \qquad (5.8 \text{ cm}^2)$$
$$r_y = 0.399 \text{ in.} \qquad (10.1 \text{ mm})$$
$$R = 3/32 \text{ in.} \qquad (2.4 \text{ mm})$$
$$F_y = 36 \text{ ksi} \qquad (250 \text{ MPa})$$
$$F = 0.6F_y = 21.6 \text{ ksi} \quad (150 \text{ MPa})$$

b. Evaluate form factor Q
   i. UCEs:

$$w_1 = 2.25/2 - 3/32 - 0.075 = 0.956 \text{ in.} \quad (24.3 \text{ mm})$$
$$w_1/t = 0.956/0.075 = 12.75$$
$$(w/t)_{\text{lim}} = 63.3/\sqrt{F_y} = 10.55$$
$$(w/t)'_{\text{lim}} = 144/\sqrt{F_y} = 24.00$$

9.8 Axially Loaded Compression Members 335

Thus,

$$F_c = F_y \left[0.767 - \frac{2.64}{1000} \left(\frac{w}{t}\right) \sqrt{F_y} \right]$$

$$= 36 \left[0.767 - \frac{2.64}{1000} (12.75)\sqrt{36} \right]$$

$$= 20.34 \text{ ksi} \quad (140 \text{ MPa})$$

and

$$Q_s = F_c/F = 20.34/21.6 = 0.942$$

ii. SCEs:

$$w_3 = 4.00 - 2(R + t) = 4.00 - 2(3/32 + 0.075)$$

$$= 3.663 \text{ in.} \quad (93.0 \text{ mm})$$

$$w_3/t = 3.663/0.075 = 48.883$$

$$(w/t)_{\text{lim}} = \frac{171}{\sqrt{f}} = \frac{171}{\sqrt{F_c}} = \frac{171}{\sqrt{20.34}} = 37.916$$

Thus,

$$\frac{b}{t} = \frac{253}{\sqrt{f}} \left[1 - \frac{55.3}{(w/t)(\sqrt{f})} \right]$$

$$= \frac{253}{\sqrt{20.34}} \left[1 - \frac{55.3}{48.833 \sqrt{20.34}} \right] = 42.012$$

and $b = 42.012 (0.075) = 3.151$ in. (80.0 mm)

$$A_{\text{eff}} = A - 2(w_3 - b)t$$

$$= 0.898 - 2(3.663 - 3.151)(0.75)$$

$$= 0.821 \text{ in.}^2 \quad (5.3 \text{ cm}^2)$$

$$Q_a = \frac{A_{\text{eff}}}{A_{\text{tot}}} = \frac{0.821}{0.898} = 0.914$$

Therefore, the form factor is $Q = Q_s Q_a = (0.942)(0.914) = 0.861$.

c. Evaluate F_{a1} and P_{all}:

  i. For $L = 4$ ft (1.22 m)

  $$KL/r = 48/0.399 = 120.30$$

  $$C_c = \left(\frac{2\pi^2 E}{F_y}\right)^{1/2} = \left(\frac{2\pi^2 (29.5 \times 10^3)}{36}\right)^{1/2} = 127.18$$

  $$(KL/r)_{\text{lim}} = C_c/\sqrt{Q} = 127.18/\sqrt{.861} = 137.06$$

  Thus, the column is intermediate and from Eq. (9.39)

  $$F_{a1} = 0.522(QF_y) - \frac{1}{75.7E}[(QF_y)(KL/r)]^2$$

  $$= 9.95 \text{ ksi} \quad (70 \text{ MPa})$$

  and

  $$P_{\text{all}} = F_{a1} A_{\text{tot}} = (9.95)(0.898) = 8.94 \text{ kip} \quad (39.8 \text{ kN})$$

  ii. For $L = 6$ ft (1.83 m) and $KL/r = 72/0.399 = 180.45 > (KL/r)_{\text{lim}}$. Thus, the column is long and from Eq. (9.40),

  $$F_{a1} = 5.15E/(KL/r)^2 = 5.15 \, (29.5 \times 10^3)/(180.45)^2$$

  $$= 4.66 \text{ ksi} \quad (30 \text{ MPa})$$

  and

  $$P_{\text{all}} = 4.66 \, (0.898) = 4.19 \text{ kip} \quad (18.6 \text{ kN})$$

Example 9.7 Point Symmetric Column (Z Section). Evaluate the allowable axial load of a $2 \times 4 \times 0.105$ in. ($51 \times 102 \times 2.7$ mm) Z section with stiffened flanges if $KL = 5$ ft (1.52 m) and $F_y = 36$ ksi (250 MPa) and if it is braced against twisting.

Solution

  a. The section properties of the point symmetric shape are taken directly from Table 3 (see Fig. 9.19).

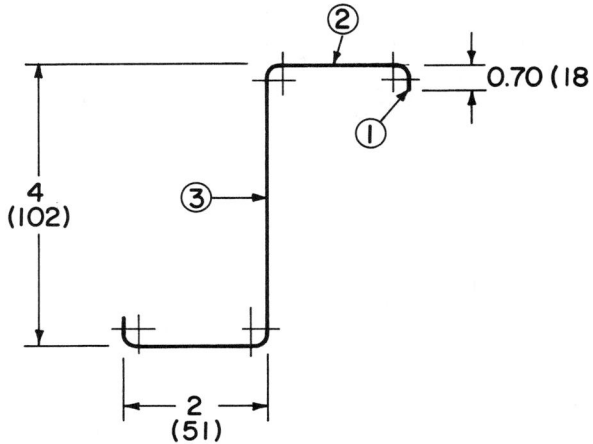

Fig. 9.19. Point symmetry Z. All dimensions are in in. (mm).

$$R = 0.188 \text{ in. } (4.8 \text{ mm})$$
$$t = 0.105 \text{ in. } (2.7 \text{ mm})$$
$$A = 0.900 \text{ in.}^2 (5.8 \text{ cm}^2)$$
$$r_{min} = 0.567 \text{ in. } (14.4 \text{ mm})$$

b. Evaluate the flat width ratios:

$$w_1/t_1 = (0.70 - 0.188 - 0.105)/0.105 = 3.881$$
$$w_2/t = [\,2 - 2(3/16 + 1.05)]/0.105 = 13.476$$
$$w_3/t = [4 - 2(3/16 + .105)]/0.105 = 32.514$$

c. Evaluate the form factors:
   i. UCEs: Since

$$(w/t)_{lim} = 63.3/\sqrt{F_y} = 10.550 > 3.881, \quad Q_s = 1.0$$

   ii. SCEs: Since

$$(w/t)_{lim} = \frac{171}{\sqrt{F}} = \frac{171}{\sqrt{21.6}} = 36.79 > \frac{w_2}{t} \text{ or } \frac{w_3}{t}$$

then $Q_a = 1.0$ and $Q = Q_a Q_s = 1.0$

d. Evaluate the allowable stress and load

$$(KL/r)_{\lim} = C_c = \sqrt{2\pi^2 E/F_y} = 127.18$$

Since $KL/r = 60/0.567 = 105.82 < C_c$, $Q = 1.0$, and $t > 0.09$,

$$F_{a1} = \frac{\left[1 - \frac{1}{2}\left(\frac{KL/r}{C_c}\right)^2\right] F_y}{\frac{5}{3} + \frac{3}{8}\left(\frac{KL/r}{C_c}\right) - \frac{1}{8}\left(\frac{KL/r}{C_c}\right)^3}$$

$$= \frac{[1 - \frac{1}{2}(105.82/127.18)^2]\, 36}{\frac{5}{3} + \frac{3}{8}(105.82/127.18) - \frac{1}{8}(105.82/127.18)^3}$$

$$= \frac{23.54}{1.91}$$

$F_{a1} = 12.32$ ksi (85 MPa) and

$P_{\text{all}} = 12.32\,(0.900) = 11.1$ kip (49.3 kN)

9.8.3 Torsionally Unstable Sections—Singly Symmetric or Unsymmetric Sections

The allowable axial compressive stress of singly symmetric or nonsymmetric shapes of open cross section shall be the smaller of F, F_{a1}, or F_{a2}, where F and F_{a1} are as defined earlier and F_{a2} is defined by Eqs. (9.46) and (9.47) in any consistent set of units,

$$F_{a2} = 0.522(QF_y) - \frac{(QF_y)^2}{7.67\sigma_{TF0}} \quad \text{if } \sigma_{TF0} > 0.5(QF_y) \quad (9.46)$$

or

$$F_{a2} = 0.522(QF_y) \quad \text{if } \sigma_{TF0} \leq 0.5(QF_y) \quad (9.47)$$

where σ_{TF0} is the elastic torsional–flexural buckling stress under concentric loading, ksi (MPa).

The elastic torsional–flexural buckling stress for shapes whose axis of symmetry is the x axis is defined in Eq. (9.48) for any consistent set of units.

$$\sigma_{TF0} = (1/2\beta)\left[(\sigma_{ex} + \sigma_t) - \sqrt{(\sigma_{ex} + \sigma_t)^2 - 4\beta\,\sigma_{ex}\sigma_t}\,\right] \quad (9.48)$$

where

$$\sigma_{ex} = \pi^2 E/(KL/r_x)^2$$

in ksi (MPa) is the allowable stress for flexural buckling about the symmetry axis;

$$\sigma_t = (1/Ar_0^2)\{GJ + [\pi^2 EC_w/(KL)^2]\}$$

in ksi (MPa) is the allowable stress for torsional buckling of the section;

$\beta = 1 - (x_0/r_0)^2$;

A = cross-sectional area in in.^{2} (mm^2);

r_0 = polar radius of gyration of the cross section about the shear center, in. (mm),

$r_0 = \sqrt{r_x^2 + r_y^2 + x_0^2}$;

r_x, r_y = radius of gyration about x and y axes in in. (mm);

E = 29.5 × 10^3 ksi (200 GPa)

G = shear modulus, 11.3 × 10^3 ksi (80 GPa);

x_0 = distance from the shear center* to the centroid along the x-axis in in. (mm);

J = St. Venant torsion constant; for sections composed of thin elements:

$$J = \frac{1}{3}\sum_{i=1}^{n}(L_i t_i^3), \text{ in.}^4 \text{ (mm}^4\text{)};$$

t_i = thickness of element i in in. (mm)

L_i = midline length of element i in in. (mm)

C_w* = warping constant of torsion of the cross section in in.^{6} (mm^6).

---

*For a discussion of and derivation of the shear center and the warping constant, consult a text on advanced mechanics and the 1972 AISI edition of the AISI Specification Part III.

340 Cold-Formed Steel Design

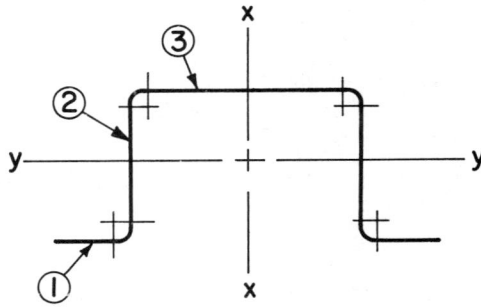

Fig. 9.20. Singly symmetric open section.

Values of J, C_w, r_0, and x_0 for typical channel sections are in Tables 9.1 and 9.2. Similar properties for hat sections are given in Table 9.9. With the exception of r_0, the same properties for angle sections are in Tables 9.7 and 9.8. Equations for evaluating J, C_w, and r_0 for stiffened and unstiffened angle, channel, hat, and Z sections not found in the tables are in Part III of the 1972 edition of the AISI Specification. Equations are available for singly symmetric I and T sections in the same reference.

Example 9.8. Singly Symmetric Open Sections. Evaluate the allowable axial compressive load of the following sections (Fig. 9.20) if $KL = 8$ ft (2.44 m) and $F_y = 33$ ksi (230 MPa).

a. A 4 × 4 × 0.105 in. (102 × 102 × 2.7 mm) hat section.
b. A 4 × 6 × 0.105 in. (102 × 152 × 2.7 mm) hat section.

Solution

a. Analyze the 4 × 4 × 0.105 in. (102 × 102 × 2.7 mm) section. The section properties are obtained from Table 9.9. In Fig. 9.20 the x axis is the axis of symmetry for consistency in the definition of σ_{TFO}. Thus, the x and y axes in Table 9.9 must be interchanged. The subscripts in this example are referenced to Figure 9.20 throughout.

$A = 1.45$ in.^{2} (9.4 cm^2), $J = 0.00534$ in.^{4} (0.22 cm^4)

$r_y = 1.53$ in. (39 mm), $C_w = 6.24$ in.^{6} (1676 cm^6)

$r_x = 1.91$ in. (48 mm), $x_0 = -3.54$ in. (-90 mm)

$r_0 = 4.31$ in. (109 mm),

9.8 Axially Loaded Compression Members 341

i. Evaluate the form factor, Q. Since $w_1/t = (1.34 - 0.105 - 0.188)/0.105 = 9.97 < 63.3/\sqrt{F_y} = 11.02$, $Q_s = 1.00$. Since $w_2/t = w_3/t = [4 - 2(0.105 + 0.188)]/0.105 = 32.51 < 171/\sqrt{.6F_y} = 38.42$, $Q_a = 1.00$. Thus, $Q = 1.0$.

ii. Evaluate F_{a2} for flexural–torsional buckling about the axis of symmetry.

$$\beta = 1 - \left(\frac{x_0}{r_0}\right)^2 = 1 - \left(\frac{-3.54}{4.31}\right)^2 = 0.325$$

$KL/r_x = 96/1.91 = 50.25$ (slenderness ratio about the axis of symmetry)

$$\sigma_{ex} = \frac{\pi^2 E}{(KL/r_x)^2} = \frac{\pi^2(29.5 \times 10^3)}{(50.25)^2} = 115 \text{ ksi} \quad (790 \text{ MPa})$$

$$\sigma_t = \frac{1}{Ar_0^2}\left[GJ + \frac{\pi^2 E C_w}{(KL)^2}\right]$$

$$= \frac{1}{(1.45)(4.31)^2}\left[(11,300)(0.00534) + \frac{\pi^2(29.5 \times 10^3)(6.24)}{(96)^2}\right]$$

$$= 9.56 \text{ ksi} \quad (70 \text{ MPa})$$

From Eq. (9.48)

$$\sigma_{TF0} = \frac{1}{2(0.33)}[(115 + 9.56) - \sqrt{(115 + 9.56)^2 - 4(.33)(115)(9.50)}]$$

$$\sigma_{TF0} = 8.91 \text{ ksi} \quad (61 \text{ MPa})$$

Since $0.5(QF_y) = 0.5(33) = 16.5$ ksi (110 MPa) $> \sigma_{TF0}$,

$$F_{a2} = 0.522(QF_y) = 0.522(16.5)$$
$$F_{a2} = 8.61 \text{ ksi} \quad (59 \text{ MPa})$$

iii. Evaluate F_{a1} for flexural buckling about the weak axis

342 Cold-Formed Steel Design

$$KL/r_y = 96/1.53 = 62.75$$

Since $(KL/r)_{\lim} = C_c = \sqrt{2\pi^2 E/F_y} = 132.84 > KL/r_y$ and $t > 0.09$ in. (2.3 mm),

$$F_{a1} = \frac{[1 - \tfrac{1}{2}(62.75/132.84)^2]\,33}{\tfrac{5}{3} + \tfrac{3}{8}(0.47) - \tfrac{1}{8}(0.47)^3} = \frac{29.36}{1.83}$$

$$= 16.04 \text{ ksi} \quad (110 \text{ MPa})$$

iv. Thus $P_{\text{all}} = F_{a2} A = 8.61(1.45)$

$$P_{\text{all}} = 12.48 \text{ kip} \quad (55.5 \text{ kN})$$

b. Analyze the $4 \times 6 \times 0.105$ in. ($102 \times 152 \times 2.7$ mm) hat section.
   i. From Table 9.9, the section properties are

$$A = 1.66 \text{ in.}^2 \ (10.7 \text{ cm}^2) \quad C_w = 16.4 \text{ in.}^6 \ (4404 \text{ cm}^6)$$
$$r_y = 1.54 \text{ in.} \ (39 \text{ mm}) \quad r_0 = 4.60 \text{ in.} \ (117 \text{ mm})$$
$$r_x = 2.69 \text{ in.} \ (68 \text{ mm}) \quad x_0 = -3.40 \text{ in.} \ (-86 \text{ mm})$$
$$J = 0.00612 \text{ in.}^4 \ (0.25 \text{ cm}^4)$$

Note again that the subscripts are referenced to the axes in Fig. 9.20.

ii. Evaluate the form factor, Q. Q_s is 1.0 since the UCE is identical to the part a section. Since $w_3/t = [6 - 2(0.105 - 0.188)]/0.105 = 51.56 > 171/\sqrt{0.6 F_y} = 38.42$, $Q_a < 1$. From Eq. (9.13),

$$b = (253/\sqrt{f}\,[1 - 55.3/(w/t)\sqrt{f}\,](t) = 4.51 \text{ in.} \ (114 \text{ mm})$$

Thus $A_{\text{eff}} = A - (w_3 - b)t = 1.66 - (5.41 - 4.51)(0.105)$
$= 1.57 \text{ in.}^2 \ (10.1 \text{ cm}^2)$ and $Q_a = 1.57/1.66 = 0.944$ and $Q = Q_s Q_a = 0.944$.

iii. Evaluate F_{a2} for flexural–torsional buckling about axis of symmetry

$$\beta = 1 - \left(\frac{-3.40}{4.60}\right)^2 = 0.454$$

$$\frac{KL}{r_x} = \frac{96}{2.69} = 35.69$$

$$\sigma_{ex} = \frac{\pi^2(29.5 \times 10^3)}{(35.69)^2} = 228.6 \text{ ksi} \quad (1580 \text{ MPa})$$

$$\sigma_t = \frac{1}{(1.66)(4.60)^2}\left[(11.3 \times 10^3)(6.12 \times 10^{-3})\right.$$
$$\left. + \frac{\pi^2(29.5 \times 10^3)(16.4)}{(96)^2}\right]$$
$$= 16.72 \text{ ksi} \quad (120 \text{ MPa})$$

$$\sigma_{TF0} = \frac{1}{2(0.454)}$$
$$\times [(228.6 + 16.72)$$
$$- \sqrt{(228.6 + 16.72)^2 - 4(.454)(228.6)(16.72)}]$$
$$= 16.04 \text{ ksi} \quad (111 \text{ MPa})$$

Since $0.5(QF_y) = 0.5(0.944)(33) = 15.58$ ksi (107 MPa) $< \sigma_{TF0}$ from Eq. (9.46)

$$F_{a2} = 0.522(0.944)(33) - \frac{(0.944 \times 33)^2}{7.67 \, (16.04)}$$
$$= 8.37 \text{ ksi} \quad (60 \text{ MPa})$$

iv. Evaluate F_{a1} for flexural buckling about the weak axis. Since $KL/r_y = 96/1.54 = 62.34 < C_c$,

$$F_{a1} = [1 - \tfrac{1}{2}(62.34/132.84)^2]33$$
$$= 16.1 \text{ ksi} \quad (110 \text{ MPa}) > F_{a2}$$

v. Thus $P_{\text{all}} = F_{a2}A = 8.37(1.66) = 13.90$ kip (61.8 kN).

9.8.4 Torsionally Unstable Point Symmetric Sections

The procedure for evaluating the allowable stresses for point symmetric column sections is similar to that for singly symmetric sections. The allowable stress is the smaller of F_{a1} or F_{a2} as defined, by Eqs. (9.49) and (9.50).

$$F_{a2} = 0.522QF_y - (QF_y)^2/7.67\sigma_t \quad \text{if} \quad \sigma_t > 0.5QF_y \quad (9.49)$$

$$F_{a2} = 0.522\sigma_t \quad \text{if} \quad \sigma_t \leq 0.5QF_y \quad (9.50)$$

The torsional constant J and the warping constant C_w are required for evaluating the allowable stress. These constants are not given in Table 9.9 for point symmetric Z sections and will have to be evaluated. Procedures for evaluating these constants for a variety of sections are outlined in Part III of the 1968 edition of the AISI Specification.

9.8.5 Cylindrical Tubular Members

The allowable compressive stress for tubular columns having a mean diameter-to-wall thickness ratio D/t less than $3,300/F_y$ is the smaller of F_{a1} from Eqs. (9.38) to (9.40) or F_a from Eq. (9.51).

$$F_a = 0.6F_y \quad \text{or} \quad 0.6F_{ya} \quad (9.51)$$

When $3,300/F_y < D/t \leq 13,000/F_y$, then the allowable compressive stress is the smaller of F_{a1} from Eqs. (9.38) to (9.40) or F_a from Eq. (9.52). In Eqs. (9.51) and (9.52), F_a and F_y both have units of ksi.

$$F_a = 662/(D/t) + 0.399 F_y \quad (9.52)$$

9.9 BEAM-COLUMNS

For sections which are not subject to torsional–flexural buckling, interaction equations (9.53), (9.54), and (9.55) must be satisfied when proportioning cold-formed sections used as beam-columns. If the ratio of axial stress to allowable axial stress, $f_a/F_{a1} > 0.15$, then both Eqs.

(9.53) and (9.54) must be satisfied. If $f_a/F_{a1} \leq 0.15$, then only Eq. (9.55) need be satisfied.

$$\frac{f_a}{F_{a0}} + \frac{f_{bx}}{F_{blx}} + \frac{f_{by}}{F_{bly}} \leq 1.0 \qquad (9.53)$$

$$\frac{f_a}{F_{a1}} + \frac{C_{mx}f_{bx}}{[1-(f_a/F'_{ex})]F_{bx}} + \frac{C_{my}f_{by}}{[1-(f_a/F'_{ey})]F_{by}} \leq 1.0 \qquad (9.54)$$

$$\frac{f_a}{F_{a1}} + \frac{f_{bx}}{F_{bx}} + \frac{f_{by}}{F_{by}} \leq 1.0 \qquad (9.55)$$

where f_a = actual axial stress, ksi (MPa); f_{bx}, f_{by}, actual maximum flexural stresses, ksi (MPa); F_{a0}, allowable compressive stress under centric loads and $KL = 0$, ksi (MPa); F_{a1}, allowable compressive stress under centric load adjusted for KL/r, ksi (MPa); $F'_{ex} = 12\pi^2 E/23 \, (KL/r)^2_x$, ksi (MPa); $F'_{ey} = 12\pi^2 E/23(KL/r)^2_y$, ksi (MPa); F_{bx}, F_{by} are allowable compressive flexural stresses in pure bending (lateral buckling considered), ksi (MPa); F_{blx}, F_{bly} are allowable compressive flexural stresses in pure bending excluding possibility of lateral buckling, ksi (MPa); $C_m = 0.6 - 0.4(M_1/M_2) \geq 0.4$. (See Section 8.9 for limitations and methods for evaluating this parameter.)

The requirements for torsionally stable cold-formed sections are nearly identical to those for hot-rolled sections. However, the load table design aids of hot-rolled standard sections are not available in cold-formed design. The analysis of existing members requires substitution of actual and allowable stresses into the appropriate interaction equations. Design of cold-formed beam-columns requires a trial and error solution.

One approach to beam column design is to select as a first trial a member slightly larger than that required for pure compression or for pure flexure. Then check its adequacy by substituting into either the simpler of the interaction equations for cases where $f_a/F_{a1} > 0.15$ or into the interaction equation for $f_a/F_{a1} \leq 0.15$. When the simpler interaction equations are satisfied, then refine the design by using the more complex interaction equation if $f_a/F_{a1} > 0.15$.

Of course another approach is to utilize the digital computer to analyze and design beam-columns. Certainly, the designer who is faced with many combined loading situations should contact AISI or AISC for available computer programs.

The behavior of beam-columns of nonsymmetric open shapes and unsymmetrically loaded singly symmetric shapes is very complex because they are subject to torsional–flexural buckling. The AISI requires that the behavior of these sections be determined by appropriate testing.

The behavior of singly symmetric shapes loaded in the plane of symmetry and unbraced against flexural–torsional buckling is also very complex. Analytical design procedures are available and are presented in Section 3.72 of Part I of the 1980 edition of the AISI Specification. The procedures are quite cumbersome, requiring numerous design aids from Part IV of 1968 AISI Specifications. The engineer who contemplates designing or analyzing cold-formed steel beam-columns of singly symmetric shapes should consult the AISI Specifications and Design Aids.

9.10 CONNECTORS

9.10.1 Introduction

Connectors for fastening cold-formed steel include welds, bolts, cold rivets, and self-tapping screws. The methods for estimating loads in cold-formed connections are presented in Chapter 7 on Connector Fundamentals. This section will deal with the specific strengths and requirements for typical adequate cold-formed connectors.

The primary difference between connections for cold-formed steel and hot-rolled steel are those related to the relatively thinner elements. In bolted connections, the ratio of bolt diameter to thickness of the connected part is large in many cold-formed steel joints and care must be exercised to prevent failure of the connected part by tearing out or piling due to bearing. Figure 9.21 illustrates these two modes of failure. Also, precautions are necessary to properly weld the thin sheets without burning the steel.

Both fusion or resistance (spot) welds are acceptable as cold-formed connectors. Resistance (also called spot or puddle) welds are normally confined to shop welding, whereas fusion welding predominates in erection practices. Fusion welding is used to connect cold-formed members and to connect cold-formed members to heavier hot-rolled members.

9.10.2 Welded Connections

If the thickness of the welded parts exceeds 0.18 in. (4.6 mm), the weld may be designed in accordance with the AISC Specification for hot-rolled steel. The 1980 edition of the AISI Specification for various welds for thinner members are summarized herein. Groove welds in butt

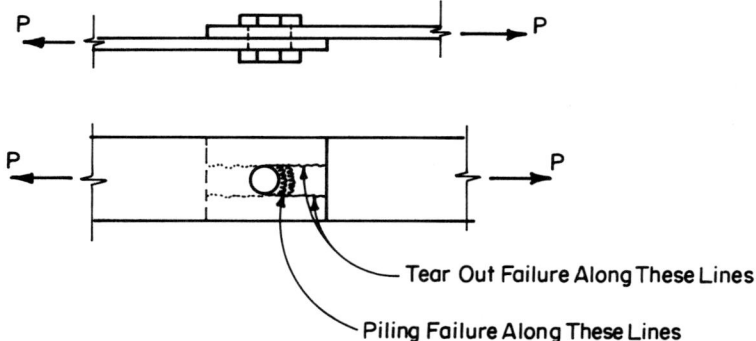

Fig. 9.21. Piling and tearout failure modes in bolted, cold-formed steel connectors.

joints are designed on the basis of the strength of the weakest part joined provided the effective throat thickness is equal to or greater than the thinner part joined. The allowable loads for arc spot (puddle) welds where the thinnest part connected is less than 0.15 in. (3.8 mm) are defined, in English units, by Eqs. (9.56) to (9.59) and Fig. 9.23. The allowable load is the smaller of the loads obtained from Eq. (9.56) or the applicable one of Eqs. (9.57), (9.58), or (9.59).

$$P = d_a^2 F_{xx}/4 \tag{9.56}$$

$$P = 0.88 t d_a F_u \quad \text{if} \quad d_a/t \leq 140/\sqrt{F_u} \tag{9.57}$$

$$P = 0.122 \left[1 + \frac{960t}{(d_a\sqrt{F_u})} \right] t d_a F_u \tag{9.58}$$

if

$$\frac{140}{\sqrt{F_u}} < \frac{d_a}{t} < \frac{200}{\sqrt{F_u}}$$

$$P = 0.56 t d_a \quad \text{if} \quad d_a/t \geq 200/\sqrt{F_u} \tag{9.59}$$

where P is load, kip; d is outside diameter of weld, in.; d_a is average diameter of weld, $d_a = d - t$ for single sheet, and $d - 2t$ for double sheets, in.; $d_e = 0.7d - 1.5t$ but $\leq 0.55d$, in.; t is total thickness of all thin sheets connected, in.; F_{xx} is designated strength of the AWS electrode used, ksi; F_y is minimum yield point of the steel, ksi; F_u is minimum tensile strength of the steel, ksi.

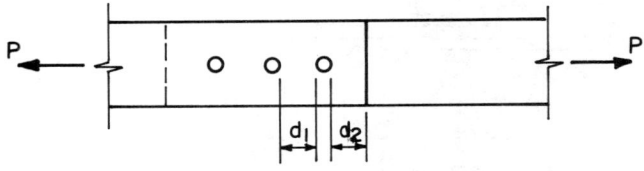

Fig. 9.22. End bolt spacings for cold-formed bolted connectors.

The allowable loads as estimated in Eqs. (9.56) to (9.59) are satisfactory provided the edge distance, e, is equal to or greater than e_{min} as calculated with Eqs. (9.60) and (9.61).

$$e_{min} = P/(0.5\, F_u t) \quad \text{if} \quad F_u/F_y \geq 1.15 \tag{9.60}$$

and

$$e_{min} = P/(0.45\, F_u t) \quad \text{if} \quad F_u/F_y < 1.15 \tag{9.61}$$

The allowable loads for fillet welds similar to those in Fig. 9.24 connecting sheet to sheet or sheet to a thicker member is specified by AISI by Eqs. (9.62) and (9.63) for longitudinal loading and by Eq. (9.64) for transverse loading. Equations (9.62) to (9.65) may be used with any consistent set of units.

$$P = 0.4[1 - 0.01\, L/t]tLF_u \quad \text{if} \quad L/t < 25 \tag{9.62}$$

$$P = 0.3tLF_u \quad \text{if} \quad L/t \geq 25 \tag{9.63}$$

$$P = 0.4tLF_u \tag{9.64}$$

For members thicker than 0.150 in. (3.8 mm), the allowable load from Eq. (9.65) may be used.

$$P = 0.3\, t_w L\, F_{xx} \tag{9.65}$$

where L is fillet weld length, t_w, effective throat thickness = $0.707w_1$ or $0.707w_2$ in Fig. 9.24; w_1 and w_2, weld leg lengths.

The allowable loads for resistance spot welds are summarized in Table 9.15. The permissible loads are for individual spots, for prescribed thickness of connected parts, and have a factor of safety of approximately 2.5.

9.10 Connectors 349

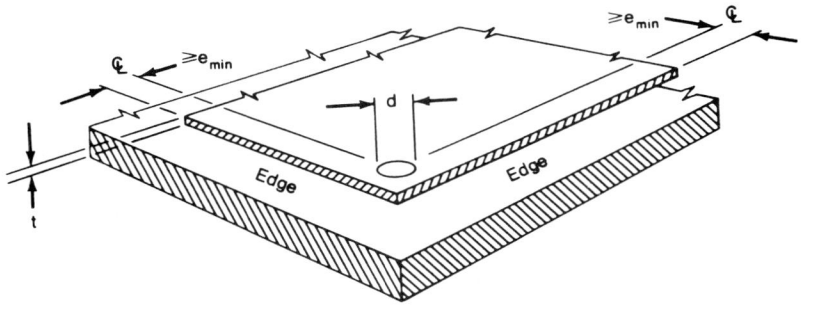

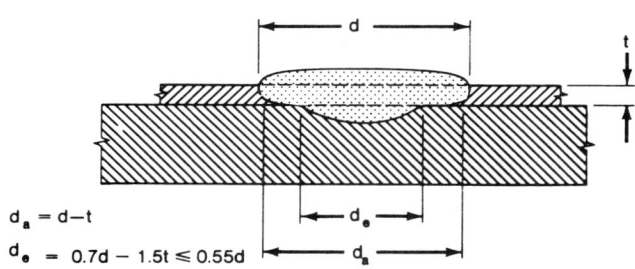

$d_a = d - t$
$d_e = 0.7d - 1.5t \leq 0.55d$

Arc Spot Weld—Single Thickness of Sheet

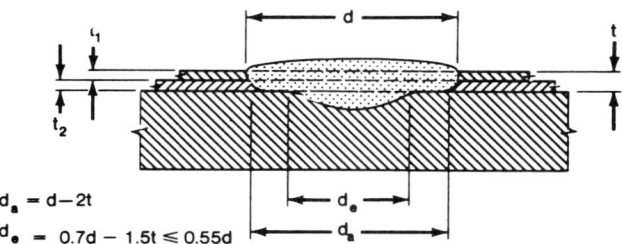

$d_a = d - 2t$
$d_e = 0.7d - 1.5t \leq 0.55d$

Arc Spot Weld—Double Thickness of Sheet

Fig. 9.23. Arc spot welds—definition sketches. Reproduced with permission from *Cold Formed Steel Design Manual: Part I—Specification for the Design of Cold Formed Steel Structural Members,* 1980 ed., American Iron and Steel Institute, Washington, DC.

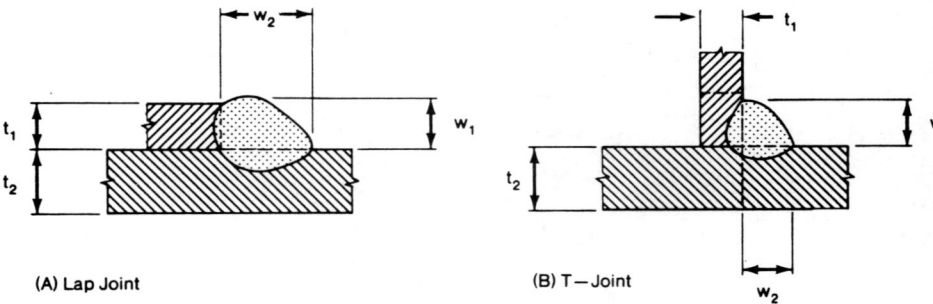

Figure 9.24. Fillet welds.

9.10.3 Bolted Connections

The allowable load for bolted connections in cold-formed steel is dependent upon the allowable shear stress of the bolts, the bearing stress exerted by the bolt on the projected area of the connected parts, the tensile stress acting on the net section of the connected parts, and the spacing of the bolts in the connector (dimensions d_1 and d_2 in Fig. 9.22). The AISI (5th ed.) requirements are summarized herein.

The allowable bolt shear stresses for a variety of approved structural bolts are summarized in Table 9.16. The allowable bearing stresses acting on the projected bolt areas ($d \times t$) must be less than the values in Tables 9.17 and 9.18 in order to prevent piling. In order to prevent tearout type failures (see Fig. 9.21), the distance, d_2 (see Fig. 9.22), between the center line of a bolt and the edge of the connected part measured in the direction of the applied force must be greater than

Table 9.15. Allowable Load in Resistance Spot Welds.

Thickness of Thinnest Outside Sheet (in.)	Allowable Shear Strength per Spot (kip)	Thickness of Thinnest Outside Sheet (in.)	Allowable Shear Strength per Spot (kip)
0.010	0.050	0.080	1.075
0.020	0.125	0.094	1.375
0.030	0.225	0.109	1.65
0.040	0.350	0.125	2.00
0.050	0.525	0.188	4.00
0.060	0.725	0.250	6.00

Reproduced with permission from *Cold Formed Steel Design Manual: Part 1—Specification for the Design of Cold Formed Steel Structural Members,* 1980 ed. American Iron and Steel Institute, Washington, DC.

Table 9.16. Allowable Shear Stress on Bolts.

Type Bolt	Threads Excluded from Shear Plane (ksi)	Threads in the Shear Plane (ksi)
ASTM A307-78, Type A	10	10
ASTM A325-79	30	21
ASTM A354-79, Grade B, D $< \frac{1}{2}$ in. dia.	40	24
ASTM A449-78a, $<\frac{1}{2}$ in. dia.	30	18
ASTM A490-79	40	28

Reproduced with permission from *Cold Formed Steel Design Manual: Part 1—Specification for the Design of Cold Formed Steel Structural Members,* 1980 ed. American Iron and Steel Institute, Washington, DC.

the distance calculated with Eqs. (9.66) or (9.67). Equations (9.66) and (9.67) are valid for any consistent set of units.

$$(d_2)_{\min} = P/0.5F_u t \quad \text{if} \quad F_u/F_y \geq 1.15 \quad (9.66)$$

$$(d_2)_{\min} = P/0.45F_u t \quad \text{if} \quad F_u/F_y < 1.15 \quad (9.67)$$

where t is thickness of thinnest part connected; P, force transmitted by the bolt; F_u, minimum tensile strength of the connected part; F_y, yield stress of the connected part.

Also, the spacing between bolts, d_1 (see Fig. 9.22), must be greater than three times the bolt diameter; and the distance between the center

Table 9.17. Allowable Bearing Stresses for Bolted Connections with Washers under Both Bolt Head and Nut.

Thickness of Connected Part (in.)	Type of Joint	F_u/F_y Ratio of Connected Part	Allowable Bearing Stress, F_p (ksi)
≥ 0.024 but $< 3/16$	Inside sheet of double shear connection	≥ 1.15	$1.50F_u$
		< 1.15	$1.35F_u$
	Single shear and outside sheets of double shear connection	No limit	$1.35F_u$
$\geq 3/16$	Use AISC Bolt Specification		

Reproduced with permission from *Cold Formed Steel Design Manual: Part 1—Specification for the Design of Cold Formed Steel Structural Members,* 1980 ed. American Iron and Steel Institute, Washington, DC.

Table 9.18. Allowable Bearing Stresses for Bolted Connections without Washers under Both Bolt Head and Nut, or with Only One Washer.

Thickness of Connected Part (in.)	Type of Joint	F_u/F_y Ratio of Connected Part	Allowable Bearing Stress, F_p (ksi)
≥0.036 but <3/16	Inside sheet of double shear connection	≥1.15	$1.35F_u$
	Single shear and outside sheets of double shear connection	≥1.15	$1.00F_u$
≥3/16	Use AISC Bolt Specification		

Reproduced with permission from *Cold Formed Steel Design Manual: Part 1—Specification for the Design of Cold Formed Steel Structural Members,* 1980 ed. American Iron and Steel Institute, Washington, DC.

of a bolt and any boundary must be greater than 1.5 times the bolt diameter.

The tension stress acting on the net area of a bolted connection may not exceed $0.6F_y$ nor may it exceed the stresses calculated by Eqs. (9.68) or (9.69) when washers are used under both the bolt head and nut and by Eq. (9.70) when fewer washers are used. Equations (9.68) to (9.70) are valid for any consistent set of units and when the thickness of the thinnest part connected is less than $\frac{3}{16}$ in. (4.8 mm). When the thickness exceeds $\frac{3}{16}$ in. (4.8 mm) the net section requirements of the AISC specification are adequate (see Chapter 8).

$$F_t = (1.0 - 0.9r_1 + 3r_1 d/s)(0.5F_u) \leq 0.5F_u \qquad (9.68)$$

$$F_t = (1.0 - 0.9r_1 + 3r_1 d/s)(0.45F_u) \leq 0.45F_u \qquad (9.69)$$

$$F_t = (1.0 - r_1 + 2.5r_1 d/s)(0.45F_u) \leq 0.45F_u \qquad (9.70)$$

where r_1 is force transmitted by the bolt, or bolts, at the section considered divided by the tension force in the member at that section; s, spacing of bolts normal to the line of stress, or the width of sheet if only one row of bolts; F_t, allowable tension on the net section; d, bolt diameter.

PROBLEMS

9.1 Show whether a simple lip stiffener with a flat width of 0.35 in. (9 mm) is adequate to stiffen the flanges of the 5 × 1.125 × 0.060 in. (127 × 29 × 1.5 mm) Z section in Table 9.4.

9.2 Evaluate the magnitude of the allowable compressive stress for the flanges and the webs of the 3 × 1.125 × 0.048 in. (76 × 29 × 1.2 mm) unstiffened channel in Table 9.2 with (a) $F_y = 33$ ksi (230 MPa); (b) $F_y = 40$ ksi (280 MPa); (c) $F_y = 50$ ksi (340 MPa).

9.3 Evaluate the effective width of the compression elements for strength and deflection determinations in the following sections if $F_y = 33$ ksi (230 MPa).

    a. A stiffened 3.5 × 2 × 0.135 in. (89 × 51 × 3.4 mm) channel.
    b. A stiffened 8 × 3 × 0.060 in. (203 × 89 × 1.5 mm) channel.
    c. An unstiffened 3 × 1.125 × 0.048 in. (89 × 29 × 1.0 mm) channel.
    d. An unstiffened 6 × 1.50 × 0.105 in. (152 × 38 × 2.7 mm) Z.
    e. A 4 × 4 × 0.075 in. (102 × 102 × 1.9 mm) hat section.

9.4 Repeat Problem 9.3 for $F_y = 40$ ksi (280 MPa). For $F_y = 50$ ksi (340 MPa).

9.5. Evaluate the effective area, moment of inertia, and section modulus for the following sections which are fabricated from steel with $F_y = 33$ ksi (230 MPa). Compare your results to those in Tables 9.1 to 9.9.

    a. A 6 × 1.50 × 0.060 in. (152 × 38 × 1.5 mm) unstiffened channel used as a flexural member bent about the strong axis.
    b. Repeat Problem 9.5a if the member is used as a compression member.
    c. A 6 × 5 × 0.105 in. (152 × 127 × 2.7) stiffened built up I section (Table 9.5) when the section is used as a flexural member bent about its strong axis.
    d. Repeat Problem 9.5c if the section is used as a compression member.
    e. A 6 × 6 × 0.105 in. (152 × 152 × 2.7 mm) hat section when used as a flexural member in the orientation shown in Fig. 9.15 and bent about the horizontal centroidal axis of the section.

9.6. Repeat Problem 9.5 for $F_y = 40$ ksi (280 MPa). For $F_y = 50$ ksi (340 MPa).

9.7. Evaluate the moment capacity of the following sections if $F_y = $

33 ksi (230 MPa) and the sections are bent about their strong axis.

- a. An adequately braced 6 × 5 × 0.105 in. (152 × 127 × 2.7 mm) stiffened built-up I section.
- b. An adequately braced 6 × 2.5 × 0.105 (152 × 64 × 2.7 mm) stiffened Z section. How frequently should this section be braced?
- c. An adequately braced 6 × 3.00 × 0.060 in. (152 × 76 × 1.5 mm) unstiffened built-up I section.
- d. An adequately braced 6 × 3 × 0.048 in. (152 × 76 × 1.2 mm) hat section when bent about an axis parallel to the lips.

9.8. Repeat Problem 9.7 for F_y = 40 ksi (280 MPa). For F_y = 50 ksi (340 MPa).

9.9. Evaluate the moment capacity of the following sections when fabricated from steel with F_y = 33 ksi (230 MPa) and bent about their strong axis.

- a. An unstiffened 4 × 1.125 × 0.060 in. (102 × 29 × 1.5 mm) Z section used in a laterally unbraced simply supported beam with a span of 4 ft. (1.22 m); 6 ft (1.83 m); 8 ft (2.44 m); 10 ft (3.05 m).
- b. Repeat Problem 9.9a for a stiffened 5 × 2 × 0.075 in. (127 × 51 × 1.9 mm) Z section.

9.10. Repeat Problem 9.9 for F_y = 40 ksi (280 MPa). F_y = 50 ksi (340 MPa).

9.11. Evaluate the magnitude of the uniformly distributed load which can be carried by a stiffened 4 × 2 × 0.060 in. (102 × 51 × 1.5 mm) channel when used over an unbraced simply supported span of 5 ft (1.52 m) if the section has a yield stress of 33 ksi (230 MPa). The beam ends rest on a 2 in. (51 mm) long bearing surface and the allowable deflection equals (1/180)(span).

9.12. Evaluate the form factor, Q, for the sections in Problem 9.3 when used as a compression member.

9.13. Evaluate the magnitude of the allowable compressive load for the following columns when fabricated with steel having a yield strength of 33 ksi (230 MPa).

- a. An unstiffened 5 × 2.5 × 0.075 in. (127 × 64 × 1.9 mm) I section with KL = 2 ft (0.61 m). KL = 4 ft (1.22 m). KL = 6 ft (1.82 m). KL = 8 ft (2.44 m). KL = 10 ft (3.05 m).

b. A stiffened 4 × 4 × 0.105 in. (102 × 102 × 2.7 mm) I section with $KL = 5$ ft (1.52 m); $KL = 8$ ft (2.44 m); $KL = 10$ ft (3.05 m).
   c. A stiffened 4 × 2 × 0.105 in. (102 × 51 × 2.7 mm) channel section with $KL = 5$ ft (1.52 m); $KL = 8$ ft (2.44 m); $KL = 10$ ft (3.05 m).
   d. An unstiffened 4 × 1.125 × 0.105 in. (102 × 29 × 2.7 mm) channel section with $KL = 5$ ft (1.52 m); $KL = 8$ ft (2.44 m); $KL = 10$ ft (3.05 m).

9.14. Repeat Problem 9.13 if $F_y = 40$ ksi (280 MPa). $F_y = 50$ ksi (340 MPa).

9.15. Determine the maximum moment the column in Example 9.6 can carry superimposed on an axial compressive load of 2.0 kip (8.9 kN).

NOMENCLATURE FOR CHAPTER 9

A	Cross-sectional area, in.² (mm²)
A_w	Area of the webs, in.² (mm²)
B	Width of the flanges of a cross section, in. (mm)
C_b	$1.75 + 1.05\ (M_1/M_2) + 0.3\ (M_1/M_2)^2$
C_c	Limiting value of slenderness ratio $\sqrt{2\pi^2(E/F_y)}$
C_w	Warping constant for a cross section
D	Overall depth of a cross section; diameter of round tubes, in. (mm)
E	Modulus of elasticity of steel, ksi (GPa)
F	Basic design stress for a section or element, ksi (MPa)
F_{a0}	Allowable compressive stress under centric loading and with $KL = 0$, ksi (MPa)
F_{a1}	Allowable average axial compressive strength, ksi (MPa)
F_{a2}	Allowable axial compressive strength for torsionally unstable sections, ksi (MPa)
F_b	Allowable flexural stress, ksi (MPa)
F_{b1}	Allowable compressive flexural stress in pure bending excluding possibility of lateral buckling, ksi (MPa)
F_{bw}	Allowable flexure stress in web of a beam, ksi (MPa)
F_c	Allowable compressive stress for an element of a cross section, ksi (MPa)
F'_{ei}	Euler column buckling stress about the i axis with a factor of safety (F.S.) of 1.92, ksi (MPa)

F_t	Allowable tensile stress, ksi (MPa)
F_u	Ultimate tensile strength, ksi (MPa)
F_v	Allowable shearing stress, ksi (MPa)
F_y	Yield strength, ksi (MPa)
F_{ya}	Average yield strength of an entire section considering cold working, ksi (MPa)
F_{yc}	Yield strength of corners, ksi (MPa)
G	Shear modulus; modulus of rigidity, ksi (GPa)
I	Second moment of area, in.^{4} (cm^4)
I_{yc}	Moment of inertia of the compressive portion of a section about the centroidal axis of the entire section parallel to the web, in.^{4} (cm^4)
J	Torsional constant for a cross section, in.^{4} (cm^4)
K	Effective length factor for columns
L	Unsupported length of columns; beam span; length of a weld, in. (mm)
L_b	Distance between lateral support of the compression flange of a beam, in. (mm)
M	Moment capacity, or bending moment, k · ft (kN · m)
P	Applied load, kip (kN)
Q	$Q_a Q_s$; form factor for local buckling of sections with both stiffened and unstiffened compression elements
Q_a	A_{eff}/A is form factor for local buckling of stiffened compression elements
Q_s	$F_c/0.6F_y$ is form factor for local buckling of unstiffened compression elements
R	Inside radius of the corners of a section, in. (mm)
R'	Midline radius of corners of a section, in. (mm)
S	Section modulus, in.^{3} (cm^3)
S_{xc}	Compression section modulus of the entire section about the axis of bending (I_x divided by the distance from the centroidal axis to the outside fiber of the compression flange), in.^{3} (cm^3)
b	Effective width of stiffened compression elements, in. (mm)
b_e	Reduced effective width of stiffened compression elements, in. (mm)
c	Distance between neutral axis (N.A.) and outer fiber of flexural element, in. (mm)
d	Length of midline of webs; depth of element stiffener; diameter of a weld; bolt diameter, in. (mm)
e	End, edge distance for connectors, in. (mm)
f	Actual stress level, ksi (MPa)

f_a	Actual axial compressive stress level in an element, ksi (MPa)
f_b	Actual flexural stress, ksi (MPa)
b_{bw}	Actual flexural stress in the web, ksi (MPa)
f_v	Average shear stress, ksi (MPa)
h	Clear distance between flanges, in. (mm)
k_v	Shear buckling coefficient
m	Distance between an element and the shear center of a cross section, in. (mm)
r	Radius of gyration of a cross section, in. (mm)
r_0	Polar radius of gyration of section about the shear center, in. (mm)
r_1	Ratio of force transmitted by bolts at a section to the total tensile load acting at the section.
s	Bolt spacing along line of stress, in. (mm)
t	Thickness of the elements of a cross section, in. (mm)
w	Flat width of an element; width of the legs of a fillet weld, in. (mm)
x	Subscript usually denoting the strong axis of a section; distance between the web and centroidal axis of some shapes, in. (mm)
x_0	Distance between the shear center and centroidal axis along the x axis, in. (mm)
y	Subscript usually denoting the weak axis of a section; distance between the web and centroidal axis of some shapes, in. (mm)
ε	Unit strain
σ_{ex}	Allowable stress for flexural buckling about the symmetry axis, ksi (MPa)
σ_t	Allowable stress for torsional buckling of the section, ksi (MPa)
σ_{TF0}	Elastic torsional—flexural buckling stress under concentric loading, ksi (MPa)

10

Light Timber Design

10.1 INTRODUCTION:

10.1.1 General Considerations

Timber is an important structural material in agricultural production systems. A large proportion of livestock and storage facilities, temporary facilities, and auxiliary pieces of handling equipment are constructed of timber. Portions of farm machinery, such as wagon beds and framing, are also of timber construction.

Timber is unique as a structural material because it is a biological material. Consequently, it is more variable in its strength and stiffness properties than materials such as concrete or steel. The structural properties of timber vary between species (pine, maple, oak, etc.) between trees of the same species (high or low density), and indeed even within a single tree of the same species. One of the purposes of this chapter is to describe how the timber design code copes with the high degree of variability.

Timber is anisotropic; i.e., its physical properties are direction dependent. The reason for this behavior is easily understood by considering the structure of the wood section in Fig. 10.1. The longitudinal (L) axis represents a line parallel to the tree trunk, the radial (R) axis is directed outward from the center of the trunk, and the tangential axis (T) is normal to both the R and L axes. In the T–R plane the annual growth rings are visible. The grain, which is visible in the T–L and L–R plains, is the side view of the annual rings. The wood structure varies with direction. Likewise, the timber strength varies

360 Light Timber Design

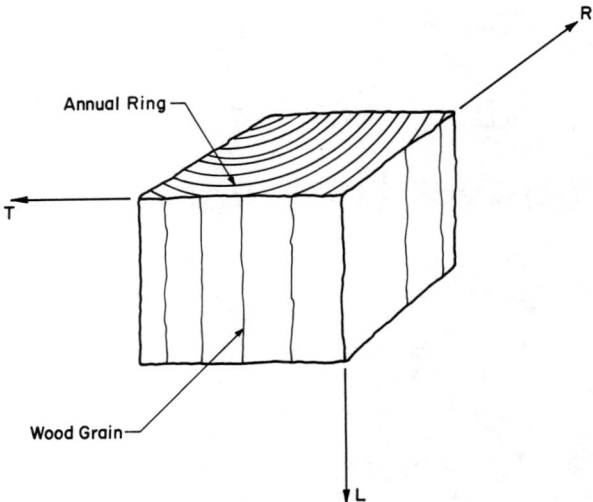

Fig. 10.1. Wood section.

with direction. Specifically, the tensile or compressive strength parallel to the grain is significantly greater than in directions oblique to the grain, the weakest direction being perpendicular to the grain. Contrariwise, the timber resistance to shear-type failure is greatest in the direction perpendicular to the grain and least in the direction parallel to the grain. It is easier to shear the bonds between annual rings (fibers) than to shear the fibers themselves.

Wood is not homogeneous. Features such as knots, changes in grain direction, and separations between and within annual rings occur to some degree in nearly all commercially available lumber, and most certainly, in all construction grade lumber. These features, some of which are illustrated in Fig. 10.2, are collectively termed wood characteristics and influence wood strength.

The primary specification for timber design in the United States is the National Design Specification for Wood Construction (NDS). Timber construction specifications are available through various groups including the American Institute of Timber Construction (AITC), the Truss-Plate Institute (TPI), and the American Plywood Association (APA). Most of the discussion in this text will be based upon the provision of the National Design Specification which will hereafter be designated as NDS.

There are many texts, handbooks, and references on timber and timber design. Among them are the following:

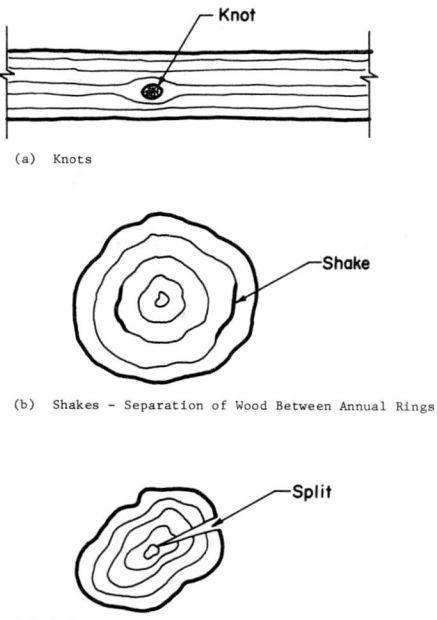

(a) Knots

(b) Shakes - Separation of Wood Between Annual Rings

(c) Splits - Separation of Wood Across Annual Rings

Fig. 10.2. Wood characteristics.

1. AITC. 1985. *Timber construction manual.* Wiley: New York.
2. Breyer, D. E. 1987. *Design of wood structures.* 2nd ed. McGraw-Hill: New York.
3. Dietz, A. G. A., Schaffer, E. L., and Gromala, D. S. 1982. *Wood as a structural material.* EMMSE: The Pennsylvania State University, University Park, PA.
4. Gurfinkel, G. 1981. *Wood engineering.* 2d ed. Kendall/Hunt: Dubuque, IA.
5. Hoyle, R. J., Jr. 1973. *Wood technology in the design of structures.* Mountain Press: Missoula, MT.
6. National Forest Products Association. 1981, 1986. *National design specification for wood construction.* Washington, DC.
7. Western Wood Products Association. 1973. *Western woods use book.* WWPA: Portland, OR.
8. United States Department of Agriculture. 1987. *Wood handbook: Wood as an engineering material.* Agric. Handbook No. 72. U.S. Printing Office: Washington, DC.

There are several timber design philosophies, including strength design, limit design, and reliability design. Limit and reliability design for timber are continually being developed and are beginning to be included in the design specifications. Most timber design practice is still based on strength design, and indeed, much of the NDS is based on strength design methods.

The strength design method first assumes timber behaves as a homogeneous, isotropic, and linearly elastic material. This allows member stresses to be evaluated using the simple strength of materials equations to predict member stresses. That is, member stresses are predicted by

$f_b = M/S$	(Flexure stress)	(10.1a)
$f_v = VQ/Ib$	(Shear stress)	(10.1b)
$f_t = P/A$	(Tensile stress)	(10.1c)
$f_c = P/A$	(Compressive stress)	(10.1d)

Allowable stresses are then adjusted to account for differences between theoretical and actual material behavior. Thus, the strength design criteria for flexure becomes

$$f_b < F_b = F_y/F.S. \tag{10.2}$$

where f_b is theoretical stress, F_b is allowable stress, F_y is yield stress, and $F.S.$ is the factor of safety defined by the design specification.

The purpose of this chapter is to learn how to analyze the structural adequacy of timber load carrying members. Specifically, the reader will learn how to (1) determine the allowable stresses for structural grade lumber; (2) adjust allowable stresses for service conditions; (3) analyze and design load-carrying solid sawn timber members; (4) analyze and design simple timber connections; and (5) analyze or design simple timber structural assemblies. The nomenclature used in this chapter is listed at the end of the chapter.

10.2 STRUCTURAL PROPERTIES OF TIMBER

10.2.1 Factors Influencing Properties

The structural properties of timber are influenced by many factors. Herein the more important factors are considered. For a more thorough discussion, the reader is referred to Gurfinkel's text, *Wood Engineering*, or USDA's *Wood Handbook*, listed in the preceding section.

There is a significant difference in the strength and stiffness of wood between species. For example, certain grades of Douglas fir timber are, respectively, 34% and 14% stronger than the same grade of white fir when used as a post in compression and as a beam. The strength of the two most common timber species used in agricultural buildings (southern pine and Douglas fir) are nearly equal.

Wood density influences strength within the same species. In dense wood the annual rings are more closely spaced than in lighter samples. For some species, particularly southern pine, the variation in strength warrants that strength properties be specified for normal, medium, and dense timber.

Moisture content influences the strength of structural timber in two ways. Generally, the strength of timber declines as moisture content increases from 0 to 25%. Above 25% moisture, no further decrease is observed. Wood, being hygroscopic in nature, also shrinks and swells as moisture content varies. Section properties, such as overall dimensions, section moduli, and moments of inertia will change as moisture content changes. This is of particular importance if a member is sawn to standard size when green (high moisture) and then used at 15% moisture. The final member size would be smaller than standard resulting in a change in the section's load carrying capacity.

The direction of loading and particularly the direction of the primary stress with respect to the direction of the wood grain, influences the strength of timber. In Figs. 10.3a to 10.3c, the direction of loading and the primary stress (compressive) are collinear. Of these configurations, the strength of the timber in Fig. 10.3a is greatest, whereas that in Fig. 10.3c is the least. In Figs. 10.3d, 10.3e, and 10.3f, the load is vertical, but the primary stress (flexural) is horizontal. Of these three configurations the descending order of flexural strength is 10.3d, 10.3e, and 10.3f. If the slope of the grain to the direction of primary stress is less than 1 to 12, then no significant strength reduction is observed. However, if the slope increases to 1 to 6, there is a 50 to 60% reduction in strength.

Member size significantly influences the load-carrying capacity of wood, because deviations between ideal elastic behavior and actual behavior increase with member size. Two examples will illustrate the point. In flexural members the strength design approach assumes $f_b = M/S$ and a triangular stress distribution across the beam section. In reality timber is not elastic and the stress distribution is not linear. The significance of this discrepancy becomes more pronounced with beam depth. In tensile members, simple elastic theory assumes a uniformly distributed tensile stress $f_t = P/A$. In reality timber members have knots and other discontinuities. The discontinuities give rise to

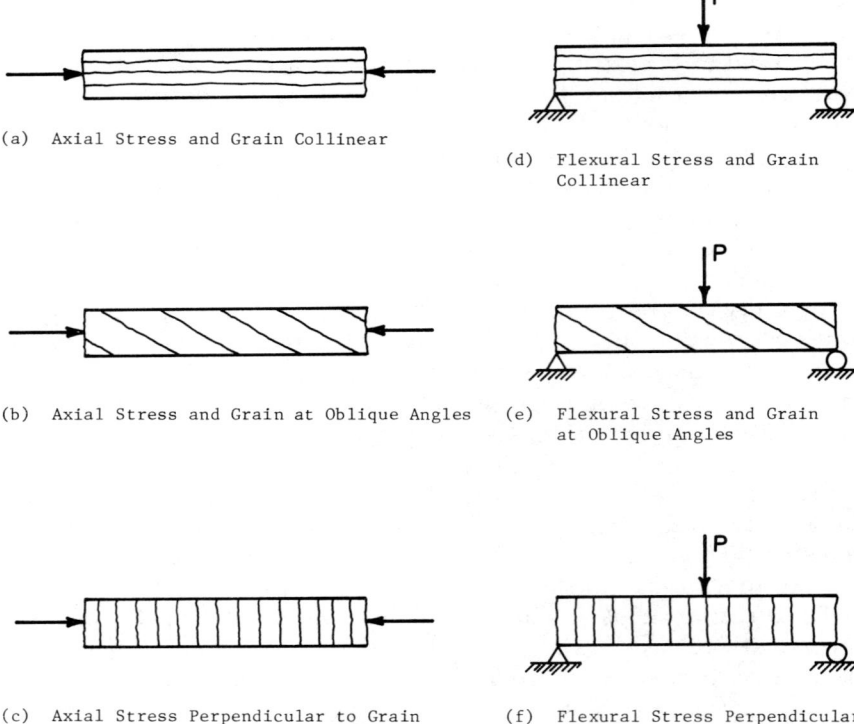

Fig. 10.3. Angles between the wood grain and the primary stress.

stress concentrations as illustrated in Fig. 10.4. Note also that the resultant internal force and the applied centric load P are not collinear. Obviously, at the discontinuity there is an internal couple Pe which must be resisted in addition to the tensile stress. The wider the member, the larger the internal couple Pe can become. Thus, the reduction in tensile load carrying capacity of timber members due to discontinuities in the wood are most severe in wider members.

Wood characteristics, such as knots, checks, shakes, and decay, influence timber strength. Strength is reduced by introduction of stress concentrations and reduction of net sections. The amount of strength reduction depends upon the wood characteristic size and location and upon the type load. For example, in a simply supported timber beam, knots near the neutral axis would be less critical than those near the outer fiber. Similarly, knots located near the end supports where

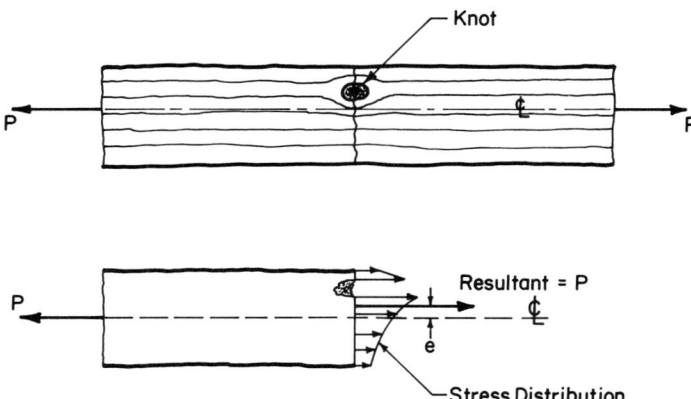

Fig. 10.4. Wood characteristics and stress concentrations.

the bending moment equals zero are less critical than those located in the middle half of the span.

The structural strength of timber is greatly influenced by the type stress, i.e., tensile, compressive, flexural, shear. For example, a construction grade southern pine 2 × 4 at 19% moisture has the following allowable stresses:

$$\text{Flexure } F_b = 1150 \text{ psi } (7.9 \text{ MPa})$$
$$\text{Tension parallel to grain } F_t = 600 \text{ psi } (4.1 \text{ MPa})$$
$$\text{Shear parallel to grain } F_v = 100 \text{ psi } (0.69 \text{ MPa})$$
$$\text{Compression parallel to grain } F_c = 1100 \text{ psi } (7.6 \text{ MPa})$$
$$\text{Compression perpendicular to grain } F_{cp} = 405 \text{ psi } (2.8 \text{ MPa})$$

The elastic limit and the ultimate strength of timber both increase as the duration of loading decreases. Thus, timber members can carry large overloads for short time periods. Similarly, the load-deformation behavior of timber is very time dependent. Thus, deflections and elongations for long-term loads are greater than for short duration loads of the same magnitude. Figure 10.5, taken from NDS, illustrates the influence of load duration upon the relative strength of timber. Note that normal loading is considered to be 10-year duration and has a relative strength of 1.0. For loads, such as dead loads, which are applied for more than 10 years, the relative strength equals 0.9. For a 1-day duration relative strength increases to 1.33, and for impact loads the strength factor is 2.0. The relative strengths for any load duration are

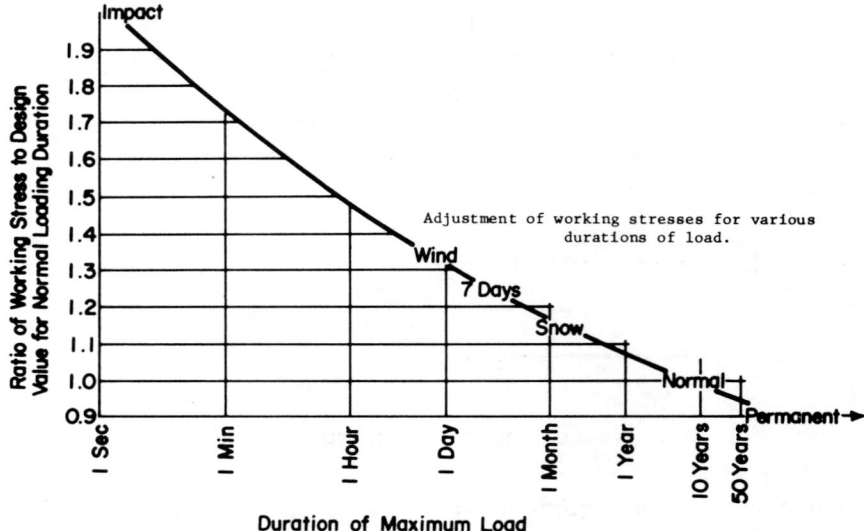

Fig. 10.5. Load duration factors. Reproduced with permission from *National Design Specification for Wood Construction,* 1977 ed., National Forest Products Association, Washington, DC.

called load duration factors and will hereinafter be abbreviated as *DF*. To illustrate the *DF* concept a timber beam which could carry a concentrated design load of 1000 lb (4.5 kN) at midspan for a period of 10 years could only carry 900 lb (4.0 kN) permanently. Also, it could carry 1250 lb (5.6 kN), 1330 lb (5.9 kN), and 2000 lb (8.9 kN) for load durations of 7 days, 1 day, and 1 hour, respectively. Table 10.1 summarizes the recommended duration factors for typical load combinations.

Often it is difficult to identify by inspection the load combinations which will control design of a timber member. Example 10.1 will amplify this statement.

Table 10.1. Load Duration Factors (*DF*).

Load Type	Duration of Load[a]	DF
Dead	Permanent	0.90
Normal	10 years	1.00
Dead and snow	2 months	1.15
Dead and wind	1 day	1.33
Dead and earthquake	1 day	1.33
Impact	1 hour	2.00

[a]Duration of load is the accumulated duration of the appropriate design load over the life of the structure.

Example 10.1. Governing Loading Condition. Consider a simply supported timber beam of span L carrying a uniformly distributed load. If the beam must carry the following load combinations during its expected life, determine which combination governs the design.

a. $DL = w_1$
b. $DL + SL = 5w_1$
c. $DL + WL + 0.5SL = 5.5w_1$

Solution

a. The required section modulus for the beam is

$$S = M_{max}/F_b = wL^2/(8F_b)$$

b. For each loading condition, $L = 10$ ft (3.05 m) and F'_b equals the allowable flexural stress adjusted for load duration.

Load condition	$F'_b = DF \times F_b$	S
DL	$0.9F_b$	$0.14w_1L^2/F_b$
D + SL	$1.15F_b$	$0.54w_1L^2/F_b$
DL + WL + 0.5SL	$1.33F_b$	$0.52w_1L^2/F_b$

Note that although condition c is the largest load, duration factor considerations show that the smaller load condition b actually controls the section size required.

Pressure preservative treatments do not reduce timber strength. Fire retardants reduce the strength by 10%.

Many of the factors just cited are considered when lumber is graded and assigned strength values. Grading rules, such as those published by the Western Wood Products Association and the Southern Forest Products Association, are available for establishing the quality, or grade, of lumber. The grade and species are then correlated to allowable stress values. Once the timber grade is assigned, allowable stresses may be found in the supplement to the NDS.

10.2.2 Timber Classification and Grades within Each Species

Structural lumber is either visually graded or machine stress rated. Visually graded lumber is assigned allowable stresses based upon the size member and visual observation of quality and wood characteristics by an inspector. Machine rated lumber is assigned allowable stress values based upon observed performance of the pieces in a standard deflection type test.

10.2.2.1 Visually Graded Lumber. Timber is classified by size and strength. A simple flow diagram of the sizing, grading, and adjustment scheme used to assign strength values to visually graded timber is shown in Fig. 10.6. The general size classifications are boards, dimension lumber, and timbers. Boards include any lumber less than two in. (51 mm) thick, i.e., 1 × 4s, 1 × 8s, etc. Dimension lumber includes lumber between 2 and 4 in. (51 and 102 mm) in thickness and greater than or equal to 2 in. (51 mm) wide. Timbers include lumber nearly square in cross section and greater than 5 in. (127 mm) on a side and are graded primarily for use as axial compression members. Boards are generally not stress rated. They are, however, graded for appearance and for sheathing purposes.

Dimension lumber is further classified as structural light framing, light framing, joists and planks, and beams and stringers. Structural light framing is relatively high grade lumber and limited in size to members 2 to 4 in. (51 to 102 mm) thick and 2 to 4 in. (51 to 102 mm) wide. Light framing is similar except it is limited to lower quality grades. Both light framing classes are graded with respect to bending about both major axes. Joists and planks include lumber 2 to 4 in. (51 to 102 mm) thick and wider than 5 in. (127 mm). Joist and plank classes are also graded with respect to bending about both major axes. Beams and stringers include members greater than 5 in. (127 mm) thick and at least 2 in. (51 mm) wider than thick. These members are graded primarily with respect to bending about the strong axis.

Within each size classification, structural lumber is graded with respect to timber density and the number, size, and location of wood characteristics. Typical grades within a size include, from highest to lowest quality, dense select structural, select structural, No. 1 dense, No. 1, No. 2 dense, No. 2, No. 3 dense, construction, standard, and utility.

Allowable tensile, flexural, compressive, and shear strengths for design with visually graded lumber are published in the NDS supplement for each grade. The allowable strength published by NDS is equal to the ultimate strength of a perfectly clear, straight grained specimen times a strength reduction factor. The strength reduction factor varies with the grade lumber, the size lumber, and the mode of loading. Pub-

10.2 Structural Properties of Timber

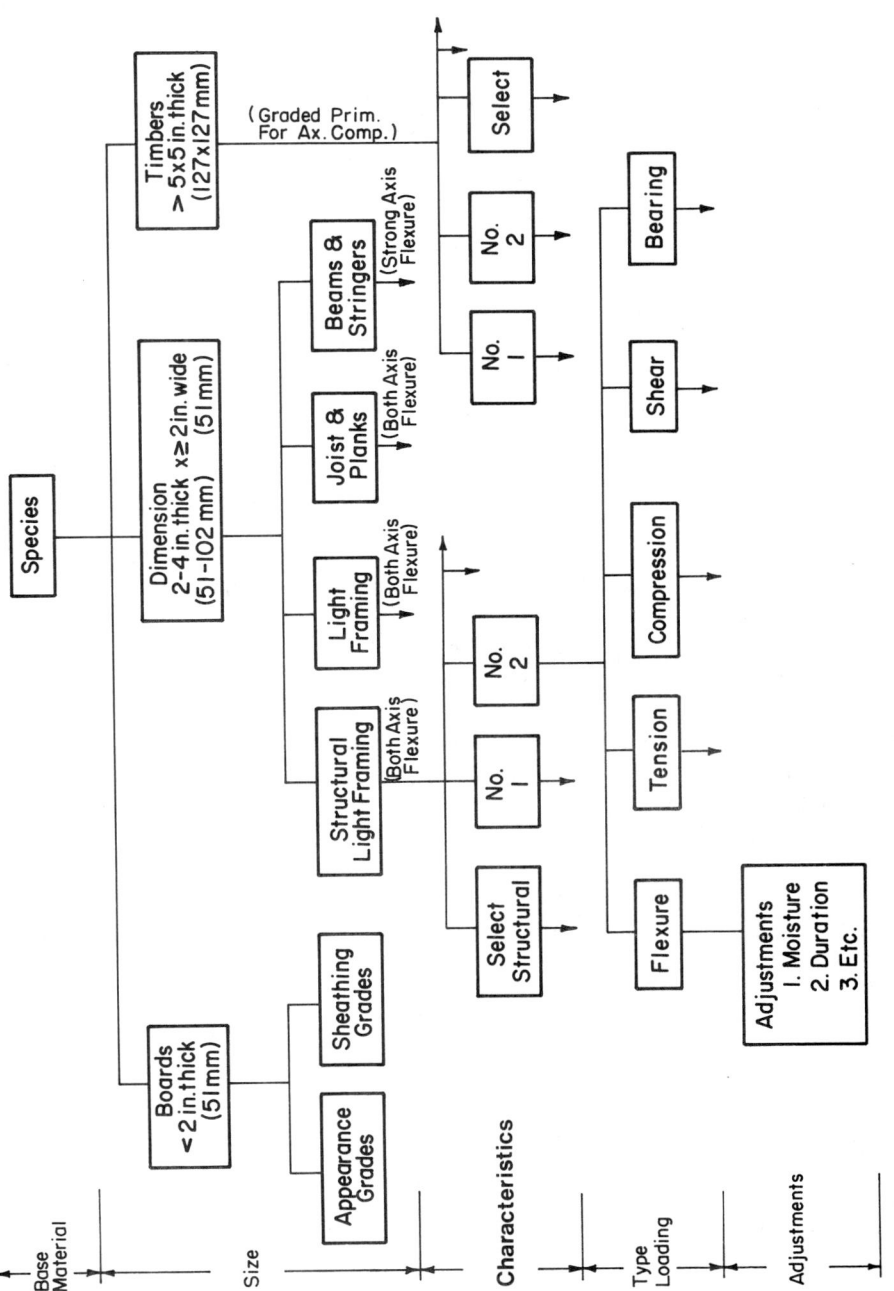

Fig. 10.6. Schematic representation of the timber grading and size classification system.

lished allowable stresses are in the order of 1/6 to 1/12 the basic strength of a "perfect" piece of lumber.

The allowable stresses for the two most common timber species, southern pine and Douglas fir, used in agricultural buildings are reproduced from NDS and are shown in Tables 10.2A and 10.2B. Note that the NDS allowable stresses have already accounted for strength and stiffness variations with moisture content, size classification, mode of loading (or primary stress), direction of loading with respect to the grain, and wood characteristics.

Allowable flexure stresses are given for single member and repetitive member use. To qualify for repetitive member use, the member must be part of the framing system with three or more adjacent members spaced 24 in. (610 mm) or less on center and connected on the compression flange by a continuous roofing or floor decking system. In such a system a higher allowable flexural stress is permitted because small localized overloads in one member can be shared by adjacent members before failure occurs.

Flexural stresses, it will be recalled, are accompanied by transverse shear stresses ($f_v = VQ/Ib$) in a beam. Further, the transverse shear stress at a point in a beam is always accompanied by an equal shear stress in the direction parallel to the longitudinal axis. In a horizontal timber beam the grain of the member will be horizontal. Since timber shear strength is lower parallel to the grain than normal to the grain, shear failures, if they occur, will be on a horizontal plane. Hence, the use of the term horizontal shear in the NDS table. Horizontal shear is simply the shear strength of the wood in the direction of the grain. The remainder of the terms in Tables 10.2A and 10.2B are self-explanatory.

Special note should be made of the footnotes to the allowable stress tables. They are very significant in some cases. For example, footnote specifications reduce allowable tensile stresses by as much as 40% in members wider than 8 in. (203 mm). The tabulated allowable stresses are for normal load durations (10 years). The tabulated values do not make adjustments for loading (stresses) at angles oblique to the grain direction and other special cases such as flexural members with depth greater than 12 in. (305 mm).

10.2.2.2 Machine Stress Rated Lumber. The allowable stresses for machine stress rated lumber are in Table 10.3. The grade designation xxxf-YYE replaces the No. 1, No. 2, etc., designations and specifies the allowable fiber stress, xxx, for a single member usage, and the modulus of elasticity, YY. The allowable stress definitions are similar to those
(text continues on page 380)

Table 10.2A. Allowable Stresses for Visually Graded Southern Pine Structural Lumber.

Species and Commercial Grade	Size Classification	Extreme Fiber in Bending F_b		Tension Parallel to Grain F_t	Horizontal Shear F_v	Compression Perpendicular to Grain F_{cp}	Compression Parallel to Grain F_c	Modulus of Elasticity E	Grading Rules Agency
		Single-Member Uses	Repetitive-Member Uses						
Surfaced at 15% moisture content, K.D. Used at 15% max MC									
Select Structural		2150	2500	1250	105	405	1800	1,800,000	
Dense Select Structural		2500	2900	1500	105	475	2100	1,900,000	
No. 1		1850	2100	1050	105	405	1450	1,800,000	
No. 1 Dense	2 to 4 in. thick,	2150	2450	1250	105	475	1700	1,900,000	
No. 2	2 to 4 in. wide	1550	1750	900	95	405	1150	1,600,000	
No. 2 Dense		1800	2050	1050	95	475	1350	1,700,000	
No. 3		850	975	500	95	405	675	1,500,000	
No. 3 Dense		1000	1150	575	95	475	800	1,500,000	
Stud		850	975	500	95	405	675	1,500,000	
Construction	2 to 4 in. thick,	1100	1250	650	105	405	1300	1,500,000	
Standard	4 in. wide	625	725	375	95	405	1050	1,500,000	
Utility		275	300	175	95	405	675	1,500,000	
Select Structural		1850	2150	1200	95	405	1600	1,800,000	
Dense Select Structural		2200	2500	1450	95	475	1850	1,900,000	
No. 1		1600	1850	1050	95	405	1450	1,800,000	SPIB (See footnotes
No. 1 Dense	2 to 4 in. thick,	1850	2150	1250	95	475	1700	1,900,000	(1, 3, 4, 5, 6,
No. 2	5 in. and wider	1300	1500	675	95	405	1200	1,600,000	12, 15, 16
No. 2 Dense		1550	1750	800	95	475	1400	1,700,000	after Table
No. 3		750	875	400	95	405	725	1,500,000	10.2B)
No. 3 Dense		875	1000	450	95	475	850	1,500,000	
Stud		800	900	400	95	405	725	1,500,000	

(continued)

Table 10.2A. Continued

Species and Commercial Grade	Size Classification	Extreme Fiber in Bending F_b		Tension Parallel to Grain F_t	Horizontal Shear F_v	Compression Perpendicular to Grain F_{cp}	Compression Parallel to Grain F_c	Modulus of Elasticity E	Grading Rules Agency
		Single-Member Uses	Repetitive-Member Uses						
Surfaced dry. Used at 19% max MC									
Select Structural		2000	2300	1150	100	405	1550	1,700,000	
Dense Select Structural		2350	2700	1350	100	475	1800	1,800,000	
No. 1		1700	1950	1000	100	405	1250	1,700,000	
No. 1 Dense	2 to 4 in. thick,	2000	2300	1150	100	475	1450	1,800,000	
No. 2	2 to 4 in. wide	1400	1650	825	90	405	975	1,600,000	
No. 2 Dense		1650	1900	975	90	475	1150	1,600,000	
No. 3		775	900	450	90	405	575	1,400,000	
No. 3 Dense		925	1050	525	90	475	675	1,500,000	
Stud		775	900	450	90	405	575	1,400,000	
Construction	2 to 4 in. thick,	1000	1150	600	100	405	1100	1,400,000	
Standard	4 in. wide	575	675	350	90	405	900	1,400,000	
Utility		275	300	150	90	405	575	1,400,000	
Select Structural		1750	2000	1150	90	405	1350	1,700,000	
Dense Select Structural		2050	2350	1300	90	475	1600	1,800,000	
No. 1	2 to 4 in. thick,	1450	1700	975	90	405	1250	1,700,000	SPIB
No. 1 Dense	5 in. and wider	1700	2000	1150	90	475	1450	1,800,000	(See footnotes
No. 2		1200	1400	625	90	405	1000	1,600,000	1, 3, 4, 5, 6,
No. 2 Dense		1400	1650	725	90	475	1200	1,600,000	12, 15, 16.)
No. 3		700	800	350	90	405	625	1,400,000	
No. 3 Dense		825	925	425	90	475	725	1,500,000	
Stud		725	850	350	90	405	625	1,400,000	

Design Values (lb/in.^{2})

Surfaced green. Used any condition

Grade	Size							
Select Structural		1600	1850	925	95	270	1050	1,500,000
Dense Select Structural		1850	2150	1100	95	315	1200	1,600,000
No. 1		1350	1550	800	95	270	825	1,500,000
No. 1 Dense	2½ to 4 in. thick, 2½ to 4 in. wide	1600	1800	925	95	315	950	1,600,000
No. 2		1150	1300	675	85	270	650	1,400,000
No. 2 Dense		1350	1500	775	85	315	750	1,400,000
No. 3		625	725	375	85	270	400	1,200,000
No. 3 Dense		725	850	425	85	315	450	1,300,000
Stud		625	725	375	85	270	400	1,200,000
Construction	2½ to 4 in. thick, 4 in. wide	825	925	475	95	270	725	1,200,000
Standard		475	525	275	85	270	600	1,200,000
Utility		200	250	125	85	270	400	1,200,000
Select Structural		1400	1600	900	85	270	900	1,500,000
Dense Select Structural		1600	1850	1050	85	315	1050	1,600,000
No. 1		1200	1350	775	85	270	825	1,500,000
No. 1 Dense	2½ to 4 in. thick, 5 in. and wider	1400	1600	925	85	315	950	1,600,000
No. 2		975	1100	500	85	270	675	1,400,000
No. 2 Dense		1150	1300	600	85	315	800	1,400,000
No. 3		550	650	300	85	270	425	1,200,000
No. 3 Dense		650	750	350	85	315	475	1,300,000
Stud		575	675	300	85	270	425	1,200,000
No. 1 SR		1350	—	875	110	270	775	1,500,000
No. 1 Dense SR	5 in. and thicker	1550	—	1050	110	315	925	1,600,000
No. 2 SR		1100	—	725	95	270	625	1,400,000
No. 2 Dense SR		1250	—	850	95	315	725	1,400,000

SPIB (See footnotes 1, 3, 4, 5, 6, 12, 14, 15, 16.)

Reproduced with permission from *Design Values for Wood Construction*, 1978 ed. National Forest Products Association, Washington, DC.

Table 10.2B. Allowable Stresses for Visually Graded Douglas Fir, Larch Structural Lumber.

Species and Commercial Grade	Size Classification		Extreme Fiber in Bending F_b		Tension Parallel to Grain F_t	Horizontal Shear F_v	Compression Perpendicular to Grain F_{cp}	Compression Parallel to Grain F_c	Modulus of Elasticity E	Grading Rules Agency
			Single-Member Uses	Repetitive-Member Uses						
Surfaced dry or surfaced green. Used at 19% max MC										
Dense Select Structural			2450	2800	1400	95	455	1850	1,900,000	
Select Structural			2100	2400	1200	95	385	1600	1,800,000	
Dense No. 1			2050	2400	1200	95	455	1450	1,900,000	
No. 1	2 to 4 in.		1750	2050	1050	95	385	1250	1,800,000	
Dense No. 2	thick,		1700	1950	1000	95	455	1150	1,700,000	
No. 2	2 to 4 in.		1450	1650	850	95	385	1000	1,700,000	
No. 3	wide		800	925	475	95	385	600	1,500,000	
Appearance			1750	2050	1050	95	385	1500	1,800,000	
Stud			800	925	475	95	385	600	1,500,000	
Construction	2 to 4 in.		1050	1200	625	95	385	1150	1,500,000	WCLIB
Standard	thick,		600	675	350	95	385	925	1,500,000	WWPA
Utility	4 in. wide		275	325	175	95	385	600	1,500,000	
Dense Select Structural			2100	2400	1400	95	455	1650	1,900,000	(See Footnotes
Select Structural			1800	2050	1200	95	385	1400	1,800,000	1–12)
Dense No. 1			1800	2050	1200	95	455	1450	1,900,000	
No. 1	2 to 4 in.		1500	1750	1000	95	385	1250	1,800,000	
Dense No. 2	thick,		1450	1700	775	95	455	1250	1,700,000	
No. 2	5 in. and		1250	1450	650	95	385	1050	1,700,000	
No. 3	wider		725	850	375	95	385	675	1,500,000	
Appearance			1500	1750	1000	95	385	1500	1,800,000	
Stud			725	850	375	95	385	675	1,500,000	
Dense Select Structural			1900	—	1100	85	455	1300	1,700,000	
Select Structural	Beams and		1600	—	950	85	385	1100	1,600,000	
Dense No. 1	stringers		1550	—	775	85	455	1100	1,700,000	
No. 1			1300	—	675	85	385	925	1,600,000	

Dense Select Structural	Posts and	1750	—	85	455	1350	1,700,000 WCLIB
Select Structural	timbers	1500	—	85	385	1150	1,600,000 (See Footnotes
Dense No. 1		1400	—	85	455	1200	1,700,000 1–12)
No. 1		1200	—	85	385	1000	1,600,000
Dense Select Structural		1900	—	85	455	1300	1,700,000
Select Structural	Beams and	1600	—	85	385	1100	1,600,000
Dense No. 1	stringers	1550	—	85	455	1100	1,700,000
No. 1		1350	—	85	385	925	1,600,000
Dense Select Structural		1750	—	85	455	1350	1,700,000 WWPA
Select Structural	Posts and	1500	—	85	385	1150	1,600,000
Dense No. 1	timbers	1400	—	85	455	1200	1,700,000
No. 1		1200	—	85	385	1000	1,600,000

Reproduced with permission from *Design Values for Wood Construction*, 1978 ed. National Forest Products Association, Washington DC.

1. Grading rules agencies listed in Table 10.2 include the following: SPIB, Southern Pine Inspection Bureau, P.O. Box 846, Pensacola, Florida 32594; WCLIB, West Coast Lumber Inspection Bureau, 6980 SW Varnes Road, P.O. Box 23145, Portland, Oregon 97223; WWPA, Western Wood Products Association, 1500 Yeon Building, Portland, Oregon 97204.

2. The design values herein are applicable to lumber that will be used under dry conditions such as in most covered structures. For 2–4-in.-thick lumber the *dry* surfaced size shall be used. In calculating design values, the natural gain in strength and stiffness that occurs as lumber dries has been taken into consideration as well as the reduction in size that occurs when unseasoned lumber shrinks. The gain in load carrying capacity due to increased strength and stiffness resulting from drying more than offsets the design effect of size reductions due to shrinkage. For 5-in. and thicker lumber, the surfaced sizes also may be used because design values have been adjusted to compensate for any loss in size by shrinkage which may occur.

3. Tabulated tension parallel to grain values for all species for 5 in. and wider, 2- to 4-in.-thick (and 2½–4-in.-thick) size classifications apply to 5 and 6 in. widths only, for grades of Select Structural, No. 1, No. 2, No. 3, Appearance, and Stud (including dense grades). For lumber wider than 6 in. in these grades, the tabulated F_t values shall be multiplied by the following factors:

	Multiply Tabulated F_t Values by		
Grade	5 and 6 in. wide	8 in. wide	10 in. and wider
Select Structural	1.00	0.90	0.80
No. 1, No. 2, No. 3, and Appearance	1.00	0.80	0.60
Stud	1.00	—	—

(*continued*)

Table 10.2B. Continued

4. Design values for all species of Stud grade in 5 in. and wider size classifications apply to 5 and 6 in. widths only.
5. Values for F_b, F_t, and F_c for all species of the grades of Construction, Standard and Utility apply only to 4-in. widths. Design values for 2-in. and 3-in. widths of these grades are available from SPIB, WCLIB, and WWPA (see note 1).
6. The values for dimension lumber 2 to 4 in. in thickness are based on edgewise use. When such lumber is used flatwise, the design values for extreme fiber in bending for all species may be multiplied by the following factors:

	Thickness		
Width	2 in.	3 in.	4 in.
2 to 4 in.	1.10	1.04	1.00
5 in. and wider	1.22	1.16	1.11

7. The design values in Table 10.2B for extreme fiber in bending for decking may be increased by 10% for 2-in.-thick decking and by 4% for 3-in.-thick decking.
8. When 2-in.- to 4-in.-thick lumber is manufactured at a maximum moisture content of 15% and used in condition where the moisture content does not exceed 15%, the design values for surfaced dry or surfaced green lumber shown in Table 10.2B may be multiplied by the following factors:

dry use

Extreme Fiber in Bending F_b	Tension Parallel to Grain F_t	Horizontal Shear F_v	Compression Perpendicular to Grain F_{cp}	Compression Parallel to Grain F_c	Modulus of Elasticity E
1.08	1.08	1.05	1.00	1.17	1.05

flatwise

9. When 2- to 4-in.-thick lumber is designed for use where the moisture content will exceed 19% for an extended period of time, the design values in Table 10.2B shall be multiplied by the following factors:

Wet use

Extreme Fiber in Bending F_b	Tension Parallel to Grain F_t	Horizontal Shear F_v	Compression Perpendicular to Grain F_{cp}	Compression Parallel to Grain F_c	Modulus of Elasticity E
0.86	0.84	0.97	0.67	0.70	0.97

10. When lumber 5 in. and thicker is designed for use where the moisture content will exceed 19% for an extended period of time, the design values shown in Table 10.2B shall be multiplied by the following factors:

Wet use

Extreme Fiber in Bending F_b	Tension Parallel to Grain F_t	Horizontal Shear F_v	Compression Perpendicular to Grain F_{cp}	Compression Parallel to Grain F_c	Modulus of Elasticity E
1.00	1.00	1.00	0.67	0.91	1.00

11. The tabulated horizontal shear values shown in Table 10.2B for lumber 4 in. and thinner should be multiplied by a factor of 0.92 when such lumber is manufactured unseasoned.
12. Stress-rated boards of nominal 1-, 1¼, and 1½-in. thickness, 2 in. and wider, of most species, are permitted the design values shown for Select Structural, No. 1, No. 2, No. 3, Construction, Standard, Utility, Appearance, Clear Heart Structural, and Clear Structural grades as shown in the 2-in.- to 4-in.-thick categories herein, when graded in accordance with the stress-rated board provisions in the applicable grading rules. Information on stress-rated board grades applicable to the various species is available from the respective grading rules agencies.
13. When Decking graded to WWPA rules is surfaced at 15% maximum moisture content and used where the moisture content will exceed 15% for an extended period of time, the tabulated design values for Decking surfaced at 15% maximum moisture content shall be multiplied by the following factors: Extreme Fiber in Bending F_b – 0.79; Modulus of Elasticity, E – 0.92.
14. Repetitive member design values for extreme fiber in bending for Southern Pine grades of Dense Structural 86, 72, and 65 apply to 2- to 4-in. thickness only.
15. When 2- to 4-in.-thick Southern Pine lumber is surfaced dry or at 15% maximum moisture content (K.D.) and is designed for use where the moisture content will exceed 19% for an extended period of time, the design values in Table 10.2A for the corresponding grades of 2½- to 4-in.-thick surfaced green Southern Pine lumber shall be used. The net green size may be used in such designs.
16. When 2- to 4-in.-thick Southern Pine lumber is surfaced dry or at 15% maximum moisture content (K.D.) and is designed for use under dry conditions, such as most covered structures, the net *dry* size shall be used in design. For other sizes and conditions of use, the net green size may be used in design.

Green

Table 10.3 Allowable Stresses for Machine Graded Structural Lumber.

Grade Designation	Size Classification	Extreme Fiber in Bending F_b [3]		Tension Parallel to Grain F_t	Compression Parallel to Grain F_c	Modulus of Elasticity E
		Single-Member Uses	Repetitive-Member Uses			
900f-1.0E	Machine rated lumber 2 in. thick or less, all widths	900	1050	350	725	1,000,000
1200f-1.2E		1200	1400	600	950	1,200,000
1350f-1.3E		1350	1550	750	1100	1,300,000
1450f-1.3E		1450	1650	800	1150	1,300,000
1500f-1.4E		1500	1750	900	1200	1,400,000
1650f-1.5E		1650	1900	1020	1320	1,500,000
1800f-1.6E		1800	2050	1175	1450	1,600,000
1950f-1.7E		1950	2250	1375	1550	1,700,000
2100f-1.8E		2100	2400	1575	1700	1,800,000
2250f-1.9E		2250	2600	1750	1800	1,900,000
2400f-2.0E		2400	2750	1925	1925	2,000,000
2550f-2.1E		2550	2950	2050	2050	2,100,000
2700f-2.2E		2700	3100	2150	2150	2,200,000
2850f-2.3E		2850	3300	2300	2300	2,300,000
3000f-2.4E		3000	3450	2400	2400	2,400,000
3150f-2.5E		3150	3600	2500	2500	2,500,000
3300f-2.6E		3300	3800	2650	2650	2,600,000
900f-1.0E	See Footnote 1	900	1050	350	725	1,000,000
900f-1.2E		900	1050	350	725	1,200,000
1200f-1.5E		1200	1400	600	950	1,500,000
1350f-1.8E		1350	1550	750	1075	1,800,000
1500f-1.8E		1500	1750	900	1200	1,800,000
1800f-2.1E		1800	2050	1175	1450	2,100,000

Design Values (lb/in^2)[5]

Reproduced with permission from *Design Values for Wood Construction*, 1978 ed. National Forest Products Association, Washington, DC.

1. Size classifications for these grades are:
    SPIB—Machine Rated Lumber; 2 in. thick or less, all widths
    WCLIB—Machine Rated Joists; 2 in. thick or less, 6 in. and wider
    WWPA—Machine Rated Lumber; 2 in. thick or less, all widths
2. Stresses apply at 19% maximum moisture content
3. Tabulated extreme fiber in bending values F_b are applicable to lumber loaded on edge. When loaded flatwise, these values may be increased by multiplying by the following factors:

Nominal width (in.)	3	4	5	6	8	10	12	14
Factor	1.06	1.10	1.12	1.15	1.19	1.22	1.25	1.28

4. Footnotes 1, 2, 9, 11, and 16 to Table 10.2 apply also to Machine-Stress-Rated Lumber.
5. Design values for horizontal shear F_v (dry) and compression perpendicular to grain F_{cp} (dry) are:

	F_v	F_{cp}
Douglas Fir South (WWPA)	90	335
Southern Pine (SPIB)	90	95 (K.D.)	405

for visually graded lumber and are subject to the conditions specified in the footnotes to Table 10.3. Note that some of the footnotes to Table 10.2 are also applicable to machine stress rated lumber.

10.2.2.3 Other Considerations. Loads are often applied at oblique angles to the grain in compression members and connections. Allowable stress of obliquely loaded timber can be calculated from the compressive strengths normal and parallel to the grain and from Hankinson's equation

$$N_\theta = \frac{F_c F_{cp}}{F_c \sin^2\theta + F_{cp} \cos^2\theta} \qquad (10.3)$$

where F_c is allowable compressive strength parallel to grain; F_{cp}, allowable compressive strength normal to grain; N_θ, allowable compressive strength at an angle θ from the direction of the grain; θ, acute angle between the direction of loading and the direction of grain.

The allowable stress values tabulated in the NDS supplement, and in tables such as Tables 10.2A, 10.2B, and 10.3 of this text, can be used with confidence. They need only be adjusted as specified earlier for factors such as load duration. That is, the published values may be used without additional factors of safety. To incorporate additional factors of safety only leads to uneconomical design.

10.3 STANDARD SIZES AND SECTION PROPERTIES

Timber member cross sections may be either full or dressed size. For example, a full size 2 × 4 has cross-sectional dimensions of 2 and 4 in. (51 and 102 mm), whereas a dressed 2 × 4 has dimensions of $1\frac{1}{2}$ in. (38 mm) and $3\frac{1}{2}$ in. (89 mm). In either case the nominal size is a 2 × 4. To identify a member accurately, both the nominal and type section must be identified. The majority of structural lumber is dressed.

The most common nominal sizes of timber are 2 (25), 4 (102), 6 (152), 8 (203), 10 (254), and 12 (305) in. (mm). In agricultural applications, both full and dressed sizes are used, with dressed sizes being the more common. The dimensions, section properties, and weights of the more common dressed lumber sizes are listed in Table 10.4. The section moduli and moments of inertia are given with respect to the cross-sectional axis parallel to the b dimension in column 1. That is, the section modulus of a dressed 2 × 4 equals 3.063 in.^{3} (50.2 cm^3), whereas the modulus of a 4 × 2 equals 1.313 in.^{3} (21.5 cm^3). When a flexural load is applied to the narrow face, bending takes place about the strong axis and the appropriate section modulus is that of a 2 × 4. This section is termed "loaded on edge." Contrariwise, when the load is applied to the wide face, bending proceeds about the weak axis and the section modulus for a 4 × 2 is used. Such an application is termed "used flat."

Table 10.4. Section Properties of Standard Dressed Rectangular Structural Lumber.

Nominal Size b(in.)d	Standard Dressed Size (S4S) b(in.)d	Area of Section A(in.^{2})	Moment of Inertia I(in.^{4})	Section Modulus S(in.^{3})	Mass in pounds per linear foot of piece when mass of wood per cubic foot equals: 30 lb	35 lb
1 × 3	¾ × 2-½	1.875	0.977	0.781	0.391	0.456
1 × 4	¾ × 3-½	2.625	2.680	1.531	0.547	0.638
1 × 6	¾ × 5-½	4.125	10.398	3.781	0.859	1.003
1 × 8	¾ × 7-¼	5.438	23.817	6.570	1.133	1.322
1 × 10	¾ × 9-¼	6.938	49.466	10.695	1.445	1.686
1 × 12	¾ × 11-¼	8.438	88.989	15.820	1.758	2.051
2 × 3	1-½ × 2-½	3.750	1.953	1.563	0.781	0.911
2 × 4	1-½ × 3-½	5.250	5.359	3.063	1.094	1.276
2 × 5	1-½ × 4-½	6.750	11.391	5.063	1.406	1.641
2 × 6	1-½ × 5-½	8.250	20.797	7.563	1.719	2.005
2 × 8	1-½ × 7-¼	10.875	47.635	13.141	2.266	2.643
2 × 10	1-½ × 9-¼	13.875	98.932	21.391	2.891	3.372
2 × 12	1-½ × 11-¼	16.875	177.979	31.641	3.516	4.102
2 × 14	1-½ × 13-¼	19.875	290.775	43.891	4.141	4.831
3 × 1	2-½ × ¾	1.875	0.088	0.234	0.391	0.456
3 × 2	2-½ × 1-½	3.750	0.703	0.938	0.781	0.911
3 × 4	2-½ × 3-½	8.750	8.932	5.104	1.823	2.127
3 × 5	2-½ × 4-½	11.250	18.984	8.438	2.344	2.734
3 × 6	2-½ × 5-½	13.750	34.661	12.604	2.865	3.342
3 × 8	2-½ × 7-¼	18.125	79.391	21.901	3.776	4.405
3 × 10	2-½ × 9-¼	23.125	164.886	35.651	4.818	5.621
3 × 12	2-½ × 11-¼	28.125	296.631	52.734	5.859	6.836
4 × 1	3-½ × ¾	2.625	0.123	0.328	0.547	0.638
4 × 2	3-½ × 1-½	5.250	0.984	1.313	1.094	1.276
4 × 3	3-½ × 2-½	8.750	4.557	3.646	1.823	2.127
4 × 4	3-½ × 3-½	12.250	12.505	7.146	2.552	2.977
4 × 5	3-½ × 4-½	15.750	26.578	11.813	3.281	3.828
4 × 6	3-½ × 5-½	19.250	48.526	17.646	4.010	4.679
4 × 8	3-½ × 7-¼	25.375	111.148	30.661	5.286	6.168
4 × 10	3-½ × 9-¼	32.375	230.840	49.911	6.745	7.869
4 × 12	3-½ × 11-¼	39.375	415.283	73.828	8.203	9.570

(continued)

Table 10.4. Continued.

Nominal Size b(in.)d	Standard Dressed Size (S4S) b(in.)d	Area of Section A(in.^{2})	Moment of Inertia I(in.^{4})	Section Modulus S(in.^{3})	Mass in pounds per linear foot of piece when mass of wood per cubic foot equals:	
					30 lb	35 lb
6 × 1	5-½ × ¾	4.125	0.193	0.516	0.859	1.003
6 × 2	5-½ × 1-½	8.250	1.547	2.063	1.719	2.005
6 × 3	5-½ × 2-½	13.750	7.161	5.729	2.865	3.342
6 × 4	5-½ × 3-½	19.250	19.651	11.229	4.010	4.679
6 × 6	5-½ × 5-½	30.250	76.255	27.729	6.302	7.352
6 × 8	5-½ × 7-½	41.250	193.359	51.563	8.594	10.026
6 × 10	5-½ × 9-½	52.250	392.963	82.729	10.885	12.700
6 × 12	5-½ × 11-½	63.250	697.068	121.229	13.177	15.373
8 × 1	7-¼ × ¾	5.438	0.255	0.680	1.133	1.322
8 × 2	7-¼ × 1-½	10.875	2.039	2.719	2.266	2.643
8 × 3	7-¼ × 2-½	18.125	9.440	7.552	3.776	4.405
8 × 4	7-¼ × 3-½	25.375	25.904	14.803	5.286	6.168
8 × 6	7-½ × 5-½	41.250	103.984	37.813	8.594	10.026
8 × 8	7-½ × 7-½	56.250	263.672	70.313	11.719	13.672
8 × 10	7-½ × 9-½	71.250	535.859	112.813	14.844	17.318
8 × 12	7-½ × 11-½	86.250	950.547	165.313	17.969	20.964
10 × 1	9-¼ × ¾	6.938	0.325	0.867	1.445	1.686
10 × 2	9-¼ × 1-½	13.875	2.602	3.469	2.891	3.372
10 × 3	9-¼ × 2-½	23.125	12.044	9.635	4.818	5.621
10 × 4	9-¼ × 3-½	32.375	33.049	18.885	6.745	7.869
10 × 6	9-½ × 5-½	52.250	131.714	47.896	10.885	12.700
10 × 8	9-½ × 7-½	71.250	333.984	89.063	14.844	17.318
10 × 10	9-½ × 9-½	90.250	678.755	142.896	18.802	21.936
10 × 12	9-½ × 11-½	109.250	1204.026	209.396	22.760	26.554
12 × 1	11-¼ × ¾	8.438	0.396	1.055	1.758	2.051
12 × 2	11-¼ × 1-½	16.875	3.164	4.219	3.516	4.102
12 × 3	11-¼ × 2-½	28.125	14.648	11.719	5.859	6.836
12 × 4	11-¼ × 3-½	39.375	40.195	22.969	8.203	9.570
12 × 6	11-½ × 5-½	63.250	159.443	57.979	13.177	15.373
12 × 8	11-½ × 7-½	86.250	404.297	107.813	17.969	20.964
12 × 10	11-½ × 9-½	109.250	821.651	172.979	22.760	26.554
12 × 12	11-½ × 11-½	132.250	1457.505	253.479	27.552	32.144

Reproduced with permission from *National Design Specification for Wood Construction*, 1977 ed. National Forest Products Association, Washington, DC.

10.4 DESIGN OF TIMBER LOAD CARRYING MEMBERS

The general procedures for analyzing or designing timber tension, compression, and flexural members follow along with procedures for

10.4 Design of Timber Load Carrying Members

analyzing members subject to combined loadings. Typical load combinations include axial tension plus flexure and axial compression plus flexure. This section discusses the governing equations for each type member, their limitations, and applications.

10.4.1 Tension Members

Tension members are most commonly encountered in lower chords and webs of trusses and in bracing members. The governing design equation is

$$F_t > f_t = P/A \qquad (10.4)$$

where F_t is allowable tensile strength; f_t, actual tensile stress; P, tensile load; A, cross-sectional area.

Care must be exercised to use the correct value for F_t since, in visually graded lumber, it is dependent upon the total width of the member. This dependence is covered in the footnotes of Tables 10.2A and 10.2B. The tensile strength of wide members is reduced because the potential for secondary stresses due to wood characteristics such as knots increases with width. Any wood characteristic or discontinuity disrupts the assumed uniform distribution of tensile stresses over the cross section. Stress concentrations result and the resultant internal tensile force no longer coincides with the centroidal axis. The eccentricity of the internal resultant introduces a secondary flexural stress, which, when coupled with stress concentrations, can produce premature brittle fracture. The magnitude of the secondary stresses is potentially greater in wider members as the eccentricity of the internal tensile force can be large.

Example 10.2. Tension Members. Evaluate the allowable tensile load that an 8 ft. (2.44 m) long No. 2 southern pine (S.P.) dressed 2 × 8 can carry if the load is induced by a wind load. The lumber is dressed and surfaced at 15% moisture.

Solution

a. The allowable stress for the member is the product of F_t from Table 10.2, the reduction factor for tensile members and the DF.

$$F_t = 675(0.80)(1.33) = 718 \text{ psi} \quad (5.0 \text{ MPa})$$

b. From Eq. (10.4) and the section properties from Table 10.4, the allowable load is simply

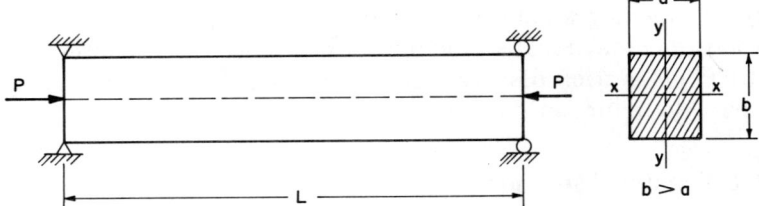

Fig. 10.7. Compression member definition sketch.

$$P = F_t(A) = 718(10.88) = 7,800 \text{ lb} \quad (34.5 \text{ kN})$$

10.4.2 Compression Members—Columns

The structural member in a state of pure compression is a rarity. That is, axial compressive stresses are usually accompanied by flexural stresses due to eccentric loading, lateral loading, and end moments. Nonetheless, the primary design of compression webs in trusses, bracing members, and some vertical framing members assumes the column is loaded in pure compression. Furthermore, the response of a timber column in pure compression is an integral component of the design procedure for beam columns (members subjected to combined axial compressive and flexural stresses).

A timber column is a straight member loaded by a centrically applied compressive force as shown in Fig. 10.7. The x and y axes of the cross section are the strong and weak axes, respectively. The strong axis has the larger moment of inertia. The slenderness ratio of a column is defined as L/r, where L equals the unsupported length of the column and r is the radius of gyration ($\sqrt{I/A}$) of the section. Two slenderness ratios are defined for the column in Fig. 10.7; namely L/r_x and L/r_y, where $r_x = \sqrt{I_x/A} = b/\sqrt{12}$ and $r_y = a\sqrt{12}$. In solid rectangular columns, the slenderness factors for the section in Fig. 10.7 are L/b about the x axis and L/a about the y axis. In general discussions the slenderness factor will be designated as L/d and will imply the largest slenderness factor for the column.

Timber columns are classified as either short, intermediate, or long depending upon their slenderness factor, L/d. A column with $L/d < 11$ is short, a column with $11 < L/d < K = 0.67\sqrt{E/F_c}$ is intermediate, and if $L/d > K$, the column is long. A short column fails by crushing of the wood fibers. An intermediate column fails by buckling inelastically; that is, buckling commences before yielding, but yielding occurs during buckling. Long columns buckle elastically; that is, they do not yield before or during buckling.

If a timber column is long or intermediate, it will buckle about the cross-sectional axis with the largest slenderness factor, L/d. Buckling does not always occur about the weak axis of the section. That is, the column unsupported length may be less than the total column length. Furthermore, the unsupported length may be different with respect to buckling about the x axis or the y axis.

The governing condition for analysis and design of timber columns is

$$F'_c > f_c = P/A \tag{10.5}$$

where F'_c is allowable compressive stress parallel to the grain adjusted for slenderness (buckling effects); $f_c, P/A$ = actual uniform compressive stress.

For a straight, centrically loaded, completely pinned end column with no lateral bracing, the NDS specifies the following allowable compressive stresses:

$$F'_c = F_c \text{ (Table 10.2 or 10.3)} \quad \text{if} \quad L/d < 11 \tag{10.6}$$

$$F'_c = F_c \left[1 - \tfrac{1}{3} \left(\frac{L/d}{K} \right)^4 \right] \quad \text{if} \quad 11 < L/d < K \tag{10.7}$$

$$F'_c = 0.3E/(L/d)^2 \quad \text{if} \quad K < L/d < 50 \tag{10.8}$$

$$F'_c = 0 \quad \text{if} \quad L/d > 50 \tag{10.9}$$

where $K = 0.67\sqrt{E/F_c} = (L/d)_{\lim}$ and F_c is the allowable compressive stress in the absence of buckling. The variation of F'_c with L/d is shown graphically in Fig. 10.8.

The F'_c values for short and intermediate columns may be adjusted for load duration. $F_c{}'$ for long columns may not be increased for load duration as it is dependent upon the modulus of elasticity, E, which does not increase with shorter duration loading.

The intermediate column equation is an empirical equation fitted to laboratory test data. The long column equation is a transformation of Euler's equation with $r = d/\sqrt{12}$ and a factor of safety of 2.74 included. Euler's equation is

$$F'_c = \frac{\pi^2 E}{(F.S.)(L/r)^2}$$

and upon substituting

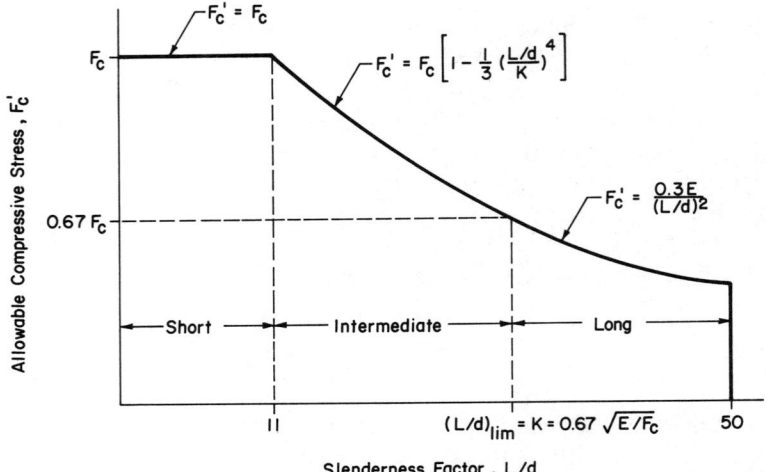

Fig. 10.8. Variation of allowable compressive stress with slenderness factor for solid rectangular timber columns.

$$F'_c = \frac{\pi^2 E}{2.74(L/d)^2(12)} = \frac{0.3E}{(L/d)^2}$$

Laboratory load test data begin to converge to Euler's equation whenever $F'_c < 0.67F_c$. By equating F_c' from Eq. (10.8) to $0.67F_c$, it can be shown that $K = (L/d)_{\lim}$ is the limiting L/d value between intermediate and long column behavior.

The allowable stresses for columns with intermediate bracing or nonpinned end restraints may also be calculated by Eqs. (10.6–10.9) provided the column length is replaced by an appropriate effective length.

The effective lengths for columns with various end constraints are summarized in Table 10.5. The influence of lateral braces upon L/d is demonstrated in Fig. 10.9 and Example 10.3.

Example 10.3. Effective Length

a. Evaluate the slenderness factor of the timber column in Fig. 10.9a. The brace prevents displacement in the plane of page, but not in any other plane. The slenderness factor about the strong axis (axis with greatest moment of inertia) is $(L/d)_x = L/b$. The slenderness factor about the weak axis (axis with least moment of inertia) is $(L/d)_y = L/(2a)$.

10.4 Design of Timber Load Carrying Members

Table 10.5. Effective Column Lengths.

End Constraints	Column Length	Theoretical Effective Length	Design Effective Length
Pinned–pinned	L	1.0L	1.0L
Fixed–fixed	L	0.5L	0.65L
Fixed–pinned	L	0.7L	0.80L
Fixed–fixed[a] (unbraced)	L	1.0L	1.20L
Cantilevered	L	2.0L	2.0L

[a]Rotation prevented.

b. Evaluate the slenderness factor of the fixed–fixed column in Fig. 10.9b. The bracing prevents buckling only in the plane of the page. The buckling mode is shown by the dashed line. The brace is essentially a pin for effective length considerations. About the strong axis, $(L/d)_x = 0.5L/b$. About the weak axis, $(L/d)_y = 0.7(L/2)/a = 0.35L/a$.

Example 10.4. Compression Member Analysis. Evaluate the allowable load for the 5 ft (1.52 m) long, dressed No. 2 southern pine 2 × 4 (surfaced dry and used at 19% moisture content) column in Fig. 10.10 if the ends are pinned with respect to buckling about both the strong and weak axes. Assume a snow load.

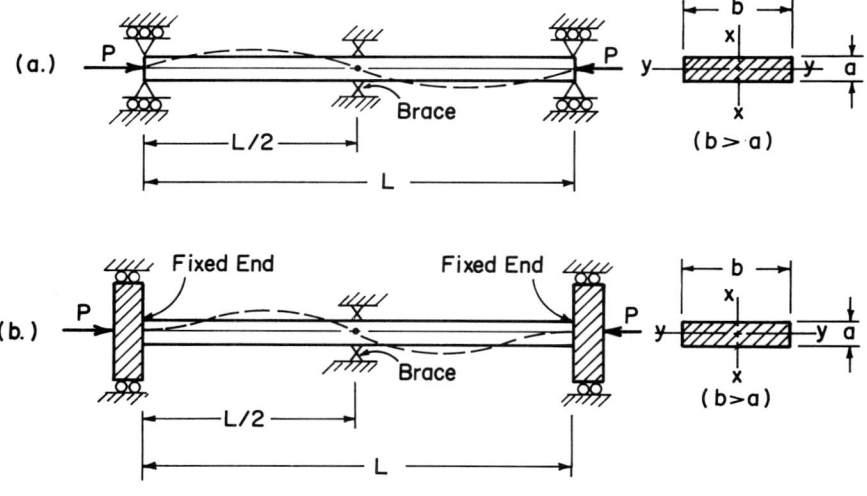

Fig. 10.9. Slenderness factors.

388 Light Timber Design

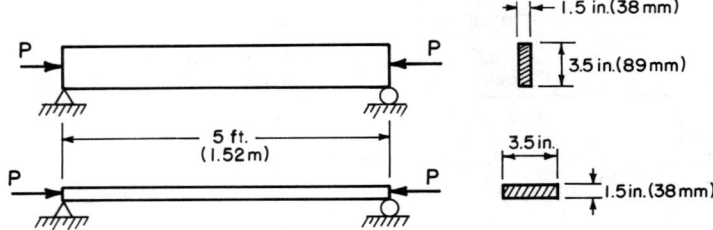

Fig. 10.10. Example 10.4—Compression member analysis.

Solution

a. Classification of the column.

$$(L/d)_{max} = 5 \times 12/1.5 = 40$$
$$K = 0.67\sqrt{E/F_c} = 0.67\sqrt{1.6 \times 10^6/975 \times 1.15} = 25.3$$

Since $L/d > K$, the column is long.

b. The allowable load for the column.

$$F'_c = 0.3E/(L/d)^2 = 0.3(1.5 \times 10^6)/(40)^2 = 300 \text{ psi} \quad (2.1 \text{ MPa})$$

Thus, $P_{all} = F'_c A = 300(5.25) = 1580 \text{ lb } (7.0 \text{ kN})$.

Example No. 5. Compression Member Design. Design the smallest square column which can support a concentric compressive load of 16,000 lb (71 kN). Use Dense No. 1 Douglas fir larch and assume the load is due to a wind load. The unsupported length of the column is 8 ft (2.44 m).

Solution

a. As a first estimate, assume a short column. For posts and timbers, $F_c = 1200 \times 1.33 = 1600$ psi (11.0 MPa). From Eq. (10.5), $A_{req'd} = 16,000/1600 = 10$ in.^{2} (64.5 cm^2) and from Table 10.4, a 4 × 4 is the smallest possible section.

b. Check the 4 × 4 for column action (buckling). Since $L/d = 8 \times 12/3.5 = 27.4 > K = 0.67\sqrt{1.7 \times 10^6/1600} = 21.8$, the column is long and from Eq. (10.8), $F'_c = 0.3E/(L/d)^2 = \dfrac{0.3(1.7 \times 10^6)}{(27.4^2)} =$

677 psi (4.7 MPa). From Eq. (10.5), $f_c = P/A = 16{,}000/12.25 = 1310$ psi (9.0 MPa) $> F'_c$ and the 4×4 is inadequate.

c. Trying a 6×6 section, the slenderness factor, $L/d = 8 \times 12/5.5 = 17.4 < K$ and the column is intermediate. Thus, from Eq. (10.7) $F'_c = 1600 \, [1 - \frac{1}{3}(17.4/21.8)^4] = 1380$ psi (9.5 MPa). Since $f_c = P/A = 16{,}000/30.25 = 512$ psi (3.5 MPa) $< F'_c$, the 6×6 is adequate.

Centrically loaded compression members are seldom found in practice. Although many columns are analyzed or designed assuming centric loading, it is better practice to assume an accidental eccentricity of the larger of 1 in. (25 mm) or $0.1b$ about the strong axis (see Fig. 10.7) and 1 in. (25 mm) or $0.1a$ about the weak axis; and to analyze or design the member as a beam-column.

The NDS prohibits the use of compression members with $L/d > 50$. In cases where $L/d > 50$, the slenderness can be reduced by installation of intermediate bracing or by increasing the minimum thickness with properly fastened scabbing in the middle $\frac{1}{2}$ to $\frac{1}{3}$ of the critical unsupported length. Another alternative is to replace the solid timber member with a spaced column. The spaced column consists of two compression members spaced approximately $1\frac{1}{2}$ in. (38 mm) apart and fastened together with special timber connectors. Slenderness ratios up to 80 are then permissible and allowable compressive stresses are $2\frac{1}{2}$ to 3 times as great as for single compression members. The reader is referred to the NDS for details.

For timber columns with nonrectangular cross sections, the design equations presented herein are easily modified. Simply replace d with $\sqrt{12}r$ in all the design equations for F'_c and evaluate the slenderness ratio instead of the slenderness factor. The limiting slenderness ratios then become $L/r < 38$ for short columns; L/r between 38 and K' for intermediate columns; and L/r between K' and 173 for long columns, where $K' = 2.32\sqrt{E/F_c}$.

10.4.3 Flexural Members—Beams

A sketch of a typical timber flexural member is found in Fig. 10.11. The length of the member is much greater than either cross-sectional dimension and the applied load is either a couple or forces with a component perpendicular to the longitudinal axis. There are two basic types of timber beams; namely, laterally braced and laterally unbraced beams. The laterally braced beam has sufficient lateral supports, such as attached decking or bridging, to permit the beam to reach its al-

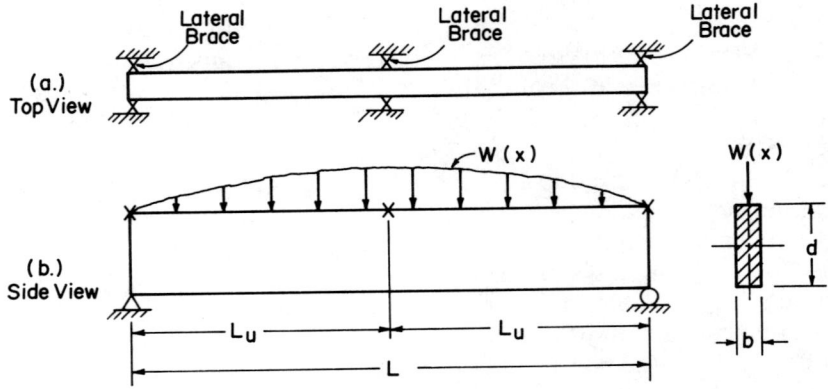

Fig. 10.11. Definition sketch for a laterally braced timber beam.

lowable flexural stress from Table 10.2 or 10.3 before buckling laterally. An unbraced beam may buckle laterally at some stress level below the allowable flexural stress from Table 10.2 or 10.3. The criteria for braced and unbraced beams will be discussed in a later section.

The design or analysis of a timber beam may be controlled by any of several factors. Chief among these are the flexural stresses, transverse shear stresses, the bearing stresses at end supports and at concentrated loads, and the beam deflection. For many timber beams in agricultural applications, flexural stresses will control the design. Thus, a first step in nearly all beam analysis is an evaluation of flexural stresses. Some exceptions to this order include very short or heavily loaded beams where shear is likely to control design and beams with severe deflection limits.

In all the following discussions it is assumed that the beams are of solid sawn cross section and that flexural loads are applied through the shear center of the section so that no torsion (twisting) of the beam occurs. The scope of this text will not include either laminated beams or torsional stresses.

10.4.3.1 Laterally Braced Beams. To qualify as a laterally braced beam, the slenderness factor C_s must be less than or equal to 10. Referring to the definition sketch in Fig. 10.11

$$C_s = \sqrt{L_e d/b^2} \qquad (10.10)$$

where d is actual dimension of the cross section in the direction normal to the axis of bending; b is actual dimension of the cross section parallel to the bending axis; L_e is effective unsupported length between bracing; $L_e = 1.61L_u$ for single span beam with a concentrated load at the center; $L_e = 1.92L_u$ for a single span beam with a uniformly distributed load; $L_e = 1.84L_u$ for a single span beam with equal end moments; $L_e = 1.69L_u$ for a cantilever beam with a concentrated load at the unsupported end; $L_e = 1.06L_u$ for a cantilever beam with a uniformly distributed load; $L_e = 1.92L_u$ for a single span or cantilever beam with any load (conservative value); L_u is distance between lateral supports.

Note that the slenderness factor increases with distance between lateral supports and with the ratio of beam depth to width. The greater the depth to breadth ratio, d/b, of the beam, the less stable the beam and the more likely it will buckle laterally.

The governing criterion for flexure stresses in beams with adequate lateral bracing is

$$F_b > f_b = Mc/I = M/S \qquad (10.11)$$

where f_b is actual flexural stress; F_b is allowable flexure stress from Table 10.2 or 10.3. The allowable flexural stress may be adjusted for load duration, single or repetitive use, edgewise or flatwise loading, cross-sectional shape, and depth. The first and second adjustments have already been discussed. The third adjustment is covered in the footnotes to Tables 10.2 and 10.3. The moment capacity of circular cross sections diverges from that predicted by elastic theory more than rectangular sections because a larger proportion of the section area is concentrated close to the neutral axis. Consequently, the NDS specifies the allowable flexure stress for such sections as $F'_b = 1.18F_b$.

As beam sections become deeper, the discrepancy between the actual stress distribution and the triangular elastic flexural stress distribution increases. The F_b values in Tables 10.2 and 10.3 adequately account for differences for beam depths to 12 in. (305 mm). In deeper beams $F'_b = F_b (12/d)^{1/9}$, where d is in inches.

The governing criterion for transverse, or horizontal shear is

$$Q = \frac{bd}{2} \cdot \frac{d}{4}$$

$$I = \frac{bd^3}{12} \qquad F_v > f_v = \frac{VQ}{Ib} \qquad (10.12)$$

where F_v is allowable horizontal shear stress from Tables 10.2 or 10.3; f_v, actual transverse shear stress; V, transverse shear force; I, second moment of area about neutral axis; b, width of beam at point shear

392 Light Timber Design

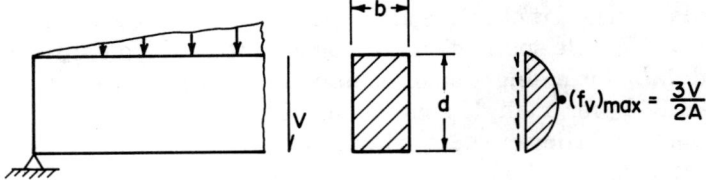

a. Shear Stress in a Solid Timber Section

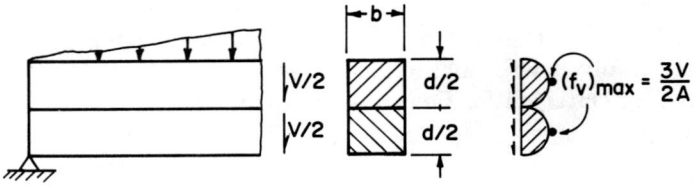

b. Shear Stress in Two Independent Beams Lying Upon One Another

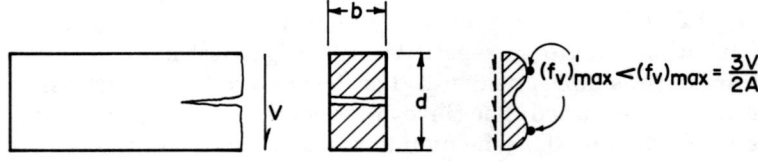

c. Shear Stress in a Checked Timber Beam

Fig. 10.12. Two-beam action in checked timber beams.

stress is evaluated; Q, first moment of the cross-sectional area between the outer fiber and the point the shear stress is evaluated taken about the neutral axis. For a rectangular section it is easily shown that $f_v = 3V/2A$. The student should verify this as a form of helpful review of mechanics.

For most solid sawn beams the shear requirement will be satisfied using the basic criterion of Eq. (10.12). However, if $f_v > F_v$, then f_v may be adjusted downwards and F_v may be adjusted upwards before the design is rejected. Both adjustments are made to account for a phenomenon known as two-beam action.

Two-beam action can be explained by reference to Fig. 10.12. The beam cross sections in Figs. 10.12a, 10.12b, and 10.12c all carry a shear force V and each has a total depth d and a width b. In Fig. 10.12a, the section is solid, and the maximum shear stress, which occurs at the neutral axis, equals $3V/2A$. In Fig. 10.12b, the section is not solid. Instead it consists of two independent beams, $d/2$ in depth, each of which carries a shear force $V/2$. The maximum shear stress in each section can be shown, by substituting $V/2$, Q for the top beam section, and I for the top beam section, into the right-hand portion of Eq. (10.12), to be $3V/2A$ where $A = b \times d$. As the shear stress in a solid timber beam approaches its allowable level, the beam begins to check at the neutral axis. That is, there is a partial separation of the upper and lower portions of the cross section. The section still carries a shear force V, but the section behaves partially as a solid beam and partially as two independent beams. The result is a stress distribution similar to that in Fig. 10.12c. Significantly, the maximum shear stress in the checked beam f_v' is less than $3V/2A$. Qualitatively, this can be visualized by recalling that the resultant of the shear stress acting over the cross section of the beam must equal the shear force V. Since some shear stress exists at the checked portion, in Fig. 10.12c, the shear stress at the centroidal axis of the top and bottom portions of the checked beam can be less than $3V/2A$, and still carry the shear force V. The design specification allows us to utilize this phenomenon.

The first level adjustment is to reduce f_v by neglecting all loads within a distance d of any supports when evaluating the internal shear force. For moving loads, the shear force may be reduced by placing single moving loads at d from the support or by placing the largest of multiple moving loads at d from the support. If the adjusted shear force is V', then the design criteria becomes:

$$F_v > f_v' = \frac{3V'}{2A} \tag{10.13}$$

Should the beam not satisfy the criterion of Eq. (10.13), then a second level adjustment of f_v and F_v is made. Actual shear forces can be modified to (see Fig. 10.13) V' as defined in Eqs. (10.14)–(10.16).

$$V_1' = \frac{10P_1(L-x_1)(x_1/d)^2}{9L[2 + (x_1/d)^2]} \quad \text{for concentrated loads} \tag{10.14}$$

and

$$V' = \tfrac{1}{2}wL[1 - (2d/L,)] \quad \text{for uniformly distributed loads} \tag{10.15}$$

For a series of n concentrated loads

$$V' = V_1' + V_2' + \cdots + V_n' \tag{10.16}$$

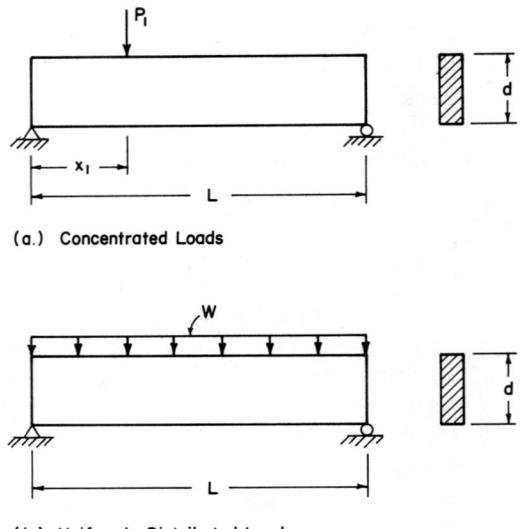

Fig. 10.13. Definition sketch for shear stress adjustments for timber beams.

The shear stress is evaluated by $f'_v = 3V'/2A$. Finally, the allowable shear stresses may be increased as shown in Table 10.6. If these stresses are designated as F'_v, the final design criteria becomes

$$F'_v > f'_v = 3V'/2A \qquad (10.17)$$

High concentrated loads at end supports or at point loads may cause compressive failures normal to the grain. To prevent this occurrence, adequate bearing area is required. The governing design criteria for end supports and point loads are $F_{cp} > f_{cp} = R/A$ at end supports and $F_{cp} > f_{cp} = P/A$ at other concentrated loads, where F_{cp} is allowable compressive stress normal to grain, R is reactive force, P is concentrated load, A is bearing area at the load or support of $b \times z$ (Fig. 10.14), and f_{cp} is actual bearing stress. At concentrated loads located at $x > 3$ in. (76 mm) from the end, F_{cp} may be increased by the factor $(z + 3/8)/z$ if $z < 6$ in. (152 mm).

Beam deflections are evaluated according to standard methods of elastic analysis. For normal loads, creep is not a factor. However, for long-term loads, creep increases deflections considerably. Thus, total

Table 10.6. Adjusted Allowable Horizontal Shear Stresses for Moving Loads and Two-Beam Action.

F'_v		Maximum Moisture Content	
	Unseasoned	19%	15%
Aspen	85	90	95
Balsam fir	85	95	95
Black cottonwood	70	75	80
California redwood	115	120	130
Coast sitka spruce	90	95	100
Coast species	90	95	100
Douglas fir, larch	130	140	145
Douglas fir, south	130	140	145
Eastern hemlock, tamarack	120	130	135
Eastern spruce	95	105	110
Eastern white pine	90	95	100
Eastern woods	85	95	95
Englemann spruce/alpine fir	95	105	110
Hemlocks, fir	105	110	115
Idaho white pine	95	100	105
Lodgepole pine	95	105	110
Mountain hemlock	130	140	150
Northern aspen	90	95	100
Northern pine	100	105	110
Northern species	90	95	100
Northern white cedar	85	95	100
Ponderosa pine/sugar pine	100	105	110
Red pine	100	110	115
Sitka spruce	105	115	120
Southern pine	125	135	145
Spruce, pine, fir	95	105	110
Western cedar	100	105	110
Western hemlock	125	135	145
Western white pine	90	100	105
White woods (Western woods, West Coast woods, mixed species)	95	100	105

Reproduced with permission from *National Design Specification for Wood Construction*, 1977 ed. National Forest Products Association, Washington, DC.

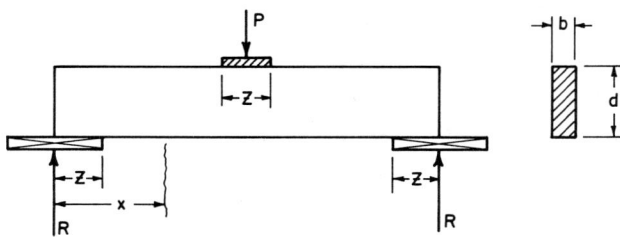

Fig. 10.14. Definition sketch for allowable bearing stresses in timber beams.

deflection Δ is usually taken as live load deflection Δ_L plus twice the dead load deflection Δ_D.

$$\Delta = \Delta_L + 2\Delta_D \tag{10.18}$$

The design criteria for deflection are

$$\Delta_{\text{all}} > \Delta = kwL^4/EI \quad \text{for distributed loads} \tag{10.19}$$

and

$$\Delta_{\text{all}} > \Delta = kPL^3/EI \quad \text{for concentrated loads} \tag{10.20}$$

where k is coefficient dependent upon the nature and distribution of loads and may be obtained by elastic beam deflection analysis or by reference to standard engineering handbooks. The allowable deflections are specified in design and building codes. Typical deflection limits include $\Delta_{\text{all}} = (1/360)$ (beam span) for beams supporting plastered ceilings and $\Delta_{\text{all}} = (1/240)$ (beam span) for beams supporting unplastered ceilings.

Example 10.6. Laterally Braced Beam. Select a No. 2 southern pine dressed member used and surfaced at 15% moisture content to carry a uniformly distributed 100 lb/ft (1.46 kN/m) snow load over a 10 ft (3.05 m) simply supported span. Assume the compression flange to be adequately braced and the allowable deflection to be (1/240)(span).

Solution

a. The maximum shear and moment for a uniformly distributed load over a simply supported span are

$V_{\max} = wL/2 = 100(10)/2 = 500$ lb (2.2 kN)
$M_{\max} = wL^2/8 = 100(10)^2(12)/8 = 15{,}000$ lb-in. (1.70 kN-m)

b. Selecting a section for flexural stresses and assuming a size greater than a 2 × 4, the allowable stress is

$$F_b = 1300(1.15) = 1500 \text{ psi} \quad (10.3 \text{ MPa})$$

where 1300 comes from Table 10.2 and 1.15 is the load duration factor. From $f_b = M/S < F_b$, $S_{\text{req'd}} > M/F_b = 15{,}000/1500 = 10$

in.^{3} (163.9 cm^3). Entering Table 10.4, a 2 × 8 is the lightest adequate member with $S > 10$ in.^{3} (163.9 cm^3). The section properties of the 2 × 8 are

$$S_{2\times 8} = 13.14 \text{ in.}^3 \quad (215.3 \text{ cm}^3)$$
$$I_{2\times 8} = 47.64 \text{ in.}^4 \quad (1983 \text{ cm}^4)$$
$$\text{Dead Load} \approx 2.6 \text{ lb/ft} \quad (38 \text{ N/m})$$

Since the beam dead load is only 2.6% of the live load it can be neglected in the design. If the dead load exceeds 5–10% of the live load, the adequacy of the selection should be rechecked.

c. Checking the horizontal shear stress

$$F_v = 95(1.15) = 109 \text{ psi} \quad (0.75 \text{ MPa})$$

where 95 comes from Table 10.2 and 1.15 is the load duration factor.

$$f_v = \frac{3V}{2A} = \frac{3(500)}{2(10.88)} = 69 \text{ psi} \quad (0.48 \text{ MPa})$$

Since $f_v < F_v$, shear strength is adequate without considering two-beam adjustments.

d. Checking the bearing requirements

$$F_{cp} = 405(1.15) = 466 \text{ psi} \quad (3.2 \text{ MPa})$$

where 405 comes from Table 10.2 and 1.15 is the load duration factor.

$$f_{cp} = R/bz = 500/(1.5z)$$

Since F_{cp} must be $\geq f_{cp}$, it follows that $z \geq 500/(466 \times 1.5) = 0.75$ in. (18 mm). Thus, a minimum bearing length of $\frac{3}{4}$ in. (19 mm) is recommended.

e. Check the bracing requirement. Since the slenderness factor $C_s = \sqrt{L_e d/b^2}$ must be less than 10, the bracing interval can be evaluated from Eq. (10.10). Since

$$C_s = \sqrt{1.92 L_u \times 7.25/1.5^2} = 10, \quad L_u = 16.2 \text{ in. (411 mm).}$$

Thus, the compression edge should be braced every 16.2 in. (411 mm).

f. Check deflection requirement from Eq. (10.18) and the deflection of a uniformly loaded simply supported beam

$$\Delta = \Delta_L + 2\Delta_D = \frac{5w_L L^4}{384EI} + 2\left(\frac{5w_D L^4}{384EI}\right)$$

$$= \left[\frac{5(100)(10)^4}{384(1.6 \times 10^6)(47.64)} + 2\left(\frac{5(2.6)(10)^4}{384(1.6 \times 10^6)(47.64)}\right)\right]1728$$

$$= 0.30 + 0.02 = 0.32 \text{ in.} \quad (8 \text{ mm})$$

Since $\Delta_{all} = (1/240)(10 \times 12) = 0.50$ in. (13 mm), the deflection requirement is met and a 2 × 8 is adequate.

10.4.3.2 Laterally Unbraced Beam. If the unsupported length of the compression edge of a timber beam is large such that $C_s > 10$, the beam will buckle laterally before reaching the full allowable bending moment $M_{all} = F_b(S)$. In the design of laterally unsupported beams, the allowable stress F'_b is reduced to a level below which lateral buckling will not occur.

Laterally unbraced beams are classified as short, intermediate, or long. As in columns, short beams fail by yielding; intermediate beams buckle inelastically, and long beams buckle elastically. The criteria for classifying beams are limiting values on the slenderness factor C_s. If $C_s < 10$, the beam is short, if $10 < C_s < C_k$ the beam is intermediate, and if $C_k < C_s < 50$, the beam is long; where $C_{s+2} = \sqrt{L_e d/b^2}$ and $C_k = \sqrt{3E/5F_b}$. The term C_k is the limiting slenderness factor between elastic and inelastic lateral buckling.

Equations (10.21)–(10.24) define the allowable flexural stresses F'_b for each class beam.

$$F'_b = F_b \qquad \text{for} \quad C_s < 10 \qquad (10.21)$$

$$F'_b = F_b\left[1 - \tfrac{1}{3}\left(\frac{C_s}{C_k}\right)^4\right] \qquad \text{for} \quad 10 < C_s \leq C_k \qquad (10.22)$$

$$F'_b = 0.4\frac{E}{C_s^2} \qquad \text{for} \quad C_k < C_s \leq 50 \qquad (10.23)$$

$$F'_b = 0 \qquad \text{for} \quad C_s > 50 \qquad (10.24)$$

Figure 10.15 graphically illustrates the variation of F'_b with C_s. F'_b may be adjusted for load duration for short and intermediate beams. However, F'_b for long beams may not be adjusted for load duration since it is dependent solely upon the stiffness parameter rather than strength. A laterally unsupported beam must satisfy the same requirements for

10.4 Design of Timber Load Carrying Members **399**

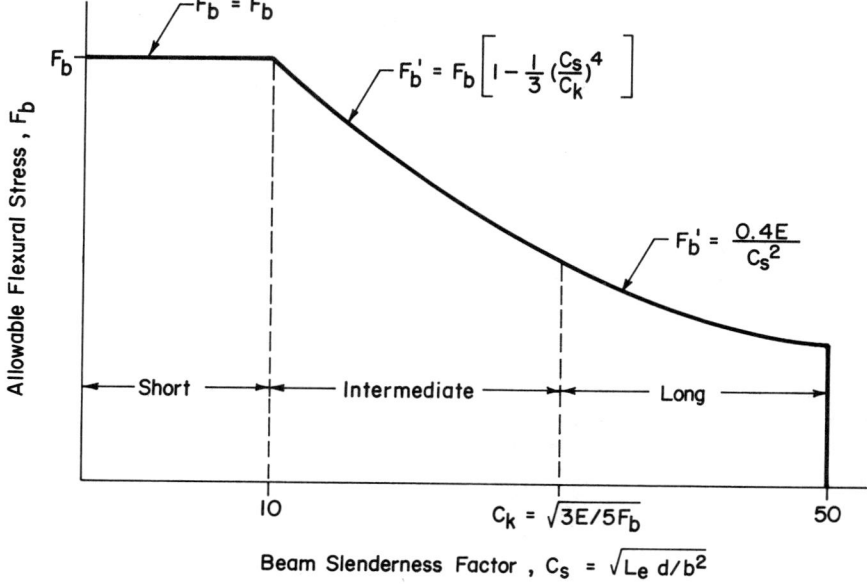

Fig. 10.15. Variation of allowable flexure stress with slenderness factor.

shear, bearing, and deflection as braced beams. The procedures for evaluating these stress levels and allowable stresses in unbraced beams are identical to those for braced beams.

Example 10.7. Laterally Unbraced Beam. Evaluate the allowable flexural stress for a dressed 2 × 10 No. 2 southern pine (surfaced and used at 15% moisture) beam which carries a uniformly distributed load over a simply supported span of 10 ft (3.05 m). Assume a snow load duration.

Solution

a. Evaluate the magnitude of the slenderness factor using Eq. (10.10). The effective length factor for the beam is

$$L_e = 1.92(10)(12) = 230.4 \text{ in.} \quad (5.85 \text{ m})$$

and

$$C_s = \sqrt{L_e d/b^2} = \sqrt{(230.4)(9.25)/(1.5)^2}$$
$$= 30.8$$

b. Evaluate the limiting slenderness factor C_k and classify the beam.

$$C_k = \sqrt{3E/5F_b} = \sqrt{3(1.6 \times 10^6)/5(1300 \times 1.15)}$$

$$= 25.3$$

Since $C_s > C_k$, the beam is classified as long.

c. Evaluate the allowable stress

$$F'_b = 0.4E/C_s^2 = 0.4(1.6 \times 10^6)/(30.8)^2$$

$$F'_b = 680 \text{ psi} \quad (4.7 \text{ MPa})$$

Notice that F'_b is less than half the allowable flexure stress without lateral buckling [$F_b = 1300 \times 1500$ psi (10.3 MPa)]. If f_b exceeds F'_b, the 2 × 10 will buckle laterally. The objective of the designer is to either brace the 2 × 10 to increase F'_b or assure that the flexural stress never exceeds $F'_b = 680$ psi (4.7 MPa) in this beam.

10.4.4 Combined Loading

Many structural members are subject to a combination of flexural, tensile, and compressive loads. The more common situations encountered include axial tension and flexure, centric axial compression and flexure, and eccentric axial compression and flexure. The combined states of stress for each are shown in Fig. 10.16.

The fundamental design criterion for timber members under combined states of loading is the interaction equation. Equation (10.25) is the generalized interaction equation for all combined loadings.

$$\frac{\text{Actual axial stress}}{\text{Allowable axial stress}} + \frac{\text{Actual flexural stress}}{\text{Allowable flexural stress}} < 1.0 \quad (10.25)$$

By referring to the stress distributions in Fig. 10.16, the generalized interaction equation is quite rational if the allowable axial and flexural stresses are equal. Reference to Tables 10.2 and 10.3, however, shows that equality generally does not exist between the allowable stresses. Thus, the interaction equation is an empirical approach. Nonetheless, it is relatively simple to use and it yields reasonable results in practice.

Equation (10.26) is the design criterion for axial tension plus flexure

$$\frac{f_t}{F_t} + \frac{f_b}{F_b} < 1.0 \quad (10.26)$$

10.4 Design of Timber Load Carrying Members 401

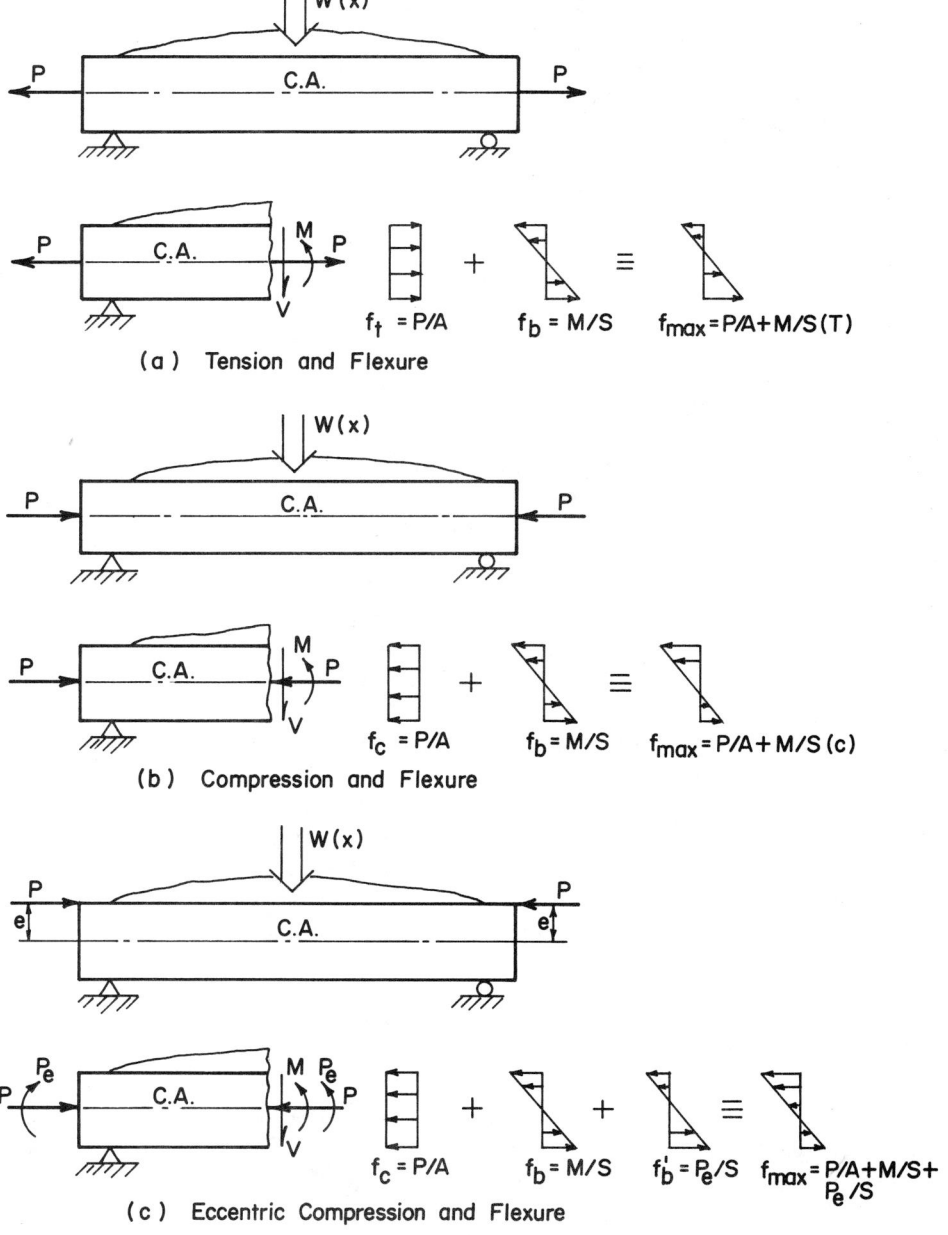

Fig. 10.16. Stress distributions for combined loadings.

Both F_t and F_b may be adjusted for load duration and F_b may need to be adjusted for lateral stability effects.

Members which carry both flexural and axial compressive loads simultaneously are called beam-columns. Equation (10.27) is the governing design criterion for beam-columns.

$$\frac{f_c}{F'_c} + \frac{f_b}{F'_b - Jf_c} < 1.0 \tag{10.27}$$

where F'_c is allowable compressive stress adjusted for slenderness effects; F'_b is allowable flexural stress adjusted for slenderness effects with respect to the axis of bending; $J = 0$ if $L/d < 11$; $J = (L/d - 11)/(K - 11)$ if $11 < L/d \leq K$; $J = 1.0$ if $L/d > K$.

The J factor in Eq. (10.27) is a moment magnification factor to compensate for secondary stresses in longer beam-columns. Figure 10.17a will help explain the phenomenon for combined flexure and centric axial compression. The total moment in the beam-column equals the moment due to flexural loads M_w plus the moment $P\Delta$. In short members, Δ is small, thus $P\Delta \ll M_w$ and therefore is negligible. As the member lengthens, Δ increases and $P\Delta$ is no longer negligible. The design specification compensates for this "P–Δ" effect by reducing the allowable flexural stress by a quantity Jf_c. For short columns, no adjustment is required. As L/d increases from 11 to $K = 0.67\sqrt{E/F_c}$, J varies from 0 to 1.0.

For the case of a beam-column with eccentric loading the maximum moment is shown in Fig. 10.17b to be $M = M_w + P(e + \Delta)$. For short members, $\Delta \ll e$ and $M \approx M_w + Pe$. For longer members Δ approaches the order of magnitude of e and $M = M_w + P(e + \Delta)$. The specification again utilizes the J factor to account for the moment magnification, or "P–Δ," effect.

Equation (10.28) is the design criterion for eccentrically loaded beam-columns.

$$\frac{f_c}{F'_c} + \frac{M_w/S + P(e + \Delta)/S}{F'_b - Jf_c} \leq 1.0 \tag{10.28}$$

If the beam-column is short ($L/d < 11$), $J = 0$, and $e + \Delta \approx e$. Hence, $P(e + \Delta)/S = Pe/(bd^2/6) = (P/A)(6e/d) = f_c(6e/d)$ for rectangular cross sections. Thus, for short beam-columns, Eq. (10.28) becomes

$$\frac{f_c}{F'_c} + \frac{M_w/S + f_c(6e/d)}{F_b} \leq 1.0 \tag{10.29}$$

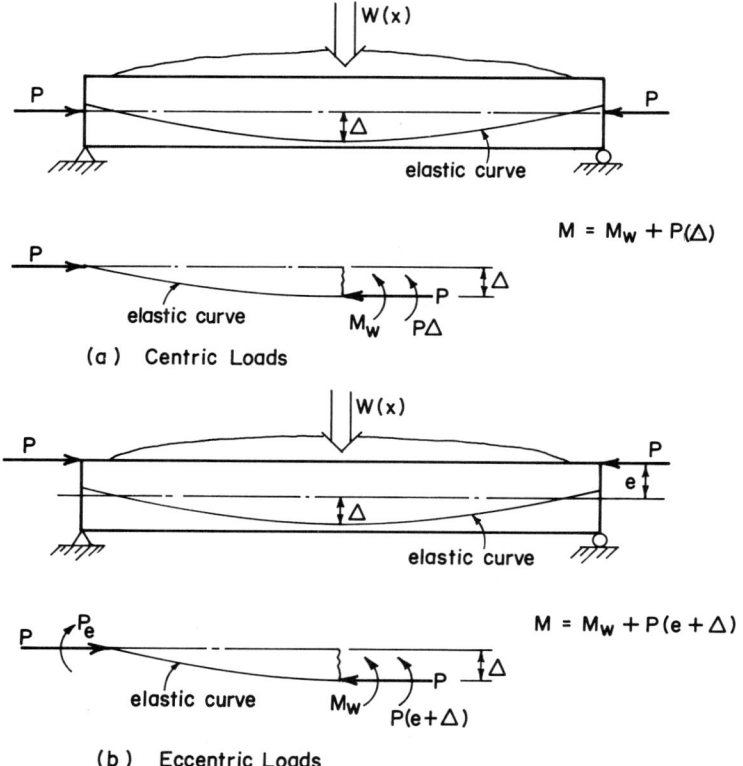

Fig. 10.17. Secondary moments in beam-columns.

If the beam-column is intermediate or long ($L/d > 11$), secondary stresses are important because $P(e + \Delta) \neq Pe$. Secondary stresses are accommodated by magnifying the coefficient on f_c in the flexure ratio and by reducing F'_b by Jf_c. Thus, the design criterion becomes Eq. (10.30) for long eccentrically loaded beam-columns.

$$\frac{f_c}{F'_c} + \frac{M_w/S + f_c(6 + 1.5J)(e/d)}{F'_b - Jf_c} \leq 1.0 \qquad (10.30)$$

Example 10.8. Beam-Column Analysis. Evaluate the magnitude of the maximum axial compression load which may be superimposed on the No. 2 southern pine dressed 2 × 4 (surfaced and used at 15% moisture) shown in Fig. 10.18. Assume normal loading and pinned ends.

404 Light Timber Design

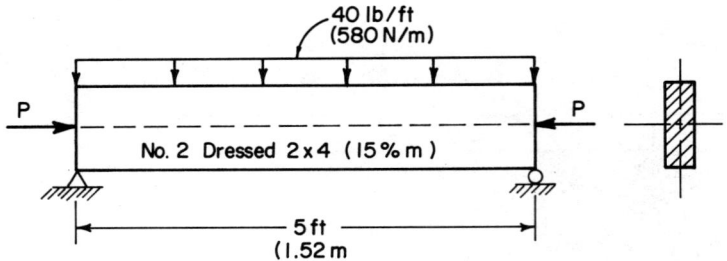

Fig. 10.18. Example 10.8—Beam-column analysis.

Solution

a. Obtain the section and structural properties from Tables 10.2 and 10.4.

$A = 5.25$ in.^{2} (33.9 cm^2) $F_b = 1550$ psi (10.7 MPa)

$b = 1.5$ in. (38 mm) $F_c = 1150$ psi (7.9 MPa)

$d = 3.5$ in. (89 mm) $E = 1.6 \times 10^6$ psi (11 GPa)

$S = 3.06$ in.^{3} (50.10 m^3)

b. Classify the column by comparing L/d to K.

$$\frac{L}{d} = \frac{5 \times 12}{1.5} = 40$$

$$K = 0.67\sqrt{1.60 \times 10^6/1050} = 25.0$$

Since $L/d > K$, the member is long and $J = 1.0$ in Eq. (10.27).

c. Evaluate the actual and allowable stresses in the member.

$$f_c = P/A = P/5.25$$

$$f_b = \frac{M_{max}}{S_x} = \left[\left(\frac{wL^2}{8}\right)\frac{(12)}{S_x}\right]$$

$$= \left[\frac{40(25)(12)/8}{3.06}\right] = 490 \text{ psi} \quad (3.4 \text{ MPa})$$

Since the column is long,

$$F'_c = \frac{0.3E}{(L/d)^2} = \frac{0.3(1.6 \times 10^6)}{(40)^2} = 300 \text{ psi} \quad (2.1 \text{ MPa})$$

From Eq. (10.10),

$$C_s = \sqrt{\frac{1.92 \times 60 \times 3.5}{1.5^2}} = 13.4 < C_k$$

$$= \sqrt{\frac{3(1.6 \times 10^6)}{5(1550)}}$$

$$= 24.9.$$

Thus,

$$F'_b = F_b\left[1 - \frac{1}{3}\left(\frac{C_s}{C_k}\right)^4\right] = 1510 \text{ psi} \quad (10.4 \text{ MPa})$$

d. Substitute the stresses into interaction equation (10.27).

$$\frac{P/5.25}{300} + \frac{490}{1510 - 1.0(P/5.25)} \leq 1.0$$

Transposing and solving for the allowable load, P,

$$P = 990 \text{ lb} \quad (4.4 \text{ kN})$$

10.5 TIMBER FASTENINGS

The basic design philosophy outlined in Chapter 7 on connectors is valid for timber connections. That is, the complex stress distributions and connector loads in timber structural joints are predicted by simple statics and elastic strength of materials methods. Discrepancies between assumed and actual load distributions are accommodated by reducing allowable connector loads to levels found acceptable in practice.

Typical timber connectors include nails, spikes, screws, bolts, lag bolts, glue, split rings, shear connectors, and metal plate connectors. This text will discuss allowable loads for nails and spikes, bolts, glue, and metal plate connectors. The discussion is intended to be representative, rather than all inclusive, of the more common connectors used in agricultural construction.

Table 10.7. Nail and Spike Sizes.

Pennyweight	Length (in.)	Box Nails	Wire Diameter (in.)		Common Wire Spikes
			Common Wire Nails	Threaded Hardened-Steel Nails	
6d	2	0.099	0.113	0.120	—
8d	2½	0.113	0.131	0.120	—
10d	3	0.128	0.148	0.135	0.192
12d	3¼	0.128	0.148	0.135	0.192
16d	3½	0.135	0.162	0.148	0.207
20d	4	0.148	0.192	0.177	0.225
30d	4½	0.148	0.207	0.177	0.244
40d	5	0.162	0.225	0.177	0.263
50d	5½	—	0.244	0.177	0.283
60d	6	—	0.263	0.177	0.283
70d	7	—	—	0.207	—
80d	8	—	—	0.207	—
90d	9	—	—	0.207	—
5/16	7	—	—	—	0.312
3/8	8½	—	—	—	0.375

Reproduced with permission from *National Design Specification for Wood Construction*, 1977 ed. National Forest Products Association, Washington, DC.

10.5.1 Nails and Spikes

Probably the most familiar timber connectors are nails and spikes. Typical nail and spike sizes may be found in Table 10.7. Among the factors affecting the strength of nails and spikes in wood are nail diameter, wood species, depth of penetration of the nail into the wood members (particularly the penetration into the joint member containing the nail point), degree of seasoning, load duration, number of shear planes in the structural joint, degree of nail clinching, service conditions, and orientation of the nail with respect to the grain direction, i.e., whether the nail penetrates the member perpendicular to (side grain) or parallel to (end grain) the wood grain.

Allowable withdrawal resistances and lateral resistances of nails and spikes driven into the side grain of southern pine and Douglas fir lumber are given in Tables 10.8 and 10.9. Allowable loads for other species may be found in the NDS. The difference between withdrawal and lateral resistance is illustrated in Fig. 10.19. The resultant force induced by the upward wind force in Figure 10.19a must be resisted by the withdrawal resistance of the nail in the purlin. The withdrawal

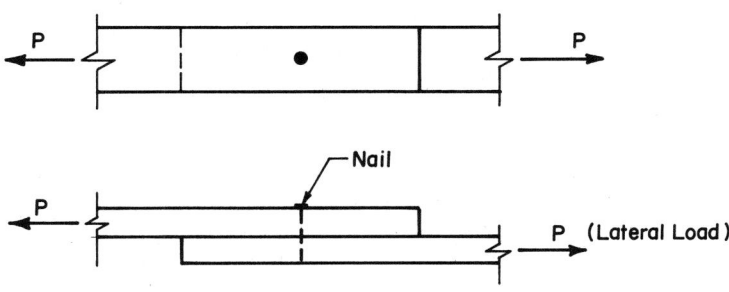

(a) Withdrawal Load And Resistance of Nails

(b) Laterally Loaded Nail

Fig. 10.19. Nail loads.

resistance is essentially induced by the frictional forces between the nail and the wood. Lateral resistance, illustrated in Fig. 10.19b, is the shear resistance in the plane normal to the nail axis.

The magnitude of the lateral resistance is controlled by crushing of the wood adjacent to the nail, by bending of the nail at the shear plane, or by tearing of the lumber in the member containing the point. Seldom do the nails fail in shear.

Table 10.8. Withdrawal Resistance of Nails and Spikes.[a]

	Specific Gravity G[b]						
	0.55	0.54	0.51	0.49	0.48	0.47	0.46
Size of Common Nail (pennyweight/diam.)							
6d/0.113	35	33	29	26	25	24	
8d/0.131	41	39	34	30	29	27	
10d/0.148	46	44	38	34	33	31	
12d/0.148	46	44	38	34	33	31	
16d/0.162	50	48	42	38	36	34	
20d/0.192	59	57	49	45	42	40	
30d/0.207	64	61	53	48	46	43	
40d/0.225	70	67	58	52	50	47	
50d/0.224	76	72	63	57	54	51	
60d/0.263	81	78	67	61	58	55	
Size of Threaded Nail[c] (pennyweight/diam.)							
30d/0.177	59	57	49	45	42	40	
40d/0.177	59	57	49	45	42	40	
50d/0.177	59	57	49	45	42	40	
60d/0.177	59	57	49	45	42	40	
70d/0.207	70	67	58	52	50	47	
80d/0.207	70	67	58	52	52	47	
90d/0.207	70	67	58	52	50	47	
Size of Box Nail (pennyweight/diam.)							
6d/0.099	31	29	25	23	22	21	20
8d/0.113	35	33	29	26	26	24	22
10d/0.128	40	38	33	30	28	27	25
12d/0.128	40	38	33	30	28	27	25
16d/0.135	42	40	35	31	30	28	27
20d/0.148	46	44	38	34	33	31	29
30d/0.148	46	44	38	34	33	31	29
40d/0.162	50	48	42	38	36	34	32
Size of Common Spike (pennyweight/diam.)							
10d/0.192	59	57	49	45	42	40	38
12d/0.192	59	57	49	45	42	40	38
16d/0.207	64	61	53	48	46	43	41
20d/0.225	70	67	58	52	50	47	45
30d/0.244	76	72	63	57	54	51	48
40d/0.263	81	78	67	61	58	55	52
50d/0.283	88	84	73	66	62	59	56
60d/0.283	88	84	73	66	62	59	56
5/16 in./0.312	97	92	80	72	69	65	62
3/8 in./0.375	116	111	96	87	83	78	74

Reproduced with permission from *National Design Specification for Wood Construction*, 1977 ed. National Forest Products Association, Washington, DC.

[a]Design values in withdrawal in pounds per inch of penetration into side grain of member holding point and are for normal load duration. d = pennyweight of nail or spike. G = specific gravity of the wood, based on weight and volume when oven dry.

[b]Specific Gravity of southern pine and Douglas fir ranges from 0.48 to 0.54.

[c]Loads for threaded, hardened steel nails, in 6d to 20d sizes, are the same as for common nails.

10.5 Timber Fastenings

The withdrawal resistances in Table 10.8 are valid for a single nail driven into the side grain, for seasoned or for unseasoned wood which will remain wet during its service life, and normal load duration. Regular duration factor increases are permissible. Note that the tabulated values are given in resistance per inch of nail penetration into the member holding the point. For nails driven into the end grain, the allowable withdrawal resistance is zero. For toenailed connections, the withdrawal resistance may be taken as two-thirds of the values in Table 10.8. While the withdrawal resistance of nails may be used in structural design, it is good practice to avoid withdrawal loads whenever possible.

The lateral resistances of a single nail or spike used in southern pine and Douglas fir wood is given in Table 10.9. The allowable loads are valid for joints loaded in single shear, normal load duration, nails driven into the side grain, seasoned lumber, and joints with nail penetration into the member containing the point equal to or greater than the 11 nail diameters. The lateral resistance does not increase if the nail penetration exceeds 11 diameters. Regular duration factor increases in allowable nail loads are permissible.

The lateral resistance values in Table 10.9 are subject to numerous adjustments which greatly expand their range of applicability. The more common adjustments follow.

If nail penetration is less than one-third the 11 nail diameters, the lateral resistance is zero. Straight-line interpolation between zero resistance at one-third (11 diameters) penetration and full tabulated resistance at 11 diameters penetration may be used for other penetrations.

When nails are used in double shear and when the nails fully penetrate all three wood members, as in Fig. 10.20, the allowable load per nail may be increased. If b, the thickness of the side members, equals $d/3$, the allowable load increases to 1.33 times the tabulated load. If $b \geq d$, the allowable load may be taken as 1.67 times the tabulated load. No further increase in load is allowed if $b > d$. Straight-line interpolation may be used for intermediate side member thicknesses. Note that when the double shear specification is used, the minimum penetration requirement is no longer applicable.

When a double shear joint is fabricated by properly clinching nails, the allowable load per nail is two times the tabulated load. A properly clinched nail requires that side member thickness be greater than or equal to $\frac{3}{8}$ in. (10 mm) and that the clinching length t be equal to at least 3 nail diameters. Clinching is limited to 12d and smaller common nails. Also, if threaded, hardened steel nails are used, the double shear provision may be used without clinching.

Table 10.9. Lateral Resistance of Single Nails and Spikes in Single Shear in Southern Pine or Douglas Fir.[a]

Nail Penetration (11D) and Allowable Load/Nail

Pennyweight (—)	Length (in.)	Box 11D (in.)	Nail P (lb)	Common 11D (in.)	Nail P (lb)	Threaded/Hardened Nails 11D (in.)	Threaded/Hardened Nails P (lb)	Common Spikes 11D (in.)	Common Spikes P (lb)
6d	2	1.09	51	1.24	63	1.32	63	—	—
8d	2½	1.24	63	1.44	78	1.32	78	—	—
10d	3	1.41	76	1.63	94	1.49	94	2.11	139
12d	3¼	1.41	76	1.63	94	1.49	94	2.11	139
16d	3½	1.49	82	1.78	108	1.63	108	2.28	155
20d	4	1.63	94	2.11	139	1.95	139	2.48	176
30d	4½	1.63	94	2.28	155	1.95	139	2.68	199
40d	5	1.78	108	2.48	176	1.95	139	2.89	223
50d	5½	—	—	2.68	199	1.95	139	3.11	248
60d	6	—	—	2.89	223	1.95	139	3.11	248

Reproduced with permission from *National Design Specification for Wood Construction*, 1977 ed. National Forest Products Association, Washington, DC.
[a] For normal load duration and for nails penetrating 11 diameters into the member holding the point.

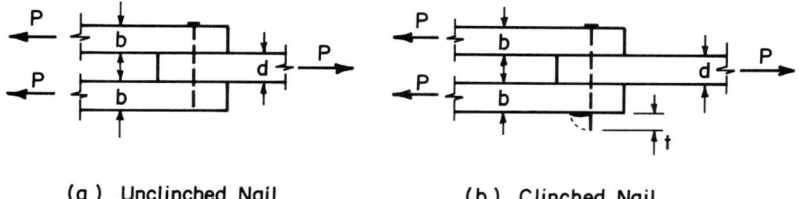

(a) Unclinched Nail (b) Clinched Nail

Fig. 10.20. Nails in double shear.

When nails are toenailed, lateral resistance is only five-sixths the tabulated values in Table 10.9.

Loads for nails used in wet or unseasoned lumber may be reduced as per factors in Table 10.10. The NDS notes that nails shall be spaced so as to prevent splitting. Some rules of thumb for nail spacing in sawn lumber are edge distance $>1\frac{1}{2}$ in. (38 mm), end distance in tension $>2\frac{1}{4}$ in. (57 mm), nail spacing parallel to grain $>2\frac{1}{2}$ in. (64 mm), and nail spacing normal to grain >1 in. (25 mm). The various distances and spacings are illustrated in Fig. 10.21.

Example 10.9. Nailed Joint. Determine the allowable load of the structural joint in Fig. 10.22 if the load is induced by a dead load plus wind load combination. The nails are driven from the gusset side of the connection.

Solution

a. Determine the allowable load per nail from Table 10.9 and increase it by 33% for wind load duration.

$$\frac{P_{\text{all}}}{\text{nail}} = 78 \times 1.33 = 104 \text{ lb} \quad (461 \text{ N})$$

if penetration into the 2×8 or 2×4 equals $11D$, or 1.44 in. (37 mm). Since the nail length equals $2\frac{1}{2}$ in. (64 mm) and the combined thickness of the 2×4 and 1×4 equals $2\frac{1}{4}$ in. (57 mm), penetration into the 2×4 is $>11D$ and the full allowable load may be used.

b. Determine the allowable load, P, for the joint based on the nail strength

$$P = \frac{P_{\text{all}}}{\text{nail}} \times \text{No. of nails in each grouping}$$

$$= 104 \times 9 = 940 \text{ lb} \quad (4.2 \text{ kN})$$

Table 10.10. Fastener Load Modification Factors for Moisture Content.

Type of Fastener	Condition of Wood[a]		Factor
	At Time of Fabrication	In Service	
Bolts	Dry	Dry	1.0
	Partially seasoned[b] or wet	Dry	c
	Dry or wet	Exposed to weather	0.75
	Dry or wet	Wet	0.67
Wire nails and spikes			
withdrawal loads	Dry	Dry	1.0
	Partially seasoned or wet	Will remain wet	1.0
	Partially seasoned or wet	Dry	0.25
	Dry	Subject to wetting and drying	0.25
lateral loads	Dry	Dry	1.0
	Partially seasoned or wet	Dry or wet	0.75
	Dry	Partially seasoned or wet	0.75
Threaded, hardened steel nails	Dry or wet	Dry or wet	1.0

Reproduced with permission from *National Design Specification for Wood Construction*, 1977 ed. National Forest Products Association, Washington, DC.

[a]Condition of wood definitions applicable to fasteners are: "Dry" wood has a moisture content of 19% or less. "Wet" wood has a moisture content at or above the fiber saturation point (approximately 30%). "Partially seasoned" wood has a moisture content greater than 19% but less than the fiber saturation point (approximately 30%). "Exposed to weather" implies that the wood may vary in moisture content from dry to partially seasoned, but is not expected to reach the fiber saturation point at times when the joint is under full design load. "Subject to wetting and drying" implies that the wood may vary in moisture content from dry to partially seasoned or wet, or vice versa, with consequent effects on the tightness of the joint.

[b]When bolts are installed in wood that is partially seasoned at the time of fabrication but that will be dry before full design load is applied, proportional intermediate values may be used.

[c]Use 1.0 for 1 or 2 bolts placed in a single line parallel to the grain or when bolts are in two or more lines with separate splice plates for each line. Otherwise use 0.4.

c. Check the adequacy of the gusset plate to carry the loads. For the 1×4,

$$P_{\text{all}} = (F_t \times 1.33)(A) = 900(1.33)(2.63)$$
$$= 3150 \text{ lb } (14.0 \text{ kN}) > 940 \text{ lb } (4.2 \text{ kN})$$

Thus, the gusset plate is adequate.

Example 10.10. Nail Joint in Double Shear. Repeat Example 10.9,

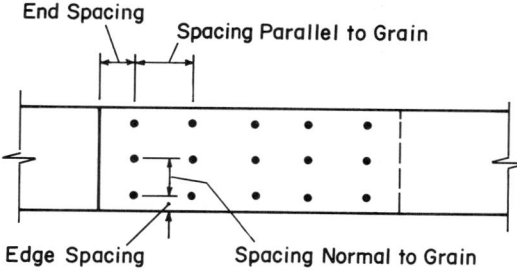

Fig. 10.21. Nail spacing and distance.

but add a second 1 × 4 gusset plate to the back side of the joint and use 10d nails.

a. Determine the allowable load per nail for double shear. The length of a 10d nail is 3 in. (76 mm) and fully penetrates the three members. Since the side member thickness b is equal to half the main member thickness, the allowable load per nail is between 1.33 and 1.67 times the tabulated values P_{tab} in Table 10.9. Interpolating between $b = d/3$ and $b = d$,

$$\frac{P_{all}}{nail} = \left[1.33 + (1.67 - 1.33)\left(\frac{d/2 - d/3}{d - d/3}\right)\right]P_{tab}$$

$$= 1.41 \, (94 \times 1.33) = 176 \text{ lb} \quad (783 \text{ N})$$

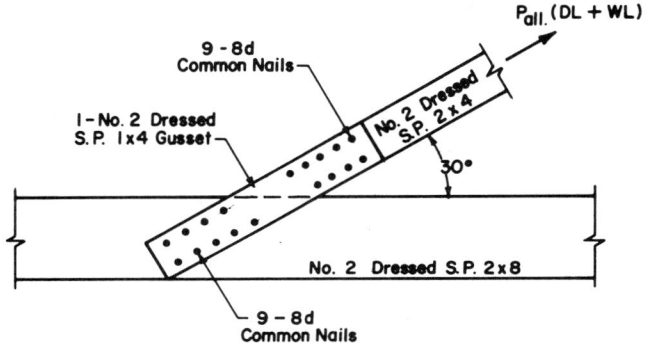

Fig. 10.22. Structural joint for Example 10.9.

b. Evaluate the joint allowable load, P.

$$P = 176(9) = 1580 \text{ lb} \quad (7.0 \text{ kN})$$

10.5.2 Bolts

The strength of bolted connections in timber depends upon bolt size, the ratio of bolt length in the main member to bolt diameter l/D, load duration, number of shear planes, types and sizes of side members, degree of seasoning, and angle between the wood grain and the direction of bolt loading.

The allowable lateral loads parallel and perpendicular to the grain for common bolts used in southern pine and Douglas fir wood are listed in Table 10.11. The tabulated values are valid for a single bolt used in a three-member joint with wood side members with dimension $b > l/2$ as shown in Fig. 10.23a. The values in Table 10.11 also assume normal load duration, seasoned wood, and adequate end, edge, and bolt spacings. Regular load duration increases are permissible in all allowable bolt loads.

When several bolts are used in a row, they are not all fully effective. That is, due to slippage and imperfect bolt placement, all bolts in a row are not loaded to the same level. When one bolt, usually one of the endmost two bolts in a row, reaches a load P, the remaining bolts are loaded to some level less than P. Consequently, the allowable load for a row of N bolts equals $K(P/\text{Bolt})(N)$ where K is less than 1.0 and is defined in Table 10.12.

When bolts are staggered, the number of bolt rows is dependent upon the pitch of the connections. If $a < b/4$ in Fig. 10.23b, the configuration would be defined as 1 row of 6 bolts. If $a > b/4$, the configuration would constitute 2 rows of 3 bolts.

Example 10.11. Rows of Bolts. Determine the allowable load for two 2 × 4s fastened together as a lap joint by 5 bolts in a row. Assume normal loading, adequate bolt spacings, and dry lumber. Let the allowable load per bolt equal P_s.

Solution. The cross-sectional areas of both the main and side members are equal and are less than 12.0 in.² (77.4 cm²). Thus, $A_1/A_2 = 1.0$ and from Table 10.12, $K = 0.85$.

Thus, $P_{\text{all}} = 0.85 P_s(5) = 4.25 P_s$ lb/bolt (N/bolt).

The allowable loads in Table 10.11 may be adjusted to accommodate a wide variety of conditions. Some of the more common follow.

When bolts are used in wet or unseasoned wood the allowable loads

Table 10.11. Allowable Bolt Loads (lb) for a Single Bolt in Double Shear.[a]

Length of Bolt in Main Member l(in.)	Diameter of Bolt D(in.)	l/D	Projected Area of Bolt $A = l \times D$(in.^{2})	Douglas Fir, Larch (Dense), Southern Pine (Dense)		Southern Pine		Douglas Fir, South	
				Parallel to Grain P	Perpendicular to Grain Q	Parallel to Grain P	Perpendicular to Grain Q	Parallel to Grain P	Perpendicular to Grain Q
$1\frac{1}{2}$	$\frac{1}{2}$	3.00	.750	1100	500	940	430	870	370
	$\frac{5}{8}$	2.40	.938	1380	570	1180	490	1090	420
	$\frac{3}{4}$	2.00	1.125	1660	630	1420	540	1310	470
	$\frac{7}{8}$	1.71	1.313	1940	700	1660	600	1530	520
	1	1.50	1.500	2200	760	1890	650	1750	570
2	$\frac{1}{2}$	4.00	1.000	1370	670	1170	570	1080	500
	$\frac{5}{8}$	3.20	1.250	1820	760	1550	650	1440	560
	$\frac{3}{4}$	2.67	1.500	2210	840	1890	720	1740	630
	$\frac{7}{8}$	2.29	1.750	2580	930	2200	790	2040	690
	1	2.00	2.000	2960	1010	2520	870	2330	750
$2\frac{1}{2}$	$\frac{1}{2}$	5.00	1.250	1480	840	1260	720	1170	620
	$\frac{5}{8}$	4.00	1.563	2140	950	1820	810	1690	710
	$\frac{3}{4}$	3.33	1.875	2710	1060	2310	900	2140	790
	$\frac{7}{8}$	2.86	2.188	3210	1160	2740	990	2530	860
	1	2.50	2.500	3680	1270	3150	1080	2910	940
3	$\frac{1}{2}$	6.00	1.500	1490	1010	1270	860	1180	750
	$\frac{5}{8}$	4.80	1.875	2290	1140	1960	970	1810	850
	$\frac{3}{4}$	4.00	2.250	3080	1270	2630	1080	2430	940
	$\frac{7}{8}$	3.43	2.625	3770	1390	3220	1190	2980	1040
	1	3.00	3.000	4390	1520	3750	1300	3460	1130
$3\frac{1}{2}$	$\frac{1}{2}$	7.00	1.750	1490	1140	1270	980	1180	870
	$\frac{5}{8}$	5.60	2.188	2320	1330	1980	1130	1830	990
	$\frac{3}{4}$	4.67	2.625	3280	1480	2800	1260	2590	1100
	$\frac{7}{8}$	4.00	3.063	4190	1630	3580	1390	3300	1210
	1	3.50	3.500	5000	1770	4270	1520	3950	1320
4	$\frac{1}{2}$	8.00	2.000	1490	1180	1270	1010	1180	960
	$\frac{5}{8}$	6.40	2.500	2330	1510	1990	1290	1840	1130
	$\frac{3}{4}$	5.33	3.000	3340	1690	2850	1440	2630	1260
	$\frac{7}{8}$	4.57	3.500	4450	1860	3800	1590	3510	1380
	1	4.00	4.000	5470	2030	4670	1730	4320	1510

(*continued*)

Table 10.11 Continued

Length of Bolt in Main Member l(in.)	Diameter of Bolt D(in.)	l/D	Projected Area of Bolt $A = l \times D$(in.^{2})	Douglas Fir, Larch (Dense), Southern Pine (Dense) Parallel to Grain P	Perpendicular to Grain Q	Southern Pine Parallel to Grain P	Perpendicular to Grain Q	Douglas Fir, South Parallel to Grain P	Perpendicular to Grain Q
$4\frac{1}{2}$	$\frac{5}{8}$	7.20	2.813	2330	1640	1990	1400	1840	1270
	$\frac{3}{4}$	6.00	3.375	3350	1900	2860	1620	2650	1410
	$\frac{7}{8}$	5.14	3.938	4530	2090	3870	1790	3580	1560
	1	4.50	4.500	5770	2280	4930	1950	4560	1700
	$1\frac{1}{4}$	3.60	5.625	7980	2670	6820	2280	6300	1990
$5\frac{1}{2}$	$\frac{5}{8}$	8.80	3.438	2330	1650	1990	1410	1840	1380
	$\frac{3}{4}$	7.33	4.125	3350	2200	2860	1880	2640	1720
	$\frac{7}{8}$	6.29	4.813	4570	2550	3900	2180	3600	1900
	1	5.50	5.500	5930	2790	5070	2380	4680	2080
	$1\frac{1}{4}$	4.40	6.875	8940	3260	7640	2790	7060	2430
$7\frac{1}{2}$	$\frac{5}{8}$	12.00	4.688	2330	1480	1990	1260	1840	1290
	$\frac{3}{4}$	10.00	5.625	3350	2130	2860	1820	2640	1800
	$\frac{7}{8}$	8.57	6.563	4560	2840	3890	2430	3600	2360
	1	7.50	7.500	5950	3550	5080	3030	4700	2800
	$1\frac{1}{4}$	6.00	9.375	9310	4450	7950	3800	7350	3310
$9\frac{1}{2}$	$\frac{3}{4}$	12.67	7.125	3360	1920	2860	1640	2640	1700
	$\frac{7}{8}$	10.86	8.313	4570	2660	3900	2270	3610	2260
	1	9.50	9.500	5950	3460	5080	2960	4700	2900
	$1\frac{1}{4}$	7.60	11.875	9300	5210	7950	4450	7340	4140
	$1\frac{1}{2}$	6.33	14.250	13410	6480	11460	5530	10590	4830
$11\frac{1}{2}$	$\frac{7}{8}$	13.14	10.062	4560	1980	3900	2060	3600	2170
	1	11.50	11.500	5950	3240	5080	2770	4690	2780
	$1\frac{1}{4}$	9.20	14.375	9300	5110	7950	4360	7340	4270
	$1\frac{1}{2}$	7.67	17.250	13410	7200	11450	6150	10580	5740
$13\frac{1}{2}$	1	13.50	13.500	5960	2410	5100	2530	4710	2680
	$1\frac{1}{4}$	10.80	16.875	9300	4860	7950	4160	7340	4130
	$1\frac{1}{2}$	9.00	20.250	13400	7070	11450	6040	10580	5920

Reproduced with permission from *National Design Specification for Wood Construction*, 1977 ed. National Forest Products Association, Washington, DC.

[a]Three (3) member joint in double shear. See Fig. 10.23a for definition sketch.

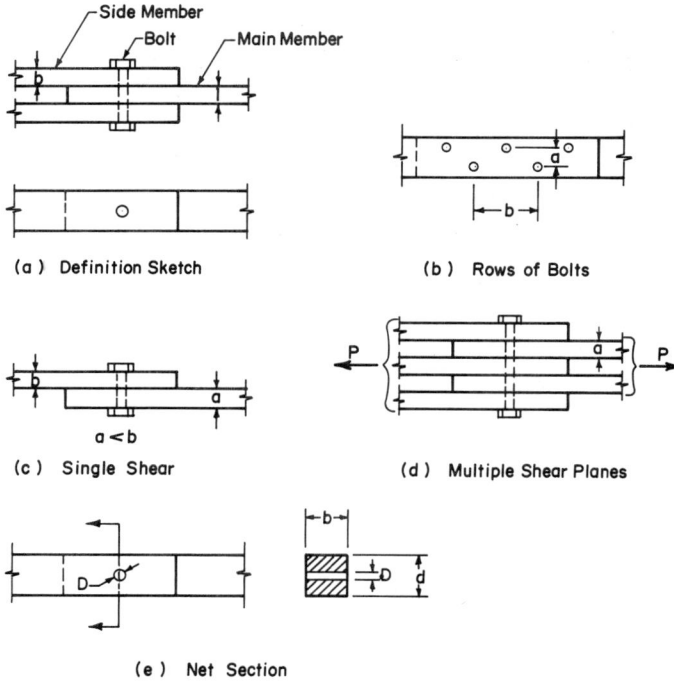

Fig. 10.23. Bolted connections.

are only 67% of those tabulated. If the side members are replaced by steel side members, the allowable loads parallel to the grain may be increased by 25% but no increase is permitted for allowable bolt loads normal to the grain.

If the side member thickness b of a three-member joint is greater than $l/2$ (see Fig. 10.23a), no increase in bolt loads is permitted. If $b < l/2$, however, the allowable bolt load is that tabulated for a main member thickness equal to $2b$.

The allowable bolt load for two-member joints (see Fig. 10.23c) is the smaller of one-half times the tabulated load for $l = a$ or one-half times the tabulated value for $l = 2b$, where $a \geq b$. If a joint is made up of more than three members each having the same thickness (Fig. 10.23d), the allowable connector load/bolt equals the number of shear planes times one-half the tabulated load for $l = a$. For the joint shown in Fig. 10.23d, $P = 4\,(^1/_2\,P_{\text{tab}}$ for $l = a)$.

When bolts are loaded at an angle to the grain, the allowable load per bolt is evaluated using Hankinson's equation. The allowable bolt

Table 10.12. K-Factor for Bolts in a Row for Two or Three Member Joints.

$A_1/A_2{}^a$	$A_1(\text{in.}^2)^b$	\multicolumn{5}{c}{Number of Fasteners in a Row}				
		2	3	4	5	6
Wood Sideplates						
$0.5^{c,d}$	<12	1.00	0.92	0.84	0.76	0.68
	12–19	1.00	0.95	0.88	0.82	0.75
	>19–28	1.00	0.97	0.93	0.88	0.82
	>28–40	1.00	0.98	0.96	0.92	0.87
	>40–64	1.00	1.00	0.97	0.94	0.90
	>64	1.00	1.00	0.98	0.95	0.91
$1.0^{c,d}$	<12	1.00	0.97	0.92	0.85	0.78
	12–19	1.00	0.98	0.94	0.89	0.84
	>19–28	1.00	1.00	0.97	0.93	0.89
	>28–40	1.00	1.00	0.99	0.96	0.92
	>40–64	1.00	1.00	1.00	0.97	0.94
	>64	1.00	1.00	1.00	0.99	0.96
Metal Side Plates						
2–12	25–39	1.00	0.94	0.87	0.80	0.73
	40–64	1.00	0.96	0.92	0.87	0.81
	65–119	1.00	0.98	0.95	0.91	0.87
	120–199	1.00	0.99	0.97	0.95	0.92
12–18	40–64	1.00	0.98	0.94	0.90	0.85
	65–119	1.00	0.99	0.96	0.93	0.90
	120–199	1.00	1.00	0.98	0.96	0.94
	200	1.00	1.00	1.00	0.98	0.97
18–24	40–64	1.00	1.00	0.96	0.93	0.89
	65–119	1.00	1.00	0.97	0.94	0.92
	120–199	1.00	1.00	0.99	0.98	0.96
	200	1.00	1.00	1.00	1.00	0.98
24–30	40–64	1.00	0.98	0.94	0.90	0.85
	65–119	1.00	0.99	0.97	0.93	0.90
	120–199	1.00	1.00	0.98	0.96	0.94
	200	1.00	1.00	0.99	0.98	0.97
30–35	40–64	1.00	0.96	0.92	0.86	0.80
	65–119	1.00	0.98	0.95	0.90	0.86
	120–199	1.00	0.99	0.97	0.95	0.92
	200	1.00	1.00	0.98	0.97	0.95
35–42	40–64	1.00	0.95	0.89	0.82	0.75
	65–119	1.00	0.97	0.93	0.88	0.82
	120–199	1.00	0.98	0.96	0.93	0.89
	200	1.00	0.99	0.98	0.96	0.93

Reproduced with permission from *National Design Specification for Wood Construction*, 1977 ed. National Forest Products Association, Washington, DC.

[a] A_1 is cross-sectional area of main member(s) before boring or grooving; A_2 is sum of the cross-sectional areas of side members before boring or grooving.
[b] When A_1/A_2 exceeds 1.0, use A_2 instead of A_1 for wood sideplates.
[c] When A_1/A_2 exceeds 1.0, use A_2/A_1 for wood sideplates.
[d] For A_1/A_2 between 0 and 1.0, interpolate or extrapolate from the tabulated values.

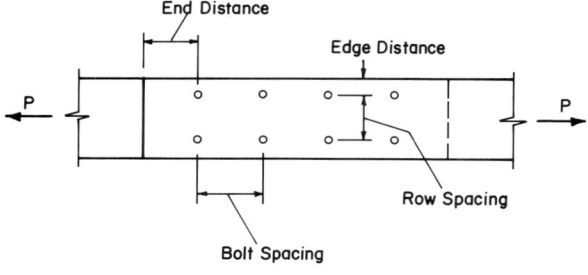

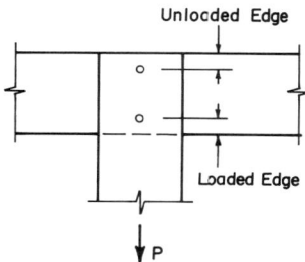

Fig. 10.24. Bolt, row, end, and edge spacing definitions.

loads parallel and perpendicular to the grain are substituted for the allowable compressive stresses in Eq. (10.3).

The net section remaining after the bolt hole has been drilled must be adequate to carry the member loads. The net section, A_{net}, in Fig. 10.23e is $bd - bD$. The design criterion for net section adequacy is $F_t > f_t = P/A_{net}$.

Minimum end, edge, and bolt spacings must be maintained in order to develop the tabulated allowable stresses for each bolt. The various distances are defined in Fig. 10.24. The criteria are as follows:

a. Bolts in a row must be placed at least four diameters apart to develop the full allowable load. Spacings between rows of bolts must be $1\frac{1}{2}$ diameters apart for parallel to grain loading and between $2\frac{1}{2}$ and 5 diameters apart for loading normal to the grain for bolt l/D ratios of 2 and 6, respectively. Bolt rows should not be greater than 5 in. (127 mm) apart.

b. End distances of 7 bolt diameters must be maintained for parallel-to-the-grain tensile loads. For parallel-to-the-grain compressive loads four bolt diameters are adequate. For loads normal to the grain, end distances of four bolt diameters are sufficient.

c. Required edge distances for parallel-to-the-grain loading in tension or compression are $1^1/_2$ bolt diameters except when $l/D > 6$, the greater of $1^1/_2$ bolt diameters or one-half the spacing between bolt rows is required. For loads normal to the grain the edge distance shall be four bolt diameters on the loaded edge and $1^1/_2$ bolt diameters on the unloaded edge.

Example 10.12. Bolted Connection. Determine the allowable load for the structural joint in Fig. 10.22 if the two-1 × 4 gussets are used and if two properly spaced $^1/_2$ in. (13 mm) diameter bolts are used in each joint member in place of the 18-8d nails. Assume normal load durations.

Solution

a. The joint is a three-member joint with $l = 1^1/_2$ in. (38 mm) and $b = l/2 = 0.75$ in. (19 mm). From Table 10.11, $P = 940$ lb (4.2 kN) and $Q = 430$ lb (1.9 kN) with $l/D = 3.0$.
b. From Table 10.12, for two bolts in a row, $K = 1.0$.
c. The 2 × 4 and gusset are loaded parallel to the grain. However, the load between the bolts and the grain of the 2 × 8 is at an angle of 30°. Using Eq. (10.3), the allowable load per bolt, N_{30}, is

$$N_{30} = \frac{PQ}{P\sin^2\theta + Q\cos^2\theta} = \frac{940(430)}{940 \sin^2 30 + 430 \cos^2 30}$$

$$= 720 \text{ lb/bolt (3.2 kN/bolt)}$$

and $P_{all} = 720(2) = 1440$ lb (6.4 kN).

d. Check the allowable load for the net section in the 2 × 4.

$$P_{all} = F_t(A_{net}) = 900 \, (5.25 - 0.750)$$

$$= 4050 \text{ lb (18.0 kN)} > 1440 \text{ lb (6.4 kN)}$$

Thus, the net section is adequate to carry the allowable bolt load.

e. Minimum spacing required to develop the allowable bolt loads.
   i. Bolt spacing in the $2 \times 4 = 4D = 2$ in. (51 mm).
   ii. Bolt spacing in the 2×8 equals, by conservatively assuming the load being perpendicular to the grain and interpolating between a row spacing of $2.5D$ and $5D$ for $l/D = 3$, $3.13D = 1.56$ in. (40 mm). The bolt spacing in the gusset is $4D = 2$ in. (51 mm). Thus, use 2 in. (51 mm).
   iii. End distance for the 2×4 measured along the center line of the 2×4 and for the gusset is $7D = 3.5$ in. (89 mm).
   iv. Edge distance in the 2×4 and gusset is $1.5D = 0.75$ in. (20 mm).
   v. Edge distance in the $2 \times 8 > 4D = 2$ in. (51 mm) to top (loaded) edge and $>1.5D = 0.75$ in. (19 mm) to the bottom (unloaded) edge.

10.5.3 Glue Joints

There are many glues and adhesives available. The most commonly used in agricultural structural applications are casein and resorcinol resin glues. Both glues are recommended for applications where moisture can be controlled. Resorcinol resin is recommended for applications in which high moisture content is likely to occur, e.g., in livestock buildings, and greenhouses.

Good glue joints require careful attention to detail. The surfaces to be glued must be dry (<15% moisture), clean and free of oil, and smooth and flat to assure intimate contact of the mating pieces. Gluing is best done when temperatures are 70°F (21°C) or greater, however, many glues can be used at temperatures as low as 40°F (4°C) if additional curing time is provided. Glues should be mixed according to manufacturer's recommendations with special care to assure the pot life is not exceeded. Glue should be spread uniformly over the mating areas. Mated parts should be held together under pressure with staples, box nails, or cement-coated nails.

The set time for casein glues varies from two hours to two days as temperature decreases from 80°F (24°C) to 40°F (4°C). Resorcinol resin glues are best when used at temperatures above 70°F (21°C) with pressure applied for 10 to 16 hours after application. Adequate pressure can be maintained with one 6d or 8d box nail for each 8 in.^{2} (52 mm^2) of glue area.

The strength of glue joints is dependent upon the orientation of the mating parts. When side grain is mated, the joint strength is as strong as the wood itself. Most glue joints are designed to transfer shear forces

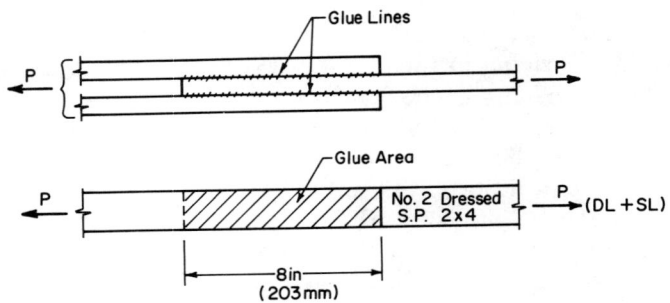

Fig. 10.25. Example 10.13—Glued joint.

in the shearing plane between the mated parts—the glue line. The shear strength of the glue line is much greater than the shear strength of the adjacent wood fibers. Thus, the wood adjacent to the glue line fails when the shear stress exceeds the allowable shear stress between the annual rings of the wood. This strength is the previously defined horizontal shear stress, F_v, of the wood. For most structural woods, F_v ranges from 65 to 105 psi (0.45–0.72 MPa). For Douglas fir and southern pine, F_v ranges from 85 to 105 psi (0.59–0.72 MPa). The shear strength of side grain mated glued joints is equal to F_v times the glue area and should be adjusted for load duration.

The strength of end-to-end grain or end-to-side grain glued joints is much lower than side-to-side grain joints. Special construction and reinforcing techniques, such as scarfing, dowels, tenons, or gussets, are required for adequate strength.

Example 10.13. Glued Joint. Determine the allowable load for the glued tension joint shown in Figure 10.25. Assume a $DL + SL$ combination.

Solution

    a. Evaluate the allowable strength of the glue line F_g after adjustment for load duration.

$$F_g = F_v \times 1.15 = 95 \times 1.15 = 109 \text{ psi} \quad (0.75 \text{ MPa})$$

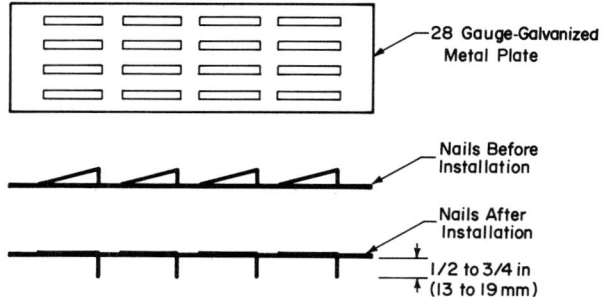

Fig. 10.26. Schematic of a metal plate connector.

b. Evaluate the allowable glue line load, P_g

$$P_g = F_g A = 109(8 \times 3.5 \times 2) = 6100 \text{ lb} \quad (27.1 \text{ kN})$$

c. Evaluate the allowable member load, P, from Eq. 10.4 and Table 10.2.

$$P = F_t A = 900(1.15)(5.25) = 5430 \text{ lb} \quad (24.2 \text{ kN})$$

Thus, the allowable load of this joint is controlled by the stress in the 2 × 4 and equals 5430 lb (24.2 kN). When a double shear joint, as in Fig. 10.25, is field constructed, it is often assumed that only one glue line is effective. This seemingly conservative approach is a hedge against the possibility of a poorly bonded glue line on one of the sides. If this practice were followed in Example 10.12, an allowable glue line load of 6100/2 = 3050 lb (13.6 kN) would be the limiting load. For an efficient utilization of the 2 × 4 strength the glue length should then be increased from 8 in. (203 mm) to 8(5430/3050) = 14.25 in. (362 mm).

10.5.4 Metal Plate Connectors

A common connector used in agricultural construction is the metal plate connector. Its most common application is in the construction of prefabricated truss joints. There are many different geometries and configurations of metal plate connectors in the market place.

A metal plate connector is essentially a battery of short nails, as shown in Fig. 10.26, created from a thin flat steel plate by a punching

424 Light Timber Design

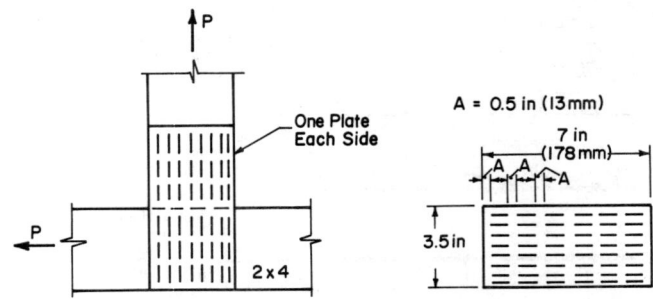

Fig. 10.27. Example 10.14—Metal plate connected joint.

process. The primary advantage of the battery of nails is a reduction of labor requirements in joint construction. Many truss joints, for example, require 40 or more nails. By using a metal plate connector and a special press, many nails can be inserted quickly with a single plate rather than with discrete fasteners.

Metal plate connectors are typically constructed of 28 gauge steel with $F_y > 23{,}000$ psi (158 MPa), $F_u > 45{,}000$ psi (310 MPa), and a minimum ductility of 20% elongation in 2 in. (51 mm) at F_u. Nails are typically $\frac{1}{2}$–$\frac{3}{4}$ in. (13–19 mm) long with lateral resistances of 40–60 lb (180–270 N) per nail. The allowable strength per nail is specified by the manufacturer. Special care must be taken to assure that the net section of the plate is adequate to carry the loads. Manufacturers usually provide this information by specifying the allowable tensile load for the plate.

Metal plate connectors are most effective for transmitting axial forces. They are seldom recommended for transfer of moments. Usually, metal plate connectors are used in pairs, one on each side of a joint, to reduce the potential for the nails to tear out when noncentric loads are applied to one or more of the connected parts.

Example 10.14. Metal Plate Connector. Evaluate the allowable load for the metal plate connection shown in Fig. 10.27. Assume the net section of the plate is adequate, a normal duration load, and an allowable load per nail of 40 lb (180 N). The metal plate has 6 rows of 8 nails each.

Solution. If a row of nails is closer than $\frac{1}{4}$ in. (6 mm) from the edge or end of a member, the row is considered to be ineffective. With careful placement of the plate, all the nails are effective. Thus, the allowable

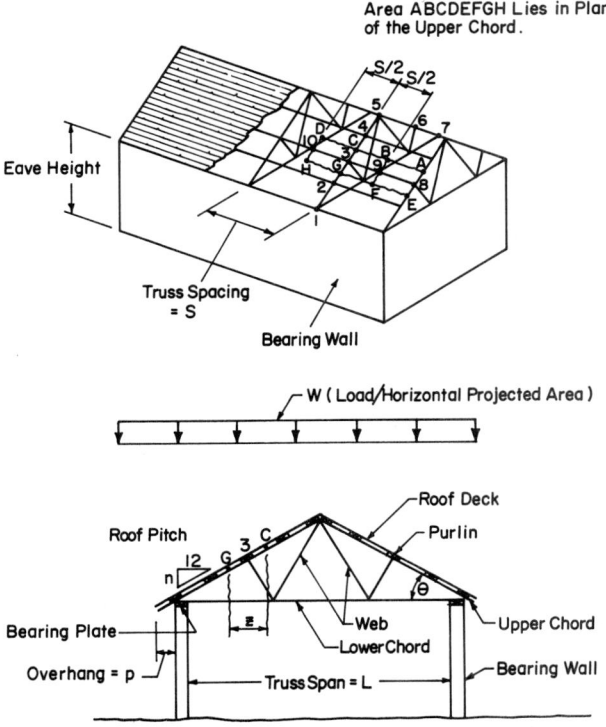

Fig. 10.28. Roof truss framing definition sketch.

load P equals the allowable load per nail times the total number of nails in each connected part.

$$P = (P/\text{nail})(N) \quad (10.31)$$
$$= 40(8 \times 3 \times 2) = 1920 \text{ lb} \quad (8.5 \text{ kN})$$

10.6 TIMBER TRUSSES

Roof trusses are an integral part of many types of timber framing, including pole-type construction, post–frame construction, and common stud framed buildings. This section will discuss the component parts of, load transfer in, and the computational procedures for analysis of typical roof truss systems.

A typical roof truss framing system resting on a bearing wall is illustrated in Fig. 10.28. The bearing walls are often masonry walls or stud walls with a 2 × 8 or 2 × 4 bearing plate on the top. The truss consists of upper and lower chords and web members connected at the joints by nails, bolts, glue, or metal plate connectors. Purlins fastened to the upper chord and running perpendicular to the plane of the trusses are nailed to the upper chords. Purlin spacings of 16 to 24 in. (406 to 610 mm) on center are typical and are usually dictated by the load capacity of the roof decking material. The purlins support the roof deck which may be plywood, steel, or aluminum roofing. Not shown in Fig. 10.28 are the lateral bracing components required for stability. Bracing requirements are the topic of a later discussion.

The function of each component is to transfer dead, live, wind, and snow loads from their point of application to the bearing wall and then to the foundation and ground. Figure 10.28 helps to illustrate how loads are transferred and also assists in classifying each member with respect to its primary stress, i.e., compression, tension, flexure, or combined stress. Purlin 3–8 carries the total load acting on the horizontal projected area of area $\overline{ABCEFG}$. If the horizontal spacing between purlins is z, each purlin carries a uniformly distributed vertical load of wz. For trusses of low slope, the purlin can be considered to be a flexural member with a uniform transverse load of $(wz) \cos \theta$. The purlin 9–10 transfers the loads acting on area $\overline{BCDFGH}$ to point 3 on the truss. The magnitude of the load at point 3 equals $(wz)s$. A load equal to wzs is transferred at points 2, 4, 5, and 6, while the loads at points 1 and 7 equal $wzs/2 + wsp$, where p is the horizontal projection of the overhang. Loads on the end trusses will be one-half of those on the intermediate trusses. It is apparent now that the upper chord of the truss in Fig. 10.28 is a beam-column since it carries the axial compression loads from truss analysis and also the flexural loads induced by the purlins located between panel points. The truss transfers the load from the area midway between adjacent trusses. Thus, each intermediate truss transfers a load $wsL + 2 wsp$ to the bearing walls. Since both the loading and truss geometry are symmetric in this illustration, the reactive forces at each bearing wall are $wsL/2 + wsp$.

During its service life, a truss is subjected to many combinations of loading. The American Institute of Timber Construction (AITC) recommends that the designer check the six load combinations listed in Table 10.13. In many agricultural buildings without storage loads it is sufficient to check the two $DL + SL$ combinations in Table 10.13 and the $DL + WL$ combination. Recent practice also suggests checking an unbalanced $DL + SL$ combination with DL on the windward slopes and $DL + 1.5SL$ on the leeward slope for gable roofs with slopes between 15° and 70°.

Table 10.13. Load Combinations for Roof Truss Analysis.

Windward Slope	Leeward Slope	Load Duration	Load Duration Factor
DL	DL	Permanent	0.90
DL + LL	DL + LL	7 days	1.25
DL	DL + LL	7 days	1.25
DL + SL	DL + SL	2 months	1.15
DL + 0.5SL	D + SL	2 months	1.15
DL + WL	DL + WL	1 day	1.33

Reproduced with permission from *Timber Construction Manual*, 2nd ed. 1974. American Institute of Timber Construction, Englewood, CO.

The designer must always be alert for the occurrence of stress reversals in a roof truss. Since wind loads can act outward and since dead loads are usually very low in agricultural buildings, it is possible for the sense (tension or compression) of one or more member forces to change with load combination. This change in sense is called stress reversal.

Stress reversals are critical to truss design because members are usually long and slender. Thus, a truss member with a slenderness ratio near 50 may be adequate to carry a moderately large tensile force, but inadequate to carry even a small compressive stress. The following example will illustrate the role of stress reversals.

Example 10.15. Truss Analysis (Stress Reversal). Assume the roof truss in Fig. 10.29a carries a $DL = 20$ lb/ft (290 N/m) of horizontal projected length on the upper chord, a full design snow load of 110 lb/ft (1.60 kN/m) of horizontal projected length, and an outward acting wind load of 30 lb/ft (440 N/m) and 50 lb/ft (730 N/m) of roof length on the left and right slopes, respectively. Assume all loads to be applied at the panel points of the truss and evaluate member loads for the balanced $DL + SL$, the unbalanced $DL + SL$, and the $DL + WL$.

Solution

a. The truss load diagrams for each load combination are shown in Fig. 10.29b, 10.29c, and 10.29d. The truss overhang is assumed to be zero in this example. The end restraints at the sills were assumed to be pinned at both reaction points. The horizontal reactive forces were assumed to be equal.

b. The member forces are summarized in Table 10.14. Notice that every member in this truss undergoes a stress reversal. Thus, each member must be analyzed as both a tension and a compression member. Member *GF* is an excellent case in point. It appears

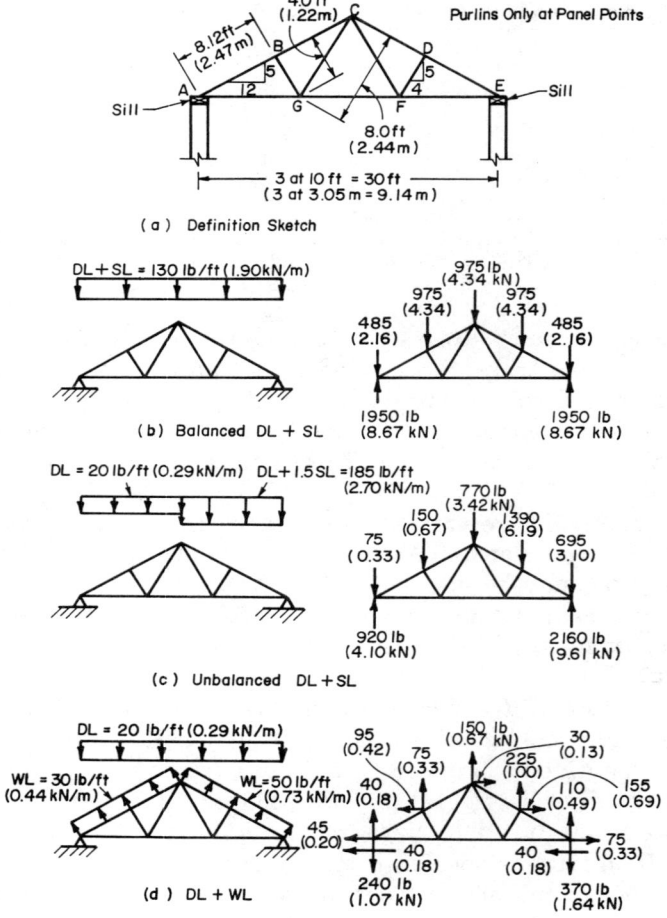

Fig. 10.29. Example 10.15—Truss analysis-stress reversal.

that the tensile force of 2340 lb (10.4 kN) will control its design. However, the relatively small compressive force of 250 lb (1.1 kN) significantly influences the final design since the slenderness ratio of GF is quite large and will require some lateral bracing to keep $L/d < 50$.

c. Since the wind may come from the right or left during the service life of the truss, design forces for both member AB and DE are the larger of the forces acting on AB and DE; i.e., 3800 lb (16.9 kN) compression and 660 lb (3.0 kN) tension. The same argument holds for other symmetrically placed members of the truss.

Table 10.14. Member Loads for Several Load Combinations.[a]

Member	DL + SL lb	(kN)	Unbalanced DL + SL lb	(kN)	DL + WL lb	(kN)
AB	−3800	(−16.9)	−2200	(− 9.8)	+510	(+2.3)
AG	+3510	(+15.9)	+2030	(+ 9.0)	−390	(−1.8)
BG	− 940	(− 4.2)	− 150	(− 0.7)	+110	(+0.5)
GC	+ 940	(+ 4.2)	+ 150	(+ 0.7)	−110	(−0.5)
BC	−3170	(−14.1)	−2100	(− 9.4)	+540	(+2.4)
GF	+2340	(+10.4)	+1850	(+ 8.2)	−250	(−1.1)
CF	+ 940	(+ 4.2)	+1330	(+ 5.9)	−260	(−1.1)
CD	−3170	(−14.1)	−2900	(−12.9)	+600	(+2.6)
DF	− 940	(− 4.2)	−1330	(− 5.9)	+260	(+1.1)
DE	−3800	(−16.9)	−3800	(−16.9)	+660	(+3.0)
FE	+3510	(+15.6)	+3500	(+15.5)	−570	(−2.5)

[a]Compressive forces are negative; tensile forces are positive.

d. Note also that the reactive force for the $DL + WL$ combination is downward due to the wind uplift. This illustrates the need for anchoring the heel joint of the truss securely to the sill plate. Toe nailing, a common practice for anchoring heel joints, is seldom sufficient to resist uplift forces in light agricultural structures.

The truss in Example 10.15 assumed the truss was loaded at panel points only. This does not occur in practice as purlins are normally spaced 16 to 24 in. (406 to 610 mm) on center. If the truss in Fig. 10.29 has purlins spaced 24 in. (406 mm) on center, member AB which is 8.12 ft (2.47 m) long would have approximately three uniformly spaced purlins between A and B, B and C, C and D, and D and E. Thus, the upper chord members have an axial stress component and a flexural stress component. The flexural stress components are called secondary stresses.

The axial stress component may be estimated by assuming all loads act at panel points. Thus, the design axial loads for the truss in Example 10.14 are those in Table 10.14. The secondary stresses may be approximated by one of several methods. Two of the more common manual methods are described herein. Computer analyses yield a more exact solution.

A conservative method is to assume the joints to be pinned and the upper chord between panel points to be a simple beam with concentrated flexural loads applied at each purlin location. The method is illustrated in Fig. 10.30 for the full $DL + SL$ acting on member AB of the truss of Example 10.15. Member AB, for a $DL + SL$ combination,

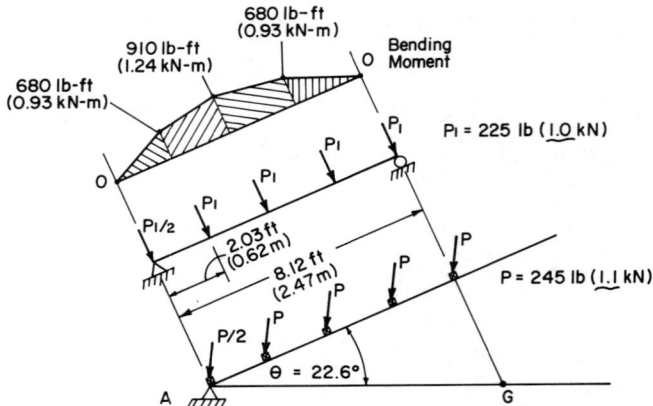

Fig. 10.30. Upper chord load distribution for a conservative estimate of the flexural (secondary) stresses in timber truss members.

must carry a compressive axial load of 3800 lb (16.9 kN) and a bending moment of 910 lb-ft (1.24 kN-m). The interaction equation for combined flexure and axial compression would be used to size the member.

An alternate, and more realistic, method of analysis is to assume some continuity, and thus some end moments, at the joints. If the upper and lower chords are continuous over the web supports, the Truss Plate Institute (TPI) recommends that secondary stresses be estimated by

$$M = \frac{1}{8}w(QL)^2 \tag{10.32}$$

where L is defined by L_i or L_a in Fig. 10.31 and Tables 10.15 and 10.16, and Q is defined for upper chords in Table 10.15 and for lower chords in Table 10.16. The distributed load term w for upper chord moments includes only the loads acting on the upper chord. Similarly, the distributed load term for lower chord moments includes only the appropriate ceiling loads acting on the lower chord.

The effective lengths recommended by TPI for evaluating allowable compressive stresses in lower chords, upper chords, and web members are summarized in Fig. 10.32. Note that purlins, bracing, sheathing, and ceiling construction all provide some lateral support in the plane normal to the truss and influence the various L/d ratios. Note also that the joint construction and member end restraints also reduce the effective L/d of the members.

To illustrate the procedures, members AB, GF, and BG of the truss

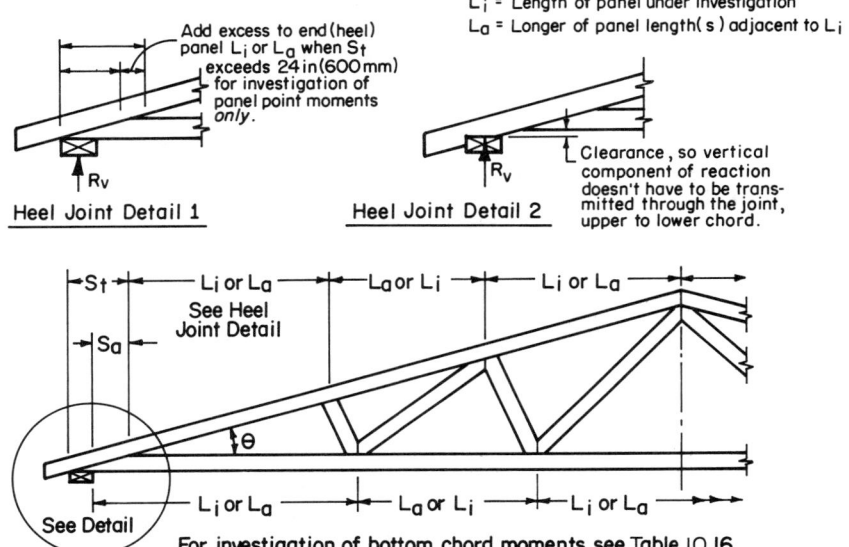

Fig. 10.31. Definition sketch for estimating moments in truss members. Reproduced with permission from *Design Specifications for Metal Plate Connected Wood Trusses* (TPI-78), 1978 ed., Truss Plate Institute, Frederick, MD.

Table 10.15. Q and L Factors for Secondary Stresses in Upper Chords of Trusses.

	Panel Point Moment		Midpanel Moment	
	Q	L	Q	L
One panel	Not applicable	Not applicable	0.90	$L_i + S_a$
Two panels[a]	0.90		$0.58(\cot\theta)^{0.23}$	
Two or three panels[b,c,d]		Largest of $0.9L_i$, $(L_i + L_a)/2$, or $0.9L_a$		Largest of $0.9(L_i + cS_a)$, $[(L_i + L_a)/2] + cS_a)$, or $0.9(L_a + cS_a)$
Three panels[a]	0.85		$0.58(\cot\theta)^{0.36}$	

Reproduced with permission from *Design Specifications for Metal Plate Connected Wood Trusses* (TPI-78), 1978 ed. Truss Plate Institute, Frederick, MD.
[a] For the midpanel moment, $Q = \alpha (\cot\theta)^\beta$, but shall not be < 0.74.
[b] For panel point moment, if $S_t > 24$ in. (610 mm), add excess to end (heel) panel L_i or L_a (see Fig. 10.31).
[c] For midpanel moment, $S_a = S_t - B$ but not < 0. cS_a shall be added only to the length of the end (heel) panel.
[d] For midpanel moment, $c = 0.5$ for two panels; $c = 0.33$ for three panels; if neither L_i nor L_a are end (heel) panels, then $cS_a = 0$.

Table 10.16. Q and L Factors for Secondary Stresses in Lower Chords of Trusses.

	Q	L
One panel	1.0	L_i
Two+ panels	1.0	L is largest of 0.9_i, $(L_i+L_a)/2$, or $0.9L_a$

Reproduced with permission from *Design Specifications for Metal Plate Connected Wood Trusses* (TPI-78), 1978 ed. Truss Plate Institute, Frederick, MD.

of Example 10.15 and Fig. 10.29 with purlins located 24.4 in. (620 mm) o.c. (on center), as in Fig. 10.30, will be designed. It will be assumed that dressed No. 2 southern pine (surface and used at 15% moisture content) is used and the nominal thickness of the lumber is 2 in. (51 mm).

Using the conservative procedure for secondary stress calculations, member AB must simultaneously carry a compressive load of 3800 lb (16.9 kN) and a moment of 910 lb-ft (1.24 kN-m). Assuming the purlins are tied into the end wall framing such that they adequately brace the upper chord in the plane normal to the truss, $C_s = (L_e d/b^2)^{1/2} = 10.7$ and 12.3, respectively, for a 2×6 and a 2×8. Thus, $F'_b \approx F_b$. For compression the slenderness ratio in the plane normal to the truss equals $24.4/1.5 = 16.3$. In the plane of the truss, from Fig. 10.32, the slenderness ratio equals $0.9 L_{AB}/h$. It is easily shown that for any upper chord equal to or larger than a 2×6, $(L/d)_{\max} = 16.2$.

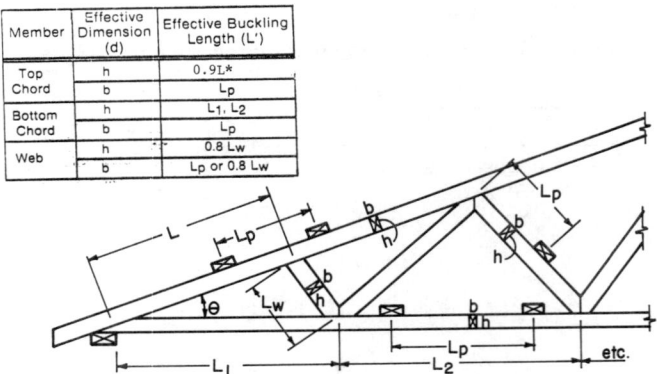

Fig. 10.32. Effective buckling lengths of truss members. Where sheathing or ceiling is nailed to a compression chord (either top or bottom), lateral support of the supported axis may be assumed to be continuous and $L'/d = L_p/b$ may be neglected. Asterisk indicates conservative values. Reproduced with permission from *Design Specifications for Metal Plate Connected Wood Trusses* (TPI-78), 1978 ed., Truss Plate Institute, Frederick, MD.

As a first estimate for sizing AB, assume the upper chord to be a 2×8. From Table 10.2 and Eq. (10.22),

$$F'_b = 1300 \times 1.15 \left[1 - \tfrac{1}{3} \left(\frac{12.3}{25.3} \right)^4 \right] = 1465 \text{ psi} \quad (10.1 \text{ MPa})$$

Also, from Table 10.2 and Eq. (10.7),

$$F'_c = 1200(1.15) \left[1 - \tfrac{1}{3} \left(\frac{16.2}{22.8} \right)^4 \right] = 1260 \text{ psi (8.6 MPa)},$$

and from Eq. (10.27), $J = (16.2 - 11)/(22.8 - 11) = 0.44$. Substituting these values into Eq. (10.27) and using the dimensions of a 2×8 yields

$$\frac{3800/10.88}{1260} + \frac{914 \times 12/13.4}{1465 - 0.44(349)} < 1.0$$

$$0.28 + 0.64 = 0.92 < 1.0$$

The left side of the interaction equation is nearly unity. It is obvious that the next smaller standard size (2×6) would not be satisfactory. Thus, a 2×8 is selected for AB. The 2×8 can carry a tensile load of $(0.8)(675)(1.15)(10.88) = 6750$ lb (30.0 kN). The 2×8, therefore is also adequate for the $DL + WL$ combination. Due to symmetry, member AB must also carry the loads acting on member DE. It is suggested that the student show that a 2×10 is indeed required to carry the unbalanced $DL + SL$ in AB.

Member BG must carry either a compressive load of 940 lb (4.2 kN) or a tensile load of 260 lb (1.1 kN). The effective length of BG equals $0.8 L_w$ from Fig. 10.32. The approximate length of L_w equals 4.0 ft (1.22 m) minus half the width of both the upper and lower chords. Thus, assuming a 2×4 lower chord and using a 2×10 upper chord $L_w = 4 \times 12 - 9.25/2 - 3.5/2 = 41.6$ in. (1.057 m). Then $0.8 L_w$ equals 33.3 in. (846 mm) yielding a maximum slenderness factor of $33.3/1.5 = 22.2$. Assuming the member is smaller than a 2×6, $K = 0.67\sqrt{E/F_c} = 23.3$, and from Eq. (10.7), $F'_c = 1150(1.15)[1 - \tfrac{1}{3}(22.2/23.3)^4] = 960$ psi (6.6 MPa). Since the allowable load for a 2×4 is $960(5.25) = 5040$ lb (22.3 kN), a 2×4 is more than adequate. Investigating a smaller size member for BG, a 1×4 web member has an $L/d = 45.5 > K$, $F'_c = 0.3(1.6 \times 10^6)/(45.5)^2 = 230$ psi (1.6 MPa) and $P_{\text{all}} = 230(2.63) = 605$ lb (2.7 kN) and is therefore inadequate for the web member.

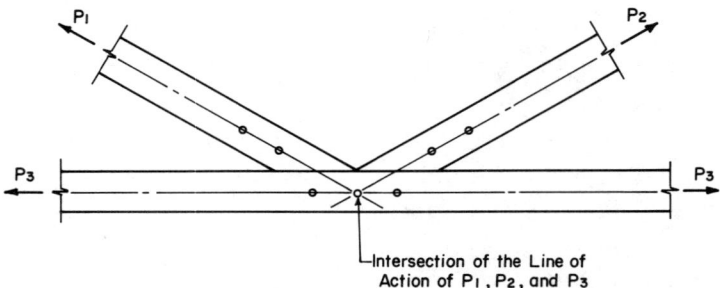

Fig. 10.33. Concentric structural joint.

Member *GF* must carry a tensile load of 2340 lb (10.4 kN) and a compressive load of 250 lb (1.1 kN). The small compressive load under $DL + WL$ may create difficulty since, for a 1.5 in. (38 mm) thick member, $(L/d)_{GF} = 120/1.5 = 80$ about the weak axis. From Eq. (10.8), $F'_c = 0$. If lateral bracing is applied at the midlength of *GF*, $L/d = 40$ and $F'_c = 260$ psi (1.8 MPa). The required cross-sectional area for *GF* equals $P/F'_c = 250/260 = 0.96$ in.^{2} (6.2 cm^2). From Table 10.4, a 2×4 is the smallest standard $2\times$ member to carry the load. Checking the 2×4 for tensile loads, $F_t = 1050(1.15) = 1210$ psi (83 MPa), the required cross-sectional area is $2340/1210 = 1.93$ in.^{2} (12.49 cm^2), and a 2×4 is adequate. A 1×4 could carry the tensile load, but requires additional bracing for the compressive load since $L/d > 50$. Thus, a 2×4 is selected for *GF*.

The entire truss can be designed by similar analysis. The upper chord is usually of one size in agricultural buildings. Thus, the entire upper chord can be sized by observing which segment carries the largest load combinations. Similarly, the lower chord is usually of constant cross section, and the cross section of all the webs are usually equal.

Once the members are sized, the joints and splices must be designed using the procedures of Chapter 7 on connectors and the allowable connector load provisions of this chapter. Care should be taken to place chord splices at approximately one-third the distance between panel points. This assures the splice to be near the inflection point where the bending moment is zero. Also, care should be exercised to assure that the connectors are placed to produce a concentric joint. That is, the line of action of the resultant force of each connector group should meet at a point. Figure 10.33 illustrates a concentric structural joint.

Lateral bracing of the truss is critical to the design. As noted in the

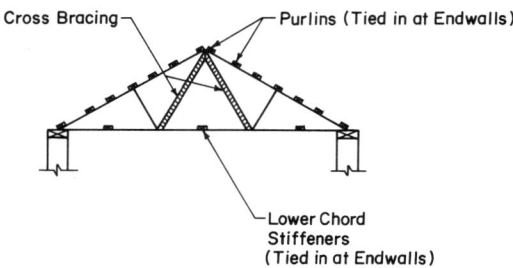

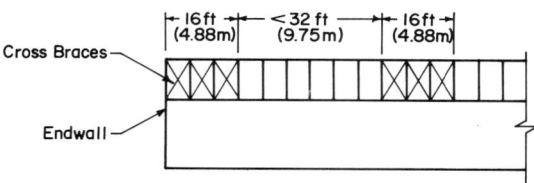

Fig. 10.34. Cross bracing and stiffeners. Reproduced with permission from *Structures and Environment Handbook,* MWPS-1, 11th ed., Midwest Plan Service, Ames, Iowa.

preceding analysis, upper and lower chords and web members must often be braced to reduce L/d ratios and to assure stability of the truss members. Lateral bracing is also required to assure stability of the entire roofing system both during and after construction. A comprehensive recommendation for bracing roof truss systems may be found in TPI publication BWT-76, *Bracing Wood Trusses: Commentary and Recommendations*. The minimum recommendation for cross bracing includes lower chord stiffeners, a diaphragm or purlins on the upper chords, and intermittent cross bracing between adjacent trusses. These braces are schematically illustrated in Fig. 10.34. Purlins are usually 1 × 4s or 2 × 4s laying flat for narrow truss spacings and 2 × 4s on edge for wider truss spacings. Lower chord stiffeners are typically 2 × 4s laying flat and attached to the top edge of the lower chord. One stiffener should be located at each lower chord panel point. Cross braces are usually 2 × 4s. Attention must be given to the L/d ratios of the stiffeners and cross bracing. A brace with $L/d > 50$ will be ineffective for resisting lateral forces if it must carry compressive forces.

A deflection analysis of the truss should be conducted to assure that under full load the lower chord panel points do not deflect below a straight line drawn between the heel joints. This requires that the lower chords be fabricated with a camber at least as great as the full load deflection of the lower chord panel points.

10.7 POLE BUILDINGS

The pole frame is probably the most common structural system in agricultural buildings in the United States. Pole frames are used in storage facilities for produce, machinery storages, shops, livestock facilities, and feed storage facilities to name but a few applications. Timber is still the most common structural material used in pole frames.

The unique feature of pole-framed structures is the multiple role played by the vertical support columns. (The vertical columns are typically called poles when of round cross section and posts when of square or rectangular cross section.) The vertical columns, which are embedded into the ground 3.5 ft (1.07 m) or more, not only transmit the service loads from the truss to the ground but they also provide resistance to overturning and sliding. In essence, the pole embedment provides the foundation and a portion of the lateral bracing and stability for the entire building.

The component parts of a typical pole-framed structure are shown in Fig. 10.35. The roof truss framing is similar to that described in the preceding section and Fig. 10.28. The trusses rest on the plates which usually consist of one or two 2 × 6, 2 × 8, or 2 × 10 members. The plate size is dependent upon the truss span, the bay length (pole spacing), and the design loads. The plates act as flexural members and transfer the truss reactive forces to the pole tops. The plates are typically connected to the poles by threaded pole barn spikes or bolts. Often the pole tops are notched to allow the bottom edge of the plate to bear on the pole. This practice greatly reduces connector requirements and improves the reliability of the plate-to-pole connection.

Since poles are typically spaced 4, 8, 12, or 16 ft (1.22, 2.44, 3.66, or 4.88 m) apart, horizontal nailing girts are required for attachment of the siding. The girts are primarily flexural members and transfer lateral wall loads to the poles.

The poles act as pure compression members for gravity loads. However, when the building is laterally loaded, as by wind or storage loads, the pole is subject to combined axial and flexural loads. In most applications the pole size requirement is controlled by the lateral loads.

The poles rest on a concrete footer pad. Typical pad sizes are 6 to 8

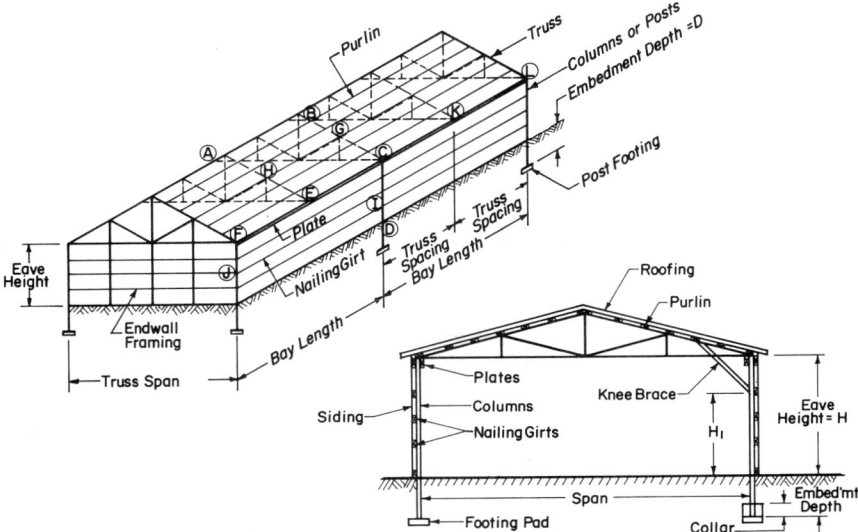

Fig. 10.35. Typical timber pole framing.

in. (152 to 203 mm) thick and 12 to 20 in. (305 to 508 mm) in diameter. Often the footer extends about 12 in. (305 mm) above the pole butt as shown in the right-hand side of the cross section in Fig. 10.35. If pole barn spikes are partially driven into the bottom of the pole, the footer becomes an integral part of the pole with increased resistance to uplift forces.

The knee brace is optional in pole-framed buildings. However, a properly installed knee brace stiffens the joints between the pole tops and the heel of the truss. The knee brace provides additional lateral stability by reducing the lateral movement and rotation of the pole top resulting in significantly reduced pole stresses and size requirements. Since the poles are the most costly component of the pole framing, addition of knee braces can significantly reduce building costs.

The embedded portion of the pole is in contact with the ground and subject to fungal attack and deterioration. Poles should, therefore, be treated with preservatives. Typical preservatives include creosote, pentachlorophenol, and nonleaching waterborne salt preservatives.

The pole frame is obviously an indeterminate structure. The frame in the cross section of Fig. 10.35 has a total of six end reactions. If the connection between the top of both poles and the truss is pinned, two equations of condition ($M = 0$) exist at the top of the pole and the frame is one degree indeterminate. If the connection between the top

438 Light Timber Design

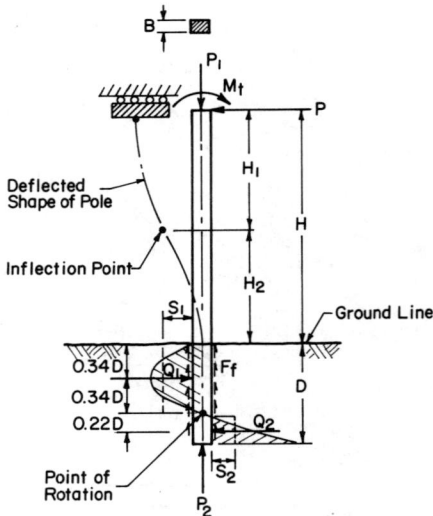

Fig. 10.36. Forces acting on a pole under action of a single lateral load P. Reproduced with permission from G. Gurfinkel, *Wood Engineering,* 1st ed., 1973, Southern Forest Products Association, New Orleans, LA.

of the pole has some moment resistance, as provided by a knee brace, then the frame is indeterminate to the third degree. The pole frame can be analyzed by one or more of the indeterminate methods of analysis. Application of several approximations, however, allows the pole frame to be analyzed as a determinate structure.

The lateral forces acting on a pole are shown in Fig. 10.36. The applied force P acts at a height H above the ground. The top of the pole, which is attached to the roof framing, has a rotational restraining moment, M_t. This restraint equals zero in the case of a pinned top and is nonzero in the case of a knee-braced pole.

The lateral force is restrained by the parabolic distribution of passive soil pressure. The resultant lateral soil resistances are Q_1 at $0.34D$ below the soil surface and Q_2 acting at $0.90D$ below the soil surface, where D equals the total embedment depth of the pole. The average lateral soil pressures above and below the point of rotation are defined in the diagram as S_1 and S_2.

The inflection point, located at H_2 above the ground surface, is the point of zero moment in the pole. If the moment restraint, $M_t = 0$ (a pinned connection between the roof frame and the pole), then $H_2 = H$. If the pole is knee braced (see Fig. 10.37 for a typical knee brace) then H_2 equals $0.66H$ for a tapered pole and approximately $0.5H$ for an

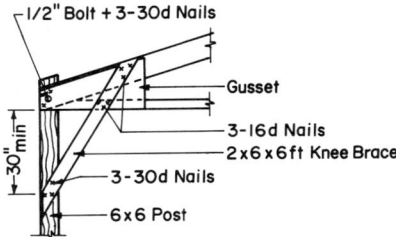

Fig. 10.37. Typical example of an adequate knee brace for pole construction. Reproduced with permission from *Structures and Environment Handbook*, MWPS-1, 11th ed., 1983, Midwest Plan Service, Ames, Iowa.

untapered pole, where H equals the distance between the ground line and the attachment of the knee bracing.

The maximum moment in a pole occurs below the ground surface or at the attachment of the knee brace. It is generally accepted that the maximum moment below ground occurs at one-fourth to one-third the pole embedment depth. The maximum moment below ground for the pole in Fig. 10.36 is obtained by drawing a free body diagram (FBD) of the pole between the inflection point and a depth $D/3$ below the ground. For the loading of Fig. 10.36 this moment equals $P(H_2 + D/3)$. In Fig. 10.36, $M_t = PH_1$.

An expression for the embedment depth can be found by drawing a FBD of the lower portion of the pole through the inflection point, by summing forces in the horizontal direction, by summing moments about the line of action of Q_2 and by noting that $Q_2 = S_1 B(0.68D)$ and $Q_2 = S_2 B(0.32D)$. The resulting expression for D is

$$D = \frac{2.37P + \sqrt{(2.37P)^2 + 10.56PH_2 S_1 B}}{2 S_1 B} \quad (10.33)$$

where S_1 is the lateral soil resistance at $D/3$; B is the diameter of a round pole, the diameter of an encasing concrete collar, or the diagonal dimension of a rectangular post. The remaining terms are as defined in Fig. 10.36.

If the pole has lateral restraint at the ground line, such as that provided by a rigid floor or concrete pavement, the embedment depth may be determined with Eq. (10.34).

$$D = \sqrt{4.25 PH_2 / S_3 B} \quad (10.34)$$

440   Light Timber Design

Table 10.17. Allowable Lateral Pressures for Soils.

Class of Material	Lateral Pressures	
	Allowable Values per Foot of Depth Below Natural Grade[a] (psf)	Maximum Allowable Values (psf)
Good[a]		
Compact well-graded sand and gravel	400	8000
Hard clay		
Well-graded fine and coarse sand		
Average[b]		
Compact fine sand	200	2500
Medium clay		
Compact sandy loam		
Loose coarse sand and gravel		
Poor[c]		
Soft clay	100	1500
Clay loam		
Poorly compacted sand		
Clays containing large amounts of salt		

Reproduced with permission from *Pole Building Design*, 6th ed. 1968. American Wood Preserver's Institute, Washington, DC.
[a]Isolated poles, such as flagpoles, or signs, may be designed using lateral bearing values equal to two times tabulated values.
[b]Drained so water will not stand.
[c]Water stands during wet season.

where S_3 is the lateral soil pressure at depth D. Typical values of lateral soil resistances are given in Table 10.17. These values are averages and should be used with caution.

The maximum moments and embedment depths for poles with more general loadings may be estimated by assuming the same general shape of soil lateral resistance, the same location of the inflection point, and that M_{max} occurs at $D/3$ below the ground surface. These procedures will be demonstrated in Example 10.16.

To illustrate the primary steps in the analysis of a pole frame consider the individual pole frame and loads shown in Figs. 10.38a and 10.38b. Using the truss FBD for Case I (Fig. 10.38c) and assuming the reactive forces R_A and R_B to be equal, the remaining truss reactive forces and member forces can be evaluated from simple statics. The analysis assumes that the wind forces acting on the windward and leeward poles are shared equally. The design loads for the two poles are shown in the FBDs in Fig. 10.38d.

For the FBD in Fig. 10.38e, the inflection point is located at H above the ground. Summing moments about 0, the maximum moment is $M_{max} = R(H + D/3) + wH(H/2 + D/3)$. M_{max} for alternate loading

10.7 Pole Buildings 441

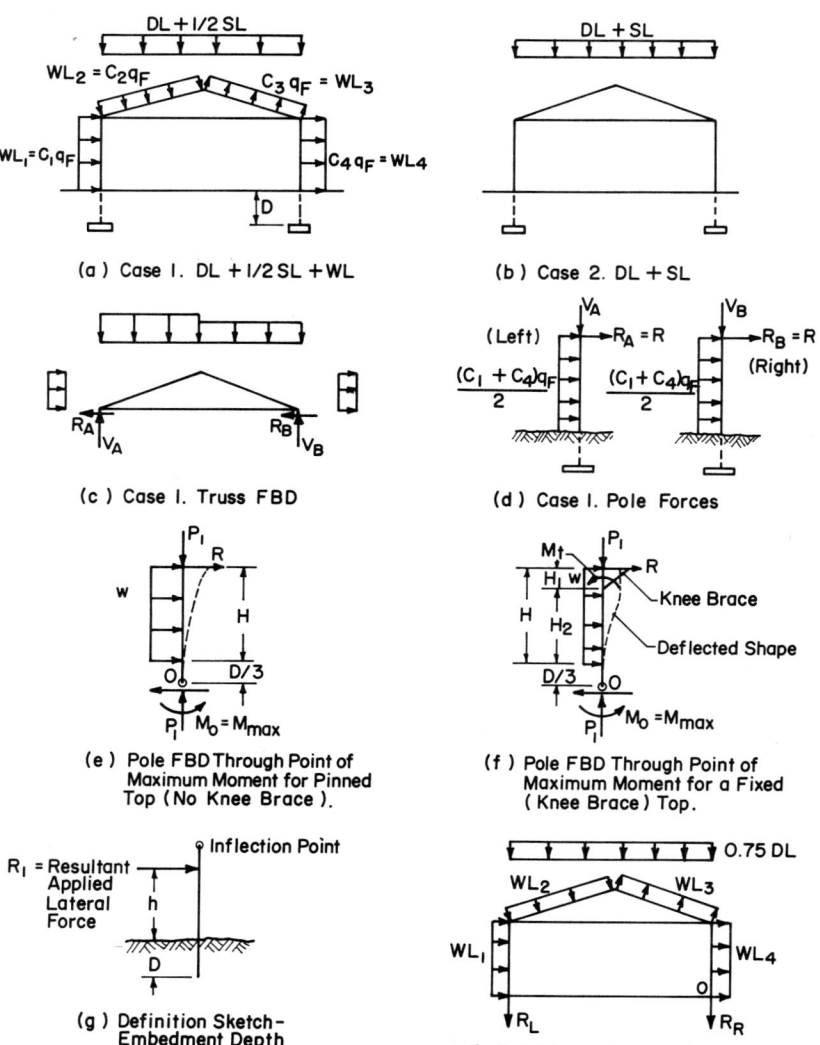

Fig. 10.38. Pole frame analysis.

schemes is easily found by sketching the appropriate free body diagram and summing moments about point 0. The maximum moment for the knee-braced pole in Fig. 10.38f is determined by drawing the FBD of the pole between 0 and the inflection point at $H_2/2$ (for a rectangular pole). With a shear force of $(R + wH - wH_2/2)$ at the inflection point and summing moments about point 0, $M_{max} = R(H_2/2 + D/3 - w(H_2^2/8 - HH_2/2 - HD/3)$.

The pole size must be large enough to satisfy Eq. (10.35).

$$\frac{P_1/A}{F'_c} + \frac{M_{max}/S}{F_b - J(P_1/A)} < 1.0 \qquad (10.35)$$

The pole size is usually controlled by either the $DL + 1/2SL + WL$ or the $DL + 1/2SL + WL + LL$ load combination. Nonetheless, it is advisable to check the adequacy of the pole selected for Case II for the full $DL + SL$. For Case II the pole is subject to pure compression and must satisfy the criterion, $F'_c > f_c$.

The appropriate depth of embedment may be estimated by either Eq. (10.33) or (10.34). In the case of more general loadings than shown in Fig. 10.36, Eqs. (10.33) and (10.34) may still be used by substituting R_1 and h, as defined in Fig. 10.38g, for P and H_2, respectively. R_1 is the resultant of all the applied horizontal loads (including the shear force) between the ground and the inflection point and h is the distance between the ground and the equivalent resultant force. To assure adequate stability it is recommended that D never be less than 4 ft (1.22 m) and that the holes be well tamped when backfilled.

The footer pads distribute the concentrated column loads over sufficient area so that the poles do not settle under design loads. The design criterion for sizing the pads is

$$F_{brg} > P_1/A_f \qquad (10.36)$$

where F_{brg} is allowable bearing stress of the soil (see Table 11.20) in psf (kPa); P_1 is maximum vertical pole design load, lb (kN); A_f is area of the footer pad, ft^2 (m^2).

Most properly embedded poles have a withdrawal resistance of at least 1000 lb (4.4 kN). The design withdrawal resistance is obtained by summing and evaluating the pole reactive forces for a $0.75\,DL + WL$ combination. By summing moments about 0 in the FBD of Fig. 10.35h, the uplift force, R_L, may be evaluated. If $R_L < 1000$ lb (4.4 kN), the withdrawal resistance is adequate. If $R_L > 1000$ lb (4.4. kN), the building may overturn unless additional anchorage is provided. Additional anchorage can be obtained by pouring an 18 to 24 in. (457 × 610 mm) concrete collar around the bottom 12 in. (305 mm) of the pole and fastening it to the pole with bolts, pins, or spikes driven partially into the pole and extending into the concrete collar. The concrete collar adds its own mass, the mass of soil above the collar, and a larger frictional surface area to the withdrawal resistance.

The purlins, nailing girts, and top plates are designed as simply supported or continuous flexural members. Care must be exercised, particularly in the top plates, to assure that the members are laterally stable.

Example 10.16. Pole Frame Analysis. The 40 × 72 ft (12.19 × 21.95 m) pole framed building shown in Figs. 10.39a and 10.39b is used to store sawdust to a depth of 8 ft (2.44 m). The poles are spaced 6 ft (1.83 m) on center and extend 10 ft (3.05 m) above the ground line. The poles are restrained at the ground and embedded and backfilled with medium clay. The frame must support the *DL, SL, WL,* and *LL* shown in Fig. 10.39c. The live load is based on Rankine's equation with material density of 12 pcf (190 kg-m^{-3}) and an angle of internal friction of 25°. The poles are adequately braced 2 ft (0.61 m) from the eave. A plywood liner is attached to the inside of the poles to transfer storage loads to the poles. Assume adequate horizontal framing is included to support the plywood between the poles: (a) Select the size No. 2 Southern Pine rectangular posts required to carry the design loads; (b) Specify the pole embedment depth, anchorage, and footer pad requirements; (c) select the top plate size for a truss spacing of 3 ft (0.91 m).

Solution

a. Design the pole for the $DL + WL + \frac{1}{2}SL + LL$
   i. The truss FBD in Fig. 10.37d and equations of statics yield the truss reactive forces V_A, V_B, and R. The magnitude of the horizontal and vertical components of wind load have the same magnitude as the normal wind load. However, the vertical wind load is based on the horizontal projected area, the horizontal wind load is based on vertical projected area of the roof, and the normal load is based on the actual roof area.

444 Light Timber Design

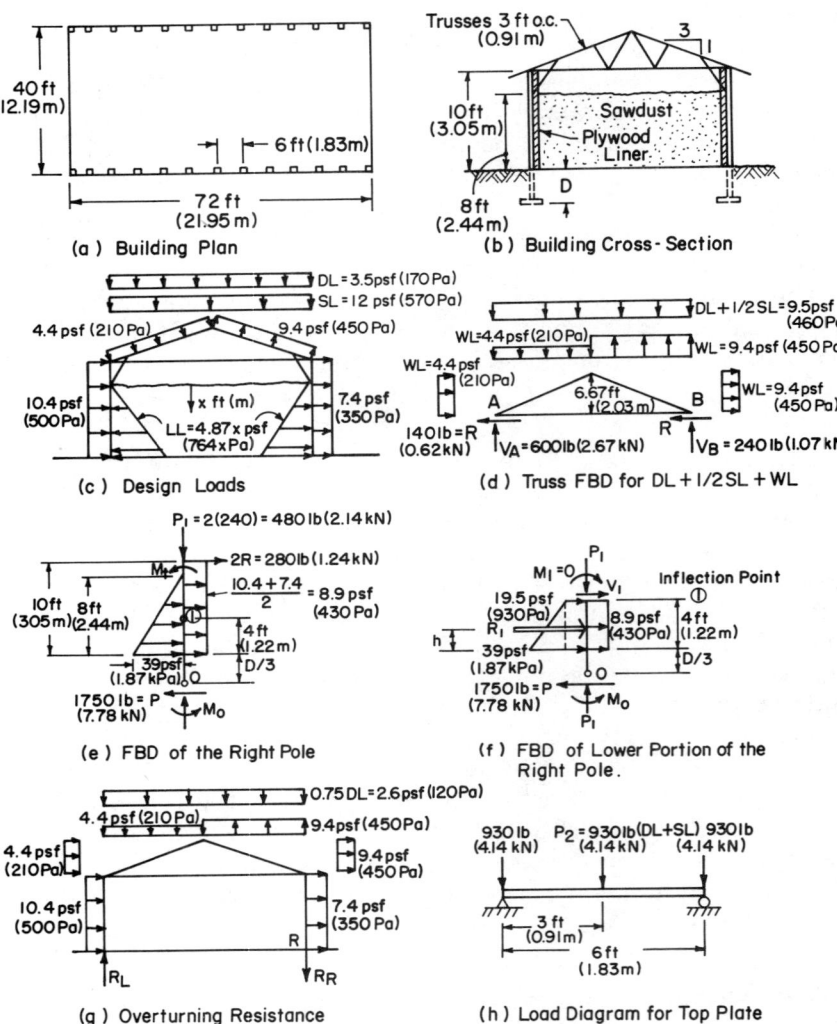

Fig. 10.39. Example 10.16—Pole framing analysis.

$$\Sigma F_h + = 0: 0 = -2R + (4.4 + 9.4)(6.67)(3)$$
$$R = 140 \text{ lb } (0.62 \text{ kN})$$

$$\overset{\curvearrowleft}{\Sigma M_A} + = 0: 0 = 9.5(40)(3)(20)$$
$$+ 4.4(20)(3)(10)$$
$$- 9.4(20)(3)(30)$$
$$+ (4.4 + 9.4)(6.67)(3)(6.67/2) - 40V_B$$
$$V_B = 240 \text{ lb } (1.07 \text{ kN}) \uparrow$$
$$+ \downarrow \Sigma F_v = 0: 0 = 9.5(40)(3) + 4.4(20)(3)$$
$$- 9.4(20)(3) - 236 - V_A$$
$$V_A = 600 \text{ lb } (2.67 \text{ kN}) \uparrow$$

ii. It is apparent that the right-hand pole carries the maximum bending due to the pressure from the stored material plus wind load. The free body diagram of the right-hand pole is shown in Fig. 10.39e. Assuming M_{max} occurs at $D/3$, an inflection point midway between the knee brace and the ground line, and an embedment depth of 5 ft (1.52 m), then the FBDs of Figs. 10.39e and 10.39f (drawn through the inflection point), and statics yield the design moment as follows:

From Fig. 10.39e, the equilibrium of horizontal forces yields

$$\Sigma F_h + \rightarrow = 0: 0 = 276 + 8.9(10)(6) + (^1/_2)(39)(8)(6) - P$$

and the reactive force $P = 1750$ lb (7.78 kN). Figure 10.39f and $\Sigma M_1 = 0$ yield the equation

$$0 = M_0 + 1750 (5.67) - (19.5 + 8.9)(6)(4)(4/2)$$
$$- (1/2)(19.5)(6)(4) (8/3),$$

from which $M = 7910$ lb-ft (10.72 kN-m)

iii. An excellent first estimate of the pole size requirement may be obtained by only considering flexure. Thus, from Table 10.2 $F_b = 1100 \ (1.33) = 1460$ psi (10.1 MPa) and from $F_b > f_b = M/S$, $S_{req'd} = M/F_b = 7910 \times 12/1460 = 64.9$ in.^{3} (1063 cm^3). A survey of Table 10.4 will show a 4×12 with $S = 73.8$ in.^{3} (1209 cm^3) to be the smallest adequate size. A more common size pole is a 6×8 with $S = 51.6$ in.^{3} (846 cm³). Since the pole bending moment is a linear function of pole spacing, a 6 × 8 could be used over a spacing of (51.6/64.9)(6) = 4.8 ft (1.46 m).

iv. Combined loading must now be checked. The plywood liner braces the 4 × 12 in the plane normal to the frame. In the plane of the frame the column buckles as an unbraced column with one end fixed and the other end restrained from rotating. Since the unbraced length equals 8 ft (2.44 m), $KL = 1.2(8) = 9.6$ ft (2.93 m) and $L/d = 9.6 \times 12/11.25 = 10.2$. Thus, $F'_c = 625 \times 1.33 = 830$ psi (5.7 MPa). Substituting into the interaction equation (10.35) with $J = 0$ and the section properties of a 4 × 12 yields

$$\frac{480/39.4}{830} + \frac{7910 \times 12/73.8}{1460} = 0.01 + 0.88 = 0.89 < 1.0$$

Note that the axial load contributed very little to the interaction sum in this case. Note also that the interaction sum is considerably less than 1.0. Nonetheless, the next smaller standard size is clearly inadequate from the flexure analysis in step iii.

b. Check the 4 × 12 post for $DL + SL + LL$. The axial load at the top of both the left and right pole equals $(12 + 3.5)(40)(6)/2 = 1860$ lb (8.27 kN). The first term of the interaction equation becomes $(1860/39.4)/(625 \times 1.15) = 0.07$. It is apparent that, since the bending moment will be smaller for the $DL + SL + LL$ than for the $DL + \frac{1}{2}SL + WL + LL$ combination, that the interaction equation criterion is satisfied. Thus, a 4 × 12 post will be used.

c. From a FBD of the portion of the pole above the inflection point in Fig. 10.39e and the summation of forces in the horizontal direction, $V_1 = 830$ lb (3.69 kN). Placing this force on the FBD in Fig. 10.39f and summing the moments of the applied forces above the ground line about the ground line, the resultant force $R_1 = 1750$ lb (7.78 kN) is found to act at $h = 2.86$ ft (0.87 m) above the ground line. Substituting into Eq. (10.34) with $P = 1750$ lb (7.78 kN), $H_2 = 2.86$ ft (0.87 m), $S_3 = 200 \times 5 = 1000$ psf (50 kPa), and $B = \sqrt{11.25^2 + 3.5^2}/12 = 0.98$ ft (300 mm),

$$D = \sqrt{4.25(1746)(2.86)/1000 \times 0.98} = 4.65 \text{ ft} \quad (1.42 \text{ m})$$

Since the required depth is less than the estimated depth of 5 ft (1.52 m), the initial assumption is conservative. By embedding the pole to only 4.67 ft (1.42 m), the net effect would be to reduce the bending moment. Thus, an embedment depth of 4.67 ft (1.42 m) can be used without further calculation.

d. The overturning stability and adequacy of anchorage under $0.75DL + WL$ can be determined by summing moments about the ground line of the right-hand pole in Fig. 10.37g.

$$\overset{\rightarrow}{\underset{+}{\Sigma}} M_R = 0: \quad 0 = (10.4 + 7.4)(10)(6)(5)$$
$$+ (4.4 + 9.4)(6.67)(6)$$
$$\times (10 + 6.67/2) + 9.4(20)(6)(10)$$
$$- 4.4(20)(6) \times (30) - 2.6(40)(6)(20)$$
$$+ 40R_L$$

$$R_L = 4340/40 = 110 \text{ lb} \quad (490 \text{ N}) \uparrow$$

and from

$$\Sigma F_v = 0$$

the right-hand reaction is found to be an uplift force.

$$R_R = 80 \text{ lb} \quad (360\text{N}) \downarrow$$

Since the maximum uplift force is $<<1000$ lb (4.45 kN), the anchorage is adequate.

e. The top plates connecting the poles may be treated as simple beams. A FBD of the plate is shown in Fig. 10.37h. The pole–plate connection is assumed pinned. The truss reactive force acting on the plate is $(15.5 \times 40 \times 3)/2 = 930$ lb (4.14 kN) for the $DL + SL$ combination. For the $DL + WL + \frac{1}{2}SL$, the maximum reactive force was 600 lb (2.67 kN) from step a. Thus, the maximum plate moment equals $P_2L/4 = 930(6)(12)/4 = 16,740$ in.-lb (1.89 kN-m). For the flexure criterion and $F_b = 1550 \times 1.15 = 1780$ psi (12.3 MPa) (no. 2 southern pine used and dressed at 15% moisture), $S_{\text{req'd}} = M/F_b = 16,740/1780 =$

9.40 in.^{3} (154.0 cm^3). A 2 × 8 with S = 13.14 in.^{3} (215.3 cm^3) is adequate to carry the moment.

f. Assuming a soil with a bearing strength of 2000 psf (100 kPa), the footer pad area required is A_f = 930 × 2/2000 = 0.93 ft^2 (6.0 cm^2) and the footer radius required is

$$r = \sqrt{A_f/\pi} = 0.54 \text{ ft} \quad (160 \text{ mm})$$

A footer 6 in. (150 mm) thick and 13 in. (330 mm) in diameter will suffice.

Example 10.16 represents a partial and preliminary design for a pole frame. A complete design would include the design of the trusses as well as consideration of all the design load combinations acting on the structure. The loads in the example assume the wind is acting normal to the ridge line. A complete design would also consider the effect of loads induced by wind moving parallel to the ridge line. Other considerations in a complete design would include connector details between trusses and plates, connector details between plates and poles, and endwall framing details. The diaphragm strength of the cladding may also be considered in the design.

PROBLEMS

10.1. Compare the allowable tensile stresses of a No. 2 southern pine (15% M) 2 × 4 to a 2 × 6, a 2 × 8, a 2 × 10, and a 2 × 12. Explain why they are different.

10.2. Compare the allowable flexural, shear, and compressive stresses for a No. 2 southern pine (15% M) 2 × 4 to a 2 × 6, a 2 × 8, and a 2 × 10.

10.3. Compare the allowable flexural, shear, compressive, and tensile stresses of a No. 2 southern pine 2 × 6 (15% M) to a No. 2 southern pine 2 × 6 (surfaced dry and used at 19% M) and to a No. 2 southern pine 2 × 6 (surfaced green). Explain any differences.

10.4. Repeat Problem 10.1 for Douglas fir lumber.

10.5. Repeat Problem 10.2 for Douglas fir lumber.

10.6. Repeat Problem 10.3 for Douglas fir lumber.

10.7. Compare the section properties (I, S, and A and cross-sectional dimensions) of full size and dressed 2 × 4s, 2 × 8s, 2 × 10s, 4 × 6s, 6 × 8s.

10.8. Evaluate the allowable compressive stress of a No. 1 (15% M) southern pine 2 × 8 when the compressive stress acts at an angle of 45° to the grain of the 2 × 8.

10.9. Determine the allowable compressive stress and load for a No. 2 2 × 4 southern pine (15% M) timber column for the following:

   a. Both ends are pinned with respect to both the strong and weak axes, there are no intermediate braces, and the column length equals 4 ft (1.22 m), 5 ft (1.32 m), 6 ft (1.83 m), 8 ft (2.44 m).
   b. Both ends are pinned with respect to both the strong and weak axes, the weak axes is braced continuously, and the column length equals 4 ft (1.22 m), 6 ft (1.83 m), 8 ft (2.44 m), 10 ft (3.05 m).
   c. The bottom end is fixed with respect to both axes and the top end is pinned with respect to buckling about the weak axis and free with respect to buckling about the strong axis, there is no intermediate bracing, and the column length equals 4 ft (1.22 m), 6 ft (1.83 m), 8 ft (2.44 m), 10 ft (3.05 m).

10.10. Repeat Problem 10.9 for a No. 2 Douglas fir dressed 4 × 6 member.

10.11. Plot a curve similar to Fig. 10.8 for an unbraced No. 2 (15% M) southern pine 6 × 6 which is 12 ft (3.66 m) long and pinned at both ends.

10.12. Determine the allowable flexural stress for the following uniformly loaded, single span, simply supported No. 2 southern pine (15% M) beams.

   a. A 2 × 4 with an unbraced span of 2 ft (0.61 m); 4 ft (1.22 m); 6 ft (1.83 m); 8 ft (2.44 m); and 10 ft (3.05 m).
   b. Repeat Problem 10.12a for a 2 × 6.
   c. Repeat Problem 10.12b for a 2 × 10.
   d. A 2 × 6 with a span of 12 ft (3.66 m) but with adequate bracing of the compression flange every 4 ft (1.22 m).

10.13. Design the lightest weight 2× or 4× No. 1 southern pine (15% M) beam to carry a uniformly distributed load of 150 lb/ft (2.2 kN · m^{-1}) over a simply supported span of 10 ft (3.05 m) if the load is due to a $DL + SL$ combination, if the allowable deflection equals (1/360)(span), and if the compression flange is ad-

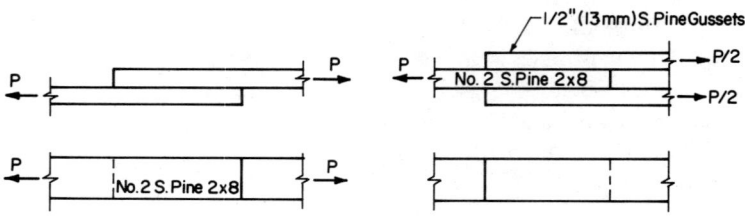

Fig. 10.40. Problem 10.18.

equately braced. How frequently must the compression flange be braced?

10.14. Repeat Problem 10.13 if the beam span equals 12 ft (3.66 m).

10.15. Repeat Problem 10.13 if the compression flange is braced only at the supports and at midspan.

10.16. Plot a curve similar to Fig. 10.15 for a No. 2 southern pine 2 × 6 simply supported beam with a span of 12 ft (3.66 m). *Hint*: Vary L_u from 0 to 12 ft (0 to 3.66 m).

10.17. Design a door header to span a 16 ft (4.88 m) sidewall opening in a 40 ft (12.19 m) wide × 100 ft (30.48 m) long machinery storage facility if trusses are spaced 4 ft (1.22 m) on center. The building is located in central Pennsylvania. Use lumber with a width of 1.5 in. (38 mm) and a maximum nominal thickness of 11.25 in. (286 mm) if

  a. The header is adequately braced laterally by the truss connections.
  b. The header is laterally unsupported between the end supports.

10.18. Evaluate the allowable load for each of the following timber joints.

  a. Eight 8d nails in the two-member joint in Fig. 10.40a.
  b. Eight 10d nails in the two-member joint in Fig. 10.40a.
  c. Two 0.5-in. (13-mm) diameter bolts in the two-member joint of Fig. 10.40a.
  d. Six 0.5-in. (13-mm) diameter bolts in a single row in the two-member joint of Fig. 10.40a.
  e. Repeat Problem 10.18a for the joint in Fig. 10.40b.
  f. Repeat Problem 10.18b for the joint in Fig. 10.40b.
  g. Repeat Problem 10.18c for the joint in Fig. 10.40b.
  h. Repeat Problem 10.18d for the joint in Fig. 10.40b.

i. Eight 12d nails in the three-member joint of Fig. 10.40b.
j. Casein glue in the two-member joint of Fig. 10.40a, if the members overlap 8 in. (203 mm).

10.19. Sketch the joint showing recommended end, edge, and connector spacings for Problems 10.18a, 10.18c, and 10.18d.

10.20. Could two rows of bolts have been used in Problem 10.18h? If so, why would it have been advantageous?

10.21. Determine if a pinned end, 10 ft (3.05 m) long dressed 4 × 4 No. 2 southern pine (15% M) timber column can carry a combined axial compressive load of 3000 lb (36 kN) and a maximum bending moment of 4000 lb-in. (56.5 N·m).

10.22. Determine the maximum axial compressive load a 12 ft (3.66 m) long simply supported 6 × 8 No. 2 southern pine (15% M) beam-column can carry if it also carries a maximum bending moment about the strong axis of 10,000 lb-ft (1.130 kN·m).

10.23. Determine the size No. 2 southern pine (15% M) member with a thickness of 1.5 in. (38 mm) required to carry the $DL + SL$ and $DL + WL$ combinations in the top chord of the roof truss in Fig. 10.29a if the truss geometry stays the same, but the truss span increases to 36 ft (10.97 m), the trusses are spaced 4 ft (1.22 m) on center, and the loads are SL = 25 psf (1.20 kPa), DL = 5 psf (240 Pa), WL = 10 psf (480 Pa) suction on the left slope, and 15 psf (720 Pa) suction on the right slope.

a. Solve if the loads are all applied at the panel points.
b. Solve by the conservative method if the loads are applied by purlins spaced 24 in. (610 mm) on center along the upper chord.
c. Solve by the TPI method if the loads are applied by purlins located every 24 in. (610 mm) on center along the upper chord.

10.24. Determine the size No. 2 southern pine (15% M) member required for members BG, AG, and GC to carry the $DL + SL$ and the $DL + WL$ combinations in Problem 10.23. If L/d ratios are exceeded, recommend a solution and size the members.

10.25. Determine the embedment depth and the size No. 2 SR southern pine rectangular pole required to carry a concentrated lateral load of 800 lb (3.6 kN) at a height 12 ft (3.66 m) above the ground line if the top of the pole is free to translate with no rotational restraint and the pole is encased in compact well graded sand and gravel.

10.26. Solve Problem 10.25 if the top of the pole is provided rotational restraint with a knee brace attached 2 ft (0.61 m) from the top of the pole.

10.27. Solve Problem 10.25 if the pole is restrained by a continuous concrete floor at the ground line.

10.28. Solve Problem 10.26 if the pole is restrained by a continuous concrete floor at the ground line.

10.29. Select the size No. 2 southern pine pole required to carry the wind, dead, and snow loads in the frame of Fig. 10.39c if the poles are spaced 8 ft (2.44 m) on center, the trusses are spaced 4 ft (1.22 m) on center, the eave height is 12 ft (3.66 m), and the top of the pole has no knee braces for rotational restraint. Assume a compact well-graded sand and gravel backfill around the poles.

10.30. Solve Problem 10.29 if the poles have knee braces attached 3 ft (0.91 m) below the eave.

10.31. Solve Problem 10.30 for poles spaced 4 ft (1.22 m) on center, an eave height of 8 ft (2.44 m), a granular media with a density of 50 pcf (800 kg·m^{-3}), and an angle of internal friction of 25°, and a storage depth of 5 ft (1.52 m).

10.32. Determine the magnitude of the secondary moments and stresses in the upper chord of the truss in Fig. 10.29 and 10.30 using the less conservative TPI-78 procedures. Then determine the size upper chord required.

NOMENCLATURE FOR CHAPTER 10

A	Cross-sectional area, in.^{2} (cm^2)
A_{net}	Net cross-sectional area remaining after drilling, in.^{2} (cm^2)
C_s	$\sqrt{L_e d/b^2}$ = slenderness factor for lateral buckling of beams
C_k	$\sqrt{3E/5F_c}$ = limiting slenderness factor for inelastic lateral buckling of beams
D	Nail or bolt diameter, in. (mm)
DF	Load duration factor
E	Modulus of elasticity, psi (MPa)
F_b	Allowable flexural stress without lateral buckling, psi (MPa)
F_b'	Allowable flexural stress as adjusted for buckling and other considerations, psi (MPa)
F_c	Allowable axial compressive strength parallel to grain without buckling, psi (MPa)
F_c'	Allowble axial compressive stress parallel to the grain reduced for buckling effects, psi (MPa)

Nomenclature for Chapter 10   453

F_{cp}	Allowable compressive stress perpendicular to the grain, psi (MPa)
F_g	Allowable shear stress at a glue line, psi (MPa)
F_v	Allowable horizontal shear stress, psi (MPa)
F_v'	Adjusted allowable horizontal shear stress for two beam action, psi (MPA)
F_y	Yield stress, ksi (MPa)
I	Second moment of area of a cross section, in.^{4} (cm^4)
J	$(L/d - 11)/(K - 11)$, beam-column moment magnification factor
K	$0.67\sqrt{E/F_c}$, limiting L/d ratio for inelastic buckling of column
K'	$2.32\sqrt{E/F_c}$, limiting L/r ratio for inelastic buckling of a column
L	Length of a column; span of a beam or truss, ft (m)
L_e	Effective unsupported length for a beam, ft (m)
L_u	Distance between lateral supports along the compression flange of a beam, ft (m)
M	Bending moment; allowable bending moment, lb · in. (kN · m)
N_θ	Allowable compressive stress at an angle θ to the grain, psi (MPa)
P	Applied concentrated load; allowable bolt load parallel to grain, lb (kN)
P_s	Allowable load per connector in a joint, lb (kN)
Q	First moment of the area between an axis parallel to the axis of bending in a beam and the outer fiber taken about the neutral axis of the section, in.^{3} (cm^3). Also, allowable bolt load perpendicular to grain, lb (kN)
R	Reactive force, lb (kN)
S	Section modulus of cross section, in.^{3} (cm^3)
V	Shear force, lb (kN)
V'	Adjusted shear force for two beam action, lb (kN)
a	Cross-sectional dimension of a member; spacing between bolt rows, in. (mm)
b	Cross-sectional dimension (nominal or actual) of a member; bolt spacing, in. (mm)
c	Distance between the centroidal axis and the extreme fiber in a cross section, in. (mm)
d	Cross-sectional dimension (nominal or actual) of a member; least dimension of a column; depth of a beam, in. (mm)

e	Eccentricity of applied or internal resultant loads, in. (mm)
f_b	Actual flexural stress, psi (MPa)
f_c	Actual axial compressive stress, psi (MPa)
f_t	Actual axial tensile stress, psi (MPa)
f_v	Actual shear stress, psi (MPa)
f'_v	Adjusted shear stress for two beam action, psi (MPa)
k	Beam deflection coefficient
l	Length of bolt in main member of a joint, in. (mm)
p	Horizontal projection of roof overhang, ft (m)
r	Radius of gyration of a cross-sectional area, in. (mm)
s	Truss, frame, or pole spacing, ft (m)
w	Applied distributed load, lb/ft (kN/m)
x	Subscript relating to the strong axis of a cross section
y	Subscript relating to the weak axis of a cross section
z	Length of bearing support for a beam, in. (mm)
Δ	Beam deflection, in. (mm)
θ	Acute angle between direction of loading and the direction of the grain; roof slope (degrees)

11

Reinforced Concrete Design

11.1 INTRODUCTION

11.1.1 General

Reinforced concrete is a composite material consisting of concrete and intimately bonded steel reinforcing bars. A typical reinforced concrete cross section is shown in Fig. 11.1. The reinforcing steel is always present in the portion of the cross section which experiences tensile stresses under load. In some cases, such as in columns and doubly reinforced beams (Fig. 11.2), steel is placed in the compression zone. The purposes of such steel are to increase the load capacity over singly reinforced sections, properly reinforce sections which are subject to stress reversals, increase the load capacity of column sections, and provide resistance to tensile stresses, which are caused by accidental eccentricities, buckling, and combined axial and flexural loads in compression members.

Plain concrete, which is strong in compression but very weak in tension, cannot be used efficiently in most structural applications. Consider, for example, the beam in Fig. 11.3a which is constructed of a highly idealized concrete with the stress–strain behavior of Fig. 11.3b. If the bottom fibers are at the tensile yield point f_{ty}, the upper fibers of the homogeneous section experience a compressive stress f_c equal to f_{ty}. If yielding is the failure criteria, the moment capacity of the section is dictated by the relatively low value of f_{ty}. Thus, a large portion of the compressive strength of the concrete is not utilized. Other limitations of plain concrete are that its behavior in tension is less predictable and less ductile than in compression.

456 Reinforced Concrete Design

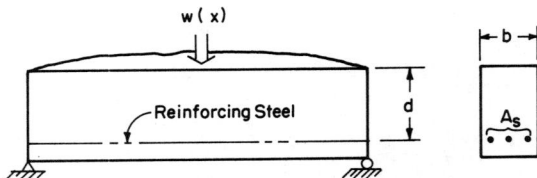

Fig. 11.1. Typical singly reinforced concrete beam.

Figure 11.4 illustrates how tensile and compressive stresses are carried by a reinforced concrete beam cross section assuming the concrete stress–strain behavior in compression is linearly elastic and the maximum flexure stress is below the yield strength. Notice that the concrete below the neutral axis (N.A.) carries no load and the total tensile force, $T = A_s f_s$, is carried by the reinforcing steel.

There are many applications of reinforced concrete in agricultural systems. Among them are foundations, footers, silo walls, retaining walls, manure storage tanks and pits, manure slats, floor slabs, and feedlot pavements.

This chapter discusses the basic strength properties of both concrete and reinforcing steel and presents the fundamental techniques required to analyze reinforced concrete beams and columns. The primary purpose of the chapter is to give sufficient background to allow the

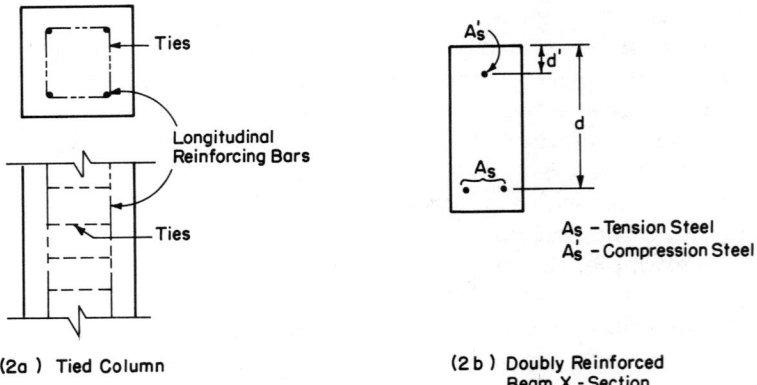

Fig. 11.2. Typical reinforced concrete columns and doubly reinforced beams.

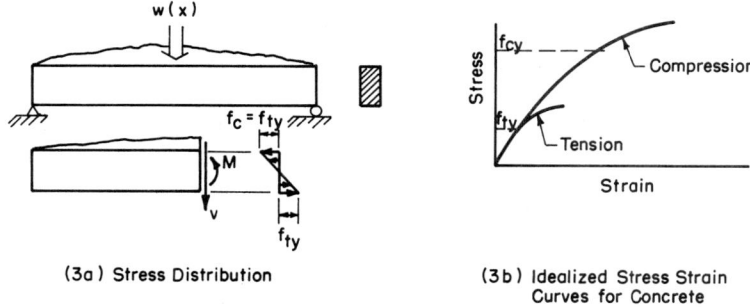

(3a) Stress Distribution

(3b) Idealized Stress Strain Curves for Concrete

Fig. 11.3. Stress distributions in a plain concrete beam.

engineer to (1) begin an independent study of reinforced concrete design; (2) intelligently select standard designs for typical applications; (3) check the adequacy of existing reinforced members; and (4) design some very simple beams and columns. Sufficient up-to-date references are also included to aid further study. Finally, most of the material presented is based upon and complies with the American Concrete Institute (ACI) 318–83 *Building Code Requirements for Reinforced Concrete*.

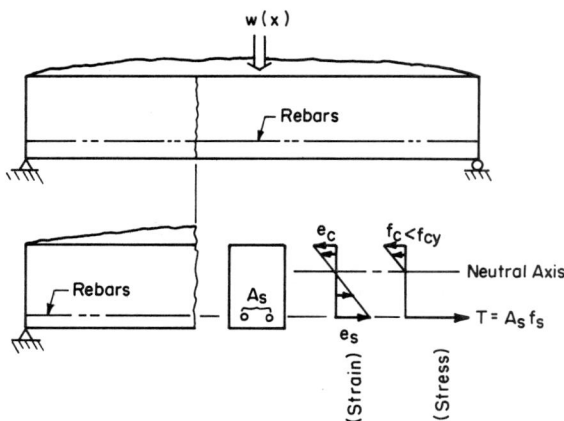

Fig. 11.4. Idealized stress and strain distributions in a reinforced concrete beam.

11.1.2 Symbols and Units

Unless noted otherwise, the symbols used in this chapter are defined at the end of the chapter. Dual units, SI, and English, are shown in the listing. Many of the equations in the text are valid for any consistent set of units. Some, however, have dimensional constants and are valid only for a specific set of units. When this occurs, it will be noted in the text and the appropriate units will be specified.

11.2 MATERIALS AND BEHAVIOR

11.2.1 Reinforcing Steel

Two suitable types of reinforcing steel are deformed bars and welded wire fabric. The several grades of reinforcing steel are given in Table 11.1. Reinforcing bar sizes are listed in Table 11.2, and some typical welded wire fabric sizes are summarized in Table 11.3.

The stress–strain behavior of reinforcing steel is assumed to be elastic–plastic as shown in Fig. 11.5. The modulus of elasticity E_s is taken as 29×10^3 ksi (200 GPa) and the ductility is in the neighborhood of 20 to 30%.

The general requirements for reinforcing steel are that bars be deformed rather than smooth, the surface be clean and free of oil, and the area be large enough to carry the tensile loads. Additionally, the ACI code imposes restrictions on steel placement tolerance, spacing, and cover.

Reinforcing steel placement is critical if the section is to perform satisfactorily. For walls, flexural members, and compression members, steel should be located within $d \pm {}^3/_8$ in. (10 mm) if $d \leq 8$ in. (203 mm), and within $d \pm {}^1/_2$ in. (13 mm) if $d > 8$ in. (203 mm). (See Fig. 11.2 for the definition of d.)

Adequate spacing is needed between reinforcing bars to ensure in-

Table 11.1. Grades and Strength of Reinforcing Steel.

Grade	Design Strength, ksi (MPa)
40	40 (280)
50	50 (340)
60	60 (410)
80	80 (550)
100	100 (690)

11.2 Materials and Behavior

Table 11.2. Reinforcing Steel Bar Sizes.

Bar No.[a]	Nominal Diameter, in. (mm)	Area, in.^{2} (mm^2)
2	$\frac{1}{4}$ (6)	0.05 (30)
3	$\frac{3}{8}$ (9)	0.11 (70)
4	$\frac{1}{2}$ (13)	0.20 (130)
5	$\frac{5}{8}$ (16)	0.31 (200)
6	$\frac{3}{4}$ (19)	0.44 (280)
7	$\frac{7}{8}$ (22)	0.60 (390)
8	1 (25)	0.79 (510)

[a]Bar no. indicates the number of eighths of an inch diameter.

timate contact between the concrete and steel. The following limits on clear space between reinforcing bars (S) as a function of bar diameter d_b are required by the ACI Code.

1. In beams, $S \geq$ the greater of d_b or 1 in. (25 mm) both between layers of reinforcement and between bars in a layer.
2. In spirally or tied reinforced columns, $S \geq$ the greater of 1.5 d_b or 1.15 in. (38 mm).
3. In walls and slabs, $S \leq$ the smaller of three times the wall thickness or 18 in. (460 mm).

Minimum cover of reinforcing steel is required to prevent spalling of concrete and to protect the steel from rust, fire, and chemical de-

Table 11.3. Typical Welded Wire Fabric Sizes and Designations.[a]

	Style Designation	Spacing, in. (mm)	Diameter, in. (mm)	Sectional Area sq in./ft (mm^2/mm)
Wire Gauge	W or D Number			
6×6–10×10	6×6–W1.4×W1.4	6 (150)	0.135 (3.4)	0.029 (0.061)
6×6–8×8	6×6–W2.1×W2.1	6 (150)	0.162 (4.1)	0.041 (0.087)
6×6–6×6	6×6–W2.9×W2.9	6 (150)	0.192 (4.9)	0.058 (0.123)
6×6–4×4	6×6–W4×W4	6 (150)	0.225 (5.7)	0.080 (0.169)
4×4–10×10	4×4–W1.4×W1.4	4 (100)	0.135 (3.4)	0.043 (0.091)
4×4–6×6	4×4–W2.9×W2.9	4 (100)	0.192 (4.9)	0.087 (0.184)
4×4–4×4	4×4–W4×W4	4 (100)	0.225 (5.7)	0.120 (0.254)

Reproduced with permission from *Structures and Environment Handbook*, MWPS–1, 11th ed. 1983. Midwest Plan Service, Ames, IA.
[a]These items may be carried in sheets by various manufacturers in certain parts of the United States and Canada. The spacing, diameter, and sectional area of the welded wire fabric are the same in both the longitudinal and transverse directions.

460 Reinforced Concrete Design

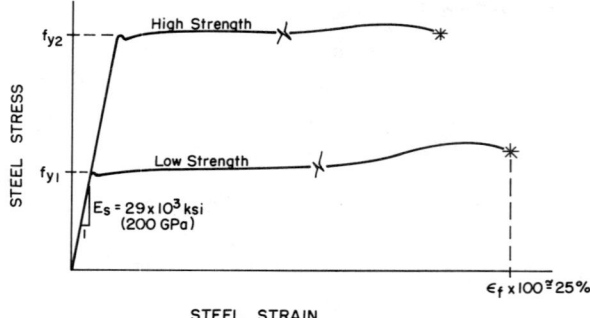

Fig. 11.5. Typical stress-strain behavior of reinforcing steel.

Table 11.4 Minimum Cover Requirements for Steel.

Placement Type	Requirements	Cover, in. (mm)
Cast-in-place	Cast against and exposed to earth	3 (76)
	Concrete exposed to earth and weather	
	$d_b \geq 3/4$ in. (19 mm)	2 (51)
	$d_b < 3/4$ in. (19 mm) W31 or D31 wire	
	or smaller	1.5 (38)
	Concrete not exposed to weather or in contact with ground	
	slabs, walls, joists [$d_b \leq 1^3/_8$ in. (35 mm)]	0.75 (19)
	primary reinforcement, ties, stirrups, spirals in beams and columns	1.5 (38)
Precast	Concrete exposed to earth and weather	
	wall panels [$d_b \leq 1^3/_8$ in. (35 mm)]	$3/4$ (19)
	other members [$3/4$ in. (19 mm) $\leq d_b \leq 1^3/_8$ in. (35 mm)]	1.5 (38)
	$d_b \leq 5/8$ in. (16 mm)	1.25 (32)
	Concrete not exposed to weather or in contact with ground	
	slabs, walls, joists [$d_b \leq 1^3/_8$ in. (35 mm)]	$5/8$ (16)
	primary reinforcement in beams	Greater of d_t or $5/8$ (16)
	ties, stirrups, or spirals in beams or columns	$3/8$ (10)

Table 11.5. Minimum Reinforcement for Temperature and Shrinkage Control.

Construction Type	Steel Ratio $\rho_s = A_s/A_g$
Slabs with Grade 40 or 50 deformed bars	0.0020
Slabs with Grade 60 deformed bars or welded wire fabric	0.0018

terioration. The cover requirements for cast-in-place concrete are greater than for precast concrete components because of better control of construction tolerances in shop fabrication. The ACI requirements for minimum cover are listed in Table 11.4.

Finally, concrete has a tendency to shrink as it cures and to change dimensions with temperature. Both of these factors cause concrete to crack. To prevent shrinkage and temperature cracks, minimum amounts of reinforcing steel are recommended by ACI. These minimums are summarized in Table 11.5 in terms of the ratio of the area of steel to the gross concrete area. ACI requires that minimum reinforcement be provided in the direction normal to the longitudinal reinforcement in structural floor and roof slabs where flexural reinforcement extends in only one direction. Further, the spacing of temperature and shrinkage reinforcement should be the smaller of five times the slab thickness or 18 in. (460 mm).

11.2.2 Concrete

Concrete is a mixture of water, cement, and aggregate, which, when properly cured, forms a bonding matrix between the particles by the chemical process of hydration. Properly cured concrete does not dry. Instead the mixing water is chemically fixed by the hydration process. The primary mechanical properties of concrete are the compressive strength and the workability.

Workability is a measure of the plasticity, or flow properties, of fresh, unhardened concrete. Workability influences the ease or difficulty of placing concrete into forms and around reinforcement and is measured by the slump test. Recommended slumps for typical constructions are given in Table 11.6.

The strength of concrete is determined by axial compression tests of 6 in. (150 mm) diameter by 12 in. (300 mm) cylinders. Concrete strength is usually specified as the ultimate strength acquired after 28 days of curing. A typical compressive stress–strain curve is shown in Fig. 11.6. Note that the curves are nonlinear and that fracture occurs at a strain of only 0.3 to 0.5%.

Table 11.6. Typical Slump Ranges for Various Types of Construction.

Types of Construction	Slump[a], in. (mm) Maximum	Minimum
Reinforced foundation walls and footings	3 (80)	1 (30)
Unreinforced footings, caissons, and substructure walls	3 (80)	1 (30)
Reinforced slabs, beams, and walls	4 (100)	1 (30)
Building columns	4 (100)	1 (30)
Bridge decks	3 (80)	2 (50)
Pavements	2 (50)	1 (30)
Sidewalks, driveways, and slabs on ground	4 (100)	2 (50)
Heavy mass construction	2 (50)	1 (30)

Reproduced with permission from *Design and Control of Concrete Mixtures*, 11th ed. 1968. Portland Cement Association, Skokie, IL.
[a]When high frequency vibrators are *not* used, the values may be increased by about 50% but in no case should the slump exceed 6 in. (150 mm)

The range of the compressive strength, f_c', is 2.5 ksi (17 MPa) to over 8 ksi (55 MPa). The compressive strength is the ultimate strength in compression and is dependent upon many factors, the most important of which are the water-to-cement ratio, specimen age, type curing, and the ratio of fine to coarse aggregates. Other factors influencing concrete quality include the quality of aggregate and water, rate of loading, and the type specimen used. The Portland Cement Association (PCA) Manual, *Guide to Proportioning, Mixing, and Placing Cement*, details many of these influences.

The higher the *water–cement* (w/c) ratio, the lower the strength of concrete. The water–cement ratio, which is probably the most important factor influencing strength, is usually expressed as the ratio of the mass of water to the mass of cement in the mix or gallons (liters)

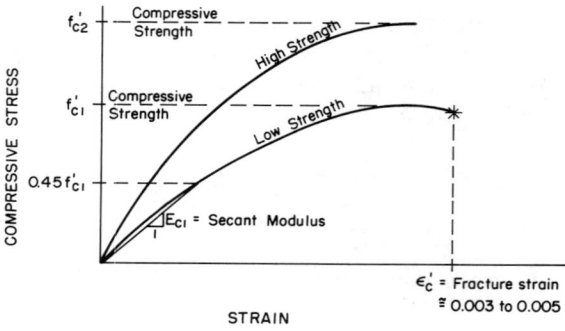

Fig. 11.6. Typical compressive stress-strain behavior of concrete.

11.2 Materials and Behavior   463

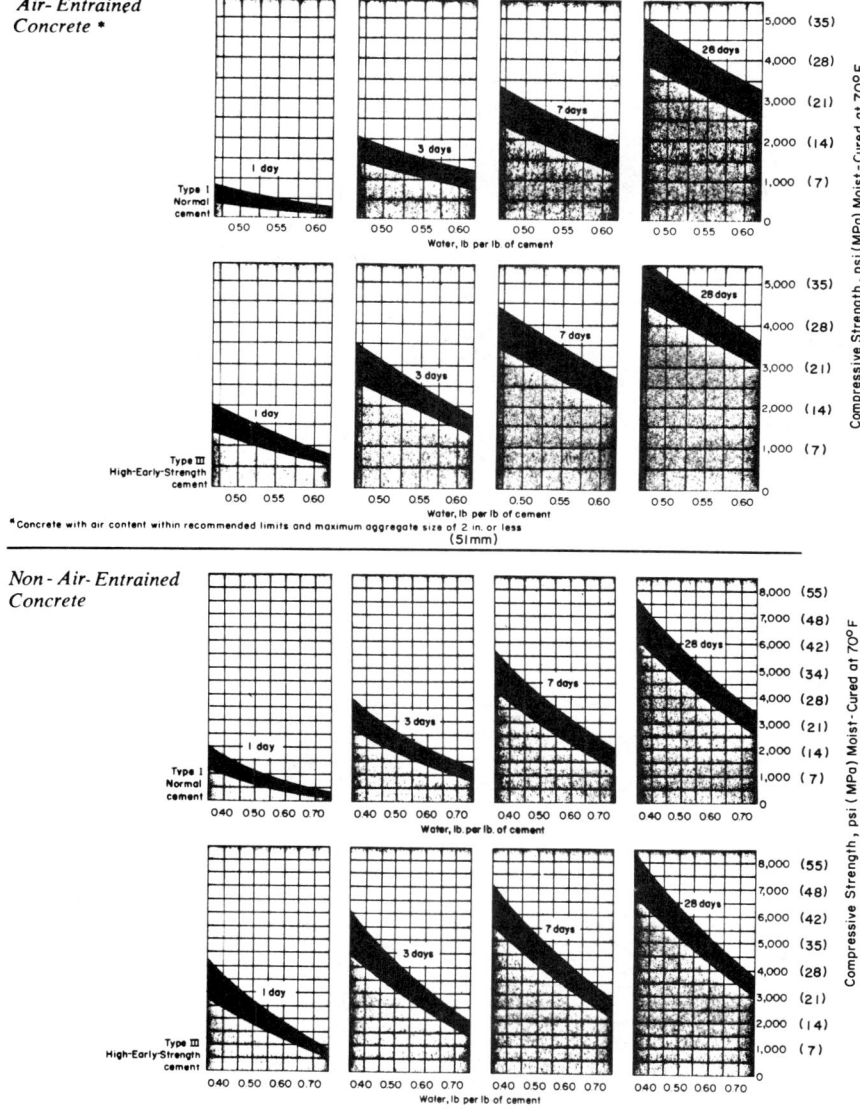

Fig. 11.7. Relationship between water-cement ratio and compressive strength for portland cements at different ages. These relationships are approximate and should be used only as a guide in lieu of data on job materials. Reproduced with permission from *Design and Control of Concrete Mixtures,* 11th ed., Portland Cement Association, Skokie, IL.

of water per bag of cement [one bag of cement equals 94 lb (43 kg)]. Typical ranges of w/c ratios are 0.4 to 0.7 or 4.5 to 8 gallons (17 to 30 L) of water per bag of cement. The workability of concrete increases as w/c ratios increase. A common error in construction is to add water to a mix to make it flow better. This practice, obviously, reduces the concrete strength, wastes money, and may yield an unsafe structure. If workability of a mix is a problem, never solve it by adding more water. Solve the problem by decreasing the quantity of aggregate used in the mix. Figure 11.7 illustrates the influence of w/c ratios upon strength. Table 11.7 recommends w/c ratios for typical types of construction.

The ratio of fine (sand) to coarse aggregate (gravel) influences the degree of bonding between individual particles of aggregate. Suggested

Table 11.7. ACI Recommended Maximum Permissible Water–Cement Ratios for Different Types of Structures and Degrees of Exposure[a]

	Exposure Conditions[b]						
	Severe wide range in temperature, of frequent alternations of freezing and thawing (air-entrained concrete only)			Mild temperature rarely below freezing or rainy or arid			
		At water line or within range of fluctuating water level or spray			At water line or within range of fluctuating water level or spray		
Type of Structures	In Air	In fresh water	In sea water or in contact with sulfates[c]	In Air	In fresh water	In sea water or in contact with sulfates[c]	
A. Thin sections such as reinforced piles and pipe	0.49	0.44	0.40	0.53	0.49	0.44	
B. Bridge decks	0.44	0.44	0.40	0.49	0.49	0.44	
C. Thin sections such as railings, curbs, sill, ledges, ornamental or architectural concrete, and all sections with less than 1-in. (25-mm) concrete cover over reinforcement	0.49	—	—	0.53	0.49	—	
D. Moderate sections, such as retaining walls, abutments, piers, girders, beams	0.53	0.49	0.44	[d]	0.53	0.44	
E. Exterior portions of heavy (mass) sections	0.58	0.49	0.44	[d]	0.53	0.44	
F. Concrete deposited by tremie under water	—	0.44	0.44	[d]	0.44	0.44	
G. Concrete slabs laid on the ground	0.53	—	—	[d]	—	—	
H. Pavements	0.49	—	—	0.53	—	—	
I. Concrete protected from the weather, interiors of buildings, concrete below ground	[d]	—	—	[d]	—	—	
J. Concrete which will later be protected by enclosure or backfill but which may be exposed to freezing and thawing for several years before such protection is offered	0.53	—	—	[d]	—	—	

Reproduced with permission from *Design and Control of Concrete Mixtures*, 11th ed. 1968. Portland Cement Association, Skokie, IL.

[a]Adapted from Recommended Practice for Selecting Proportions for Concrete (ACI 613-54).
[b]Air-entrained concrete should be used under all conditions involving severe exposure and may be used under mild exposure conditions to improve workability of the mixture.
[c]Soil or groundwater containing sulfate concentrations of more than 0.2%. For moderate sulfate resistance, the tricalcium aluminate content of the cement should be limited to 8% and for high sulfate resistance to 5%; CSA Sulfate-Resisting cement limits tricalcium aluminate to 4%. At equal cement contents, air-entrained concrete is significantly more resistant to sulfate attack than nonairentrained concrete.
[d]Water–cement ratio should be selected on basis of strength and workability requirements, but minimum cement content should not be less than 470 lb per cubic yard (210 kg/m^3).

ratios (by mass) range from 0.53 for ⅜ in. (10 mm) maximum aggregate size to 0.32 for 1.5 in. (38 mm) maximum aggregate size.

The effect of *curing method* upon concrete strength is illustrated in Figs. 11.8 and 11.9. Moist curing generally produces concrete of higher strength than dry curing. The difference in strength is most pronounced at 7 days of curing. Moist curing assures that all the mixing water is available for hydration to take place. In dry curing some of the mixing water evaporates before the hydration process is completed.

The effect of *length of cure* upon compressive strength is illustrated in Figs. 11.8 and 11.9. Generally, strength increases with length of cure. The strength of dry-cured concrete does not increase much after 7 days. However, moist-cured concrete strength increases at a decreasing rate with length of cure. Moist-cured, 28-day strength is the standard by which concrete strength is usually specified.

Some recommended trial mixes for concrete used in agricultural structures are in Table 11.8. Additional guidelines for trial mixes may be found in the PCA Manual, *Guide to Proportioning, Mixing, and Placing Concrete.*

A summary of the primary mechanical properties of hardened concrete is in Table 11.9. Note that the tensile strength is approximately one-tenth as great as the compressive strength. As noted earlier this is the primary reason why reinforced, rather than plain concrete, is used in construction. In Table 11.9, E_c is the secant modulus of the concrete at $f_c = 0.45f'_c$. (See Fig. 11.6.)

11.3 FUNDAMENTAL CONCEPTS

11.3.1 Flexural Behavior of Plain Concrete

The compressive stress–strain behavior of concrete under flexural loading is considerably different from that in pure compression. An idealized version of the flexural compressive stress–strain behavior is shown in Fig. 11.10. Note that the ultimate flexural compressive stress f_c'' only equals $0.85f'_c$ and that the ultimate strain ε_u equals approximately 0.003 at fracture. Also note that the stress–strain diagram is nonlinear even at low stress levels.

11.3.2 Design Methods

The two basic procedures for reinforced concrete design are the working stress design (WSD) and the ultimate stress design (USD) methods.

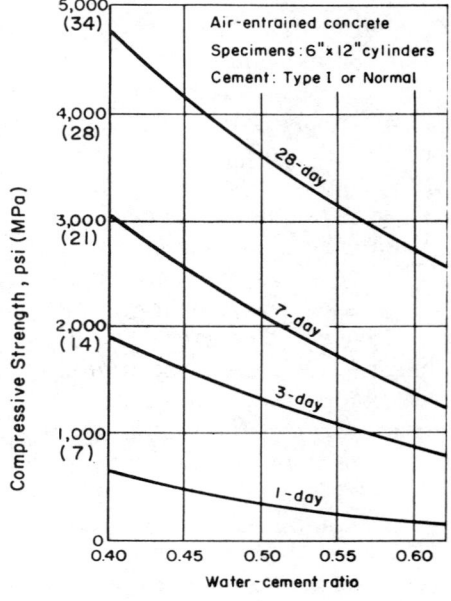

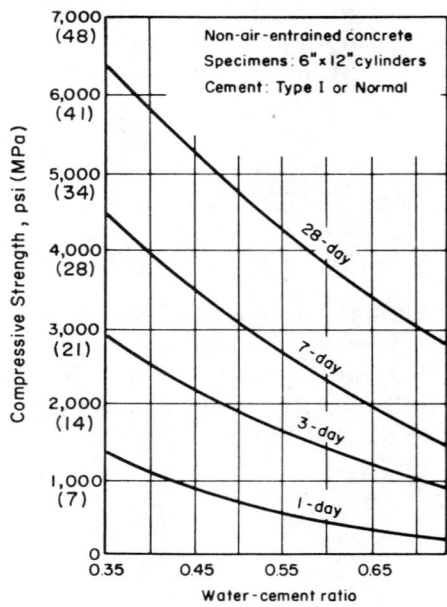

Fig. 11.8. Typical age-strength relationships based on compression tests of 6 × 12 in. (150 × 300 mm) type I normal cement. Moist cured at 70°F (21°C). Reproduced with permission from *Design and Control of Concrete Mixtures*, 11th ed., Portland Cement Association, Skokie, IL.

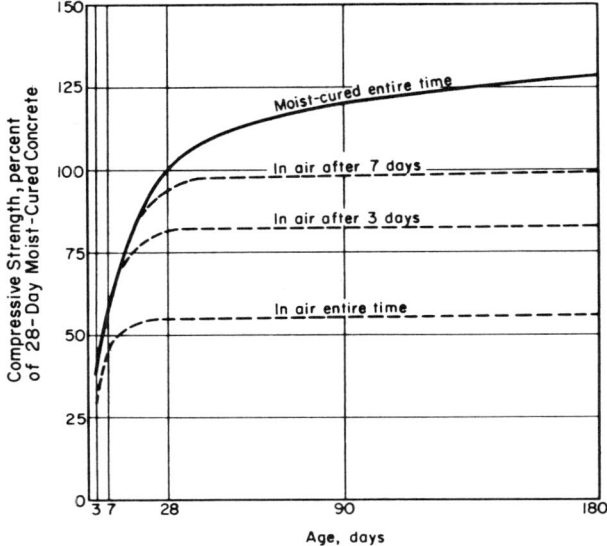

Fig. 11.9. Effect of cure length and type upon compressive strength of concrete. Reproduced with permission from *Design and Control of Concrete Mixtures,* 11th ed., Portland Cement Association, Skokie, IL.

The primary differences between the two methods are summarized in Table 11.10 and the following discussion.

The working stress method uses actual estimated service loads and assumes a linear strain distribution over a flexural member cross section, a linear stress–strain relationship for concrete, the concrete carries no tension, and the allowable concrete stress in flexure equals 0.45 f'_c. The essence of the method is to use actual service loads, assume elastic behavior, and then arbitrarily reduce the allowable stress in the concrete to compensate for discrepancies between elastic and real behavior.

The design philosophy of the ultimate stress method is to use the actual stress–strain behavior of concrete and to apply probability factors to both the service loads and the load, or moment, carrying capacity of a reinforced concrete section. The method assumes a linear strain distribution over the flexural cross section with $\varepsilon_c = \varepsilon_u = 0.003$ at the outer compressive fiber and negligible concrete tensile strength. The maximum allowable flexural compressive stress in concrete is $0.85 f'_c$ which corresponds to laboratory findings. Service loads are factored by amounts dependent upon the variability of the load. For example, if

Table 11.8 Suggested Trial Mixes for Concrete.

		Gallons (L) of Water for Each Sack of Cement,[a] Using:			Suggested Mixture for 1-Sack Trial Batches[b]			
						Aggregates		
	Max. Size Aggregate in. (mm)	Damp sand[c]	Wet (average) sand[d]	Very wet sand[e]	Cement, sacks	Fine	Coarse	Ready-Mix Sacks Cement[f]
					ft³ (m³)	ft³ (m³)	ft³ (m³)	yd³ (m³)
5-gallon mix Use for concrete subjected to severe wear, weather, or weak acid and alkali solutions	¾ (19)	4½ (17)	4 (15)	3½ (13)	1 (0.028)	2 (0.057)	2¼ (0.064)	7¾ (10¼)
6-gallon mix Use for floors (home, barn), driveways, walks, septic tanks, structural concrete	1 (25)	5½ (21)	5 (19)	4½ (17)	1 (0.028)	2¼ (0.064)	3 (0.085)	6¼ (8¼)
	1½ (38)	5½ (21)	5 (19)	4½ (17)	1 (0.028)	2½ (0.071)	3½ (0.099)	6 (7⅞)
7-gallon mix Use for foundation walls, footings, mass concrete, etc.	1½ (38)	6¼ (24)	5½ (21)	4¾ (18)	1 (0.028)	3 (0.085)	4 (0.113)	5 (6½)

Reproduced with permission from *Structures and Environment Handbook,* MWPS-1, 11th ed. 1983. Midwest Plan Service, Ames, IA.

[a]Increasing the proportion of water to cement reduces the strength and durability of concrete. Adjust the proportions of trial batches without changing the water-cement ratio. Reduce gravel to improve smoothness; reduce both sand and gravel to reduce stiffness. One sack equals 94 lb (43 kg).

[b]Proportions will vary slightly depending on gradation of aggregates.

[c]Damp sand will fall apart after being squeezed in the palm of the hand.

[d]Wet sand will ball in the hand when squeezed, but leaves no moisture on the palm.

[e]Very wet sand has been recently rained on or pumped.

[f]Medium consistency [3 in. (76 mm) slump]. Order air-entrained concrete for outdoor use.

Table 11.9. Summary of Concrete Properties.

Property	Magnitude
Ultimate compressive strength, f'_c, ksi (MPa)	2.5–8.0 (17–55)
Tensile strength, f_t, ksi (MPa)	$0.1 f'_c$
Modulus of elasticity,[a] E_c, psi	$33 w^{1.5} \sqrt{f'_c}$
Mass density, w (normal concrete), pcf (kg/m³)	145 (2320)

[a] Relationship is valid for w in pcf and f'_c in psi and yields E_c in psi.

the dead load and live load moments are denoted by M_D and M_L, respectively, the factored loads would be $1.4 M_D$ and $1.7 M_L$. The live load moment is factored upward more than the dead load moment because it is more variable, and, thus, its magnitude is less predictable. Finally, the load capacity of a section is reduced by a capacity reduction factor ϕ. The magnitude of ϕ is dependent upon the inherent variability, or unpredictability, of the section under load. For example, the magnitude for ϕ is 0.90 for the moment capacity of a reinforced concrete beam and 0.65 for a plain concrete beam. The behavior under load of a reinforced concrete beam is more predictable, and thus more reliable, than that of a plain concrete beam. Consequently, the theoretical moment capacity of a reinforced section is reduced by a smaller amount than that of a plain concrete beam.

The advantages of USD are that it uses more realistic levels of loads and section capacities than WSD; and fortunately, many USD design procedures are computationally much simpler than those in WSD.

Prior to 1963, working strength design (WSD) was the accepted method in the ACI Code and ultimate strength design (USD) was not included. In 1963 the ACI Code included USD as an alternate design method. In the 1971, 1977, and 1983 editions of the ACI Code, USD is the primary design method with WSD the alternate method. Most modern reinforced concrete practice is based upon USD, and practically all new literature and texts are based on USD. In keeping with modern trends, the coverage in this text will be limited to USD methods.

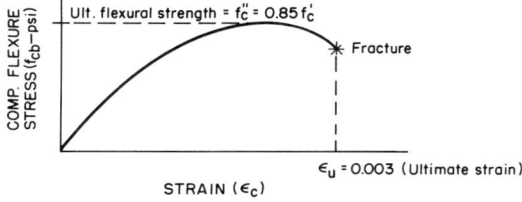

Fig. 11.10. Compressive stress-strain behavior of plain concrete under flexural loading.

Table 11.10. Illustration of Conceptual Differences between WSD and USD for a Beam with Tensile Steel Only.

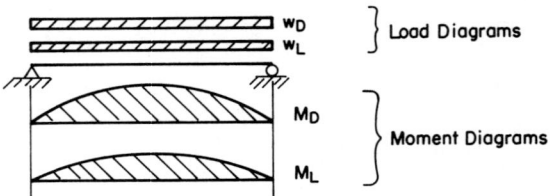

WORKING STRESS (WSD)	ULTIMATE STRESS (USD)
(a.) Service Loads	(a.) Service Loads

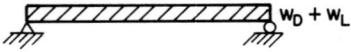 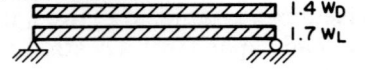

(b.) Req'd Moment Capacity	(b.) Req'd Moment Capacity
$M_{req'd} = M_D + M_L$	$M_{req'd} = M = 1.4 M_D + 1.7 M_L$
(c.) Strain & Stress Distribution	(c.) Strain & Stress Distribution

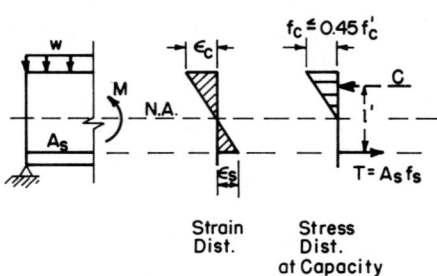 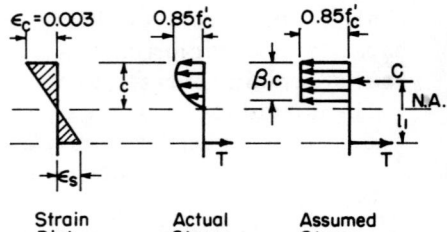

Strain Stress	Strain Actual Assumed
Dist. Dist.	Dist. Stress Stress
at Capacity	at Capacity Dist. Dist.
(d.) Moment Capacity of Section	(d.) Moment Capacity of Section
From $\Sigma F = 0$, $C = T$ and	$\Sigma F = 0 \Rightarrow C = T$
Mom. Cap. $= M = Cl' = Tl'$	Mom. Cap. $= M_n = Cl_1 = Tl_1$
(e.) The Design Criterion	(e.) The Design Criterion
$M \geq M_{req'd} = M_D + M_L$	$\phi M_n \geq M_{req'd} = 1.4 M_D + 1.7 M_L$

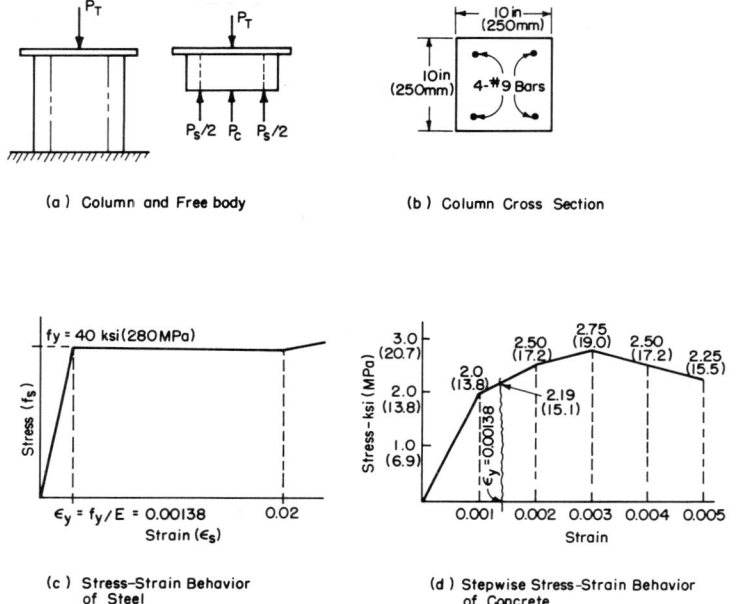

Fig. 11.11. Reinforced column composite behavior.

11.3.3 Composite Behavior of Concrete and Steel

The composite behavior of concrete and steel can be illustrated by considering the behavior of a short axially loaded square compression member, 10 in. (250 mm) on a side with four #9 reinforcing bars symmetrically spaced as shown in Fig. 11.11. The stress–strain behavior of the steel is elastic–plastic with a yield stress of 40 ksi (280 MPa) (see Fig. 11.11). The stress–strain behavior of the concrete is stepwise linear as in Fig. 11.11c. By using the free body diagram of Fig. 11.11a, the equilibrium equation, and Hooke's law, the total load carried by the column, P_T, and the portion carried by both the steel, P_s, and concrete, P_c, are as tabulated in Table 11.11.

When the steel yields at a strain of $\varepsilon_{sy} = 0.00138, P_s = 160$ k, $P_c = 210$ k, and $P_T = 370$ k. At strains beyond ε_{sy}, P_s remains constant while both P_c and P_T continue to increase until the strain exceeds 0.003. The load carrying capacity at ultimate load is $[(400 - 370)/370](100) = 8.1\%$ greater than that when the steel first yields. This behavior, in which fracture or collapse does not occur when the steel first yields, is characteristic of reinforced concrete sections and the concept is used throughout the USD method.

Table 11.11. Load Capacity of a Short Reinforced Column at Various Levels of Uniform Strain.

ε	f_s ksi (MPa)	f_c ksi (MPa)	P_c k (kN)	P_s k (kN)	P_T k (kN)
0.001	29 (200)	2 (13.8)	192 (854)	116 (510)	308 (1370)
0.00138	40 (280)	2.19 (15.1)	210 (934)	160 (712)	370 (1646)
0.002	40 (280)	2.50 (17.2)	240 (1068)	160 (712)	400 (1779)
0.003	40 (280)	2.75 (19.0)	264 (1174)	160 (712)	424 (1886)
0.004	40 (280)	2.50 (17.2)	240 (961)	160 (712)	400 (1779)
0.005	40 (280)	2.25 (15.5)	216 (961)	160 (712)	376 (Failure)

11.3.4 Fundamentals of the Ultimate Strength Design Method

11.3.4.1 Required Strength. All reinforced concrete sections must be proportioned to carry factored service loads. That is, service loads, such as dead, live, and wind, are estimated in accordance with building code requirements and then are factored upward an amount dependent upon their inherent variability. The ACI Code requires the following factoring of service loads.

1. For dead and live loads

$$U = 1.4D + 1.7L$$

2. For dead, live, and wind load combinations, the factored load is the greater of

$$U = 0.75(1.4D + 1.7L + 1.7W),$$
$$U = 0.9D + 1.3W,$$
$$U = 1.4D + 1.7L$$

3. For dead, live, and lateral earth loads, the factored load is the greater of

$$U = 1.4D + 1.7L + 1.7H \quad \text{if } D \text{ and } L \text{ in same direction as } H$$
$$U = 0.9D + 1.7H \quad \text{if } D \text{ or } L \text{ oppose } H$$
$$U = 1.4D + 1.7L$$

4. If fluid pressures F are involved, use the same criteria as for earth pressure combinations except substitute $1.4F$ for $1.7H$ in expressions for U.

In items 1–4, U is the required load or moment the section must carry; D, dead load or moment; L, live load or moment; W,

11.3 Fundamental Concepts

Table 11.12. Strength Reduction Factors, φ.

Type Load and/or Section	φ
Flexure with or without axial tension	0.90
Axial tension	0.90
Axial compression with or without flexure[a]	
spiral columns	0.75
tied columns	0.70
Shear and torsion	0.85
Bearing	0.70
Flexure in plain concrete	0.65

[a] For special cases of lightly loaded columns in which $f_y \leq 60$ ksi (410 MPa), reinforcement is symmetrically placed, $(h - d' - d_s)/h \geq 0.70$. φ may be increased linearly to 0.90 as ϕP_n decreases from $0.10 f'_c A_g$ to 0.

wind load or moment; H, lateral earth pressure load or moment; F, fluid pressure load or moment.

11.3.4.2 Design Strength. The design moment or load capacity of a section is the product of the theoretical capacity times a strength reduction factor φ. For example, if the theoretical moment capacity of a section is M_n, the design moment capacity is ϕM_n. The value of φ depends upon the type section and the predictability of the section behavior under load. Typical values of φ are in Table 11.12.

11.3.4.3 Design Criteria. The basic design criterion is that the design strength be greater than or equal to the required strength. For example, in a flexural member subject to dead and live loads the required moment capacity is $M_u = 1.4 M_D + 1.7 M_L$, the theoretical moment capacity is M_n, and the design moment capacity is ϕM_n. The resulting design criterion is

$$\phi M_n \geq M_u = 1.4 M_D + 1.7 M_L \tag{11.1}$$

11.3.4.4 Strain and Stress Distributions. In all but deep flexural members with a depth/span ratio greater than 2/5 for continuous spans and 4/5 for simple spans, the strain variation over the cross section varies linearly with distance from the neutral axis (see Fig. 11.12b); and at ultimate capacity of the section, the maximum compressive strain in the concrete equals 0.003.

The actual compressive stress distribution in the concrete at ultimate capacity is shown in Fig. 11.12c. The tensile stress in the concrete is assumed to be zero and all tensile forces are carried by the steel. The force T in the steel equals $A_s f_s$, where A_s equals area of steel reinforcing

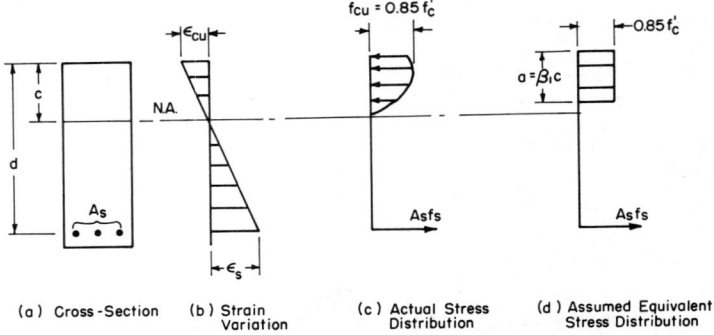

(a) Cross-Section (b) Strain Variation (c) Actual Stress Distribution (d) Assumed Equivalent Stress Distribution

Fig. 11.12. Variation of strain and stress for reinforced concrete flexural and compression members.

and f_s equals steel stress. Note that the actual compressive stress distribution is parabolic with a maximum stress of $0.85f'_c$. Evaluation of the moment capacity of a reinforced concrete section using the actual stress distribution is a mathematically tedious process. The ACI Code allows the designer to assume any compressive stress distribution (rectangular, parabolic, or trapezoidal) which yields design strengths in substantial agreement with test results.

Fortunately for the designer, the rectangular stress distribution (called the Whitney Stress Block) of Fig. 11.12d yields design strengths within a few percentage points of the parabolic stress block. The rectangular distribution is most commonly used in modern practice. The Whitney Stress Block assumes the stress to be uniform at $0.85f'_c$ over a compression zone of depth $a = \beta_1 c$, where c is the distance from outer compression fiber to the N.A.; $\beta_1 = 0.85$ if $f'_c \leq 4$ ksi (28 MPa); and $\beta_1 = 0.85 - 0.05 (f'_c - 4)$ if $f'_c > 4$ ksi (28 MPa), but c is not less than 0.65. In the latter expression for β_1, f'_c must be in ksi for the relationship to be valid.

11.4 FLEXURAL MEMBERS—ULTIMATE STRENGTH DESIGN

11.4.1 Moment Capacity of Composite Sections

The moment capacity M_n of a reinforced concrete cross section cannot be evaluated by the elastic flexure equation $M = f_S$. The composite section violates the assumptions of homogeneity, linearly elastic ma-

11.4 Flexural Members—Ultimate Strength Design 475

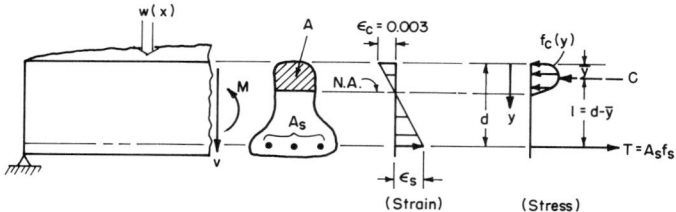

Fig. 11.13. Moment capacity of a general flexural cross section.

terial behavior, and identical stiffness properties in tension and compression.

The moment capacity of the reinforced section is obtained by evaluating the location and magnitude of the resultant compressive and tensile forces acting on a beam cross section and then evaluating the magnitude of the couple. Figure 11.13 helps illustrate the procedure.

At the cut section, the strain distribution is assumed to be linear and the general stress distribution is as shown. The steel carries all the tensile forces. The N.A. is located at the point of zero strain and usually does not coincide with the centroidal axis of the section. The resultant tensile force is located at the centroid of the steel area and equals the steel area times the steel stress, or $T = A_s f_s$. Since $\Sigma F_h = 0$, the resultant compressive force C is necessarily equal to T for the case of no applied axial forces. The compressive resultant can also be evaluated by $C = \int_A [f_c(y)] \, dA$. The location of the resultant force C is simply the center of pressure at $\bar{y}$ and is defined by $\bar{y} = \int y f_c(y)_y \, dA / C$. The moment capacity, which is the magnitude of the couple created by C and T, equals $C(\ell)$ or $T(\ell)$ where $\ell = d - \bar{y}$.

11.4.2 Balance Conditions—USD

The balance condition for USD exists in a reinforced concrete section if the steel in the tension zone reaches its yield strength, $f_s = f_y = E_s \varepsilon_y$ at the same time the concrete in the compression zone reaches a strain, ε_{cu}, of 0.003. If the steel yields before the concrete reaches ε_{cu}, the section is underreinforced. Conversely, if the concrete reaches $\varepsilon_{cu} = 0.003$ before the steel yields, the section is overreinforced. These general states of stress and strain are illustrated in Fig. 11.14.

The ACI Code requires that all reinforced concrete sections be underreinforced. The requirement is that the steel ratio $\rho = A_s/A_g$ shall not exceed 75% of the steel ratio for balance conditions. That is,

$$\rho = A_s/A_g \leq 0.75\rho_b \qquad (11.2)$$

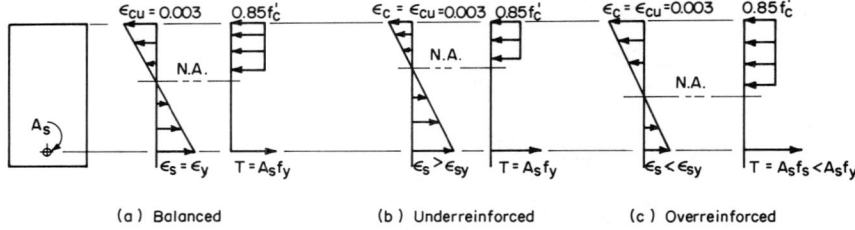

(a) Balanced (b) Underreinforced (c) Overreinforced

Fig. 11.14. Balanced, underreinforced, and overreinforced stress and strain distributions in reinforced concrete at ultimate capacity.

where ρ = steel ratio; A_s, area of tension steel; A_g, area of concrete between the outer compression fiber and the centroid of the tension steel; ρ_b, steel ratio at balance conditions.

This requirement ensures that if the section is overloaded, the steel will yield before the concrete reaches its ultimate strain, and the section will experience a ductile rather than a brittle failure. In a ductile failure, the steel yields excessively, warning of impending failure, long before collapse occurs.

If balanced or overreinforced sections are overloaded, the concrete strain exceeds ε_{cu}. Since concrete is brittle, very little additional strain is required for the section to collapse. Thus, collapse can occur almost explosively with no prior warning.

11.4.3 Analysis of Rectangular Beams with Tension Steel

The singly reinforced beam cross section in Fig. 11.15 has cross-sectional dimensions $b \times d'$ and a steel area A_s less than that required for balance A_{sb}. Thus, the section is underreinforced and the steel stress equals f_y.

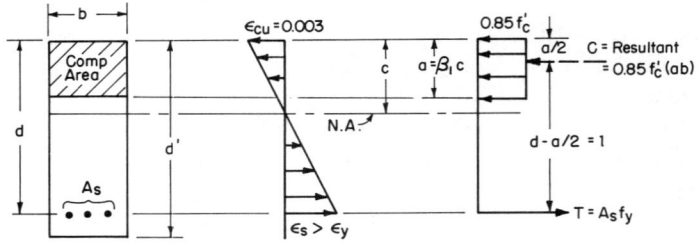

Fig. 11.15. Stress and strain distributions in an underreinforced rectangular beam at ultimate capacity.

11.4 Flexural Members—Ultimate Strength Design

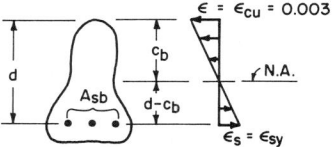

Fig. 11.16. Strain distribution of an arbitrary section at balance conditions.

The theoretical moment capacity of the section is the magnitude of the couple $T(\ell)$ or $C(\ell)$. Thus

$$M_n = T(d - a/2) = A_s f_y (d - a/2) \tag{11.3}$$

or

$$M_n = C(d - a/2) = 0.85 f_c' (ab)(d - a/2) \tag{11.4}$$

In an existing section all the terms in Eqs. (11.3) and (11.4) are known except the depth of the compression block. For flexural members with no axial loads equilibrium yields $T = C$ and $A_s f_y = 0.85 f_c'(ab)$, from which

$$a = A_s f_y / 0.85 f_c' b \tag{11.5}$$

The analysis thus far assumes that the section is underreinforced and that the steel yields. When analyzing a beam, it is necessary to check that the steel indeed does yield and that $\rho < 0.75\,\rho_b$. Utilizing the definition of a balanced section and the assumption of a linear strain distribution helps define the location of the N.A. at balance and subsequently yields a tool for checking the assumption of an underreinforced section.

The generally shaped section of Fig. 11.16 has a balanced steel ratio and is loaded to its design strength. Thus, the top fibers have a compressive strain of 0.003 and the steel has just yielded so $\varepsilon_s = f_y/E_s$. From the strain diagram and similar triangles

$$\frac{c_b}{0.003} = \frac{d - c_b}{\varepsilon_y} \tag{11.6}$$

where c_b = distance to N.A. at balance.

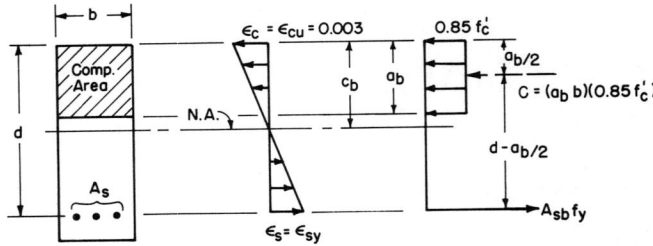

Fig. 11.17. Definition sketch for a singly reinforced rectangular section at balance conditions.

Rearranging Eq. (11.6) and substituting $\varepsilon_y = f_y/E_s$ yields an expression for c_b,

$$c_b = \frac{0.003 d E_s}{0.003 E_s + f_y} \tag{11.7}$$

Equation (11.8) yields the depth of the rectangular stress block at balance.

$$a_b = \beta_1 c_b = \beta_1 \frac{0.003 d E_s}{0.003 E_s + f_y} \tag{11.8}$$

Equations (11.7) and (11.8) were derived for an arbitrary singly reinforced section and may be used to define balance conditions for any singly reinforced beam.

The steel area and steel ratio at balance can now be derived for a rectangular section. Figure 11.17 shows a singly reinforced rectangular section along with the stress distributions at balance conditions. From $\Sigma F_h = 0$, $T = C$, and $A_{sb} f_y = 0.85 f'_c a_b b$. Rearranging and substituting Eq. (11.8) for a_b,

$$A_{sb} = (0.85 \beta_1 f'_c/f_y)\,[0.003 E_s/(0.003 E_s + f_y)] b d \tag{11.9}$$

Directly, the steel ratio at balance is

$$\rho_b = A_{sb}/bd = (0.85 \beta_1 f'_c/f_y)[0.003 E_s/(0.003 E_s + f_y)] \tag{11.10}$$

11.4 Flexural Members—Ultimate Strength Design

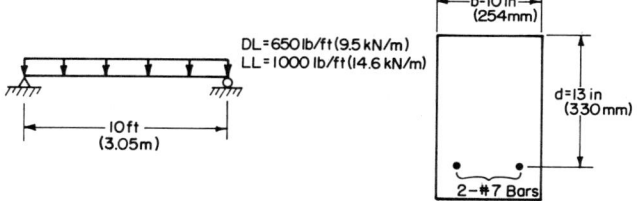

Fig. 11.18. Example 11.1—Singly reinforced beam.

The maximum allowable steel ratio is 0.75 ρ_b. Thus

$$\rho_{\text{all}} = (0.638\beta_1 f'_c/f_y)[0.003E_s/(0.003E_s + f_y)] \quad (11.11)$$

and

$$(A_s)_{\text{all}} = \rho_{\text{all}}(bd) \quad (11.12)$$

To prevent cracking of lightly loaded beams, the minimum steel ratio for a section of any shape is

$$\rho_{\min} = 200/f_y \quad (11.13)$$

where f_y is in pounds per square inch.

Example 11.1. Singly Reinforced Beam Analysis. The reinforced concrete beam of Fig. 11.18 is to be used in a confinement beef housing system. The beam spans 10 ft (3.05 m) and carries the uniformly distributed dead and live loads shown in Fig. 11.18. The reinforced concrete section (rectangular) in the figure has been designed to carry the flexural stresses. Use the ultimate design specification to determine if the section can carry the flexural stresses. The yield stress of steel is 36 ksi (250 MPa) and the compressive strength of the concrete is 3.5 ksi (25 MPa).

Solution

a. Evaluate the steel ratio ρ,

$$\rho = A_s/bd = 2(0.6)/(13 \times 10) = 0.0092$$

b. Evaluate the balanced steel ratio ρ_b with Eq. (11.10)

$$\rho_b = (0.85\beta_1 f'_c/f_y)[0.003E_s/(0.003E_s + f_y)]$$

$$\rho_b = [0.85(0.85)(3.5)/36][0.003 \times 29 \times 10^3/$$

$$(0.003 \times 29 \times 10^3 + 36)]$$

$$= 0.0497$$

Since $\rho < \rho_b$, beam is *underreinforced*.

c. Evaluate the magnitude of the allowable steel ratio, $0.75\rho_b = 0.0373 > \rho$. Thus the steel ratio is okay.

d. Check if $\rho > \rho_{min} = 200/f_y$

$$\rho_{min} = 200/36{,}000 = 0.0056 < \rho$$

Therefore, the steel ratio exceeds the minimum requirements, the steel yields, and the steel stress f_s equals the steel yield stress f_y.

e. Evaluate the depth of the compressive stress block a and the theoretical moment capacity M_n with Eqs. (11.5) and (11.3), respectively.

$$a = A_s f_y / 0.85 f'_c b = 1.2(36)/0.85(3.5)(10)$$

$$= 1.45 \text{ in.} \quad (37 \text{ mm})$$

$$M_n = A_s f_y (d - a/2)$$

$$M_n = 1.2(36{,}000)(13 - 1.45/2)$$

$$= 530 \text{ k} \cdot \text{in} \quad (59.9 \text{ kN} \cdot \text{m})$$

f. Allowable moment capacity is

$$\phi M_n = 0.9(530{,}000) = 477 \text{ k} \cdot \text{in} \quad (53.9 \text{ kN} \cdot \text{m})$$

g. Evaluate the required moment capacity M_u and compare to M_n. The factored uniformly distributed load for a dead plus live load combination is

$$w = 1.4w_D + 1.7 w_L$$

$$= 1.4(650) + 1.7(1000) = 2610 \text{ lb/ft} \quad (38.1 \text{ kN/m})$$

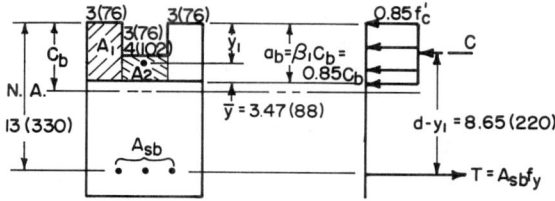

Fig. 11.19. Example 11.2—Analysis of a nonrectangular cross section. All dimensions are in in. (mm).

Using the factored load in the expression for the maximum moment in a simply supported beam yields

$$M_u = wL^2/8 = 2610(10)^2/8 \times 12 = 392 \text{ k-in} \quad 44.3 \text{ kN-m})$$

Since $M_u < \phi M_n$, the section is adequate.

11.4.4 Analysis of Singly Reinforced Nonrectangular Beams

Analysis of nonrectangular sections parallels that of rectangular sections. The stress and strain distributions are similar. The primary difference is that the compression resultant C is not located at $a/2$. Instead its location is found by locating the centroid of the nonrectangular compression zone. The technique will be demonstrated in Example 11.2.

Example 11.2. Analysis of Nonrectangular Beams. Evaluate the moment capacity of the section in Fig. 11.19 for a balanced area of steel. Also evaluate the design moment capacity for three No. 7 bars. Let $f_y = 36$ ksi (250 MPa) and $f_c' = 3.5$ ksi (25 MPa).

Solution for Balance Conditions

a. Locate N.A. at balance (assume N.A. below the notch) using Eq. (11.7).

$$c_b = 0.003 dE_s/(0.003E_s + f_y) = 0.003(13)(29 \times 10^3)/$$

$$(0.003 \times 29 \times 10^3 + 36)$$

$$= 9.2 \text{ in. } (234 \text{ mm})$$

and $a_b = 0.85(9.2) = 7.82$ in. (199 mm).

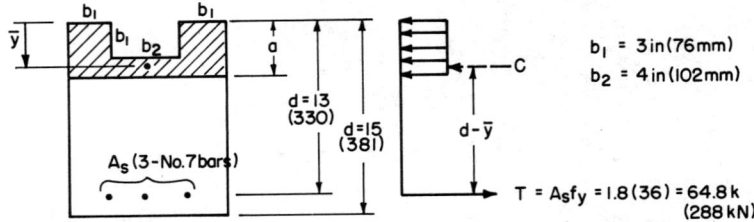

Fig. 11.20. Section and force diagram for part 2 of Example 11.2.

Thus, the N.A. is below the notch as assumed.

b. Evaluate C the compressive resultant:

$$C = 0.85f'_c \, (ba_b - 4 \times 3) = 0.85(3.5)[10 \times 7.82 - 12]$$
$$= 197 \text{ k} \quad (876 \text{ kN})$$

c. Locate the distance between the bottom of the compression zone and the centroid of the compression zone y by using the first moment of areas.

$$A_T \bar{y} = 2A_1(a_b/2) + A_2(a_b - 3)/2$$

Substituting for A_T, a_b, A_1, and A_2,

$$(7.82 \times 10 - 12)\bar{y} = 2(7.82)(3)(7.82/2) + 4(7.82 - 3)(4.82/2)$$
$$\bar{y} = 3.47 \text{ in.} \quad (88 \text{ mm})$$
$$y_1 = 7.82 - 3.47 = 4.35 \text{ in.} \quad (110 \text{ mm})$$

d. Evaluate the balance design strength of ϕM_n

$$M_n = C(d - y_1) = 197(13 - 4.35) = 1700 \text{ k} \cdot \text{in.}$$
$$(192.1 \text{ kN} \cdot \text{m})$$

and

$$\phi M_n = 0.9(1700) = 1530 \text{ k} \cdot \text{in.} \quad (172.9 \text{ kN} \cdot \text{m})$$

Solution for the Moment Capacity for Three No. 7 Bars

a. Sketch the section and force diagrams as in Fig. 11.20.

b. Evaluate the magnitude of a by assuming $a > 3$ in. (76 mm) and summing forces in the horizontal direction

$$\Sigma F_h = 0 : T = C$$

$$64.8 = 0.85 f'_c(10a - 12)$$

$$a = 3.38 \text{ in. (86 mm)} > 3 \text{ in. (76 mm)} \quad \text{as assumed}$$

Since $a < a_b$ as evaluated in the previous solution, the section is underreinforced.

c. Locate the centroid of the compression zone by taking the first moment of the compression area about the top edge of the section.

$$\bar{y}(3.38 \times 10 - 12) = (3.38/2)(10 \times 3.38) - 1.5(12)$$

$$\bar{y} = 1.79 \text{ in. (45 mm)}$$

d. Evaluate design capacity ϕM_n,

$$M_n = T(d - \bar{y}) = 64.8(13 - 1.79) = 730 \text{ k} \cdot \text{in.}$$

$$(82.0 \text{ kN} \cdot \text{m})$$

and

$$\phi M_n = 0.9(730) = 660 \text{ k} \cdot \text{in} \quad (74.2 \text{ kN} \cdot \text{m})$$

11.4.5 Design of Singly Reinforced Rectangular Beams

For design purposes it is convenient for Eq. (11.3) to be rewritten by defining the parameter $\omega = \rho(f_y/f'_c)$. Then Eq. (11.5) for the depth of the compression zone a can be redefined in terms of ω as follows:

$$a = A_s f_y / 0.85 f'_c b = (\rho b d) f_y / 0.85 f'_c b = 1.18 \rho d(f_y/f'_c) = 1.18 \omega d$$

Substituting this expression for a into Eq. (11.3) yields

$$M_n = A_s f_y (d - 0.59 \omega d) \tag{11.14}$$

Since $\omega = (A_s/bd)(f_y/f'_c)$, $A_s f_y = \omega(bd)f'_c$. Substituting this term into Eq. (11.14), yields Eq. (11.15), which is a very useful form of Eq. (11.3) for designing sections.

$$M_n = f'_c(bd^2)(\omega - 0.59 \omega^2) \tag{11.15}$$

484 Reinforced Concrete Design

In one common design situation b and d are known and it is required to evaluate the steel area required. One systematic approach is outlined herein:

1. Determine the quantity $\omega - 0.59\omega^2$ by rearranging Eq. (11.15) and substituting the required moment capacity M_u for M_n

$$(\omega - 0.59\,\omega^2) = M_u/(f_c'\,bd^2)$$

2. Solve for ω by evaluating the quadratic equation or by trial and error.
3. Evaluate the required steel ratio by $\rho = \omega f_c'/f_y$.
4. Find the area of steel from $A_s = \rho(bd)$.
5. Check if $\rho_{min} \leq \rho \leq 0.75\rho_b$.

Example 11.3. Design When b and d Are Known. The section of Example 11.1 and Fig. 11.18 is obviously overdesigned. Design an efficient section with $b = 8$ in. (203 mm) and $d = 10$ in. (254 mm) which will carry the required moment $M_u = 392$ k $\cdot$ in. (44.3 kN $\cdot$ m).

Solution

a. Evaluate the parameter $\omega - 0.59\,\omega^2$

$$(\omega - 0.59\,\omega^2) = M_u/(f_c'\,bd^2) = 392/(3.5(8)(10)^2) = 0.14$$

b. Solve for ω

$$0.59\,\omega^2 - \omega + 0.14 = 0$$
$$\omega^2 - 1.69\omega + 0.24 = 0$$
$$\omega = 1.54,\ 0.16$$

c. Evaluate the steel ratio required to carry the moment

$$\rho = \omega f_c'/f_y$$
$$= 1.54(3.5/36) = 0.150$$

or

$$\rho = 0.16(3.5/36) = 0.016$$

d. Check ρ against the maximum allowable steel ratio from Eq. (11.11)

$$\rho_{all} = (0.638(.85)(3.5)/36)\,(0.003 \times 29 \times 10^3/(0.003 \times 29 \times 10^3 + 36)) = 0.037$$

Thus $\rho = 0.150$ would result in an overreinforced section and the required steel ratio ρ equals 0.016.

e. Evaluate the required steel area

$$A_s = \rho bd = 0.016(8)(10) = 1.28 \text{ in.}^2 \quad (830 \text{ mm}^2)$$

Thus, the required steel located at $d = 10$ in. (254 mm) below the top fiber is two No. 7 bars with $f_y = 36$ ksi (250 MPa).

There are design situations where b, d, and A_s are unknown. In such cases the following steps yield a fairly direct solution:

1. Evaluate the balance steel ratio ρ_b using Eq. (11.10).
2. Assume $\rho \leq 0.75\,\rho_b$.
3. Evaluate $\omega = \rho f_y/f'_c$ and the parameter $(\omega - 0.59\omega^2)$.
4. Rearrange Eq. (11.15) to get an expression for the required (bd^2).

$$(bd^2)_{req'd} = M_u/(f'_c\,(\omega - 0.59\,\omega^2))$$

5. Select, by trial and error, b and d such that $bd^2 \geq (bd^2)_{req'd}$
6. Evaluate the steel area using $A_s = \rho(bd)$.

Example 11.4. Beam Design. Design a singly reinforced rectangular beam to carry a factored moment of 100 k · in. (11 kN · m). Assume $f'_c = 4$ ksi (28 MPa) and $f_y = 40$ ksi (280 MPa).

Solution

a. Evaluate ρ_b using Eq. 11.10.

$$\rho_b = (0.85)(0.85)(4)/(40)\,[0.003 \times 29 \times 10^3/(0.003 \times 29 \times 10^3 + 40)] = 0.049$$

b. Evaluate ρ_{all}.

$$\rho = 0.75\,\rho_b = 0.037$$

c. Evaluate ω and $(\omega - 0.59\,\omega^2)$.

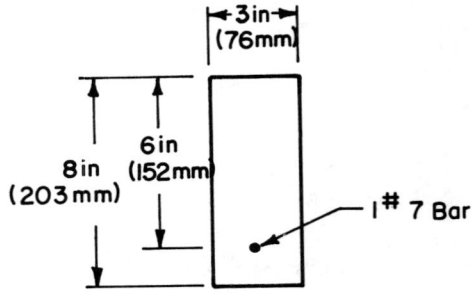

Fig. 11.21. Example 11.4—Section design.

$$\omega = 0.037(40/4) = 0.37$$
$$(\omega - 0.59\,\omega^2) = 0.29$$

d. Evaluate $(bd^2)_{\text{req'd}}$.

$$(bd^2) = 100/[(0.29)(4)] = 86.2 \text{ in.}^3 \ (1412 \text{ cm}^3)$$

e. Select b and d. Suppose $d = 2b$, then $b(2b)^2 = 86.2$ and $b = \sqrt[3]{21.55} = 2.76$ in. (70 mm).

f. Thus, select a beam with $b = 3$ in. (76 mm), $d = 6$ in. (152 mm), and $A_s = \rho bd = 0.037(3)(6) = 0.67$ in.² (430 mm²).

g. Since b and d are somewhat larger than required, try a No. 7 bar with $A_s = 0.60$ in.² (390 mm²) and check the moment capacity. Since the section is underreinforced, the depth of the compression area is found by Eq. (11.5) and the theoretical moment capacity M_n is calculated by Eq. (11.3).

$$a = A_s f_y/0.85 f'_c b = 0.6(40)/(0.85(4)(3)) = 2.35 \text{ in.} \quad (60 \text{ mm})$$

$$M_n = A_s f_y\,(d - a/2) = 0.6(40)(6 - 2.35/2)$$
$$= 116 \text{ k} \cdot \text{in.} \ (13.1 \text{ kN} \cdot \text{m})$$

Thus, $\phi M_n = 104$ k · in. (11.8 · kN · m) > 100 k · in. (11.3 kN · m) and the section as shown in Fig. 11.21 is adequate.

11.5 REINFORCED CONCRETE COLUMNS

11.5.1 General Requirements

The term reinforced column, in the context of this text, includes any reinforced concrete section subject to either pure axial compression or

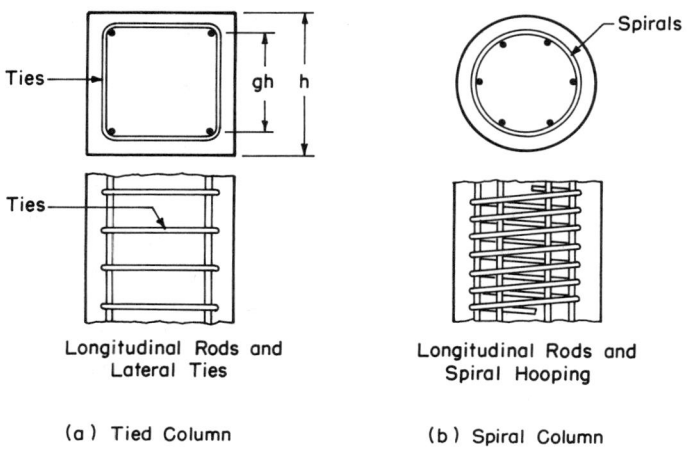

Fig. 11.22. Reinforced column types.

combined axial compression and bending moments. The moments may be induced by lateral loading, or eccentricity of axial loading.

There are two primary types of reinforced concrete columns; namely, tied and spiral columns. In tied columns the longitudinal bars are enclosed in a series of closed ties as in Fig. 11.22a. In spiral columns the longitudinal bars are enclosed by a continuous closely spaced spiral as in Fig. 11.22b.

The lateral ties serve several functions. They hold the longitudinal bars in place during construction, they prevent the highly stressed longitudinal bars from buckling outward and breaking off the outer layer of concrete, and they carry flexural shear stresses induced by lateral forces.

In tied columns the ties should be at least No. 3 bars when longitudinal bars are less than or equal to No. 10 bars. The tie size may be larger if shear forces are large. (Shear reinforcement will be covered later.) Tie spacings should be less than or equal to the smaller of 16 longitudinal bar diameters, 48 tie bar diameters, or the least dimension of the column. Every corner and alternate longitudinal bar must be supported by a tie with an included angle of not more than 135° and the maximum allowable distance between a corner tied bar and alternate bar is 6 in. (152 mm) on both sides.

The amount of lateral reinforcement required in spiral columns is specified by the spiral column steel ratio ρ_s as defined in the following:

$$\rho_s \geq 0.45 \, (A_g/A_c - 1) \, f'_c/f_y$$

where ρ_s is (volume of spiral steel/loop)/(volume of enclosed concrete per spiral loop); A_g, gross column area; A_c, concrete area enclosed by outside of spiral. Spirals must be No. 3 bars or larger and spacing between spirals must be between 1 and 3 in. (25 and 76 mm) but greater than or equal to 4/3 × (maximum aggregate size). Spiral size may also be dictated by shear requirements.

The limits on longitudinal reinforcement in all columns are as follows: (1) The steel ratio, based on gross column area, is to be between 0.01 and 0.08. The larger value is a practical limit on placement while the lower limit protects against unexpected flexure. (2) The minimum number of longitudinal bars is four for rectangular arrangements and six for circular arrangements. The capacity reduction factors ϕ for reinforced columns are 0.75 for spiral columns and 0.70 for tied columns except in lightly loaded columns ϕ may be increased as follows.

1. If g, as defined in Fig. 11.22, is greater than or equal to 0.7, if $f_y \leq 60$ ksi (410 MPa), and if the reinforcement is symmetrically placed

$$\phi = 0.9 - (P_n/0.1f'_c A_g)(0.90 - \phi_c)$$

where P_n is design axial load strength at required eccentricity; ϕ_c is 0.70 for tied columns and 0.75 for spiral columns.

2. For columns where $g < 0.7$ or for eccentrically placed reinforcement

$$\phi = (0.90 - P_b/F)(0.9 - \phi_c)$$

where F is the smaller of $0.1f'_c A_g$ or P_b; P_b is design axial load strength at balanced strain conditions and will be defined in a later section.

11.5.2 Short Centrically Loaded Column Analysis

The basic design criterion for analysis or design of reinforced columns is that the factored axial load $P_u = 1.4P_D + 1.7 P_L$ must be less than the allowable axial load capacity ϕP_n. The theoretical ultimate load capacity of a centrically loaded short compression member is the internal resistance provided when the entire concrete area is uniformly strained to $\varepsilon_c = 0.003$. At this strain level the steel has already yielded

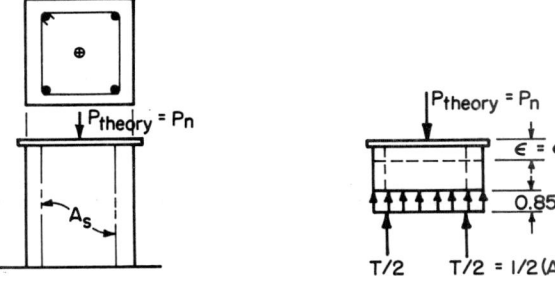

Fig. 11.23. Theoretical axial ultimate load capacity of a short reinforced concrete column.

and $f_s = f_y$. By summing forces in the longitudinal direction in the free body of Fig. 11.23, the theoretical load capacity P_n is

$$P_n = (0.85 f'_c)(A_g - A_s) + f_y A_s \qquad (11.16)$$

The theoretical load capacity can never be reached because accidental loading eccentricities cannot be avoided. Also, the maximum design strength of a short column must be further reduced by the capacity reduction factors to account for unpredictable variations in material behavior. Thus, the maximum allowable load capacity for axially loaded short columns is defined by Eqs. (11.17) and (11.18) for spiral and tied columns, respectively.

For spiral columns

$$(\phi P_n)_{max} = 0.85\phi[0.85 f'_c(A_g - A_s) + A_s f_y] \qquad (11.17)$$

For tied columns

$$(\phi P_n)_{max} = 0.80\phi[0.85 f'_c(A_g - A_s) + A_s f_y] \qquad (11.18)$$

Past ACI code specifications required that axially loaded columns be analyzed with a minimum eccentricity. The current specification simplifies the procedure by accounting for minimum eccentricity with the capacity reduction factors of 0.85 and 0.80.

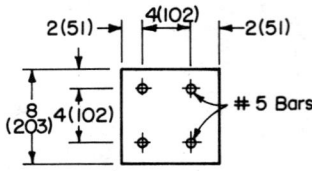

Fig. 11.24. Example 11.5—Column cross section. All dimensions are in in. (mm).

Example 11.5. Analysis of a Short Axially Loaded Column. Evaluate the maximum load capacity of a short-tied square column, with the dimensions shown in Fig. 11.24 and reinforced with four No. 5 bars. Assume f_y = 40 ksi (280 MPa), f'_c = 4 ksi (28 MPa), and that the section is properly tied.

Solution

a. The properties of the cross section in Fig. 11.24 are

$$A_s = 0.31 \times 4 = 1.24 \text{ in.}^2 \text{ (800 mm}^2\text{)}$$

$$A_g = 64 \text{ in.}^2 \text{ (413 cm}^2\text{)}$$

$$A_c = 64 - 1.24 = 62.76 \text{ in.}^2 \text{ (405 cm}^2\text{)}$$

b. Check the limits of ρ.

$$\rho = A_s/A_g = 1.24/64.0 = 0.019$$

and falls within the acceptable range of 0.01 and 0.08.

c. Evaluate the maximum design strength.

$$(\phi P_n)_{max} = 0.80(0.70)[0.85(4)(64 - 1.24) + 40(1.24)]$$

$$(\phi P_n)_{max} = 147 \text{ k} \quad (654 \text{ kN})$$

For a column to be loaded axially the axial load must pass through the plastic centroid. The plastic centroid is that load point required to produce a uniform strain of 0.003 at fracture. Its location is simply at the center of pressure of all the internal forces. The following example illustrates how it is evaluated.

Example 11.6. Location of the Plastic Centroid. Find the plastic centroid of the column cross section shown in Fig. 11.25. Assume f'_c = 3 ksi (21 MPa), f_y = 50 ksi (340 MPa), and A_s = 8 in.² (51.7 cm²).

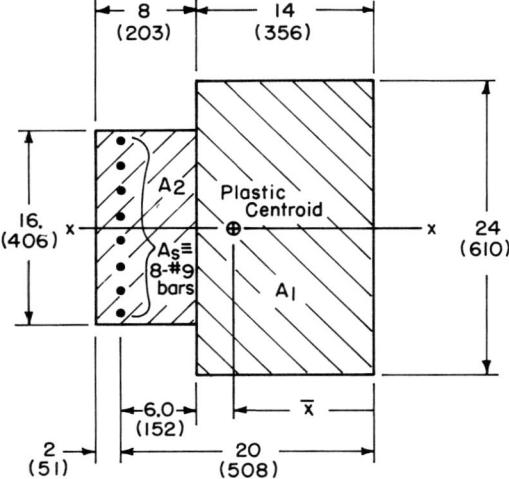

Fig. 11.25. Example 11.6—Column section. All dimensions are in in. (mm).

Solution

a. The center of pressure is defined from statics by

$$\bar{x} = \frac{\Sigma xF}{\Sigma F} = \frac{(A_1\bar{x}_1 + A_2\bar{x}_2)f_c + A_s\bar{x}_sf_y - A_s\bar{x}_sf_c}{(A_1 + A_2 - A_s)f_c + A_sf_y}$$

b. Substituting $f_c = 0.85 f'_c = 2.55$ ksi (18 MPa) and evaluating $\bar{x}$,

$$\bar{x} = \frac{[336(7) + 128(18) - 8(20)]\,2.55 + 8(20)(50)}{(336 + 128 - 8)(2.55) + 8(50)}$$

$$\bar{x} = 12.46 \text{ in.} \quad (316 \text{ mm})$$

11.5.3 Short Eccentrically Loaded Column Analysis

Most columns in practice are subject to eccentric loads or to lateral loads and end restraints which induce flexural stresses in the column. The basic analysis technique is similar for any of these situations and involves the use of equivalent column loading schemes.

In statics and strength of materials it was shown that an eccentric

492 Reinforced Concrete Design

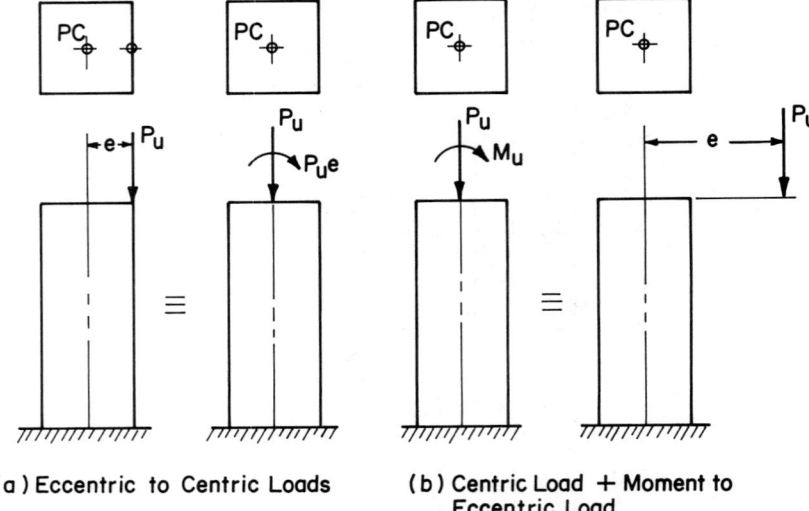

Fig. 11.26. Equivalent loadings for columns. PC = plastic centroid of the section.

compressive load on a column can be replaced by an equal load acting through the member centroid and a couple equal to $P_u e$ (see Fig. 11.26a). Similarly, if a column is subject to an axial load P_u and a bending moment M_u, it can be replaced by an equivalent system of forces P_u at an eccentricity e such that $M_u = P_u e$ (see Fig. 11.26b). In eccentrically loaded columns and beam columns, the analysis technique will be to transform any bending moments and/or axial loads into an equivalent axial load P_u at an eccentricity e from the plastic centroid of the column cross section.

The relationship between the eccentricity of loading and the stress and strain distributions in the column section are of paramount importance in eccentrically loaded column analysis. Figures 11.27 and 11.28 and the ensuing discussion summarize the salient points.

The steel in an eccentrically loaded column is loaded in tension on the side opposite the direction of the eccentricity and in compression on the side of the eccentricity. All parameters related to compression steel will be designated by primes (A'_s, d', f'_s), whereas their tension steel counterparts will be unprimed (see Fig. 11.27). At the limiting axial load at a given eccentricity the concrete will have just reached its ultimate strain of 0.003. Dependent upon the magnitude of eccentricity, the tension steel will yield simultaneously as $\varepsilon_c = 0.003$ (bal-

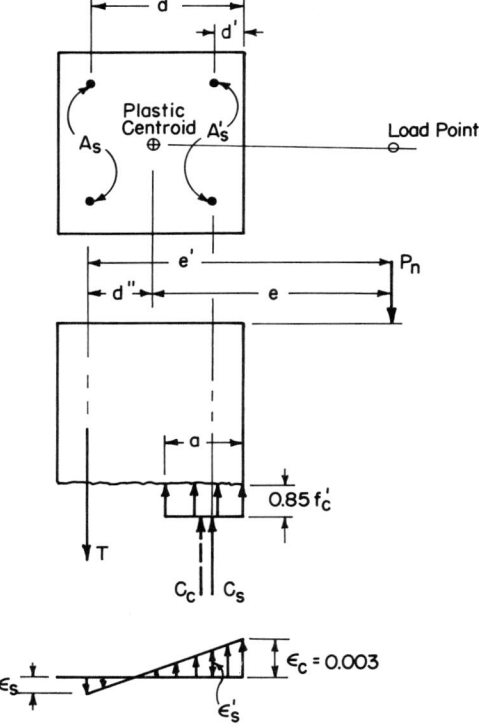

Fig. 11.27. Definition sketch for eccentrically loaded columns.

ance condition) or the tension steel will have already yielded (underreinforced-tension control), or the steel will not have yielded and $f_s = E_s \varepsilon_s$ (overreinforced-compression control). In any of the three situations, the compression steel A'_s may or may not have yielded.

The balance condition in the column is produced by a unique combination of load and eccentricity. The combination denoted by P_{nb} and e_b, respectively, is called the balanced load and eccentricity.

The stress and strain distributions for the various possibilities of eccentricity are shown in Fig. 11.28. The stress distribution at balance ($e = e_b$) is shown in Fig. 11.28b. Note that A_s has just yielded, but A'_s may or may not have yielded. The magnitude of the compression block depth at balance a_b is calculated exactly as it was earlier for beams.

When $e > e_b$, the steel yields further but the tension steel resultant T remains constant at $A_s f_y$ and the compression zone decreases ($a <$

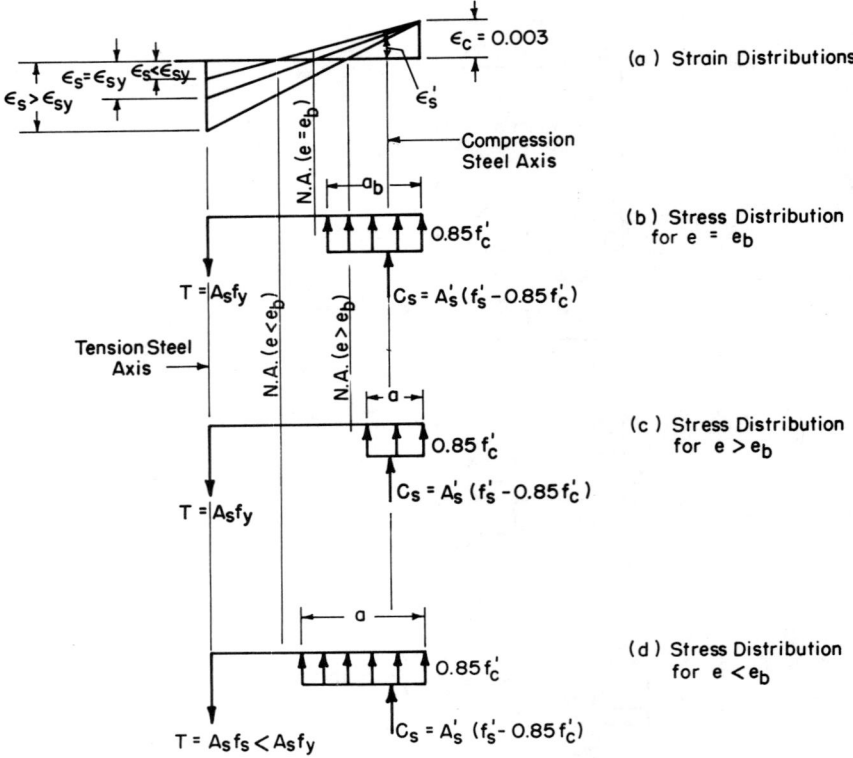

Fig. 11.28. Stress and strain distributions for reinforced columns for $e = e_b$, $e < e_b$, and $e > e_b$.

a_b). Also ε'_s decreases, so the compression steel resultant C_s decreases. The net result is that as e increases, the axial load capacity P_n of the section decreases.

When $e < e_b$, A_s no longer yields, the compression block size increases ($a > a_b$) and ε'_s increases, but may still be less than ε_y. The net result is that when $e < e_b$, $P_n > P_{nb}$ and increases to a limiting value of $(P_n)_{\max}$ as computed by Eq. (11.16) for a centrically loaded column ($e = 0$).

The eccentrically loaded column problem thus reduces to evaluating the combinations of P and e which it can carry and ensuring that $0.01 \leq \rho \leq 0.08$. A handy design aid for analyzing load capacities is the interaction diagram. An interaction diagram is a plot of the combinations of axial load ϕP_n and moment $\phi P_n e$ which induce failure. Any combination which falls within the interaction diagram envelope can

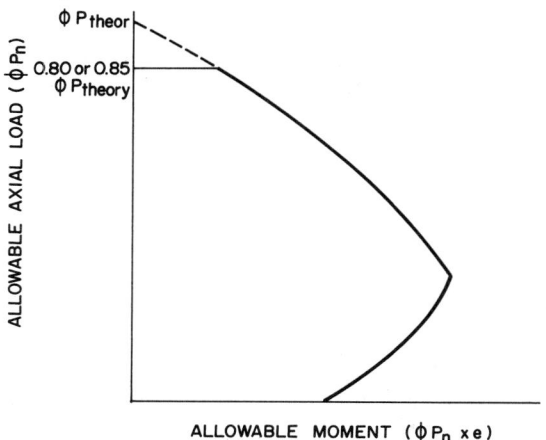

Fig. 11.29. Interaction diagram for an eccentrically loaded column.

safely be carried by the column. Figure 11.29 is a general interaction diagram.

Equilibrium of the reinforced column section (Figs. 11.27 and 11.28) is more complex than that for singly reinforced beams. The introduction of the axial load P_u is a state of stress where the internal compression and tensile resultants C and T are no longer equal and are no longer a simple couple. The main equilibrium conditions used in analysis are the summation of forces and the summation of moments about the tensile steel ($\Sigma M_T = 0$).

$$\Sigma F = 0 : P_n = C_c + C_s - T \qquad (11.19)$$

$$\Sigma M_T = 0 : P_n e' = C_c(d - a/2) + C_s(d - d') \qquad (11.20)$$

An alternate equilibrium equation in which moments are summed about the plastic centroid (PC) is useful when $e < e_b$ and A_s does not yield.

$$\Sigma M_{PC} = 0 : P_n e$$
$$= Td'' + C_s(d - d'' - d') + C_c(d - d'' - a/2) \qquad (11.21)$$

The following illustration shows how to use these principles to evaluate the load capacity of an eccentricity loaded column and will show how to generate an interaction diagram.

Example 11.7. Eccentrically Loaded Column. Evaluate the allowable axial load ϕP_n for the short-tied column cross section of Example 11.5 (Fig. 11.24) for eccentricity $e = 0, e_b, 0.5e_b, 2e_b$, and ∞. Then sketch the interaction diagram for the section.

Solution

a. Evaluate the load capacity ϕP_n when $e = 0$. The maximum load for zero eccentricity from Example 11.5 equals $(\phi P_n)_{max} = 147$ k (654 kN). Note that the plastic centroid and centroidal axis of concrete coincide since all steel yielded and the steel is symmetrically placed.

b. Evaluate the load capacity, $\phi P_n = \phi P_{nb}$, when $e = e_b$ (see Fig. 11.28b). At balance conditions A_s yields and A'_s will be assumed to yield. (This assumption will have to be checked.)

    i. Using Eq. (11.8), the depth of the compression block and the N.A. are calculated.

$$a_b = 0.85[0.003E_s/(0.003E_s + f_y)]d$$

$$= 0.85[0.003 \times 29 \times 10^3/(0.003 \times 29 \times 10^3 + 40)](6)$$

$$= 3.49 \text{ in. } (89 \text{ mm})$$

$$c_b = 3.49/0.85 = 4.11 \text{ in. } (104 \text{ mm})$$

    ii. Evaluate the strain in the compression steel, ε'_s, using the strain distribution of Fig. 11.30a and similar triangles and compare it to the yield strain of the steel, ε_{sy}.

$$\varepsilon'_s = 2.11/4.11 \,(0.003) = 0.00154$$

$$\varepsilon_{sy} = f_y/E = 40/(29 \times 10^3) = 0.00138$$

Since $\varepsilon'_s > \varepsilon_{sy}$, the compression steel A'_s yields and $f'_s = f_y$.

    iii. Evaluate the balance load capacity, P_{nb}, by summing forces in the axial direction in Fig. 11.28b.

$$P_{nb} = C_c + C_s - T$$

$$= 0.85f'_c(a_b \times 8) + (f'_s - 0.85f'_c)A'_s - f_yA_s$$

$$= 0.85(4)(3.49 \times 8) + (40 - 0.85 \times 4)(0.62)$$

$$- 40(0.62)$$

$$= 92.8 \text{ k } (412 \text{ kN})$$

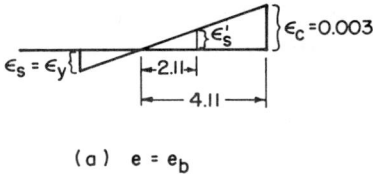

(a) e = e_b

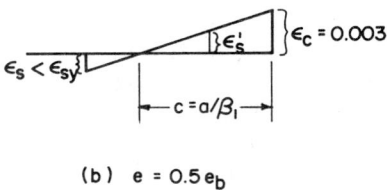

(b) e = 0.5 e_b

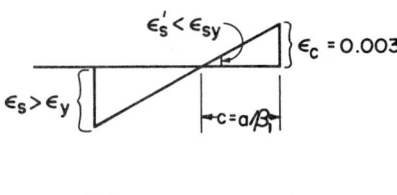

(c) e = 2e_b and $\epsilon'_s < \epsilon_{sy}$

Fig. 11.30. Example 11.7—Strain distributions. All dimensions are in in. (mm).

iv. Evaluate the magnitude of e_b by summing moments about the tension steel in Fig. 11.28b [Eq. (11.20)].

$$P_{nb}e'_b = C_c(d - a_b/2) + C_s(d - d')$$

$$92.8e'_b = 3.4(3.49 \times 8)(6 - 3.49/2) + 40(.62)(6 - 2)$$

$$e'_b = 5.42 \text{ in. } (138 \text{ mm})$$

$$e_b = e'_b - 2$$

$$e_b = 3.42 \text{ in. } (87 \text{ mm})$$

v. Evaluate the allowable load capacity ϕP_{nb} at e_b from the plastic centroid.

$$(P_{nb})_{\text{all}} = \phi P_{nb} = 0.70(92.8)$$

$$\phi P_{nb} = 65.0 \text{ k} \quad (289 \text{ kN}) \quad \text{at } e_b = 3.42 \text{ in. (87 mm)}$$

c. Evaluate the load capacity ϕP_n when the eccentricity equals $0.5e_b = 1.71$ in. (43 mm).
  i. The tension steel A_s does not yield and A'_s may or may not yield. Assume A'_s yields and check the assumption. Consequently, the steel stresses are $f'_s = f_y = 40$ ksi (280 MPa) and $f_s = E_s \varepsilon_s$. Using the strain diagram of Fig. 11.30b, similar triangles and Hooke's Law, an expression for f_s can be derived, as follows, in terms of the dimensions of the section.

$$0.003/c = \varepsilon_s/(d - c)$$

$$\varepsilon_s = (d/c - 1)(0.003)$$

$$= (\beta_1 d/a - 1)(0.003) = [0.85(6)/a](0.003)$$

$$= 0.015/a - 0.003$$

$$f_s = (29 \times 10^3)\varepsilon_s = 444/a - 87 \text{ (ksi)}$$

  ii. An expression for P_n is obtained by summing forces in the axial direction for the conditions illustrated in Fig. 11.28d.

$$P_n = C_c + C_s - T$$

$$C_s = A'_s f_y = (40 - 3.4)(0.62) = 22.7 \text{ k}$$

$$C_c = 0.85 f'_c (8a) = 27.2a \text{ k}$$

$$T = A_s f_s = 0.62 f_s = 275/a - 53.94 \text{ k}$$

$$P_n = 27.2a + 22.7 - (275/a - 53.94) \text{ k}$$

$$P_n = 27.2a - 275/a + 76.64 \text{ k} \tag{11.21a}$$

  iii. A second expression for P_n as a function of a is obtained by summing the moments of the forces in Fig. 11.28d about the tension steel.

$$P_n e' = C_c(d - a/2) + C_s(d - d')$$
$$P_n(2 + 1.71) = 27.2a(6 - a/2) + 22.7(6 - 2)$$
$$3.71 P_n = -13.6a^2 + 163.2a + 90.8 \text{ k}$$
$$P_n = -3.67a^2 + 44.0a + 24.47 \text{ k} \qquad (11.21b)$$

iv. Solve for the dimension a by equating Eqs. (11.21a) and (11.21b).

$$-3.67a^2 + 44.0a + 24.47 = 27.2a - 275/a + 76.64$$

Upon rearranging and solving, $a = 4.84$ in. (123 mm). Using Fig. 11.30b and similar triangles, it is now possible to check if ε_s' has yielded.

$$\varepsilon_s' = 0.003 \, [1 - 2/c] = 0.003 \, [1 - 2/(4.84/0.85)]$$
$$= 1.95 \times 10^{-3} > \varepsilon_{sy}$$

Since $\varepsilon_s' > \varepsilon_{sy}$, A_s' yields as assumed.

v. Evaluating P_n from Eq. (11.21a),

$$P_n = 27.2(4.84) - 275/4.85 + 78.74 = 154 \text{ k (685 kN)}$$

and

$$\phi P_n = 0.70(154)$$
$$\phi P_n = 108 \text{ k (480 kN)} \quad \text{at} \quad e = 0.5e_b = 1.71 \text{ in. (43 mm)}$$

d. Evaluate the load capacity ϕP_n when the eccentricity equals $2e_b = 6.84$ in. (174 mm)

i. Since $e > e_b$, A_s yields but A_s' may or may not yield. Assume A_s' yields and check. From Fig. 11.28c,

$$C_c = 0.85 f'_c(ab) = 27.2a \text{ k}$$
$$C_s = A'_s f'_s = 0.62(40 - 3.4) = 22.7 \text{ k}$$
$$T = A_s f_y = 0.62(40) = 24.8 \text{ k}$$

ii. From $\Sigma F = 0$ of the section in Fig. 11.28c, an expression for P_n as a function of a is

$$P_n = 27.2a + 22.7 - 24.8 = 27.2a - 2.10 \text{ k} \qquad (11.21c)$$

iii. By summing moments about the tension steel in Fig. 11.28c, a second expression is obtained for P_n

$$(P_n)(6.84 + 2) = 27.2a(6 - a/2) + 22.7(6 - 2)$$
$$P_n = -1.54a^2 + 18.46a + 10.27 \text{ k} \qquad (11.21d)$$

iv. Equating Eqs. (11.21c) and (11.21d) and solving for a yields $a = 1.20$ in. (30 mm) and $c = 1.41$ in. (36 mm). By sketching the strain diagram, it is apparent that $\varepsilon'_s < \varepsilon_{sy}$ and the compression steel does not yield.

v. From the strain distribution of Fig. 11.30c and similar triangles, ε'_s can be written in terms of the dimension a.

$$\varepsilon'_s/0.003 = 1 - d'/c$$
$$\varepsilon'_s = 0.003 [1 - (\beta_1/a)d'] = 0.003 [1 - 0.85(2)/a]$$
$$\varepsilon'_s = 0.003 - 0.005/a$$

and from Hooke's law $f'_s = E\varepsilon'_s = 87 - 148/a$ and

$$C_s = A'_s (f'_s - 0.85f'_c) = 0.62[87 - 148/a - 3.4]$$
$$= 51.83 - 91.76/a \text{ k}$$

vi. Rewriting the equilibrium equations yields two equations for P_n as a function of a

$$\Sigma F = 0 : P_n = 27.2a + 51.83 -$$
$$91.76/a - 24.8 \qquad (11.21e)$$
$$P_n = 27.2a - 91.76/a + 27.03$$
$$\Sigma M_T = 0 : P_n (8.84) = 27.2a(6 - a/2)$$
$$+ (51.83 - 91.76)(4) \qquad (11.21f)$$
$$8.84\, P_n = -13.6a^2$$
$$+ 163.2a - 367.04/a$$
$$+ 207.32\, P_n = -1.54a^2$$
$$+ 18.46a - 41.52/a$$
$$+ 23.45$$

By equating the Eqs. (11.21e) and (11.21f), the solution for a and c is

$$a = 1.93 \text{ in. } (49 \text{ mm})$$
$$c = 1.93/0.85 = 2.27 \text{ in. } (58 \text{ mm})$$

Note that since $c > d'$, A'_s is indeed in compression as assumed.

vii. Substituting $a = 1.93$ in. (49 mm) into Eq. (11.21e) yields the value of the theoretical axial compressive load P_n at $2e_b$

$$P_n = 27.2(1.93) - 91.76/1.93 + 27.03$$
$$= 32 \text{ k} \quad (142 \text{ kN})$$

and since $\phi P_n = 22.4 < 0.1\, f'_c A_g = 25.6$ and $g < 0.7$, from Section 11.5.1

$$\phi = 0.90 - 22.4/25.6(0.9 - 0.7) = 0.73$$

and

$$\phi P_n = 0.73(32) = 23.2 \text{ k} \quad (102 \text{ kN})$$

Thus, $\phi P_n = 23.2$ k (102 kN) at $e = 2e_b = 6.84$ in. (174 mm).

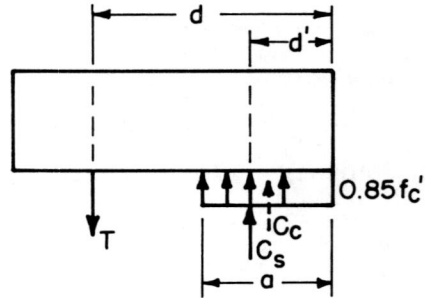

Fig. 11.31. Definition sketch for the moment capacity of a doubly reinforced flexural member.

e. Evaluate the moment capacity of the section when $P_n = 0$ (see Fig. 11.31).

  i. The force resultants in Fig. 11.31 equal

  $$C_c = 0.85 f'_c(8a) = 27.2a \text{ k}$$
  $$C_s = A'_s (f'_s - 0.85 f'_c) \text{ k}$$
  $$T = 0.62(40) = 24.8 \text{ k}$$

  ii. If the tension steel A_s is assumed to yield but the compression steel A'_s is assumed to be below yield, then, from subsection d, part v of this example $f'_s = 87 - 148/a$ ksi and $C_s = 51.83 - 91.76/a$ k. From equilibrium of forces in the axial direction $a = 1.41$ in. (36 mm) and $c = 1.65$ in. (42 mm). Since $c < d'$, the area of steel A'_s is in tension rather than compression, as assumed.

  iii. Using the stress and strain diagrams for the cross section as in Fig. 11.32 and from similar triangles, ε'_s and f'_s can be written as functions of a.

  $$\varepsilon'_s/(d' - c) = 0.003/c$$

  $$\varepsilon'_s = 0.003(d'/c - 1)$$
  $$= 0.003 [0.85(2)/a - 1] = 0.005/a - 0.003$$
  $$f'_s = E_s \varepsilon'_s = 148/a - 87$$
  $$T' = A'_s f'_s = 91.76/a - 53.94 \text{ k}$$

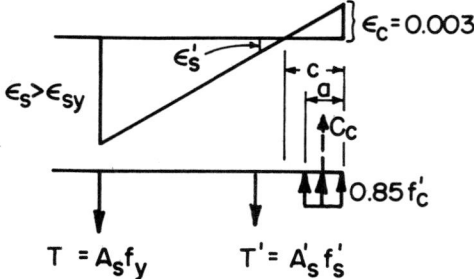

Fig. 11.32. Example 11.7—Stress and strain distribution for all steel in tension.

By summing forces in axial direction, the magnitude of a is evaluated.

$$\Sigma F = 0 : = T + T' - C_c$$

$$0 = (24.8) + (91.76/a - 53.94) - (27.2a)$$

and

$$a = 1.38 \text{ in. } (35 \text{ mm})$$

iv. The design moment capacity M_n can now be determined by summing moments about the axis of A_s.

$$\Sigma M_T = 0 : M_n = C_c(d - a/2)$$
$$- T'(d - d')$$
$$= (27.2)(1.38)(6 - 1.38/2)$$
$$- (91.76/1.38 - 53.94)$$
$$\times (6 - 2)$$
$$= 149 \text{ k-in. } (16.8 \text{ kN-m})$$

and

$$\phi M_n = 0.9(149.1) = 134 \text{ k-in.}$$
$$(15.1 \text{ kN-m}) \text{ at } P_n = 0$$

The interaction diagram for the rectangular column of Example 11.7 is shown in Fig. 11.33. From this diagram one can easily evaluate the

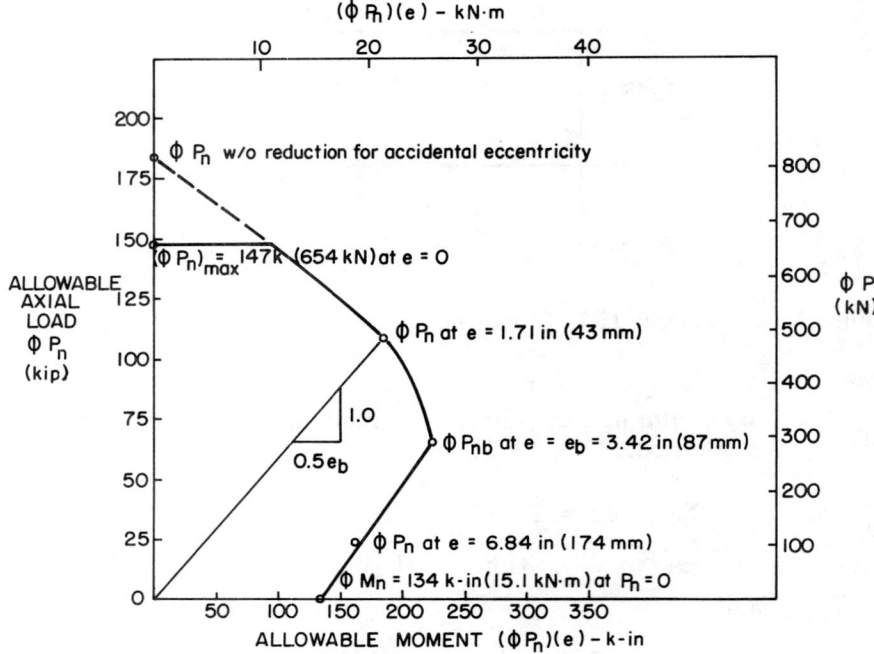

Fig. 11.33. Example 11.7—Interaction diagram for column.

combinations of axial load ϕP_n and moment $M_n = (\phi P_n)(e)$ the column can safely carry. If the column-load combinations fall within the envelope, the column can carry the load. Interaction diagrams for a wide variety of spiral and tied rectangular and circular columns are available in the American Concrete Institute publication, *Ultimate Strength Design Handbook: Part II, Columns.*

The analysis technique, as presented, is applicable only to short columns of solid square or rectangular cross section with all steel located at d and d' from the compression face. The technique is easily adapted to nonrectangular shapes. The only variations, as was the case with nonrectangular beam cross sections, are as follows:

1. The compression resultant C_c is not equal to $0.85 f'_c (ab)$. Instead it is replaced by $C_c = (0.85 f'_c)$ (compression area to depth a).
2. The location of the centroid of the compression area, and thus C_c, is not necessarily at $a/2$.

A reinforced concrete column may be treated as a short column if, in an unbraced frame where sidesway is permitted, $kl_u/r \leq 22$. In a braced frame without sidesway, the column may be considered short if $kl_u/r \leq 34 - 12M_1/M_2$. In these criteria, l_u is the unsupported column length, k is the column effective length factor, $r = \sqrt{I_g/A_g}$ equals the radius of gyration of the gross section, and M_1 and M_2 are the end moments acting on the column. The end moment with the maximum absolute value is always M_2, M_2 is always assumed to be positive, and M_1 is positive if the column is bent in single curvature and negative when bent in double curvature.

11.6 SHEAR REINFORCEMENT

11.6.1 Flexural Members Without Torsion

Shear reinforcement is required in reinforced concrete beams to carry the tensile principal stresses, also called diagonal tensile stresses, induced by the combined normal and shear stresses acting on a beam cross section. A review of the stress states in Fig. 11.34 and strength of materials will verify that even in the compression zone of the section, significant tensile stresses can occur. The orientation angle, θ, of the tensile stresses varies from 0° to nearly 90° depending upon the relative magnitude of the shear V and the moment M acting on the section and depending upon the distance y from the N.A.

The total stress and strain distribution in a reinforced concrete beam at ultimate load capacity is more complex than that of Fig. 11.34. Nonetheless, combined shear and normal forces on the beam cross section still induce tensile stresses σ_T in the web of the cross section. If σ_T exceeds the allowable tensile stress f_t of the concrete ($f'_t \approx 0.1 f'_c$), a beam without web reinforcement will develop tension cracks in the direction normal to σ_T (see Fig. 11.35). Once these cracks develop and the load is not removed the beam will be rendered unserviceable due to collapse or excessive deformation.

Diagonal tension cracks can be controlled by increasing the web size (a very uneconomical alternative) or by providing web reinforcement as shown in Fig. 11.36. Web reinforcement is also called shear reinforcement, or stirrups, and must be No. 2 bars or larger. When web reinforcement is used, it is proportioned to carry the diagonal tensile stress in excess of what the concrete can carry. It also is spaced to assure that at least one stirrup crosses each potential tension crack.

Equation (11.22) represents the governing criterion for designing shear reinforcement.

$$V_u \leq \phi V_n = \phi(V_c + V_s) \qquad (11.22)$$

506   Reinforced Concrete Design

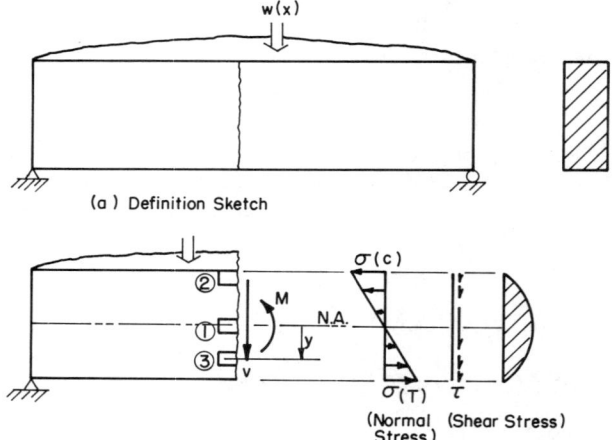

(a) Definition Sketch

(b) Stress Distributions at Arbitrary Section

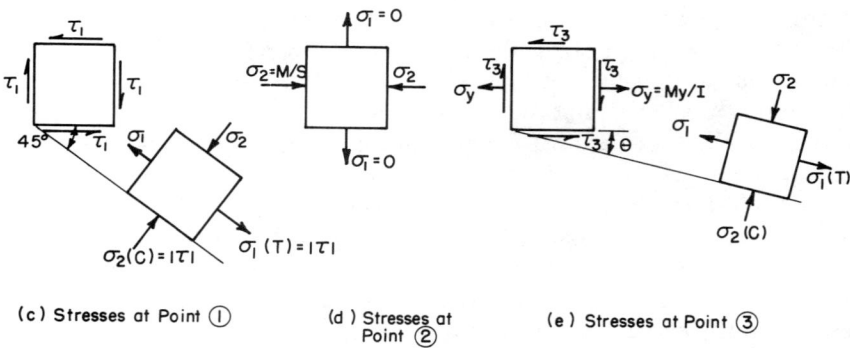

(c) Stresses at Point ①   (d) Stresses at Point ②   (e) Stresses at Point ③

Fig. 11.34. Stress distribution and biaxial stress states in a homogeneous flexural member.

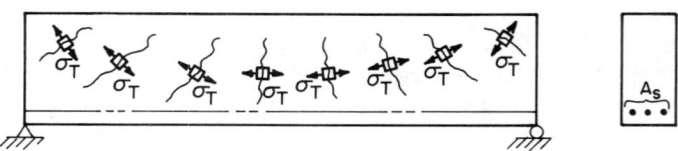

Fig. 11.35. Effect of shear induced diagonal tension stresses in reinforced concrete beams without web reinforcement.

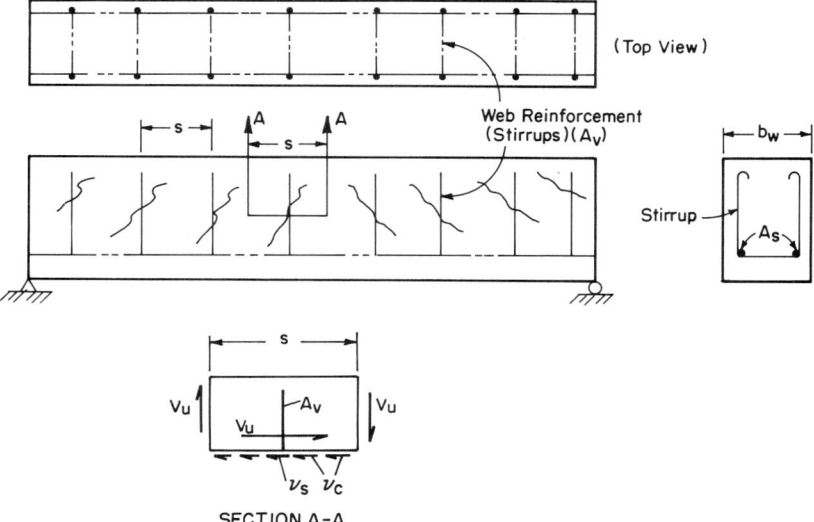

Fig. 11.36. Definition sketch for shear reinforcement in beams.

where V_u is factored shear load at station x along the beam length, k (kN), $V_u = 1.4 V_D + 1.7 V_L$; V_n, shear capacity of the section, k (kN); ϕ, strength reduction factor, $\phi = 0.85$; V_c, shear capacity of plain concrete, k (kN); V_s, shear capacity of web steel, k (kN).

The average shear capacity of the concrete V_c is simply equal to the shear strength of plain concrete times the web area and can be estimated with either Eq. (11.23) or (11.24). Equation (11.23), which is the simpler estimate, yields conservative results for V_c. The shear strength of the steel is simply the nominal steel area A_v times the steel yield strength; i.e., $V_s = A_v f_y$.

$$V_c = (2\sqrt{f'_c}) b_w d \qquad (11.23)$$

or

$$V_c = (1.9 f'_c + 2500 \rho_w V_u d / M_u) b_w d \leq (3.5 \sqrt{f'_c}) b_w d \qquad (11.24)$$

where $\rho_w = A_s/(b_w d)$; V_u, M_u are factored shear and moment at station x along beam; f'_c is compressive strength of concrete in psi.

There are some special cases where no shear reinforcement is required. Among them are footers, slabs, and beams with overall depth less than 10 in. (254 mm) and beams with $V_u \leq 0.5(\phi V_c)$.

When $0.5(\phi V_c) \leq V_u \leq \phi V_c$, a minimum steel area $(A_v)_{min}$ is required. For these relatively low loads, the minimum steel area is designed to carry an average shear stress v_{min} of 50 psi (340 kPa) over the beam web area $(b_w s)$ (see Fig. 11.36) between two adjacent stirrups. By assuming that v_{min} between the two adjacent stirrups in section A-A of Fig. 11.36 is carried by the single stirrup with area $(A_v)_{min}$ and equating the resultant forces, $(A_v)_{min} f_y = v_{min}(b_w s)$. Rearranging this relationship and solving for the minimum shear reinforcement yields Eq. (11.25).

$$(A_v)_{min} = v_{min}(b_w s)/f_y \qquad (11.25)$$

When $V_u \geq \phi V_c$, the shear reinforcement area A_v must carry an average shear stress of $(v_u/\phi - v_c)$ over the beam area $(b_w s)$ between two adjacent stirrups. Equating the shear force carried by the web steel between two adjacent stirrups to the excess vertical shearing force between two stirrups which is not carried by the concrete yields Eq. (11.26).

$$A_v f_y = (v_u/\phi - v_c)(b_w s) \qquad (11.26)$$

where $v_u = V_u/(\phi\, b_w d)$ is average factored shear *stress,* psi (kPa); $v_c = V_c/(b_w d)$ is average concrete shear *strength,* psi (kPa).

Rewriting Eq. (11.26) in terms of the shear forces and solving for A_v yields an expression relating web steel area and stirrup spacing as shown in Eq. (11.27).

$$A_v = (V_u - \phi V_c)s/(\phi f_y d) \qquad (11.27)$$

In cases where $V_u \geq \phi V_c$, A_v is always taken as the larger of $(A_v)_{min}$ from Eq. (11.25) or A_v from Eq. (11.27).

Limits are placed upon spacing between stirrups to assure that each diagonal crack is crossed by one or two stirrups. For lightly loaded beams where $V_s = (V_u/\phi - V_c) \leq 4\sqrt{f'_c}(b_w d)$ with f'_c in psi, s_{max} must be less than the smaller of $d/2$ or 24 in. (610 mm) to assure one stirrup through each crack. When $4\sqrt{f'_c}(b_w d) < V_s \leq 8\sqrt{f'_c}(b_w d)$ with f'_c in psi, s_{max} equals the smaller of $d/4$ or 12 in. (305 mm) to assure that two stirrups pass through each diagonal crack. If $V_s > 8\sqrt{f'_c}(b_w d)$ with f'_c in psi, the beam section is unsatisfactory and needs to be redesigned.

At reaction points where there is a compressive bearing stress, the factored shear force V_u may be reduced for sections within a distance

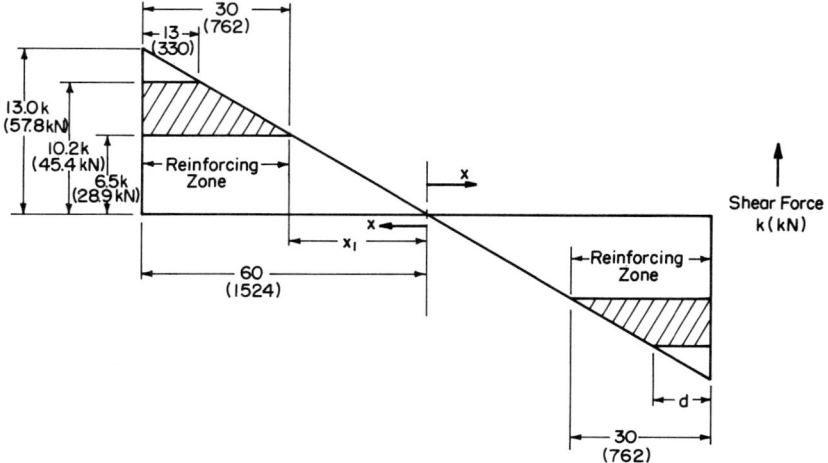

Fig. 11.37. Shear force diagram for the beam in Fig. 11.8. All dimensions are in in. (mm).

equal to d from the end of the support. Sections within d of the support may be designed for the shear force at distance d from the support.

Example 11.9. Web Reinforcement in Beams. Determine the region and quantity of web reinforcement required for the beam and loads of Example 11.1 and Fig. 11.18.

Solution

a. Sketch the shear diagram as in Fig. 11.37.
b. Using similar triangles the shear force V'_u at a distance d from each end support can be calculated.

$$V'_u = (60 - 13)/60(13.0) = 10.2 \text{ k} \quad (45.4 \text{ kN})$$

V'_u is the required shear capacity of the beam for the first 13 in. (330 mm).

c. Evaluate the allowable shear force ϕV_c with Eq. (11.23).

$$V_c = 2\sqrt{f'_c} \, (b_w d) = 2\sqrt{3500}(10 \times 13) = 15.4 \text{ k} \quad (68.5 \text{ kN})$$

and

$$\phi V_c = 0.85(15.4) = 13.1 \text{ k} \quad (58.3 \text{ kN}) > V_u$$

d. Determine the zone of reinforcement. The limiting value of shear force for which no reinforcement is required is $0.5(\phi V_c) = 6.5$ k (28.9 kN). Thus, $A_v = 0$ when $V_u < 6.5$ (28.9 kN) or from the center of the beam to the distance x_1 as defined in Fig. 11.37.

$$x_1/6.5 = 60/13.0 \quad \text{and} \quad x_1 = 30 \text{ in.} \quad (762 \text{ mm})$$

When $0.5V_c \leq V_u < \phi V_c$, or for $x = 30\text{–}60$ in. (762–1524 mm), $A_v = (A_v)_{min}$.

e. Determine the spacing of the stirrups. By rearranging Eq. (11.25), and using No. 2 reinforcing bars for A_v, the stirrup spacing can be evaluated.

$$s = (A_v)_{min} f_y / [(v)_{min} b_w] = 2(.055)(36{,}000)/50(10)$$

$$= 7.9 \text{ in.} \quad (201 \text{ mm})$$

To assure at least one reinforcing bar at each potential tension crack, the minimum stirrup spacing must be checked. The maximum force carried by the steel, $V_s = 10.2 - 6.5 = 3.7$ k (16.5 kN). The limiting value of V_s requiring a stirrup every $d/2$ equals $4\sqrt{f'_c}\ b_w d = 4\sqrt{3500}\ (10 \times 13) = 7.1$ k (31.6 kN). Since $V_s < 4\sqrt{f'_c} b_w\ d$, $s_{max} = d/2 = 6.5$ in. (165 mm). Thus, place No. 2 stirrups 6.5 in. (165 mm) on center for $x = 30\text{–}60$ in. (762–1524 mm).

11.6.2 Torsional Shear

In some applications torsional moments act on a cross section and induce shear stresses. The ACI Code has extensive provisions for treating torsion and should be consulted where torsional loads are significant. In a prismatic beam which can be subdivided into i-rectangular areas torsional shear may be neglected if the torsional moment T_u satisfies the condition in Eq. (11.28).

$$T_u \leq \phi(0.5\sqrt{f'_c}\ \Sigma\ x_i^2 y_i) \tag{11.28}$$

where T_u is the torsional moment, lb· in; x_i, the smaller dimension of component rectangle i, in.*; y_i, the larger dimension of component rectangle i, in.; f'_c, compressive strength of the concrete, psi.

*The term $\Sigma x_i^2 y_i$ is evaluated by subdividing the cross section into i rectangles with dimensions x_i any y_i.

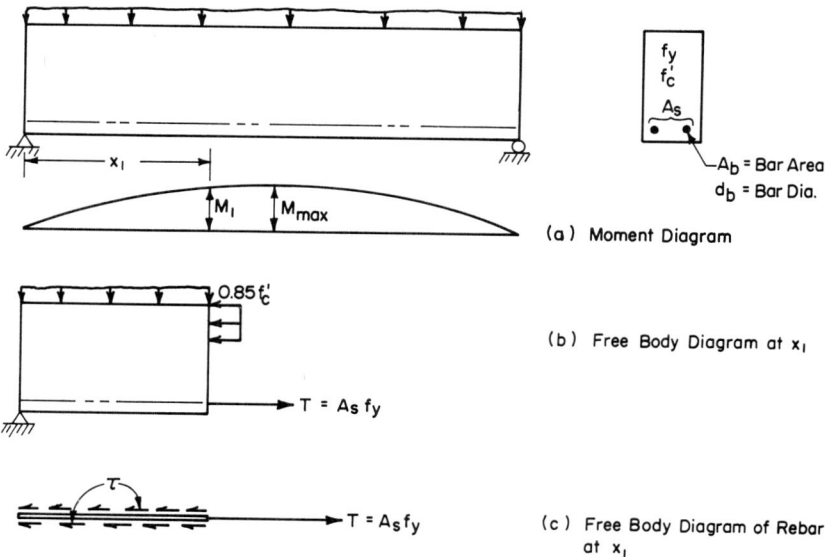

Fig. 11.38. Definition sketch for development length l_d.

11.7 DEVELOPMENT OF REINFORCEMENT

The development length l_d of a reinforcing bar is the embedment depth required for the reinforcing steel to transfer its full design load $A_s f_y$ to the concrete without either being pulled out of the concrete or damaging the concrete. The distance between the end of a reinforcing bar and a point along the bar where $f_s = f_y$ must exceed l_d.

The beam and simplified free body diagrams of Fig. 11.38 help clarify the development length concept. At midspan of this beam the moment is maximum and the steel in a properly designed reinforced beam will have yielded. However, since $\rho_{\text{all}} \leq 0.75\,\rho_b$, the steel will yield at some moment less than $M_{\max}$, say, M_1, at x_1 from the left end. The free body at section x_1 shows a tensile force $T = A_s f_y$ acting on the steel. The only forces which keep the steel and concrete from pulling apart are the bond stresses τ between the rebar and concrete. If τ is assumed constant (which it is not) for clarity, then the tensile force T must be less than or equal to $\tau(x_1)$ (bar perimeter) to prevent pullout. Thus the steel design force $A_s f_y$ can be developed in A_s only if Eq. (11.29) is satisfied.

$$x_1 \geq A_s f_y / [\tau \text{ (bar perimeter)}] = l_d \tag{11.29}$$

Table 11.13. Basic Development Lengths for Reinforcement.

Type Load in A_s or A_s'	Reinforcement Type and Size	l_d (in.)[a]
Tension	≤No. 11 bars	Larger of $0.04A_b f_y/\sqrt{f_c'}$ or $0.0004 d_b f_y$ but ≥12 in.
	Deformed wire	$0.03 d_b f_y/\sqrt{f_c'}$ but ≥12 in.
Compression	Any	Larger of $0.02 d_b f_y/\sqrt{f_c'}$ or $0.0003 d_b f_y$ but ≥8 in.

Reproduced with permission from *Building Code Requirements for Reinforced Concrete* (ACI 318–77), 1977 ed. American Concrete Institute, Detroit, MI.
[a] A_b in in.^{2}; f_y in psi; f_c' in psi.

The development length depends upon the type loading, the reinforcement size, the concrete quality, the grade steel, and the presence or absence of hooks. The basic development lengths for straight deformed bars are summarized in Table 11.13 for tension and compression reinforcement.

In tensile reinforcement, standard hooks, which provide an equivalent development length l_e, are often used. The equivalent development length for a hook is obtained by substituting $f_h = \zeta\sqrt{f_c'}$ for f_y in the expressions for l_d in Table 11.13. ζ is defined in Table 11.14 for a variety of bar sizes, strengths, and applications. Thus, for a standard hook in tension reinforcement, l_e is the smaller of

$$l_e = 0.04 A_b (\zeta\sqrt{f_c'})/\sqrt{f_c'}, \quad f_c' \text{ in psi}$$

or

$$l_e = 0.0004 d_b (\zeta\sqrt{f_c'}), \quad f_c' \text{ in psi}$$

where l_e has dimensions in in.

Table 11.14. ζ Values for Standard Hooks.

Bar Size	$f_y = 60$ ksi (410 MPa)		$f_y = 40$ ksi (280 MPa)
No.	Top Bars	Other Bars	All Bars
3–5	540	540	360
6	450	540	360
7–9	360	540	360
10	360	480	360
11	360	420	360

Reproduced with permission from *Building Code Requirements for Reinforced Concrete* (ACI 318-77), 1977 ed. American Concrete Institute, Detroit, MI.

Table 11.15. Minimum Bend Diameters.

Bar Size	Minimum diameter
3–8	$6d_b$
9–11	$8d_b$

Reproduced with permission from *Building Code Requirements for Reinforced Concrete* (ACI 318-77), 1977 ed. American Concrete Institute, Detroit, MI.

Any of the following definitions meet the requirements for a standard hook:

1. A 180° bend plus an extension of the greater of $4d_b$ or $2\frac{1}{2}$ in. (64 mm) at the free end of the bar.
2. A 90° bend plus an extension of $12d_b$ at the free end of the bar.
3. In stirrups and ties, either a 90° or a 135° bend plus an extension of the larger of $6d_b$ or $2\frac{1}{2}$ in. (64 mm) at the free end of the bar.

The minimum inside bend diameters for reinforcing bars, except ties and stirrups, are summarized in Table 11.15. Bend diameters as low as $5d_b$ are permissible in 180° bends in Grade 40 bars of size No. 3–11. Stirrups and ties up to No. 5 bars may be bent to a diameter of $4d_b$. Larger bars must conform to the requirements in Table 11.15.

The critical sections for checking development length in flexural members are at points of maximum moment and points where adjacent reinforcing bars terminate. Bars should extend the larger of d or $12d_b$ beyond the beam section where it is required by exact analysis.

Web reinforcement should be placed as close to the compression and tension surfaces as minimum cover requirements permit. Stirrups are adequately anchored if they meet one of three conditions:

1. Embedment equivalent to a standard hook plus $\frac{1}{2}l_d$ measured from $d/2$ to the beginning of the hook.
2. Embedment equivalent to the greater of l_d or $24d_b$ for bars or 12 in. (305 mm) for wire and measured from $d/2$.
3. A 135° hook around compression steel A'_s plus 0.33 l_d when $f_y \geq 40$ ksi (280 MPa) and less than or equal to No. 5 bars and measured from $d/2$.

11.8 DEFLECTION CONTROL IN BEAMS

Reinforced concrete beam deflections must be checked against the maximum allowable deflections. Table 11.16 summarizes ACI recommen-

Table 11.16. Maximum Permissible Computed Deflections.

Type of Member	Deflection to Be Considered	Deflection Limitation
Flat roofs not supporting or attached to nonstructural elements likely to be damaged by large deflections	Immediate deflection due to live load L	$\dfrac{l^a}{180}$
Floors not supporting or attached to nonstructural elements likely to be damaged by large deflections	Immediate deflection due to live load L	$\dfrac{l}{360}$
Roof or floor construction supporting or attached to nonstructural elements likely to be damaged by large deflections	That part of the total deflection occurring after attachment of nonstructural elements (sum of the long-time deflection due to all sustained loads and the immediate deflection due to any additional live load)[c]	$\dfrac{l^b}{480}$
Roof or floor construction supporting or attached to nonstructural elements not likely to be damaged by large deflections		$\dfrac{l^d}{240}$

Reproduced with permission from *Building Code Requirements for Reinforced Concrete* (ACI 318-77), 1977 ed. American Concrete Institute, Detroit, MI.

[a] Limit not intended to safeguard against ponding. Ponding should be checked by suitable calculations of deflection, including added deflections due to ponded water, and considering long-time effects of all sustained loads, camber, construction tolerance, and reliability of provisions for drainage.

[b] Limit may be exceeded if adequate measures are taken to prevent damage to supported or attached elements.

[c] Long-time deflection may be reduced by amount of deflection calculated to occur before attachment of nonstructural elements. This amount shall be determined on basis of accepted engineering data relating to time-deflection characteristics of members similar to those being considered.

[d] But not greater than tolerance provided for nonstructural elements. Limit may be exceeded if camber is provided so that total deflection minus camber does not exceed limit.

dations for allowable deflections in several types of reinforced concrete construction.

If the depth d of a reinforced beam exceeds the limits in Table 11.17, deflections will not be a problem and do not have to be computed. For example, if the depth of a simply supported beam of span l is greater than $l/16$, the deflections will be within the allowable limits of Table 11.16.

If d is less than the limits in Table 11.17, then deflections must be

11.8 Deflection Control in Beams

Table 11.17. Minimum Thickness of Nonprestressed Beams or One-Way Slabs Unless Deflections Are Computed.[a]

	Minimum Thickness, h^b			
	Simply Supported	One End Continuous	Both Ends Continous	Cantilever
Member	Members not supporting or attached to partitions or other construction likely to be damaged by large deflections.			
Solid one-way slabs	$l/20$	$l/24$	$l/28$	$l/10$
Beams or ribbed one-way slabs	$l/16$	$l/18.5$	$l/21$	$l/8$

Reproduced with permission from *Building Code Requirements for Reinforced Concrete* (ACI 318-77), 1977 ed. American Concrete Institute, Detroit, MI.
[a] Span length l is in inches.
[b] Values given shall be used directly for members with normal weight concrete ($w_c = 145$ pcf) and Grade 60 reinforcement. For other conditions, the values shall be modified as follows: (a) For structural lightweight concrete having unit weights in the range 90–120 lb per cu ft, the values shall be multiplied by $(1.65 - 0.005 w_c)$ but not less than 1.09, where w_c is the unit weight in lb per cu ft. (b) For f_y other than 60,000 psi, the values shall be multiplied by $(0.4 + f_y/100,000)$.

computed and compared to allowable values. Deflection calculations for reinforced concrete are somewhat more complex than those for homogeneous members because (1) the cross section is nonhomogeneous; (2) creep is significant during long-term loading; (3) the section cracks progressively in the tension zone as loading proceeds thus altering the section's effective moment of inertia; and (4) the beam material is not linearly elastic and is loaded beyond its elastic limit at design loads.

The basic procedure for deflection calculation is to use standard methods of elastic analysis to evaluate the immediate deflections using an equivalent section moment of inertia and to add to it additional long-term deflections due to creep. The basic equation is

$$\Delta = \Delta_{LP} + \lambda_\infty \Delta_D + \lambda \Delta_{LS} \qquad (11.30)$$

where Δ is design deflection; Δ_{LP}, immediate live load deflection ($k_L w_L L^4/E_c I_c$); Δ_D, immediate dead load deflection ($k_D W_D L^4/E_c I_e$); Δ_{LS}, immediate long-term (12 months) live load deflection ($k_s w_{LS} L^4/E_c I_e$); λ_∞, creep factor for dead loading, $= [2 - 2(A'_s/A_s)] \geq 0.60$; λ, creep factor

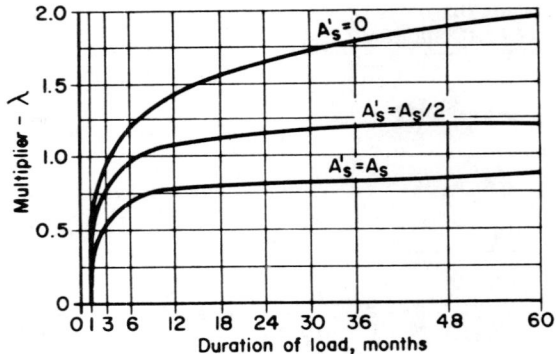

Fig. 11.39. Multipliers for long-term deflections. Reproduced with permission from *Building Code Requirements for Reinforced Concrete,* ACI 318-77, 1977 ed., American Concrete Institute, Detroit, MI.

for long-term live loading (defined in Fig. 11.39); k, constant dependent upon geometry of beam loading and supports; $E_c = 33w^{1.5} \sqrt{f'_c}$ for w in pcf, f'_c in psi, and E_c in psi; w_L, w_D, w_{LS}, *service* loads; I_e, equivalent area moment of inertia.

The expression for evaluating I_e is empirical and equals

$$I_e = (M_{cr}/M_a)^3 I_g + [1 - (M_{cr}/M_a)^3] I_{cr} \qquad (11.31)$$

where M_{cr} is cracking moment of a plain concrete section, $M_{cr} = f_r I_g/y_T$; $f_r = 7.5\sqrt{f'_c}$ in psi for f'_c in psi; $I_g = bD^3/12$; D, total beam depth; y_T, distance from N.A. of plain concrete section to the tension face; I_{cr}, moment of inertia of the composite steel and cracked concrete area assuming the allowable concrete tensile strength f'_{ct} equals 0; M_a, maximum service moment in the beam.

Example 11.10 clarifies the definition of the terms in Eq. (11.31).

Example 11.10. Calculation of the Equivalent Moment of Inertia I_e of a Beam Section for Deflection Evaluations. Evaluate the equivalent moment of inertia I_e for the beam section of Fig. 11.40, if $A_s = 1.32$ in.² (80 mm²), $f'_c = 2500$ psi (17 MPa), $f_y = 40$ ksi (280 MPa), $E_c = 2.85 \times 10^6$ psi (20 GPa), and the maximum service moment equals 320 k in. (36 kN · m).

Solution

a. Evaluate f_r

$$f_r = 7.5\sqrt{f'_c} = 7.5\sqrt{3500} = 375 \text{ psi} \quad (2.6 \text{ MPa})$$

11.8 Deflection Control in Beams 517

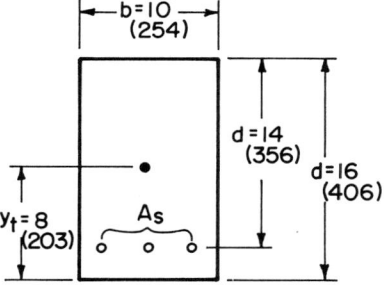

Fig. 11.40. Example 11.10—Beam section. All dimensions are in in. (mm).

b. Evaluate the moment of inertia of the gross concrete section I_g

$$I_g = bD^3/12 = (10)(16)^3/12 = 3410 \text{ in.}^4 \quad (1.43 \times 10^5 \text{ cm}^4)$$

c. Evaluate cracking moment for a plain concrete section M_{cr}.

$$M_{cr} = f_r I_g/y_T = [375(3410)]/(16/2) = 160 \text{ k} \cdot \text{in. } (18.1 \text{ kN} \cdot \text{m})$$

d. Evaluate the moment of inertia of the cracked concrete section by considering the compression zone concrete and the tension steel areas.
   i. Replace steel area by equivalent area of concrete A_{eq} located at the centroid of the steel. See Fig. 11.41.
$A_{eq} = (E_s/E_c)(A_s) = nA_s = 10A_s = 13.2 \text{ in.}^2 \quad (85.2 \text{ cm}^2)$
   ii. Locate the centroid and neutral axis of cracked section. Taking the first moment of the transformed area in Fig. 11.41 about the N.A. yields $y_1(n A_s) = (y/2)A_c$. Substituting values yields $13.2(14 - y) = (y/2)(10y)$ from which $y = 4.9 \text{ in. } (124 \text{ mm})$ and $d - y = 9.1 \text{ in. } (231 \text{ mm})$.

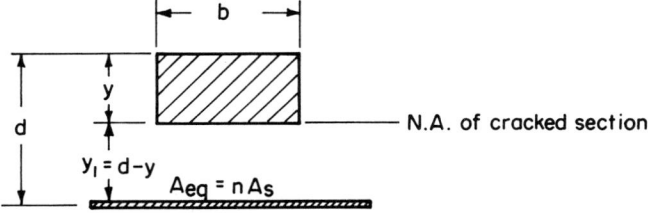

Fig. 11.41. The equivalent, or transformed, cracked section.

iii. Evaluate the moment of inertia of the cracked section about the N.A.

$$I_{cr} = 1/3 \, by^3 + (n \, A_s)(d - y)^2$$
$$= 1/3 \, (10)(4.9)^3 + 13.2(9.1)^2$$
$$= 1480 \text{ in.}^4 \quad (6.16 \times 10^4 \text{ cm}^4)$$

e. Evaluate the equivalent moment of inertia I_e using Eq. (11.31).

$$I_e = (160/320)^3 \, (3410) + [1 - (160/320)^3](1480)$$
$$I_e = 1720 \text{ in.}^4 \quad (7.16 \times 10^4 \text{ cm}^4)$$

11.9 PRESTRESSED CONCRETE BEAMS

11.9.1 Basic Behavior

The load carrying capacity of a reinforced concrete beam or column can be increased by 15 to 20% by selectively precompressing the portions of the cross section which experience tensile strains and stresses under service loads. The precompression, or prestressing, is accomplished by placing highly stressed, high strength steel strands with strengths of 180 to 250 ksi (1.24–1.72 GPa) in the tensile zone of the beam and then transferring the prestressing load to the concrete. The various stages of prestressing and associated stress distributions are illustrated in Fig. 11.42.

One fabrication method for prestressed concrete is called pretensioning. Steel strands are stressed by a force F and held in place by abutments. Concrete is then placed. At this stage, the concrete mass is supported by the bulkwork (see Fig. 11.42b). After several days of accelerated curing, the bulkwork is removed, the strands are anchored at the beam ends, and the strands are cut. Figures 11.42c and 11.42d, respectively, illustrate the beam and stress distributions after the bulkwork is removed and the prestressing force F is transferred to the concrete.

The prestress force F is usually applied at some eccentricity e from the centroidal axis. The force F at eccentricity e is equivalent to a force F and a couple $F(e)$ acting through the centroid of the section. These equivalent forces are shown by dashed vectors in Fig. 11.42d.

The stress distributions in Fig. 11.42d assume that the concrete has

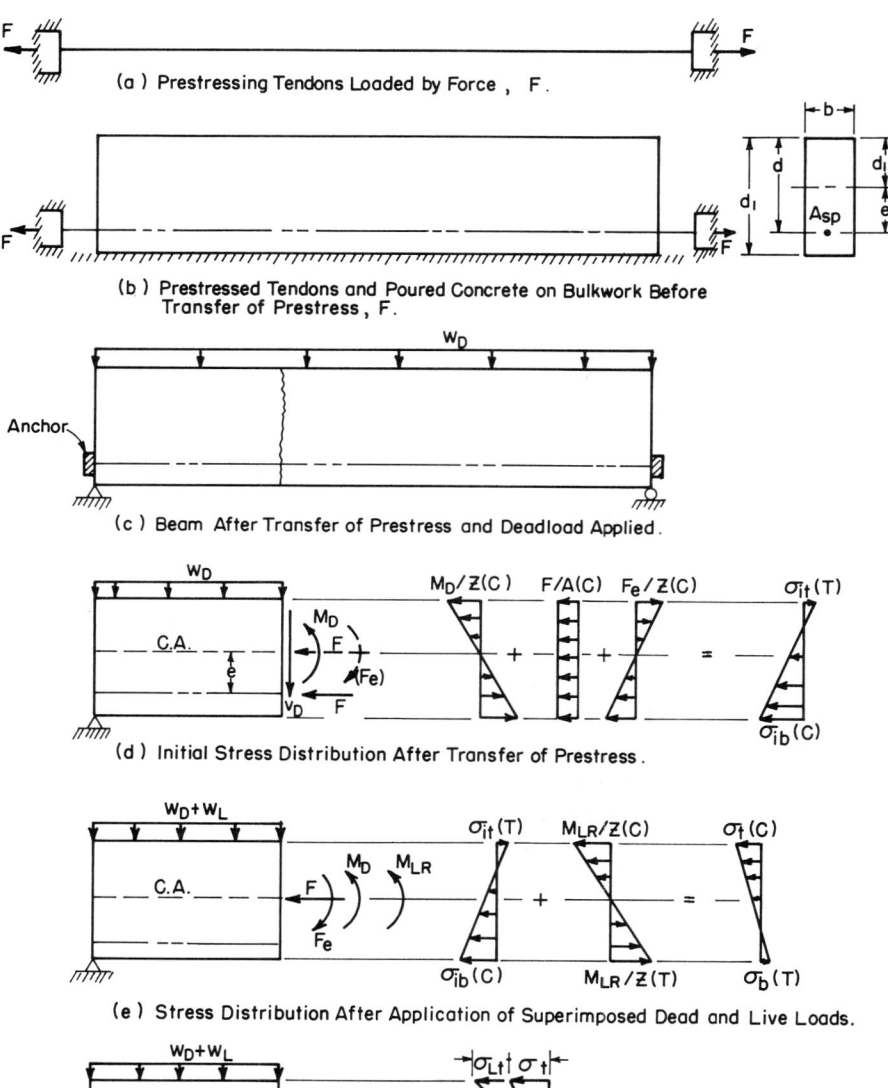

Fig. 11.42. Stages of prestressing and associated stress distributions.

not reached its rupture strength and has not cracked. Thus, the entire concrete section is effective. Also, the distributions assume that the bulkwork is removed when the prestress force is transferred to the concrete. The magnitude of the stresses in the top and bottom fibers at transfer of prestress are defined by Eqs. (11.32) and (11.33), respectively.

$$\sigma_{it} = Fe/Z - F/A - M_D/Z \tag{11.32}$$

$$\sigma_{ib} = -Fe/Z - F/A + M_D/Z \tag{11.33}$$

In Eqs. (11.32) and (11.33), tensile stresses are positive, Z is the section modulus, and M_D is the bending moment due to the beam's dead load w_D.

For a given beam cross section, the location and magnitude of F influences the final stress distribution. For example, if F is applied at $d_1/3$ from the bottom of a rectangular beam and M_D is not applied, then $e = d_1/2 - d_1/3$ and the stress in the top fiber equals zero; i.e.,

$$\sigma_{\text{top}} = -F/A + Fe/Z = F[-1/bd_1 + (d_1/2 - d_1/3)/(6\ bd_1^2/6)] = 0$$

The point at which the prestress force produces zero stress on the outer fiber is called the kern point. If the prestress force is applied above the kern point, the top fiber will be in compression with M_D removed. Contrariwise, the top fiber experiences tensile stresses if F is applied below the kern point with M_D removed. The usual practice is to apply F at an eccentricity such that neither the compressive stress at the bottom nor the tensile stress at the top fiber exceeds the initial allowable stresses in the concrete.

After the prestress is transferred, the beam experiences losses in prestress. Prestress losses are the result of elastic shortening of the concrete at transfer of prestress, creep shortening of the concrete, shrinkage of concrete during curing, tension creep of the steel, slippage between the anchors and strands, and friction and wobble losses between the strands and concrete. Friction losses are most commonly encountered in posttensioned beams and wobble losses are found in beams with curved prestressing tendons. Each of these losses has the effect of shortening the steel strands and reducing the prestressing force F. These losses result in approximately an 18% reduction of F in pretensioned beams with straight tendons. The ACI Code commentary suggests that a prestress loss of 35 ksi (240 MPa) is satisfactory for most pretensioning applications. For posttensioned cases prestress losses of 25 ksi (170 MPa) are typical. Since much of the prestress losses are due to time dependent phenomena, they are significant only after service loads are applied.

Figure 11.42e illustrates a prestressed beam after application of the service loads (superimposed dead load w_D plus live load w_L) and before prestress losses. Using the equivalent centroidal loading and stress distributions of Fig. 11.42e, the top and bottom fiber stresses are calculated by Eqs. (11.34) and (11.35).

$$\sigma_t = \sigma_{it} \text{ (Tension)} - M_{LR}/Z \text{ (Compression)} \tag{11.34}$$

$$\sigma_b = -\sigma_{ib} \text{ (Compression)} + M_{LR}/Z \text{ (Tension)} \tag{11.35}$$

where M_{LR} is superimposed dead plus live load bending moment and σ_{ib} and σ_{it} are defined in Eqs. (11.32) and (11.33).

When prestress losses are considered, the prestress force decreases an amount ΔF to F'. The consequence is a decrease in the concrete prestress at both the top and bottom fibers. That is, the top fiber tensile prestress is reduced (becomes more compressive) by $\sigma_{Lt} = -(\Delta Fe)/Z + \Delta F/A$ and the bottom fiber compressive prestress is reduced (becomes more tensile) by $\sigma_{Lb} = +(\Delta Fe)/Z + \Delta F/A$. These prestress losses increase the final service load stresses in the top and bottom fibers to the levels predicted by Eqs. (11.36) and (11.37), respectively.

$$\sigma_{ft} = \sigma_t + \sigma_{Lt} \tag{11.36}$$

$$\sigma_{fb} = \sigma_b + \sigma_{Lb} \tag{11.37}$$

The signs of Eqs. (11.36) and (11.37) must be carefully chosen. σ_t will be a compressive stress and the prestress loss σ_{Lt} will tend to make the top fiber stress more compressive. Thus, if compressive stresses are taken as negative, both σ_t and σ_{Lt} must have negative signs. Contrariwise, both σ_b and σ_{Lb} require positive signs in Eq. (11.37).

11.9.2 Materials

Prestressed concrete requires both high strength steel and concrete for satisfactory performance. Steels with yield stresses of 180 to 250 ksi (1.24–1.72 GPa) assure that the concrete is adequately prestressed after prestress losses. If ordinary steels were used the shortening of the tendons due to shrinkage, etc., would reduce F to a level near zero. [Recall that prestress losses are in the range of 35 ksi (240 MPa).] High strength steels have significant prestress force remaining after losses.

High strength concrete is required to reduce flexural cracking around the prestressing tendons and to increase the diagonal tension strength of the webs. Control of flexural cracking is critical since excessive cracking exposes the small diameter tendons to potentially corrosive environments. Since the tendons are highly stressed, small reductions in cross sections could be catastrophic. The increased diagonal tension of high strength concrete allows designers to utilize more economical sections, such as Is and Ts, which have thin webs.

11.9.3 Basic Analysis Procedure for Beams

The usual design procedure is to assume a fully effective section with elastic behavior up to service loads; and then to check the ultimate capacity at factored loads. Maximum stresses must be evaluated and compared to allowable stresses at two stages: (1) at transfer of prestress with the gravity load in place and no prestress losses; and (2) after application of the services loads and prestress losses. The ultimate capacity check is to assure that the section can safely carry typical overloads without collapsing. Finally, to protect against abrupt flexural failure due to rupture of tendons immediately after cracking of the concrete, the prestressing steel must be adequate to carry 1.2 times the cracking load based upon the concrete modulus of rupture, $f_r = 7.5\sqrt{f'_c}$, where both f_r and f'_c are in psi.

11.9.4 Allowable Stresses

The allowable stresses and/or strains in steel and concrete depend upon the stage of construction. Prestress is usually transferred after a relatively few hours of curing. Thus, the compressive strength is less than the 28-day strength f'_c. Service loads are usually applied only after the concrete has cured and has a compressive strength f'_c. The allowable stresses for steel and concrete for various stages are summarized in Tables 11.18 and 11.19. These stresses are used in the elastic analyses at prestress transfer and at service loads. The ultimate capacity is based upon the usual factored procedures for reinforced concrete beams with the following exception. The steel yield strength f_y is replaced by

$$f_{ps} = f_{pu}(1 - 0.50\rho_p \, f_{pu}/f'_c)$$

where f_{ps} is allowable stress in prestressing steel at ultimate capacity

Table 11.18. Allowable Concrete Stresses.

Loading Stage	Load Type	Allowable Stress (psi)
Transfer of prestress	Fiber stress in compression	$0.60 f'_{ci}$ [a]
	Fiber stress in tension	$3\sqrt{f'_{ci}}$
	Fiber stress in tension at ends of simply supported beams	$6\sqrt{f'_{ci}}$
After-service loads applied and prestress losses considered	Fiber stress in compression	$0.045 f'_c$
	Fiber stress in tension in precompressed tensile zone	$6\sqrt{f'_c}$

Reproduced with permission from *Building Code Requirements for Reinforced Concrete* (ACI 318-77), 1977 ed. American Concrete Institute, Detroit, MI.
[a] f'_{ci} = initial compressive strength at time of transfer of prestress in psi.

of beam; f_{pu}, ultimate strength of steel; ρ_p, ratio of prestressed reinforcement; ρ_p, A_{sp}/bd; A_{sp}, area of prestressing steel.

11.9.5 Ultimate Strength of Prestressed Beams

If design loads are exceeded, the concrete section will crack and only a portion of the section will be effective. The method for predicting the ultimate capacity is similar to that for ordinary reinforced concrete beams. The procedure is modified only to the account for differences in the stress–strain behavior of ordinary and high strength steels. The procedure for rectangular prestressed beam sections is as follows:

1. If the section is underreinforced, then the ultimate capacity of the section from Eq. (11.3) and Fig. (11.15) is $\phi M_n = \phi[T(d - a/2)]$ with $T = A_{ps}f_{ps}$ in place of $A_s f_y$. The depth of the compression zone a is calculated by Eq. (11.5); i.e., $a = A_{ps}f_{ps}/(0.85 f'_c b)$.
2. If the section is overreinforced and rectangular, then the ultimate capacity is estimated by $\phi M_u = \phi(0.25 f'_c b d^2)$.

Table 11.19. Allowable Stresses in Prestressing Steel.

Loading Stage	
Preloading tendon before transfer of prestress	Smaller of $0.80 f_{pu}$ [a] or $0.94 f_{py}$ [b]
Immediately after transfer of prestress	$0.70 f_{pu}$

Reproduced with permission from *Building Code Requirements for Reinforced Concrete* (ACI 318-77), 1977 ed. American Concrete Institute, Detroit, MI.
[a] f_{pu} = ultimate strength of prestressing steel.
[b] f_{py} = yield strength of prestressing steel.

524   Reinforced Concrete Design

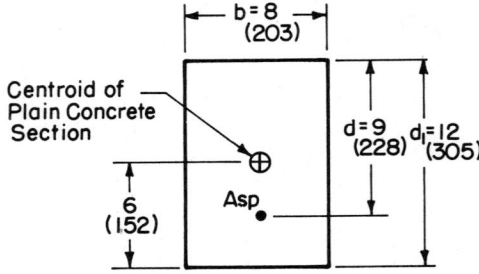

Fig. 11.43. Prestressed concrete beam section. All dimensions are in in. (mm).

3. A section with only prestressing steel may be considered underreinforced if the ratio $\rho_p f_{ps}/f'_c \leq 0.30$.

Example 11.15. Prestressed Beam Analysis. The rectangular section of Fig. 11.43 is used as prestressed beam over a simply supported span of 20 ft (6.10 m). The steel area A_{sp} equals 0.50 in.² (320 mm²), $f'_c = 4$ ksi (28 MPa), $f'_c = 6$ ksi (41 MPa), $f_{pu} = 250$ ksi (1.72 GPa), and $f_{py} = 200$ ksi (1.38 GPa).

1. Check the stresses at transfer of prestress.
2. Evaluate the allowable live load plus superimposed dead load moment the section can carry after prestress losses.
3. Evaluate the ultimate moment capacity and compare to the service moment capacity.
4. Check if the steel can carry 1.2 times the cracking moment for a modulus of rupture f_r equal to $7.5\sqrt{f'_c}$.

Solution

a. Evaluate the allowable stresses from Tables 11.18 and 11.19. The allowable concrete stresses at transfer of prestress

$$0.60\, f'_c = 0.6(4) = 2.4 \text{ ksi } (16.6 \text{ MPa}) \quad \text{in compression}$$
$$3\sqrt{f'_c} = 3\sqrt{4000} = 190 \text{ psi } (1.31 \text{ MPa}) \quad \text{in tension}$$

The allowable concrete stress at service loads

$$0.45\, f'_c = 0.45(6) = 2.7 \text{ ksi } (18.6 \text{ MPa}) \quad \text{in compression}$$
$$6\sqrt{f'_c} = 6\sqrt{6000} = 465 \text{ psi } (3.21 \text{ MPa}) \quad \text{in tension}$$

The allowable steel stress at transfer of prestress

$$0.70 f_{pu} = 0.7(250) = 175 \text{ ksi} \quad (1.20 \text{ GPa})$$

b. Estimate the prestress losses for the pretensioned case as 35 ksi (240 MPa). Thus, the effective steel prestress at service loads is 175 − 35 = 140 ksi (965 MPa).
c. Evaluate the flexural stresses at transfer of prestress. Determine the gravity load and midspan moment.

$$w_D = (150 \text{ lb/ft}^3)(8 \times 12/144) = 100 \text{ lb/ft} \ (150 \text{ kg/m})$$
$$M_D = w_D L^2/8 = 100(20)^2/8 = 5000 \text{ lb} \cdot \text{ft} \ (6.8 \text{ kN} \cdot \text{m})$$

Determine the extreme fiber stresses at transfer of prestress

$$F = 175(0.5) = 87.5 \text{ k} \ (390 \text{ kN})$$
$$Z = bd_1^2/6 = 8(12)^2/6 = 192 \text{ in.}^3 \ (3146 \text{ cm.}^3)$$

From Eqs. (11.32) and (11.33) and the limits of Table 11.18

$$\sigma_{it} = 87{,}500(3)/192 - 87{,}500/96 - 5000(12)/192$$
$$= 143 \text{ psi} \ (990 \text{ kPa}) < 3\sqrt{f'_{ci}}$$
$$\sigma_{ib} = -87{,}500(3)/192 - 87{,}500/96 + 5000(12)/192$$
$$= -1965 \text{ psi} \ (13.6 \text{ MPa}) < 0.6 f'_c$$

Thus, the strength at transfer of prestress is adequate.

d. Evaluate the magnitude of the allowable superimposed live plus dead loads. The allowable effective prestress force is $F' = 140 A_{sp} = 140(0.5) = 70 \text{ k} \ (310 \text{ kN})$. Using the effective prestress force and Eq. (11.36), the top fiber stresses may be expressed as

$$\sigma_{ft} = F'e/Z - F'/A - M_D/Z - M_{LR}/Z$$
$$\leq -2{,}700 \text{ psi} \ (18.6 \text{ MPa})$$
$$= 70{,}000(3)/192 - 70{,}000/96 - 60{,}000/192$$
$$- M_{LR}/192 \leq -2700$$
$$M_{LR} = 528 \text{ k} \cdot \text{in} \ (59.7 \text{ kN} \cdot \text{m})$$

Similarly from Eq. (11.37), the bottom fiber stress is

$$\sigma_{fb} = -F'e/Z - F'/A + M_D/Z + M_{LR}/Z$$
$$\leq +465 \text{ psi } (3.21 \text{ MPa})$$
$$= -70,000(3)/192 - 70,000/96 + 60,000/192$$
$$+ M_{LR}/192 \leq +465$$
$$M_{LR} = 380 \text{ k} \cdot \text{in } (42.9 \text{ kN} \cdot \text{m})$$

Thus, tension in the bottom fiber controls the allowable moment and $(M_{LR})_{\text{all}} = 380$ k · in. (42.9 kN · m). Over the 20 ft (6.10 m) span the allowable uniformly distributed load w_{LR} in addition to w_D can now be calculated from the maximum moment relationship $w_{LR}L^2/8 \leq 380,000$ which yields an allowable load of 630 lb/ft (9.2 kN · m).

e. Determine the ultimate moment capacity. Check if the section is over- or underreinforced; that is, determine if $\rho_P f_{ps}/f'_c$ is less than, greater than, or equal to 0.30. Since $\rho_p = A_{sp}/bd = 0.50/(8 \times 9) = 0.0069$ and (from Section 11.9.4)

$$f_{ps} = f_{pu} (1 - 0.5 \rho_p f_{pu}/f'_c)$$
$$= 250 [1 - 0.5(0.0069)(250/6)]$$
$$= 214 \text{ ksi } (1.48 \text{ GPa})$$

the parameter $\rho_p(f_{ps}/f'_c) = 0.0069(214/6) = 0.25$. Since $\rho_p(f_{ps}/f'_c) < 0.3$, the beam is underreinforced and the ultimate moment capacity can be determined by Eq. (11.3).

$$\phi M_n = \phi A_{sp}f_{ps}[d - (A_{ps} f_{ps}/(0.85f'_c b)]$$
$$= (0.85)(0.50)(214)[9 - 0.5(214)/((0.85 (6) (8))]$$
$$\phi M_n = 580 \text{ k} \cdot \text{in} \quad (65.5 \text{ kN} \cdot \text{m})$$

The allowable service moment must be calculated by comparing

ϕM_n to $M_u = 1.4M_D + 1.7M_L$; i.e.,
$$1.4M_D + 1.7 M_L < \phi M_n = 580 \text{ k·in} \quad (65.5 \text{ kN} \cdot \text{m})$$

Assuming the total dead load to be equal to the total live load, then $1.4M_L + 1.7M_L = 580$ k · in. (65.5 kN · m) from which the allowable live load equals 187 k · in. (21.1 kN · m). Since the dead load moment due to the mass of the beam M_D equals 60 k · in. (6.8 kN · m), the *allowable superimposed* dead plus live load moment M_{LR} equals $187 + 187 - 60 = 314$ k · in. (35.5 kN · m).

Thus, to satisfy the ultimate capacity requirements the superimposable dead plus live load moment must be reduced by 380,000 − 314,000 = 66 k · in. (7.5 kN · m). The section could be made more efficient by increasing the area of steel. A_{sp} could be increased by a factor of 1.2 to 0.60 in.² and still be underreinforced.

f. Check the cracking moment

$$M_{cr} = f_r Z = 7.5 \sqrt{f'_c} Z = 7.5 \sqrt{6000}(192)$$
$$= 112 \text{ k} \cdot \text{in.} \quad (12.7 \text{ kN} \cdot \text{m})$$
$$1.2 M_{cr} = 134 \text{ k} \cdot \text{in.} \ (15.1 \text{ kN} \cdot \text{m})$$

Since the moment capacity of the reinforcing steel and concrete is greater than $1.2 M_{cr} = 134$ k · in. (15.1 kN · m), the beam satisfies all the requirements except the ultimate capacity requirement.

11.10 CLOSURE FOR REINFORCED CONCRETE

The purpose of the preceding sections was to introduce the basic concepts of reinforced and prestressed concrete behavior. The coverage is by no means complete. The student who wishes to explore the subject in detail is referred to the design specifications, codes, commentaries, and design aids of the Prestressed Concrete Institute, the American Concrete Institute, and the Portland Cement Association. Additionally, the student should consult any one of the many available texts on reinforced and prestressed concrete design.

11.11 FOUNDATIONS FOR AGRICULTURAL BUILDINGS

11.11.1 Introduction

One of the most important components of an agricultural structure is its foundation. Foundations are required to distribute concentrated column loads or uniformly distributed wall loads, such as those encountered in silo wall footers or continucus wall footers under stud or masonry walls. The functions of foundations are to (1) prevent excessive settlement of the entire structure by distributing loads over areas larger than the column or wall bearing area; (2) prevent differential settle-

528 Reinforced Concrete Design

ment of various portions of the structure by equalizing or reducing the ground pressure under the walls or columns; (3) provide lateral support for poles in pole-type structures; (4) prevent heaving due to freezing and thawing of soil; and (5) keep moisture and vermin out of the building.

The primary factors which control foundation length and width are the allowable bearing capacity of the soil and the magnitude of the loads transmitted to the soil. The primary factors controlling the thickness of concrete foundations are the shear strength and moment capacity of the plain or reinforced footer.

11.11.2 Foundation Loads

Foundation loads are evaluated from the load and stress analysis of the structure. In a livestock, plant, or storage building the foundation load is the largest combination of dead, snow, wind, and live loads the structure may be expected to encounter. The foundation loads induced by overturning moments due to wind or due to uneven loading of tall storage structures must be carefully evaluated. The foundation loads may be a combination of concentrated column loads and distributed wall loads. The distributed loads may or may not be uniform.

11.11.3 Allowable Bearing Capacity of Soils

The allowable bearing capacity for a foundation is dependent upon the type soil and upon the depth of the bottom of the footer below grade. If large design loads are anticipated, soil bearing capacities should be obtained from soil tests. In the absence of soil tests and when the foundation is lightly loaded, the bearing capacities in Table 11.20 may be used as a guide. These bearing capacities assure that excessive settlement will not occur and have a factor of safety of 2.5 to 3.0 with respect to exceeding the ultimate soil bearing capacity. Bearing capacities may be increased by 33% for wind loading.

11.11.4 Design Criteria

11.11.4.1 Footer Area. The basic design criterion for establishing the size footer is that the pressure exerted by the foundation on the soil be less than the bearing capacity of the soil. The actual pressure distribution between the soil and footer varies nonlinearly and is fairly complex. However, several simplifying assumptions are used in foundation design for light structures.

11.11 Foundations for Agricultural Building

Table 11.20. Approximate Bearing Capacities of Soils.

Class of Material	Minimum Depth of Footing below Grade ft (m)	Brg. Capacity at Minimum Depth psf (kPa)	Increase in Min. Brg. Cap., per foot (305 mm) psf (kPa)	Maximum Brg. Capacity psf (kPa)
Rock	0 (0)	20% of crushing strength	0 (0)	20% of crushing strength
Compact sand–gravel mix	1 (0.30)	4,000 (190)	600 (30)	12,000 (570)
Compact coarse sand	1 (0.30)	1,500 (70)	300 (15)	8,000 (380)
Compact fine sand	1 (0.30)	1,000 (50)	200 (10)	8,000 (380)
Loose sand	2 (0.61)	500 (20)	100 (5)	3,000 (140)
Hard clay or sandy clay	1 (0.30)	4,000 (190)	800 (40)	8,000 (380)
Medium stiff clay	1 (0.30)	2,000 (100)	200 (10)	6,000 (290)
Soft sandy clay	2 (0.61)	1,000 (50)	50 (2)	2,000 (100)
Compact inorganic sand–silt mixtures	1 (0.30)	1,000 (50)	200 (10)	4,000 (190)
Loose inorganic sand–silt mixtures	2 (0.61)	500 (20)	100 (5)	1,000 (50)
Loose organic sand, silt mixtures, mud	0 (0)	0 (0)	0 (0)	0 (0)

The bearing pressure distribution for a centrically loaded footer is nonuniform as shown in Figs. 11.44a and 11.44b for granular and cohesive soils. However, adequate results are often obtained by assuming the pressure distribution to be uniform. Thus, the design criterion for sizing a centrically loaded footer is

$$P/A = q \leq q_a \qquad (11.38)$$

or

$$A \geq P/q_a \qquad (11.39)$$

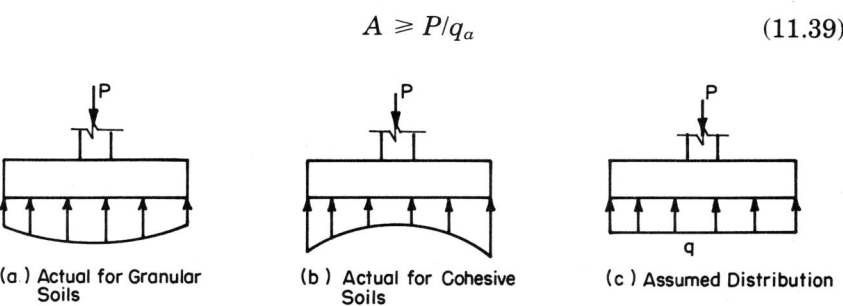

(a) Actual for Granular Soils   (b) Actual for Cohesive Soils   (c) Assumed Distribution

Fig. 11.44. Pressure distribution under centrically loaded footers.

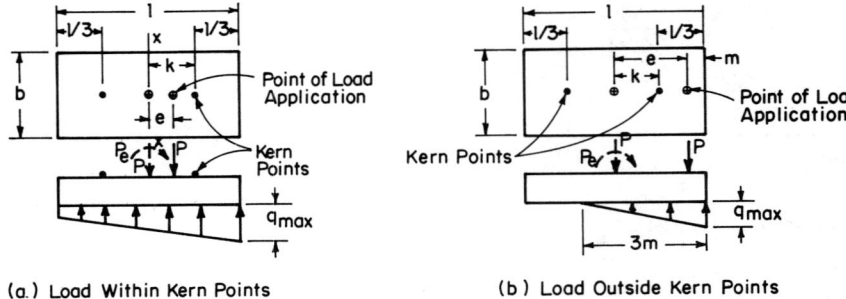

(a.) Load Within Kern Points (b) Load Outside Kern Points

Fig. 11.45. Pressure distribution under eccentrically loaded footers.

where q is bearing stress; q_a, allowable bearing stress; A, footer area; P, footer concentrated load.

The assumed bearing pressure distributions under eccentrically loaded footings are shown in Fig. 11.45. The eccentricity may be the result of an eccentric gravity load or due to a fixed or partially fixed end restraint at the base of the wall; e.g., the end reaction of a restrained rigid frame. If the footer is loaded by a centric load and a moment M, it can be replaced by an equivalent vertical load at an eccentricity from the centroidal axis. If the eccentricity is within the kern points ($\ell/3$ from either edge for a rectangular footer), the soil stress may be assumed to be similar to that shown in Fig. 11.45a and the maximum soil pressure is predicted by Eq. (11.40).

$$q_{max} = P/A + Pec/I_x = P/A(1 + ec/r^2) \qquad (11.40)$$

where I_x is the moment of inertial of the footer area about the centroidal axis; $r_x = \sqrt{I_x/A}$; and the design criterion for footer sizing the footer is Eq. (11.41).

$$q_{max} \leq q_a \qquad (11.41)$$

When the eccentricity of loading falls outside the kern points, then only a portion of the footer resists the vertical load. Part of the pressure distribution becomes negative, but no tension forces can be transferred between the soil and concrete. For rectangular footings, the pressure distribution of Fig. 11.45b may be assumed and the maximum soil pressure may be estimated by Eq. (11.42).

$$q_{max} = 2P/3bm \leq q_a \qquad (11.42)$$

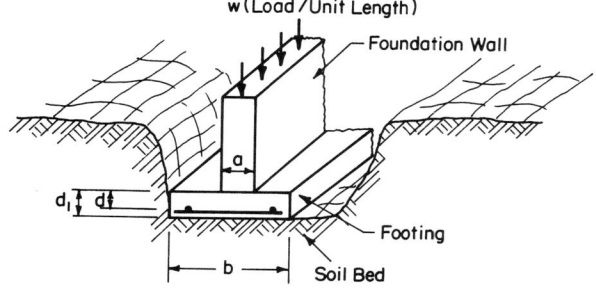

Fig. 11.46. Continuous wall footing.

11.11.4.2 Footer Size. The design criteria for determining the thickness and reinforcement requirements for the footer depend upon the type footer and are based upon the ultimate design stress theory of reinforced beams. The footer section must be large enough such that diagonal shear capacity, punching shear capacity, and moment capacity of the concrete are not exceeded. Since the criteria vary somewhat with the type footer, specific criteria will be discussed separately for each type.

11.11.5 Footer Types

The two main footer types encountered in agricultural buildings are continuous wall footers and single column footers. Both reinforced and nonreinforced concrete footers of both types are common. For example, silo foundations require reinforced wall footings, whereas footer pads for pole structures are usually unreinforced column footings. Typical wall and column footings are illustrated in Figs. 11.46 and 11.47.

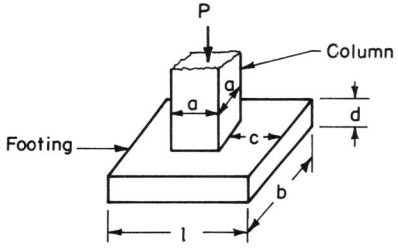

Fig. 11.47. Rectangular column footer.

11.11.6 Design of Wall Footings

Wall footings usually support a foundation wall which may be either solid concrete or masonry supporting stud framing. The load on such footings in agricultural structures may usually be considered as being uniformly distributed. If the footer load per unit length is w, the footer width b is estimated by Eq. (11.43).

$$b = w/q_a \qquad (11.43)$$

A flexural analysis determines the thickness and reinforcement requirements for the footing. By assuming the footer in Fig. 11.46 to be two cantilever beams of length $b/2$ the maximum shear and moment would occur at the center of the wall. However, the rigidity provided by the wall shifts the critical section for the bending moment to the faces of the wall and the critical section for shear to a distance d from the wall faces. At these points, the expressions for the moment and shear per unit length of foundation wall, M_u and V_u, are

$$M_u = q_u (1)(b - a)^2/8 \qquad (11.44)$$
$$V_u = (q_u)(1)(b/2 - a/2 - d) \qquad (11.45)$$

where q_u is the factored ground pressure per unit length of footer.

Since shear reinforcement is usually avoided in footings, footer depth d_1 and reinforcement depth d are often controlled by the shear strength of the concrete, $V_c = 2\sqrt{f'_c}(bd)$ with f'_c in psi, and are estimated by equating the factored shear force of Eq. (11.45) to the allowable concrete shear stress $\phi V_n = \phi V_c$. Solving the equality for d yields

$$d = [(b - a)/2]/[\phi(2\sqrt{f'_c}b/q_u + 1)] \qquad (11.46)$$

Once d is known, the steel area required in the direction normal to the wall is calculated by simply setting $M_u = \phi M_n$ as in any reinforced concrete beam. Longitudinal steel is required for temperature and shrinkage control. The ACI code recommends a minimum steel area of $0.0020(bd)$ for Grade 40 or 50 deformed bars.

In many agricultural buildings the foundation loads are so light that footers between 10 and 20 in. (250 and 500 mm) wide are adequate. In such cases reinforcement is generally not required in the lateral

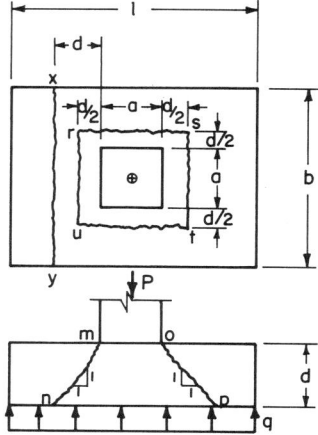

Fig. 11.48. Critical sections for column footers.

direction and two No. 4 bars are sufficient in the longitudinal direction. A "rule of thumb" for lightly loaded footers is to construct them twice as wide as the foundation wall and as deep as the foundation wall is wide.

11.11.7 Design of Column Footings

Column footings in agricultural structures are usually square or round. Unreinforced round footers are frequently used under pole-type structures. When larger loads occur and footer reinforcement is necessary, square footers are most common.

If the column load is P, then the required footer area is $A = P/q_a$. The thickness of the footer is usually dictated by the flexural shear (diagonal tension) or the punching shear. Footer reinforcement requirements are determined by the moment capacity in the two directions parallel to the footer sides.

The column in Fig. 11.48 has a tendency to punch through the footer line along lines $\overline{mn}$ and $\overline{op}$ which make an approximate angle of 45° with the horizontal. This behavior is called punching shear and must be resisted by the concrete. The average punching shear stress can be estimated by assuming the fracture line to occur at a distance $d/2$ from all the column faces. From equilibrium of the column and the footer free body cut at a vertical section through $\overline{rstu}$, the average punching shear stress v_{up} is

$$v_{up} = P_u/l_o d \qquad (11.47)$$

where l_o is the perimeter of $\overline{rstu} = 4(a + d)$. The allowable shear stress in punching shear, in psi, is $v_{cp} = 4\sqrt{f'_c}$ for f'_c in psi. This is greater than the allowable shear stress in flexural members subject to one-way action and is the consequence of increased concrete strength under the triaxial state of stress induced in the vicinity of the column. The design requirement for punching shear is

$$v_{up} \leq \phi v_{cp} \qquad (11.48)$$

or

$$P_u/l_o\, d \leq \phi\, (4\sqrt{f'_c}) \qquad (11.49)$$

The critical section for flexural shear is at a distance d from the faces of the column. Section $\overline{xy}$ in Fig. 11.48 taken along the shorter dimension of a rectangular footer is the critical section. If the distance $\overline{xy}$ is b, the design criteria for flexural shear of column footers is identical to Eqs. (11.45) and (11.46) for wall footers with the unit length term replaced by the column dimension normal to b.

The critical sections for the moment capacity of the footer are at the column faces for concrete columns resting directly on concrete footers and halfway between the column face and steel baseplate for steel columns resting on concrete footers. The moment capacity, and reinforcement requirements, are equal in both directions in a square column. In a rectangular footer the factored bending moment and reinforcement needs will be greatest on a cross section parallel to the short side of the footer. The procedure for evaluating the reinforcement requirements is identical to that for wall foundations. Equation (11.44) estimates the factored moments in both directions and application of Eq. (11.1) yields the reinforcement requirement in both directions.

Punching shear or flexural shear usually control the footer thickness. Thus, the design procedure is to evaluate d based on shear requirements and then to evaluate the steel area needs.

For lightly loaded structures, reinforcement is often unnecessary in column footers. The following rules of thumb serve as guidelines for such cases. Footers up to 3 ft (0.91 m) square generally require no reinforcement. Referring to Fig. 11.47, the footer thickness d is usually the larger of 8 in. (203 mm) or $1.5(c)$. Adequacy of plain column footers

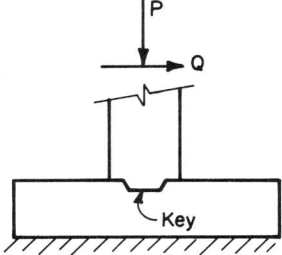

Fig. 11.49. Keys for footers.

is usually assured if the unfactored maximum flexural stress in tension, in psi, is less than $1.6\sqrt{f'_c}$, where f'_c is in psi.

11.11.8 General Requirements

Reinforced footers must be checked for adequate development length, cover, and spacing of reinforcement. In footers the minimum cover is 3 in. (76 mm) and the maximum reinforcement spacing is the smaller of $3d$ or 18 in. (451 mm). Development length requirements are the same as for beams.

Foundation walls must often resist the horizontal thrust at the base of a structural wall. Resistance to horizontal movement of the footer and foundation wall is provided by the lateral resistance of the soil. The lateral soil resistance may be conservatively estimated as 250 psf (12 kPa) for most foundations. Alternatively it may be predicted by Rankine's equation for passive pressures or by reference to Table 10.17 in the timber design chapter. Keying is recommended to prevent sliding between the wall and footer (see Fig. 11.49). The bottom of all footers should be placed on undisturbed or carefully compacted soil and should extend below the frost line.

Example 11.12. Footer Design. A steel bin, 15 ft (4.57 m) in diameter by 20 ft (6.10 m) high, is supported by four equally spaced columns. If the stored material has a density of 80 pcf (1280 kg/m³) and the average bin wall thickness is 0.0625 in. (1.6 mm), design a square concrete footing for one of the columns. Assume the bottom of the footer is 1 ft (305 mm) below grade and rests on a hard clay soil. Design the footer for gravity loads only. Do not consider the effect of wind loads. Assume the columns are steel sections with an 8 × 8 in. (203 × 203 mm) square base plate and that $f'_c = 4,000$ psi (28 MPa).

Solution

a. Estimate the loads.

$$DL = \text{(volume of bin wall)} \times \text{(mass density of steel)}$$
$$= [2\pi(7.5)(20)(0.0625/12) + 2(\pi)(7.5)^2(0.0625/12)](491)$$
$$= 3300 \text{ lb} \quad (14.7 \text{ kN})$$
$$LL = \text{(bin volume)} \times \text{(mass density of stored material)}$$
$$= (\pi R^2 h)\gamma_g = \pi(7.5)^2(20)(80)$$
$$= 283 \text{ k} \quad (1.26 \text{ MN})$$
$$P = DL + LL = 286 \text{ k} \quad (1.27 \text{ MN})$$

Evaluating the factored column load,

$$P_u = 1.4DL + 1.7LL = 486 \text{ k} \quad (2.16 \text{ MN})$$

b. Determine the footer area A and dimension b.

$$A = \tfrac{1}{4}(DL + LL)/q_a = \tfrac{1}{4}(286,000)/4000$$
$$= 17.88 \text{ ft}^2 \quad (1.66 \text{ m}^2)$$
$$b = \sqrt{A} = 4.25 \text{ ft (1.3 m)} = 51 \text{ in. (1.3 m)}$$

c. Determine the required footer thickness.
   i. For punching shear, assume that the critical section is at $d/2$ from the edge of the base plate. Then the factored punching shear stress may be calculated with Eq. (11.47).

$$v_{up} = [(486,000/4)/(4(8 + d)d)] = 30,375/[d(8 + d)]$$

The allowable concrete shear stress, $v_{cp} = 4\sqrt{f'_c} = 4\sqrt{4000} = 253$ psi (1.74 MPa). By equating v_{up} to ϕv_{cp}, the required distance d between the top of the footer and the tension steel equals 8.5 in. (216 mm).

   ii. For flexural shear, assume that the critical section is at d from the edge of the base plate, then the average ultimate shear force may be calculated by Eq. (11.45).

$$V_u = (P_u/4A)(b/2 - a/2 - d)(b)$$
$$= [486,000/(4 \times 51^2)][51/2 - 8/2 - d][51]$$
$$= 2382 \ (21.5 - d)$$

and directly the average ultimate shear stress becomes $v_u = V_u/bd = 45.81 \ (21.5/d - 1)$ The ultimate shear strength ϕv_n equals

$$\phi v_c = 0.85(2)\sqrt{f'_c} = 0.85(2)\sqrt{4000} = 108 \text{ psi} \quad (740 \text{ kPa})$$

Equating $v_u = \phi v_n$, the required depth d for flexural shear equals 6.4 in. (163 mm). Thus punching shear controls and the required depth d equals 8.5 in. (216 mm).

d. Determine steel area requirements. Assuming the effective column width to be 7 in. (152 mm), the factored moment M_u can be evaluated with Eq. (11.44).

$$\begin{aligned} M_u &= q_u(b)(b-a)^2/8 \\ &= [486/4 \times 51^2][51][51-7]^2/8 \\ &= 576 \text{ k-in.} \quad (65.1 \text{ kN-m}) \end{aligned}$$

Using the procedures discussed in Section 11.4.5 for beam design with b and d known, evaluate the parameters $M_u/(f'_c bd^2)$ and $\omega - 0.59\omega^2$.

$$\begin{aligned} \omega - 0.59\omega^2 &= M_u/f'_c bd^2) \\ &= 576,000/[4000(51)(8.5)^2] = 0.039 \end{aligned}$$

Thus $\omega = 0.040$.

By definition of ω and for Grade 40 steel reinforcement, evaluate the steel ratio, ρ

$$\rho = f'_c/f_y \, \omega = (4000/40,000)(0.040) = 0.0040$$

From the definition of ρ, the steel area, $A_s = \rho(bd) = 0.0040(51)(8.5) = 1.73$ in.^{2} (12.1 cm^2). Thus, use six No. 5 bars (both ways) with $A_s = 6(0.31) = 1.86$ in.^{2} (12.0 cm^2) spaced 9.2 in. (230 mm) apart with 3-in. (76-mm) cover on all sides and bottom. The footer design is illustrated in Fig. 11.50.

Since the bars in the transverse direction are above those in the longitudinal direction, the top bars are placed at 8.5 in. (216 mm) below the top surface and the bottom bars are 9.25 in. (235 mm) below the top. The total footer depth is 12.5 in. (318).

PROBLEMS

11.1. For a rectangular reinforced concrete beam section with dimensions, $b = 8$ in. (203 mm) and $d = 10$ in. (254 mm), $f'_c = 4$ ksi (28 MPa), and $f_y = 40$ ksi (280 MPa):

538 Reinforced Concrete Design

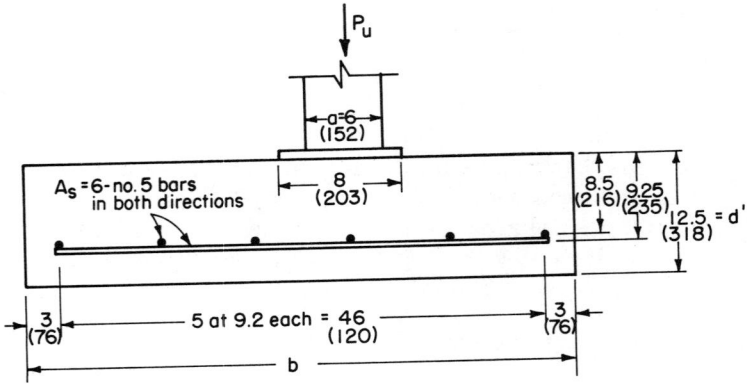

Fig. 11.50. Example 11.12—Footer design. All dimensions are in in. (mm).

    a. Determine the balanced steel area.
    b. Determine the allowable steel area.
    c. Select the reinforcing bars required for the allowable steel area.
    d. Determine the balanced moment capacity.
    e. Determine the allowable moment capacity.
    f. If the section is used as a simply supported beam with a span of 10 ft (3.04 m), determine the magnitude of the factored allowable uniform live plus dead load combination.
    g. What allowable live load can the section carry in addition to its own dead load?

11.2. If the 3 in. (76 mm) × 4 in. (102 mm) notch in the section of Fig. 11.19 were replaced with concrete and if the 3 in. (76 mm) × 3 in. (76 mm) solid portions at the top of the section were removed:
    a. Determine the depth of the compression block at balance.
    b. Determine the allowable steel ratio.
    c. Determine the moment capacity at balance conditions.
    d. Determine the allowable moment capacity at the allowable steel ratio.
    e. Determine the moment capacity for the three No. 7 reinforcing bars.
    f. Determine the allowable uniform live load the section can carry over a 10 ft (3.04 m) simply supported span

in addition to its own dead load if $A_s = 0.75A_{sb}$, $f'_c = 4$ ksi (28 MPa), and $f_y = 50$ ksi (340 MPa).

g. How would the solution to 2f differ if $f_y = 40$ ksi (280 MPa)?

11.3. Design an economical rectangular reinforced section to carry a uniformly distributed live load of 500 lb/ft (680 N/m) over a simply supported span of 12 ft (3.66 m). Use $f'_c = 4$ ksi (28 MPa) and $f_y = 40$ ksi (280 MPa).

11.4. Determine whether the column section of Fig. 11.24 is long or short if it is used as a single independent section completely pinned at each end and is 10 ft (3.05 m) long.

11.5. Determine the location of the plastic centroid of the section of Fig. 11.25 if the steel area is replaced with seven No. 6 reinforcing bars and if $f'_c = 4$ ksi (28 MPa) and $f_y = 40$ ksi (280 MPa).

11.6. For a rectangular reinforced concrete beam with $b = 4$ in. (102 mm), $d = 6$ in. (152 mm), two No. 5 bars located at d, an overall depth of 7.5 in. (191 mm), $f'_c = 3$ ksi (21 MPa), and $f_y = 40$ ksi (280 MPa):

a. Show whether the steel yields and explain why.
b. Show whether the steel reinforcement limitations are met.
c. Sketch the tensile and compressive stresses and strains acting on the cross section at ultimate capacity.
d. Calculate the magnitude of ultimate moment capacity.
e. Calculate the magnitude and the allowable uniformly distributed live load the beam can carry over an 8 ft (2.44 m) simple span.

11.7. Repeat Problem 11.6 if the steel reinforcement is three No. 5 bars.

11.8. A reinforced concrete section with $b = 6$ in. (152 mm), $d = 6$ in. (152 mm), $d' = 8$ in. (203 mm), one No. 7 reinforcing bar located at d, $f'_c = 4$ ksi (28 MPa), and $f_y = 40$ ksi (280 MPa) is used as a 10 ft (3.04 m) simply supported beam in a beef confinement facility. If the design live load equals 250 lb/ft (340 N/m):

a. Determine if the section is over- or underreinforced.
b. Determine the allowable moment capacity of the section.

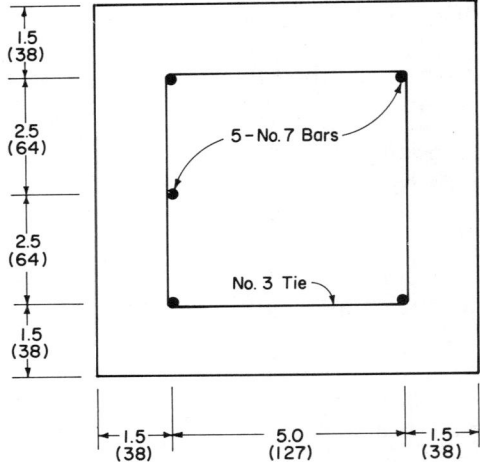

Figure 11.51. Problem 11.12—Reinforced column section. All dimensions are in in.

    c. Determine the required moment capacity of the section.
    d. Determine if the section is adequate to carry the design loads.

11.9. Repeat Example 11.5 if the overall section size decreases to 6 in. (152 mm) on a side and the steel is located 2 in. (51 mm) from each corner as in Fig. 11.24.

11.10. Determine whether the short column cross section of Examples 11.5 and 11.7 can safely carry an axial compressive live load of 50 k (222 kN) when the load is applied (a) at the plastic centroid; (b) at 2 in. (51 mm) to the right of the plastic centroid. (Solve analytically, then use the interaction diagram of Fig. 11.33 to check your solution.)

11.11. If the column section of Example 11.5 and Fig. 11.24 is subjected to an axial compressive load of 20 k (89 kN) and a bending moment of 40 k · in. (4.5 kN · m), determine:

    a. The equivalent eccentric load for the column.
    b. The allowable axial load for the eccentric load.
    c. Whether the section is adequate to carry the loads.

11.12. For the short column cross section of Fig. 11.51 in which $f'_c = 4$ ksi (28 MPa) and $f_y = 40$ ksi (280 MPa), determine:

    a. The location of the plastic centroid.
    b. The magnitude of the allowable axial compressive load

capacity for a load applied through the plastic centroid.
c. How frequently ties are needed in the column.
d. The magnitude of the allowable combined axial compressive load and moment capacity when the axial load is applied at a point 2 in. (51 mm) to the right of the plastic centroid.

11.13. Determine the location and spacing of No. 2 vertical stirrups for shear reinforcement for the beam section of Problem 11.1 when it carries its maximum allowable uniformly distributed live load over a 10 ft (3.04 m) simple span.

11.14. Determine the diagonal shear reinforcement required for the beam section and loads of Problem 11.3. Use No. 2 vertical stirrups.

11.15. Determine the diagonal shear reinforcement required for the beam section and loads of Problem 11.6 when carrying its maximum allowable live load over a simple span of 8 ft (2.44 m). Use No. 2 vertical stirrups.

11.16. Determine the diagonal shear reinforcement required for the beam section and loads of Problem 11.8. Use No. 2 vertical stirrups.

11.17. Evaluate the long-term design deflection for the beam in Problem 11.1g. Is this an acceptable deflection? Why?

11.18. Evaluate the design deflection for the beam in Problem 11.6e.

11.19. Evaluate the design deflection for the beam in Problem 11.8.

11.20. Analyze the prestressed concrete beam section of Example 11.11 and Fig. 11.43 if the area of prestressing steel is increased to 0.6 in.² (390 mm²).

11.21. Design a square reinforced concrete footer for a 6 in. (152 mm) square column which carries an axial load of 80 k (356 kN). The footer rests on a medium stiff clay and is 2 ft (610 mm) below grade. Use f'_c = 3 ksi (21 MPa) and f_y = 40 ksi (280 MPa).

NOMENCLATURE FOR CHAPTER 11

A_b Area of a single reinforcing bar, in. (mm)
A_g Gross concrete cross-sectional area, in.² (cm²)
A_s Area of tension steel, in.² (mm²)
A'_s Area of compression steel, in.² (mm²)
A_{sb} Area of steel at balance, in.² (mm²)
A_{sp} Area of prestressing steel, in.² (mm²)
A_v Area of web, or shear, steel reinforcement, in.² (mm²)

C_c, C	Resultant compressive force in the concrete, k (kN)
C_s	Resultant compressive force in the steel, k (kN)
D	Dead load, k (kN)
E_c	Concrete modulus of elasticity, ksi (GPa)
E_s	Steel modulus of elasticity, ksi (GPa)
F	Prestressing force, k (kN)
F'	Prestressing force after losses, k (kN)
ΔF	$F - F'$ = prestressing loss, k (kN)
L	Live load, k (kN)
M	Bending moment, k · in. (kN · m)
M_D	Dead load moment, k · in. (kN · m)
M_L	Live load moment, k · in. (kN · m)
M_n	Theoretical moment capacity of a section, k · in. (kN · m)
M_{LR}	Superimposed dead plus live load moment, k · in. (kN · m)
M_u	Required moment capacity of a section, k · in. (kN · m)
P	Concentrated load, k (kN)
P_b	Theoretical axial compressive load capacity at balance conditions, k (kN)
P_D	Dead axial load for a column, k (kN)
P_L	Live axial load for a column, k (kN)
P_n	Theoretical axial load capacity of a column, k (kN)
P_{nb}	Theoretical allowable axial compressive load at balance conditions, k (kN)
P_u	Factored applied column axial load, i.e., $1.4P_D + 1.7P_L$, k (kN)
S	Spacing between reinforcing bars, in. (mm)
T	Tensile force, k (kN)
U	Factored service load, k (kN)
V	Shear force, k (kN)
V_c	Shear capacity of plain concrete, lb (kN)
V_{cp}	Allowable punching shear strength of concrete, lb (kN)
V_n	Theoretical shear capacity of a section, lb (kN)
V_s	Shear capacity of web steel, lb (kN)
V_u	Factored shear load, lb (kN)
V_{up}	Design punching shear force, lb (kN)
W	Wind load, k (kN)
a	$\beta_1 c$ is distance between the compression edge and the lower extremity of the compressive stress block, in. (mm)
a_b	$\beta_1 c_b$ is distance between the compression edge and the lower extremity of the compressive stress block at balance, in. (mm)
b	Width of a cross section, in. (mm)

Nomenclature for Chapter 11

b_w — Width of the web of a cross section, in. (mm)
c — Distance from the compression edge to the neutral axis of a section, in. (mm)
c_b — Distance from the compression edge and the neutral axis of a balanced section, in. (mm)
d — Distance from compression edge of a member to the centroid of the tension steel, in. (mm)
d' — Total depth of a beam cross section, in. (mm)
d' — Distance from the compression edge of a member to the centroid of the compression steel, in. (mm)
d'' — Distance between tension steel and the plastic centroid, in. (mm)
d_b — Diameter of reinforcing bars, in. (mm)
e — Eccentricity of applied load or resultant, in. (mm)
e' — Distance between applied axial load and the tension steel in a column section, in. (mm)
e_b — Eccentricity of applied axial compressive load at balance, in. (mm)
f_c — Compressive stress of concrete, ksi (MPa)
f'_c — Ultimate compressive strength of concrete, ksi (MPa)
f'_{ci} — Initial compressive strength of concrete in prestressed applications, ksi (MPa)
f_{cu}, f''_c — Ultimate flexural compressive strength of concrete, ksi (MPa)
f_{ps} — Allowable stress in prestressing steel, ksi (MPa)
f_{pu} — Ultimate strength of prestressing steel, ksi (MPa)
f_s — Steel tensile stress, ksi (MPa)
f_t — Tensile strength of concrete, ksi (MPa)
f_y — Yield strength of steel, ksi (MPa)
g — Ratio of the distance between outer rows of longitudinal reinforcement in column sections to the overall dimension of the section
l_d — Development length for reinforcing steel, in. (mm)
l_e — Equivalent development length for standard hooks, in. (mm)
q — Bearing load, psf (kPa)
q_a — Allowable soil bearing strength, psf (kPa)
q_{max} — Maximum soil pressure, psf (kPa)
s — Spacing between web stirrups, in. (mm)
v_c — Average concrete shear strength, psi (kPa)
v_{cp} — Allowable punching shear stress, psi (kPa)
v_u — Average factored shear stress, psi (kPa)

v_{up}	Average punching shear stress, psi (kPa)
w	Uniformly distributed load, k/ft (kN/m)
w_D	Uniformly distributed dead load, lb/ft (kN/m)
w_{LR}	Uniformly distributed superimposed live plus dead load, lb/ft (kN/m)
β_1	a/c
Δ	Design deflection of a beam, in. (mm)
Δ_D	Immediate dead load deflection, in. (mm)
Δ_{LP}	Immediate live load deflection, in. (mm)
Δ_{LS}	Immediate long-term live load deflection, in. (mm)
ε_c	Compressive strain in the concrete
$\varepsilon_{cu}, \varepsilon_u$	Ultimate concrete compressive strain
ε_s	Compressive strain in the steel
ε_s'	Strain in the compression steel
ε_{sy}	Yield strain of steel
l	Distance between the compressive and tensile resultants in a beam cross section, in. (mm)
λ	Creep factor for long-term live loading
λ_x	Creep factor for dead loading
ϕ	Capacity reduction factor for a concrete section
ρ	Ratio of steel to gross concrete area
ρ_b	Ratio of steel to gross concrete area at balance conditions
ρ_p	Ratio of prestressing steel to gross concrete area above centroid of steel
ρ_s	Ratio of spiral column lateral reinforcement steel ratio $\equiv$ volume of spiral steel per loop $\div$ volume of concrete enclosed by each spiral loop
ρ_w	$A_s/b_w d$ = steel ratio of shear reinforcement
σ	Normal stress, ksi (MPa)
σ_i	Initial flexural prestress, ksi (MPa)
σ_f	Final flexural stress after prestress losses, ksi (MPa)
σ_L	Prestress losses, ksi (MPa)
τ	Shear stress, ksi (MPa)
ζ	Constant for a standard hook

Index

Page numbers in *italic* indicate illustrations or tables.

A36 steel, 213–14
  design aids for columns of, 222–23, 224–29
  stress-strain curves for, *215*
Adhesive connectors, 188, *189*
Agricultural Engineers Yearbook of Standards, 136
Allowable stress, 8
  on cold-formed steel, 302–5, 314
  on connectors, 264, 265, 267, 348, *350, 351, 352*
  on hot-rolled steel, 212, 222–23, *224–28*, 234–50
  on reinforced concrete, 522, *523*
  on timber, 370, *371–79*, *395*
American Concrete Institute (ACI), 457, 469, 472, 504, 510, 520, 527
American Institute of Steel Construction (AISC), 210, 218, 243–44
American Institute of Timber Construction (AITC), 360, 426
American Iron and Steel Institute (AISI), 210, 299
American National Standards Institute (ANSI) standards, 127
  load combinations, 128–29
  snow loads, *139*, 143, 144, *145*
  wind loads, 149, 151, *152*, 155, 156, 157
American Plywood Association (APA), 360
Analysis, and design concepts, 1–5. *See also* Free body analysis; Load(s); Stress analysis
Angle, cold-formed steel, *286*
  equal, with stiffened legs, *293*
  equal, with unstiffened legs, *294*
Applied stress, 8
Arches
  area-moment equations applied to, 108
  in granular masses, 170–73, *174*, 175
Area-moment analogies and bending deformations, 80–86
Area-moment analysis of indeterminate frames, 92–108
  applications, 94–98
  concepts, 92–94
  theorem examples, 98–108
Area property of steel sections, 308
Axial loads on joints, *187*, 188–90
  combined with in-plane moments, 194–96

545

Axial loads on joints (*Cont.*):
   combined with out-of-plane moments, 196–99
Axial stress, 8, 9
   in timber, 363, *364,* 400–405
   in trusses, 33–34
Axis bending in steel sections, allowable flexural stresses in, 236

Balance conditions, reinforced concrete, 475, *476, 477*
Barre, H. J., vii
Bars, summation of, in a compound truss, 40, 41–43
Beam-columns
   cold-rolled steel, 344–46
   hot-rolled steel, 258–63
      problems examples, 260–63
   timber, 400–405
      problem examples, 403–5
      secondary moments in, *403*
   truss, *425,* 426
Beam(s). *See also* Beam-columns; Column(s)
   area-moment analysis, 108, *109, 110*
   dead-load carrying capacity of, 132
Beam(s), cold-formed steel
   allowable loads on, to prevent web crippling, *317–18*
   with lateral bracing, 314–23
   laterally unsupported beams, 325–31
Beam(s), hot-rolled steel
   allowable flexural stresses in, 242–43, *244–45*
   allowable loads in, *246–49*
   allowable moments in, 250–51, *252–53*
   buckling of, 211, *212*
   flexural design of steel, 234–58
   laterally braced and unbraced, 257–58
   laterally supported, 256–57
   selection of, using beam load tables, 249, *250*
   selection of, using design aids, 254–56
   selection of, using moment charts, 251, *254*
Beam(s), reinforced concrete, *456, 457*
   analysis of singly reinforced nonrectangular beam, 481–83
   analysis of singly reinforced rectangular beam, 476–81
   deflection control in, 513–18
   design of singly reinforced rectangular beam, 483–86
   prestressed, 518–27
   shear reinforcement for, 505–10
   working strength vs. ultimate strength methods for designing, *470*
Beam(s), timber, 389–400
   allowable horizontal shear stress for, *395*
   laterally braced, 390–98
   laterally unbraced, 398–400
   shear stress adjustments, *394*
   two-beam action, *392, 395*
Bearing stresses, on flexural steel members, 241, *242*
Bending, defined, 63
Bending deformation, 63–90
   application of deformation equations, 72–80
   area-moment analogies, 80–86
   curvature and bending, 64–65
   definitions, 63–64
   displacements, 68–70
   due to bending moment, 71–72
   elastic line deformation sketches, 86–87
   nomenclature, 90
   problems, 87–90
   rotations, 65–68
   sign conventions for, 70–71
Bending effect β(beta), 63–64, 65

Bending moment
  deformation due to, 71–72
  diagram of, *82*
Bending stress, 8, 10. *See also*
  Moment diagrams; Shear
  diagram
Bernoulli, John, 54
*BOCA Basic General Building Code/
  1975,* 135, 138
Bolted connections, 263–65
  for cold-formed steel, *348,* 350–52
    allowable stresses on, *351, 352*
  for hot-rolled steel
    allowable stresses for, *264*
    bearing type, 263, *264*
    example, 269–73
    friction type, *263*
    minimum edge distances, *265*
  for timber, 414–21
    allowable loads on, *415–16*
    example problems, 414–21
    spacing definitions, *419*
    two-, three-member joints, *417, 418*
Box-type steel sections
  allowable flexural stress in hot-rolled, 236
  cold-formed, *286*
*Bracing Wood Trusses: Commentary
  and Recommendations,* 435
Buckling of steel sections, 211–*12*.
  See also Post-buckling
  cold-formed, 284, *296,* 297–99
  local, in UCEs, 319–20
  hot-rolled, 211–12
  design for no local, 216–21
Buckling of timber sections, 385
*Building Code Requirements for
  Reinforced Concrete,* 457, 514, 523
Building codes on loads, 127–28
Building materials, dead loads of, *130–32*
Building Officials and Code
  Administrators (BOCA), 127

Butt joint, *186,* 187, *190*

Cantilevered bracket, area-moment
  bending deformation, 82, *83,* 84
Cantilevered frame, computing
  bending deformation in, 75–76, 77
Cantilevered member, computing
  bending deformation in
  bent by end moment, 74, *75*
  bent by intermediate couple, 74–75, *76*
Carry-over moment (COM), 114–15
Castigliano, Alberto, 54
Channel, cold-formed steel, *286*
  with stiffened flanges, *287*
  two, with stiffened flanges back-to-back, *291*
  two, with unstiffened flanges back-to-back, *292*
  with unstiffened flanges, *288*
Checking of design adequacy, 3
Cold-formed steel design, 283–357
  allowable stresses in, 302–5
  axially loaded compression members, 331–44
  beam-columns, 344–46
  cold-formed vs. hot-rolled steel, 209–10, 284–302
  connectors, 346–52
  effective widths of SCEs, 305–7
  flexural members, 314–31
  introduction to, 283–84
  maximum flat-width ratios for section elements, 307, *308*
  nomenclature, 355–57
  problems, 352–55
  properties of sections with thin elements, 308–14
  types of, and strength properties, 299, *300, 301*
Cold Formed Steel Design Manual,
  309, 317, 318, 349, 350, 351, 352
Collapse mechanism, *34,* 35

548 Index

Collinear members complex truss, 48–50
Column analogy analysis of indeterminate frames, 120
Column Research Council (CRC) equations, 218, 219, 220
Column(s), cold-formed steel, 331–44. *See also* Beam(s); Beam-columns
  cylindrical tubular members, 344
  introduction, 331–32
  torsionally stable, 332–38
  torsionally unstable (point symmetric sections), 344
  torsionally unstable (single symmetric/unsymmetric sections), 338–43
Column(s), hot-rolled steel, 223–34. *See also* Beam(s); Beam-columns
  allowable compressive stress, *229*
  allowable loads, extra-strong pipe, *226*
  allowable loads, standard pipe, *227*
  allowable loads, weak axis, *224–25*
  axial load, *228*
  buckling strength of, 216–21
  design aids for A36, 222–23, *224–29*
  design illustrations, 231–32, 233–34
  effective slenderness ratios, 221–22
  selection illustration, 223–31
Column(s), reinforced concrete, 486–505
  composite behavior, *471, 472*
  foundation footers, *531, 533*–35
  general requirements, 486–88
  short centrically loaded analysis, 488–91
  short eccentrically loaded analysis, 491–505
  types, *487*

Column(s), timber, 384–89. *See also* Beam(s), timber; Beam-columns, timber; Pole buildings
  allowable compressive stress with slenderness factor, *386*
  definition sketch, *384*
  effective length of, *387*
  problem examples, 387–89
  slenderness factors, *387*
Combined loading. *See* Beam-columns
Compact steel sections, allowable flexural stresses on, 235–36, *239*
Complex truss, 36, 40–*41*
  stress analysis of
    collinear members method, 48, *49, 50*
    simultaneous equations method, 47
    undetermined member stress method, 47, *48*
Components and cladding, wind pressures on, 156–57, 160
Compound truss, 36, 38, *39*, 40
  stress analysis of, 46, 47
Compression (C)
  on reinforced concrete, *462, 463, 464, 469*
  on trusses, 45
Compression elements, stiffened/unstiffened. *See* Stiffened compression elements (SCE); Unstiffened compression elements (UCE)
Compression member design. *See also* Column(s)
  cold-formed, 331–44
  hot-rolled steel, 216–34
  timber, 362, 365, 384–89
Computer-based commercial design systems, 53
Concentric structural joint, *434*
Concrete, materials and behavior, 461–69
  allowable stresses on, *523*

Index 549

curing, 465, *466, 467*
slump ranges, *462*
summary of properties, *469*
trial mixes for, *468*
water-cement ratio, 462, *463, 464*
Conjugate Frame Analogy, 81
Connectors. *See* Structural connectors
Continuous connectors, 186
Coplanar structure, 7. *See also* Statically indeterminate coplanar frames, analysis of
force/stress equilibrium equations for static, 13–15
kinds of stresses in, 8, *9,* 10
Corners, cold-formed steel, 285, 299, *301*
Corrugated sheet, cold-formed steel, *286*
Coulomb's equation, 162, 164
Critical form, 48–50, *51*
Cross bracing and stiffeners for timber trusses, *435*
Curvature
bending and, 64–*65*
defined, 63
Cut-and-try process, 3, 4

Dead loads, 128, 129–34
concrete slat example, *129,* 130
specific material, *130–32*
steel beam example, 132
truss, 132, *133,* 134
Deflection, control of, in reinforced concrete beams, 513–18
equivalent moment of inertia, 516–18
maximum permissible computed, *514*
minimum thickness of nonprestressed beams or one-way slabs, *515*
multipliers for long-term deflection, *516*

Deflection, flexural steel member, 241–43
Deflection, truss, 54–59
calculations, *58,* 59
real and ideal trusses, 54
virtual work, 54–59
Deformation equations,
applied to areas/moments of areas, 80–86
applied to bending effects, 72–80
Degree of indeterminacy, 91
Design, 1–5. *See also* Design aids
guides, aids, and shortcuts, 4–5
objectives of, and analysis, 1–2
philosophy, 3–4
procedures, 2–3
Design aids, 4–5
for A36 steel columns, 222–23, *224–29*
computerized, 53
for structural steel, 243–51
Design and Control of Concrete Mixtures, 462, 463, 464
Design drawings, final, 3
Design Specifications for Metal Plate Connected Wood Trusses, 431, 432
Design Values for Wood Construction, 373, 375, 378
Design of wood structures, 361
Development length, for concrete reinforcement bars, 511–13
basic lengths, *512*
definition sketch, *511*
minimum bend diameters, *513*
values for standard hooks, *512, 513*
Direction of a force, 10
Discrete connectors, 186, 188, *189*
Displacement (geometrical) analysis of indeterminate frames, 120
Distributed moment, 113. *See also* Moment distribution analysis of indeterminate frames

Distribution factor, 114
Dynamic equilibrium, 13

Edge-stiffened, cold-formed steel
    sections, 297, *298*
Effective slenderness ratios, of
    columns, 221–22
Elastic line
    bending deformations, *68*–*70*, *80*,
        81
    bending deformation sketches, *73*,
        86–*87*
    defined, 63
    direction of, defined, 63
    effects of bending on (rotation),
        66–*68*
Euler buckling in noncompact steel
    sections, *237*, 238
Euler buckling stress equations,
    217–21, 386
External load, 8, 13

Farm Structures, vii
Fink truss configuration, *53*
Fixed end bending moment (FEM),
    111
Fixed-end members, area-moment
    equations applied to, 108, *109*,
    110
Flexural member design, cold-
    formed steel, 302–5, 314–31
    beams with adequate lateral
        bracing, 314–23
    laterally unsupported beams, 325–
        31
Flexural member design, hot-rolled
    steel, 211, *212*, 234–58
    allowable stresses, 235–41
        summary of, *239*–41
    bearing stresses, 241, *242*
    combined with axial compression,
        258–63
    deflection, 241–43
    design aids, 243–51
        beam selection with, 254–56

    illustrations of, 251–58
    section properties, *244*–*45*
    shear stresses, 241
Flexural member design, reinforced
    concrete, 465–74
    flexural behavior, 465, *469*
    ultimate strength method, 474–86
    working stress vs. ultimate stress
        methods, 465–69, *470*
Flexural member design, timber,
    362, 363, *364*, 365, 389–400
    combined loading, 400–405
    laterally braced beams, 390–98
    laterally unbraced beam, 398–400
Flexural stress. *See* Flexural
    member design
Floors
    agricultural building loads, 135,
        136–37
    minimum uniformly distributed
        loads, *134–35*
Flow patterns of granular materials
    in bins, 176–79
Fluid pressure problem, 161,
    162
Footers, foundation
    area, 528–30
    keys for, *535*
    pressure distribution under
        eccentrically loaded, *530*
    problem examples, 535–37
    size, 531
    types, 531
Force. *See also* External load;
    Internal stresses; Reactions
    equilibrium of, 13–14, *15*
    signs and symbols describing, 10–
        13
Force (equilibrium) analysis of
    indeterminate frames, 120
Foundations for reinforced concrete,
    527–37
    allowable bearing capacity of soils,
        528, *529*
    column footings design, 533–35

footer design criteria
  area, 528–30
  size, 531
  types, 531
  general requirements, 535–37
  introduction, 527–28
  loads, 528
  wall footings design, 532–33
Frame(s)
  analysis of indeterminate. *See* Statically indeterminate coplanar frames, analysis of
  bending deformation of
    cantilevered, 75–77
    simply-supported, 84–86
    three-hinged, 79–80
    two-hinged, 76–79
  free body analysis of
    entire frame, 20, *21*
    parts of, *22*
Free body analysis, 20–23
  diagrams. *See* Free body diagrams
  graphical, 52
  inflection point, *27*
Free body diagrams, 15–18
  frames, *21, 22*
  granular mass, *163, 174*
  joint, 16, *17*
  one-hinge frame, *101*
  two-hinged drawbar frame, *105*
Frictionless hinge, 18
Frictionless roller, 18
Funnel flow, 176

Glued joints, 421–23
Granular materials, loads exerted by, 160–61
  depth of transition from shallow to deep bin, 173–76
  dynamic pressures, 176–79
    moist materials, 177–78
    silage, 178–79
  properties of granular materials, *168*
  with bridging, 170–73
  without bridging, 162–70
Ground snow load, 138–40
  ground-to-roof relationship, 142–43
Guide to Proportioning, Mixing, and Placing Cement, 462, 465
Gusset plates, 186–87

Handbook of Building Plans, 5
Hankinson's equation, 380
Hat sections, cold-formed steel, *286, 295,* 314, *315, 323*
Heat-treated construction alloy steels, 214
Heat-treated high strength carbon steels, 214
High strength-low alloy steels, 214
Hingleless frame, area-moment analysis of, *95*
Horizontal member, shear/moment diagrams for, *25*
Hot-rolled steel design
  bolted connections, 263–65
  vs. cold-formed, 209–10, 284–302
  combined axial compression and flexural stress (beam columns), 258–63
  compression members, 216–34
  design aids, *243–51*
  flexural member design, 234–58
  introduction to, 209–10
  mechanical properties of structural steel, 213–15
  nomenclature, 278–81
  problems, 276–78
  stability of steel sections, 211–12
  stiffened/unstiffened compression elements, *213*
  stress-strain curves, *214, 215*
  tension members, 215–16
  welded connections, 265–76
  working strength design philosophy, 210–11

Howe truss configuration, *53*
Hurricanes, 147

Ideal vs. real trusses, 52
  truss deflection, 54–59
Immovable base, 18
Importance factor (I), *139,* 140
Inflection points, 27
In-plane moments, *187*
  combined with axial forces on joints, 194–96
  joints with, 190–94
Internal stresses, 13
  applied to trusses to test for indeterminacy, 50–51
Investigation of design adequacy, 3
I-section
  stiffened, *321*
  unstiffened, *320*

Janssen's equation, *170,* 171–72, 173
Joint(s)
  axially loaded, *187,* 188–90
  classes of structural, *186–88*
  combined axial and in-plane moment forces, 194–96
  combined axial and out-of-plane moments, 196–99
  concentric structural, *434*
  evaluation of load capacity examples, 199–204
  free body diagrams, 16, *17*
  glued, 421–23
  with in-plane moments, 190–94
  lap, *186, 189, 190*
  loads transferred by, *187*
  metal plate connectors, 423–25
  multijoint structures, 116–19
  nailed, 411–12, *413*
  number of, as a stability test, 41–42
  translation, 119
  truss stress analysis, 43–52

Kern point, 520
Knee braces, for pole frame buildings, 437, *439*

Lap joint, *186, 189, 190*
Line elements, properties of, *309*
Liquids, loads exerted by confined, 160, 161–*62*
Live loads, 128, 134–37
  concentrated, *138*
  floors, agricultural, 135, *136–37*
  floors, minimum uniformly distributed, *134–35*
  roof, 137, *139*
Load(s), 2, 127–83. *See also* Live loads; Loads on timber
  allowable, on cold-formed steel beams, *317, 318*
  classification, 128
  combination of, 128–29
  dead, 128, 129–34
  exerted by confined liquids, 160–62
  exerted by granular materials, 160, 162–79
  external, 8, 13
  on foundations, 528
  nomenclature, 181–83
  problems, 180–81
  snow, 137–46
  wind, 146–60
Loads on timber, 380
  design of load-carrying members, 382–405
  duration, 365, *366,* 367
Low carbon steels, 213–14. *See also* A36 steel
Lumber. *See* Timber design; Wood

Magnitude of a force, 10
Manual of Steel Construction, 132, 210, 243, 244
Mass flow condition, 176
Maxwell diagram, 52

Mechanism, 34–35
Metal plate connectors, 423, *424,* 425
Method of joints, 43–52
Midline method for cold-formed steel section properties, 308, 310–14
  error introduced by, *314*
Midwest Plan Service (MWPS), 4–5
Moist granular materials, 177–78
Moment. *See* Area-moment analogies; Area-moment analysis of indeterminate frames; Bending moment; In-plane moments; Out-of-plane moments; Plastic moment of steel
Moment capacity of composite reinforced concrete sections, 474, *475*
Moment diagrams, *23,* 24
  for horizontal member, *25*
  moment-shear relationships, *26*
  one-hinge frame, *101*
  preparation of, 24–26
  two-hinge drawbar frame, *105*
Moment distribution analysis of indeterminate frames, 108–19
  computations, *115*
  concepts, 110–15
  four-member rigid frame example, 115–16
  joint translation, 119
  multijoint structure examples, 116–19
Moment of inertia property, calculation of equivalent, in reinforced concrete beams, 516–18
  steel sections, 308
Mountainous regions, snow loads in, 140, *141,* 142
Multijoint structures, moment distribution analysis of, 116–19
  computations, *117*

  distribution factors/FEM in frame diagram for, *118*

Nails and spikes, 406–14
  double shear, 409, *411,* 412–13
  lateral resistance, 406, 408–9, *410*
  load modification, 411, *412, 413*
  problem examples, 411–14
  sizes, *406*
  withdrawal load/resistance, 406, *407, 407–8,* 409
National Building Code of Canada, 143
National Design Specification for Wood Construction (NDS), 360, 361, 382, 408, 412
National Weather Service (NWS), 138
Nearly compact steel sections, allowable flexural stresses on, 236, *239*
Neutral axis (NA), of reinforced concrete, 456
Nomenclature
  area moment and moment distribution analysis, 124–25
  bending deformation, 90
  cold-formed steel design, 355–57
  hot-rolled steel design, 278–81
  loads, 181–83
  reinforced concrete, 541–44
  stress analysis, 30–31
  structural connections, 206–7
  timber design, 452–54
  truss stress analysis, 61–62
Noncompact steel sections, allowable flexural sections in, *236–39*

One-hinge frame, area-moment analysis of, 95, *96,* 97
  theorem application example, *99–*105

One-hinge frame (*Cont.*):
 computed values, *102*
 free body/moment diagrams for, *101*
Out-of-plane moments
 combined with axial loads on joints, 196–99

Plastic moment of steel, 209
Point of application, force, 10
Pole Building Design, 440
Pole buildings, 436–48
 allowable lateral pressures for soil, *440*
 analysis, *441, 444*
 knee brace for, *439*
 lateral forces on, *438*
 problem example, 443–48
 typical framing, *437*
Portland Cement Association (PCA), 462, 527
Post-buckling, 285, *296,* 297
Pratt truss configuration, *53*
Pressure, wind, 151, *152*
 coefficients, 152–54, *155, 156*
 general patterns of, *154*
Prestressed concrete beams, 518–27
 allowable stresses, 522, *523*
 basic analysis procedure for, 522
 basic behavior, 518–21
 materials, 521–22
 stages of prestressing and stress distributions, *519*
 ultimate strength of, 523–27
Prestressed Concrete Institute, 527
Primary stresses, 52
Professional Design Supplement, 5
Punching shear, 533–35

Rankine active and passive states, *164,* 165–68, 173
Reactions, 8, 13
 redundant, 91

statically indeterminant, 91, 92, 93
Real vs. ideal trusses, 52
truss deflection, 54–59
Recurrence intervals (RI), 140
Reinforced concrete design, 455–544
 closure for, 527
 columns, 485–505
 deflection control in beams, 513–18
 development of reinforcement, 511–13
 flexural members, 474–86
 foundations for agricultural buildings, 527–37
 fundamental concepts, 465–74
 introduction, 455–58
 materials and behavior, 458–65
 nomenclature, 541–44
 prestressed beams, 518–27
 problems, 537–41
 shear reinforcement, 505–10
Residual stresses, on steel, *218,* 219, *220*
Roof
 live loads on, 137, *139*
 snow loads
  effects of heat loss through roofs, 144–46
  ground-to-roof relationship, 142–43
  roof slope effect, 143–44
  snow drifting, 144, *145*
  wind pressures on, 154, *155, 156,* 156–57
Rotation, elastic line
 bending and, 65–68
 defined, 64

St. Venant buckling, 237
Sammet, L. L., vii
Scissors truss configuration, *53*
Secondary stresses, 52
Section modulus property of steel sections, 308

Index 555

Sections, cold-formed steel
  compression elements. *See*
    Stiffened compression
    elements (SCE); Unstiffened
    compression elements (UCE)
  geometry of, 285, *286, 287–95*
  maximum flat-width ratios for
    elements of, 307, *308*
  properties of, with thin elements,
    308–14
  strain hardening, 285
Sections, hot-rolled steel
  compression elements. *See*
    Stiffened compression
    elements (SCE); Unstiffened
    compression elements (UCE)
  stability of, 211–12
Sections, reinforced concrete,
  moment capacity of, 474, *475*
Shear diagram, *23,* 24
  for horizontal member, *25*
  moment-shear relationships, *26*
  preparation of, 24–26
Shear reinforcement for reinforced
  concrete beams, 505–10
  definition sketch for, *507*
  effects of shear stresses, *506*
  stress states, 505, *506*
  torsional shear, 510
  web reinforcement example, 509–
    10
Shear stress, 8, 9
  double, on bolts, *415–16*
  double, on nails, 409, *411,* 412–13
  on flexural members, 241
  on joints, 194–99
  punching, 533–34
  on reinforced concrete. *See* Shear
    reinforcement for reinforced
    concrete beams
  on timber, 362, 365
Sign convention, 11, *12,* 13, 14
  for area-moment analysis, 102–3
  for bending deformation, 70–71,
    74

  for moment distribution, 114–15,
    116
  for truss deflection, 57, 59
Signs and symbols, 10–13, 14. *See
  also* Sign convention
  reinforced concrete, 458
  shear/moment diagrams, 24
  stress analysis, 21–22
Silage, 178–79
Similitude in Engineering, 81
Simple truss, 36, *37–38*
  stress analysis of, 43, 44–*45, 46*
Simply supported frame, bending
  deformation of, 84, *85,* 86
SI units, viii
Sketches, valid/invalid elastic line,
  73, 86–87
Slope-deflection analysis of
  indeterminate frames, 120
Snow loads, 137–46
  ground, 138–40
  ground-to-roof relationship, 142–
    43
  heat loss through roofs and, 144–
    46
  in mountainous regions, 140–42
  roof slope effect, 143–44
  snow drifting, 144, *145*
  summary, 146
Soil Conservation Service (SCS), 138
Soil(s)
  allowable bearing capacity of, for
    foundations, 528, *529*
  allowable lateral pressures for,
    and pole buildings, *440*
Solid steel sections, allowable
  flexural stresses on, 236
*Specification for the Design,
  Fabrication, and Erection of
  Structural Steel Buildings,*
  210
*Specification for the Design of Cold-
  Formed Steel Structural
  Members,* 210
Spikes. *See* Nails and spikes

Stable truss, 34–35
Statically determinate structures, 7, 35–36
　testing for determinacy in trusses, 40–43, 50–51
　truss types, 36–40
Statically indeterminate coplanar frames, analysis of, 91–125
　area-moment method, 92–108
　characteristics, 91–92
　moment distribution method, 108–19
　nomenclature for, 124–25
　other methods, 119–20
　problems, 121–24
Statically indeterminate structures, 36
　testing for in trusses, 50–51
Static equilibrium of forces, 13–14, *15*
　sign conventions for, 11, *12,* 13, 14
Steel. *See* Cold-formed steel design; Hot-rolled steel design
Steel reinforcing bars for concrete design, 455, *456, 457*
　allowable stresses in prestressing, *523*
　materials and behavior of, 458–61
Stiffened compression elements (SCE)
　in cold-formed steel, 284, 285, *286, 296,* 297–99
　　allowable stresses in, 304–5
　　effective widths of, 305, *306,* 307
　　with NA closer to compression range, 323
　　with NA closer to tension flange, 321–23
　in hot-rolled steel, *213,* 216
Stiffness factor, 112–13
Stiffness modulus, 65
Storms and wind loads, 147, *148*
Strain hardening, of cold-formed steel, 285

Stress, statically indeterminate, 91, *93*
Stress analysis, 2, 7–31. *See also* Load(s)
Stress-strain forces on concrete, 455–56, *457, 460, 462, 469*
　rectangular beam, *476*
　ultimate stress method, 473, *474*
Stress-strain forces on steel,
　cold-formed, 299, *301*
　hot-rolled,
　　compressive, *219*
　　curves, *214, 215*
Structural connectors, 185–207
　classes of connectors, 186
　classes of joints, 186–88
　for cold-rolled steel, 346–52
　combined axial and in-plane moments (superposition), 194–96
　for hot-rolled steel, 263–76
　introduction to, 185–86
　joint load/carrying capacity (axially loaded joints), 188–90
　joints with axial loads and out-of-plane moments, 196–99
　joints with in-plane moments, 190–94
　nomenclature, 206–7
　problems, 204–6
　special considerations/examples, 199–204
　timber, 405–25
　welded. *See* Welded connectors
Structural design, 2–3
Structural Mechanics and Analysis, 120
Structures and Environmental Handbook, 5, 130, 176, 435, 439, 459
Superposition, 194
Support conditions, 18–20
　seven elements illustrated, *19*
Symbols. *See* Signs and symbols

Tension member design
  cold-formed steel, 302–3
  hot-rolled steel, 215–16
  timber, 383–84
Tension (T) and tensile stress
  on joints, 196–97, *198,* 199
  on reinforced concrete, 455–56
  on timber, 362, 363, *364,* 365
  on trusses, 45
Three-hinged frame, bending
    deformation of, *79,* 80
Timber Construction Manual, 361,
    427
Timber design, 359–454
  connectors, 405–25
  introduction to, 359–62
  load carrying member design,
    382–405
  nomenclature, 452–54
  pole buildings, 436–48
  problems, 448–52
  standard timber size and section
    properties, 380–82
  structural properties of timber,
    362–80
  trusses, 425–36
Tornadoes, 147, *148*
Total stress, 2, 7, 8
Transformation mechanism, *34,* 35
  critical form, 48–50
Transition depths in shallow to deep
    bins, 173–76
Translation, joint, 119
Transverse shear stress, *9*
  diagram of internal. *See* Shear
    diagram
Trusses
  dead load capacity of, 132, *133,*
    134
  stress analysis. *See* Trusses, stress
    analysis of coplanar statically
    determinate
  timber, 425–36
    concentric structural joint, *434*
    cross bracing and stiffeners, *435*

definition sketch, *425*
effective buckling lengths,
  432
estimating moments in, *431*
load combinations, *427, 429*
lower chords, *432*
truss analysis-stress reversal
  example, 427–36
upper chords, *430, 431*
Trusses, stress analysis of coplanar
  statically determinate, 33–62
  commercial systems for stress
    analysis of, 53
  concepts/definitions, 33–36
  configurations, *42*
  nomenclature, 61–62
  problems, 59–61
  real vs. ideal trusses, 52
  stress analysis, 43–52
  tests for stability/determinacy,
    40–43
  truss deflection, 54–59
  truss shapes, 52–53
  truss types, 36–40
Truss-Plate Institute (TPI), 360
Tube, cold-formed steel, *286*
  compressive stress on column,
    344
Two-hinged frame
  area-moment analysis of, *97, 98*
  theorem application example,
    105–8
  bending deformation of, 76–77, *78,*
    79

Ultimate Strength Design Handbook,
  504
Ultimate stress design method
    (USD), for reinforced concrete,
    472–74
  design criteria, 473
  design strength, *473*
  flexural member design, 474–86
    analysis of rectangular beam
      with tension steel, 476–81

Ultimate stress design method (USD) (*Cont.*):
  analysis of singly reinforced nonrectangular beams, 481–83
  balance conditions, 475, *476, 477*
  design of singly reinforced rectangular beams, 483–86
  moment capacity of composite sections, 474–75
  prestressed beams, 523–27
  required strength, 472–73
  stress/strain distributions, 473, *474*
  vs. working stress method, 465–69, *470*
Uniform Building Code (UBC), 127, 138
Unstable truss. *See* Mechanism
Unstiffened compression elements (UCE)
  cold-formed steel, 284, 285, *286, 296,* 297–99
  allowable stresses in, 303, *304*
  with local buckling, 319–20
  hot-rolled steel, *213,* 216

Velocity of wind, 148, *149, 150,* 151
  gust response factors, *152*
Virtual work, and truss deflection, 54–59

Wall footers, *531,* 532–33
Water-cement ratio, 462, *463, 464*
Welded connectors, 188–*89*
  cold-formed steel, 346–50
  allowable loads for fillet, 348, *350*
  allowable loads for spot, 348, *350*
  arc spot welds, *349*
  piling/tearout failure in, *347*
  hot rolled steel, 265–76

allowable stress on, 265, *267*
effective areas of, *189,* 265–66, *268*
example, 273–76
size limitations on, 266–69
types of, 265, *266*
Western woods use book, 361
Whitney Stress Block, 474
Wind, 146–60
  pressure, 151–52
  pressure coefficients, 152–56
  pressure on components and cladding, 156–57
  storms, 147, *148*
  summary/problem examples, 157–60
  velocity, 148–52
Wire fabric, for reinforced concrete, *459*
Wood
  characteristics of, 359, *360, 361*
  classification and grades, 367–80
  machine stress rated lumber, 370, *378–79,* 380
  visually graded lumber, 368, *369,* 370, *371–73, 374–77*
  standard sizes and section properties of, 380, *381–82*
Wood as a structural material, 361
Wood Engineering, 361, 362, 438
Wood Handbook: Wood as an Engineering Material, 361, 362
Wood technology in the design of structures, 361
Working strength concept, 210–11
Working stress design method (WSD), for reinforced concrete, 465–69, *470*

Z section, cold-formed steel, *286*
  axial load on column example, 336–38
  with stiffened flanges, *289*
  with unstiffened flanges, *290*